Principles of Radiation Protection

A TEXTBOOK OF HEALTH PHYSICS

EDITORS

K. Z. Morgan

School of Nuclear Engineering, Georgia Institute of Technology

&

J. E. Turner

Health Physics Division, Oak Ridge National Laboratory

ROBERT E. KRIEGER PUBLISHING COMPANY
HUNTINGTON, NEW YORK
1973

Original edition 1967
Reprint, with corrections 1973

Printed and published by
ROBERT E. KRIEGER PUBLISHING COMPANY, INC.
BOX 542, HUNTINGTON, NEW YORK, 11745

Library of Congress Catalog Number 67-22415
ISBN No. 0-88275-128X

Printed in the United States of America

Contributions

J. A. Auxier, *Health Physics Division, Oak Ridge National Laboratory*
R. D. Birkhoff, *Health Physics Division, Oak Ridge National Laboratory*
P. R. J. Burch, *Department of Medical Physics, University of Leeds, England*
A. D. Callihan, *Neutron Physics Division, Oak Ridge National Laboratory*
L. W. Cochran, *Department of Physics, University of Kentucky, Lexington, Kentucky*
D. M. Davis, *Health Physics Division, Oak Ridge National Laboratory*
F. J. Davis, *Health Physics Division, Oak Ridge National Laboratory*
L. C. Emerson, *Health Physics Division, Oak Ridge National Laboratory*
M. F. Fair, *Health Physics Division, Oak Ridge National Laboratory*
E. D. Gupton, *Health Physics Division, Oak Ridge National Laboratory*
G. S. Hurst, *Health Physics Division, Oak Ridge National Laboratory*
R. F. Kimball, *Biology Division, Oak Ridge National Laboratory*
K. Z. Morgan, *School of Nuclear Engineering, Georgia Institute of Technology*
J. Neufeld, *Health Physics Division, Oak Ridge National Laboratory*
R. H. Ritchie, *Health Physics Division, Oak Ridge National Laboratory*
W. S. Snyder, *Health Physics Division, Oak Ridge National Laboratory*
J. E. Turner, *Health Physics Division, Oak Ridge National Laboratory*
A. C. Upton, M.D., *Health Sciences Center, State University of New York, Stony Brook*
Niel Wald, M.D., *Graduate School of Public Health, University of Pittsburgh, Pittsburgh, Pennsylvania*

Preface

Although the need for radiation protection was recognized almost immediately following the discovery of X-rays, there was no organized professional effort for radiation protection until health physics had its beginning at the University of Chicago shortly before and after the first nuclear reactor was operated on December 2, 1942. This early group of health physicists consisting of E. O. Wollan, H. M. Parker, C. C. Gamertsfelder, K. Z. Morgan, J. C. Hart, R. R. Coveyou, O. G. Landsverk, and L. A. Pardue who were assembled first at Chicago and a short time later at Oak Ridge and Hanford was made up mostly of physicists with a variety of backgrounds. For example, Parker had worked in the field of medical physics and Wollan and Morgan had conducted cosmic ray research, where they learned something of the problems associated with the instrumentation, dosimetry, shielding and protection requirements of ionizing radiation; but for the most part the originators of health physics and those who joined with them in this early period had little knowledge, no formal training, and limited experience related to this work. Thus, by the process of trial and error, through extensive reading, by seminars with physicists, biologists, and chemists and as a result of their own research, they redesigned and adapted the early ion chambers, film meters, electroscopes, electrometers, proportional counters, and Geiger counters to meet the requirements of personnel monitoring and building and environmental radiation surveys. There were a few guiding principles provided by the International Commission on Radiological Protection and the National Committee on Radiation Protection, but usually these health physicists had to make up their own recommendations, rules, procedures, and codes of practice for safe work with X, γ, and neutron radiations that were given off by the reactors and for the protection of employees from external and internal hazards associated with a great variety of radionuclides, many of them new to man's environment.

From the beginning when A. H. Compton called this new group together, health physics was recognized as the study and practice of radiation protection. It was clear, too, that the health physicist had the important and practical assignment of working with the scientist and engineer in his day-to-day tasks to see that his work was done safely. Fortunately, it was equally evident to the early leaders in the field of health physics that these programs could be successful only if accompanied by efforts of similar magnitude in education, training, and research. Health physics research was defined early as the study of the effects of ionizing radiation on matter and its influence on man and his environment. The purpose of health physics was to protect man from the hazards of ionizing radiation so that he might make full use of its benefits. The long-range goal of the health physicist was set to develop a coherent theory of radiation damage; but it was recognized that before this goal could be achieved, extensive studies would have to be conducted into a wide field of research at many levels—atomic, molecular, solid, gaseous, liquid, plasma, cell (organism, animal, and ecosystem). Looking back over the last two decades, it is evident that great progress has been made by health physicists from many scientific disciplines—physicists, chemists, engineers, biologists, mathematicians, geologists, ecologists, meteorologists—all working together toward the one focal interest of health physics—radiation protection. It is recognized that before these studies will have been completed, many generations of this nuclear age will have passed.

Beginning in the fall of 1943 and with the early operation of Clinton Laboratory (now Oak Ridge National Laboratory), health physics training programs were set up for all laboratory employees and many courses were offered for the training of those who were later to become leaders in health physics in national and private research establishments, Army, Air Force, Navy, public health services, civil defense, state agencies, universities, hospitals, U.S. Atomic Energy Commission, international agencies, space programs, power industry, etc. Subjects of instruction included atomic physics, nuclear physics, radiation physics, electronics, instrumentation, units of radiation exposure and dose and their physical meaning, methods of calculating radiation exposure, absorbed dose and dose equivalent, application of instruments in radiation protection work, radiation shielding and scattering, internal dose, radiation biology and maximum permissible exposure. Later, instruction was given the health physicists in radiation biophysics, evaluation of human exposure data, prevention of criticality accidents, hazard evaluation of reactors, accelerators and associated facilities, radioactive waste disposal, environmental contamination, radiation ecology, rules, regulations, codes of practice, public relations, etc. The duration of these courses was from a few days to two years and the level of instruction ranged from that for technicians to that for Ph.D. candidates and post-doctoral students. Karl Z. Morgan was in charge of this training at Oak Ridge National Laboratory from 1943 to 1947, during which period most of the instruction was on a rather informal basis. From 1947 through 1949, Ross S. Thackeray directed this program and during this period special health physics fellowship programs were administered at ORNL by the

National Research Council. All the applied training and courses were conducted at ORNL and were taught without college credit. In 1949, Elda E. Anderson came to ORNL and was placed in charge of these training programs. It was then that the classroom instruction portion of these programs was placed in universities—first at Vanderbilt University and the University of Rochester and later at other universities (University of Washington, University of Kansas, University of Puerto Rico, Harvard University, University of California, University of Michigan, University of Tennessee, and Texas A & M), and the applied training during the summer months was conducted at ORNL and at Brookhaven National Laboratory.

At the same time, the Oak Ridge Institute of Nuclear Studies was asked to administer the AEC health physics fellowship program. Later, summer training was offered also at Argonne National Laboratory, Lawrence Radiation Laboratory, National Reactor Testing Station, Hanford, Puerto Rico Nuclear Center and Savannah River Laboratory. There was never a teacher more dedicated to helping her students than was Elda Anderson. At the time of her death, Myron F. Fair, who had been teaching many of these classes at ORNL and at Vanderbilt University, took charge of this training program. In 1963 the University of Tennessee became the second university cooperating directly with ORNL in this training. At this time, also, the fellowship grants were extended to a period of three years and particular interest was placed on the programs at the Ph.D. level.

Throughout all this period (1943 to the present), there has been a sparsity of instructional and textbook material. Morgan furnished lecture notes to his students from 1943 to 1947, and in 1947 he issued a series of chapters covering such subjects as units of radiation exposure, methods of calculating radiation exposure and dose, instruments for radiation protection, radiation shielding, time factor in radiation damage, internal dose, health physics requirements, maximum permissible exposure to ionizing radiation, and techniques of personnel monitoring. Over a thousand copies of these unbound chapters were issued in notebook form to students by Morgan, Thackeray, Anderson, and Fair and, for the most part, this material served as the principal textbook source of information for these courses. In addition, hundreds of copies of these chapters were circulated to health physics students abroad and were used by Anderson, Fair, and L. C. Emerson as support material for three overseas health physics programs—Stockholm, Sweden (1955), Mol, Belgium (1957), and Bombay, India (1958)—sponsored by ORNL, WHO, and the U.S. Atomic Energy Commission and under the direction of Elda Anderson.

During this 22-year period, various unsuccessful attempts were made to prepare a health physics textbook and when these mentioned chapters were written, the objective was to get them into textbook form and more conveniently available to graduate students in health physics. With this in mind, an editorial committee consisting of K. Z. Morgan, E. E. Anderson, M. F. Fair, and R. D. Birkhoff was formed in 1958. This resulted in the revision of Morgan's earlier chapters and in the preparation of a number of additional chapters that are included in this health physics textbook. This effort, however, never progressed to the point of final publication, but it did result in the issuance to the health physics students of laboratory-bound chapters as follows: "History of Damage and Protection from Ionizing Radiation," by K. Z. Morgan, "Interaction of Charged Particles with Matter," by J. Neufeld, "The Physical Basis of Radiation Dosimetry," by R. H. Ritchie, "Ionization Methods of Mixed Radiation Dosimetry," by G. S. Hurst, "Special Methods in Radiation Dosimetry," by J. A. Auxier, "Distance and Shielding as Protection Factors," by K. Z. Morgan, "Internal Dose," by W. S. Snyder and K. Z. Morgan, "Radiation Biology," by A. C. Upton and R. F. Kimball, "Maximum Permissible Exposure Levels-External and Internal," by K. Z. Morgan, "Health Physics Instruments," by D. M. Davis, E. D. Gupton, and K. Z. Morgan, and "Prevention of Criticality Accidents," by Dixon Callihan. All these authors are associated with ORNL. Although these chapters were very useful to the student in this form, they were more clumsy to handle than a textbook. Furthermore, they were fast becoming outdated. Unfortunately, though, their publication as a textbook was delayed because several needed chapters had never been written. Also, in 1961 E. E. Anderson, who was urging on this effort, died of leukemia, the myelogenous variety of the disease, which she believed was developed as a result of her earlier work with ionizing radiation at Los Alamos. This exposure was received during the early part of the war under working conditions she described as less than adequate for radiation protection. Ironically, this person so loved by all her students and who strove to teach others how to avoid the hazards of ionizing radiation and who tried so hard to get this health physics text available for her classes was herself a victim of these same hazards.

In 1962 Dr. J. E. Turner came to ORNL. He had been one of Dr. Anderson's students in the second health physics AEC fellowship program at Vanderbilt University in 1951. He had a keen interest in health physics education and from firsthand experience realized the need for a textbook in this new field. He joined Morgan, Fair, and Birkhoff on their health physics editorial committee and agreed to serve as co-editor with Morgan. With this, the tide was turned and for the first time someone was pushing this publication full speed ahead and keeping after the authors to get their chapters finished. To no one is owed so much credit as to Jim Turner for the final effort that has pulled together in textbook form these chapters, has made this dream come true, and brought to fruition an often frustrated but never forsaken goal. As co-editor with Jim, I would like to express my deep appreciation to him, for without his untiring effort this text never would have been published. Also, both the editors and all the authors wish to express their deepest gratitude for the inspiration and devotion to health physics we have witnessed and received from our friend and associate Dr. Elda E. Anderson.

Karl Z. Morgan

Oak Ridge, Tennessee
January, 1966

As this book reaches the final stages before publication, we acknowledge with much appreciation the help and assistance of many friends and colleagues. In particular, detailed criticisms received from Dade W. Moeller and the staff at the Northeastern Radiological Health Laboratory (Winchester, Massachusetts), from G. F. Stone, Tennessee Valley Authority (Knoxville, Tennessee), and from Klaus Becker, Kernforschungsanlage (Jülich, Germany), have eliminated some shortcomings that would otherwise exist. It is with pleasure that we acknowledge the very competent assistance of Beverly Varnadore (Mrs. C.D.), who typed major portions of the text through several drafts, handled the large volume of correspondence connected with the editing, and lent a valuable hand in the compilation of the indices. To our secretaries Jeanne Carver (Mrs. D.W.) and Norma Brashier (Mrs. J.W.), go our thanks for their help during many phases of the production of this volume. Finally, we express our appreciation to William G. Bryden, Dorothy Garbose, Maura Fitzgerald, Ruth Weine, and others on the staff of Wiley with whom it has been a pleasure to work.

Karl Z. Morgan
James E. Turner

Oak Ridge, Tennessee
May 18, 1967

Contents

CHAPTER 3 Radiation Quantities and Units

CHAPTER 4 The Physical Basis of Radiation Dosimetry

CHAPTER 5 Detection and Measurement of Ionization

CHAPTER 6 Ionization Methods of Mixed Radiation Dosimetry

C. STOPPING POWER OF MATTER FOR ELECTRONS

D. DOSE FROM ELECTRONS AND BETA RAYS IN UNIFORM INFINITE MEDIA

C. MEDICAL MANAGEMENT OF RADIATION OVEREXPOSURE

D. CONCLUSION

CHAPTER 14 **Maximum Permissible Exposure Levels—External and Internal**

CHAPTER 15 **Health Physics Instruments**

CHAPTER 16 **Prevention of Criticality Accidents**

CHAPTER 1

History of Damage and Protection from Ionizing Radiation

K. Z. Morgan

1-1 INTRODUCTION

Probably the most important and basic problems of the health physicist are (1) an understanding of the mechanisms of radiation damage, (2) the establishment of appropriate levels of maximum permissible exposure to ionizing radiation, and (3) the intelligent enforcement of these levels. In this chapter, man's exposure to natural background radiation is reviewed, and the large increase in exposure resulting from man-made sources is described. We recount the growing awareness of man to the biological consequences of exposure to ionizing radiation and indicate how his concern for radiation protection led to the development of the profession of health physics. Subsequent chapters treat the problems of understanding the mechanisms of radiation damage and evaluating the effects of exposure to given sources of ionizing radiation.

1-2 NATURAL SOURCES OF HUMAN EXPOSURE TO IONIZING RADIATION

Although a recognized need for health physics protection did not arise until the discovery of ionizing radiation at the close of the nineteenth century, man has always been exposed to naturally occurring sources of ionizing radiation. The radiation that occurs naturally in man's environment is referred to as natural background radiation. Undoubtedly, it has played an important part in the evolution of man, and, in some instances, has been known to cause radiation damage. All life, including man, has been exposed throughout its period of existence to natural sources of ionizing radiation. At the present time it has not been established whether exposure at the relatively low dose rate of average background is harmful or whether, in fact, it is beneficial to man. Some investigators estimate that from 2–10% of the natural mutations in man are caused by background radiation, and a few writers have attempted to show a relationship between the dose from natural background radiation and the frequency of genetic malformations as well as leukemia and other forms of cancer. Without doubt, lung damage in individuals resulted from the natural radon in certain mines several centuries before the discovery of X-rays by Roentgen in 1895. Although a major portion of all genetic mutations results directly or indirectly in inherited defects and in genetic death, the fact should not be overlooked that an exceedingly small but nevertheless important fraction of the natural mutations must be beneficial in that it contributes to the evolution of higher forms of life. Thus, although from a general point of view all ionizing radiation exposure to the individual is potentially harmful, and although radiation damage in man always appears to increase with the amount of exposure, man has managed to survive and evolve to his present state of development in spite of background exposure, if not because of it.

The natural background of ionizing radiation received by man is comprised of cosmic rays and radiation from naturally occurring radionuclides that are in his environment and are incorporated to a varying degree in his body. There are some 40 naturally occurring radioelements and about twice this number of naturally occurring radioisotopes.

By far the major portion of human exposure from the naturally occurring radionuclides results from ^{40}K (half-life 1.27×10^9 yr) and from the ^{238}U (4.56×10^9 yr) and ^{232}Th (1.39×10^{10} yr) series of radionuclides. Figures 1-1 through 1-4 show the four series of radionuclides ($4n + b$) in which b has values of 0, 1, 2, or 3. The numeral 4 occurs in this series because during an α-transition there is a change in the nucleus of 4 mass units. The values of b indicate the number of neutron and/or proton departures from the thorium series ($4n + 0$). The ($4n + 2$) and the ($4n + 3$) series always occur together in nature with ^{235}U (6.8×10^8 yr), comprising on a weight basis 0.714%; ^{238}U, 99.2830%; and ^{234}U (2.48×10^5 yr), 0.0054% of the uranium isotopic mixture. The ($4n + 1$) series is missing in nature because its longest-lived radionuclide ^{237}Np (2.20×10^6 yr) has a relatively short half-life compared with the approximate age of the earth's crust, 5×10^9 yr. Long-lived radionuclides ^{48}Ca ($>7 \times 10^{18}$ yr), ^{50}V (4×10^{14} yr), ^{87}Rb (4.7×10^{10} yr), ^{115}In (6×10^{14} yr), ^{129}I (1.6×10^7 yr), ^{138}La (1.1×10^{11} yr), ^{144}Nd (2.4×10^{15} yr), ^{147}Sm (1.2×10^{11} yr), ^{176}Lu (2.2×10^{10} yr), ^{187}Re (6×10^{10} yr), ^{209}Bi ($>2 \times 10^{18}$ yr), and ^{239}Pu (24, 413 yr) occur in nature in such small concentrations that they do not contribute appreciably to the human dose from

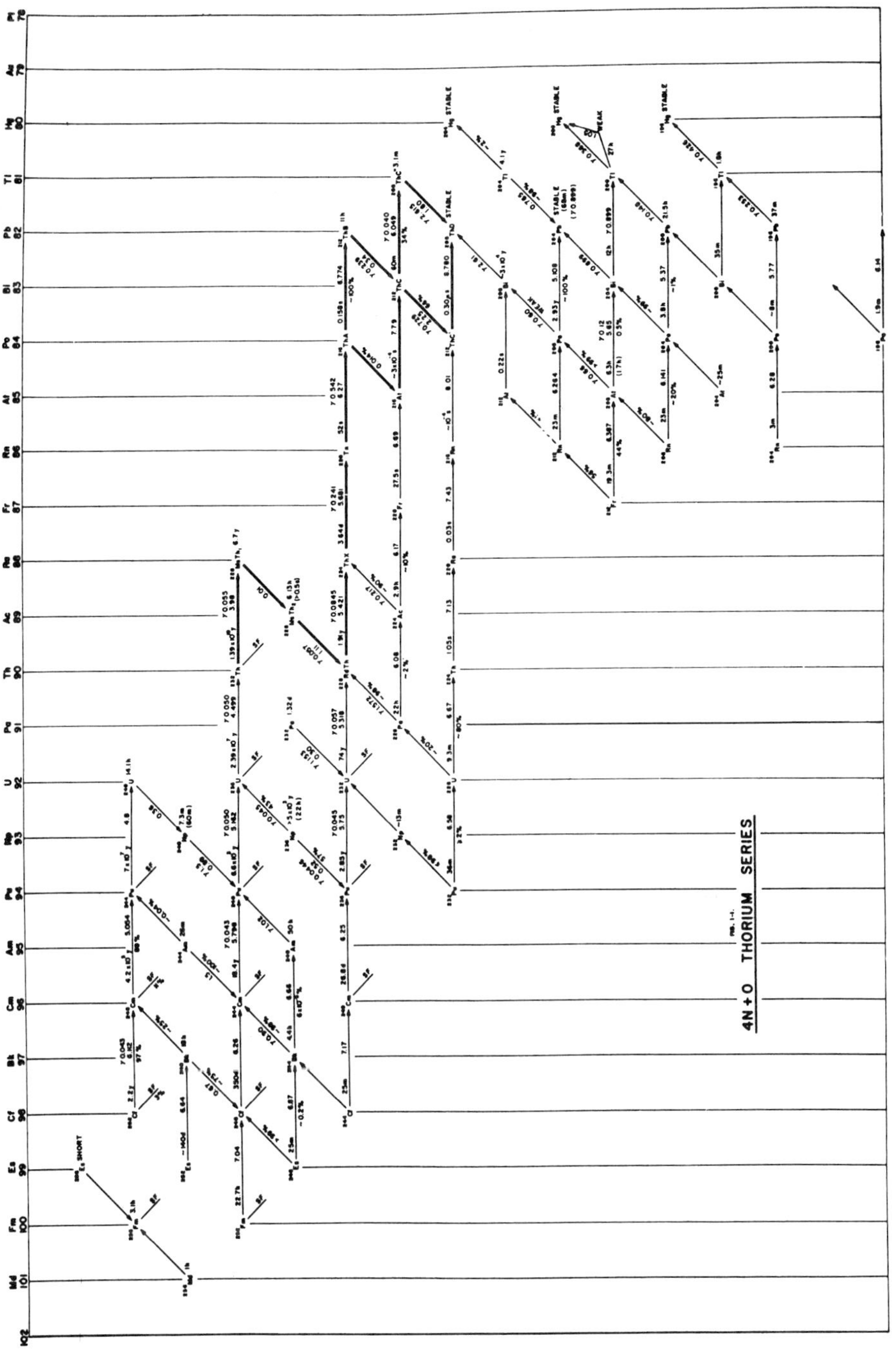

Fig. 1-1.

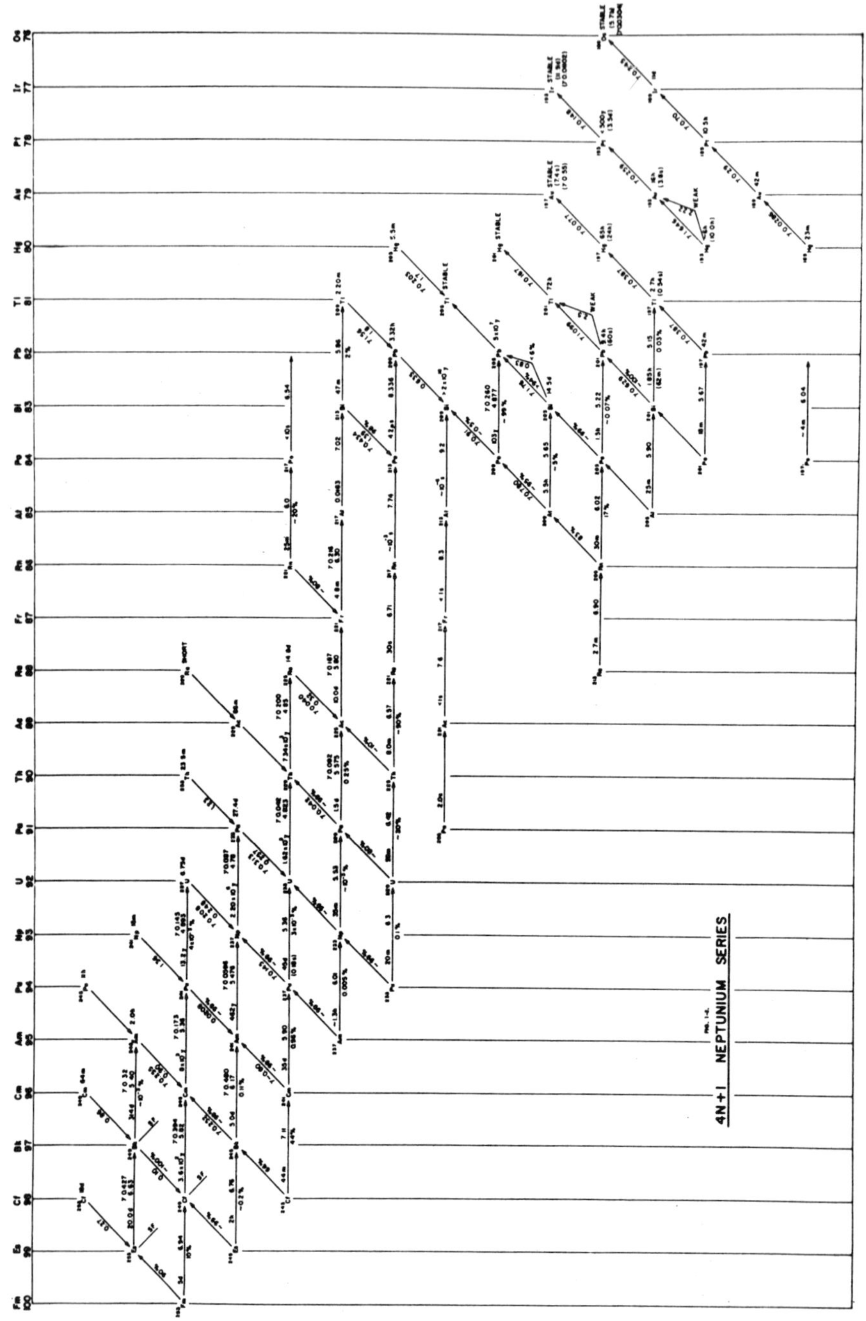

Fig. 1-2.

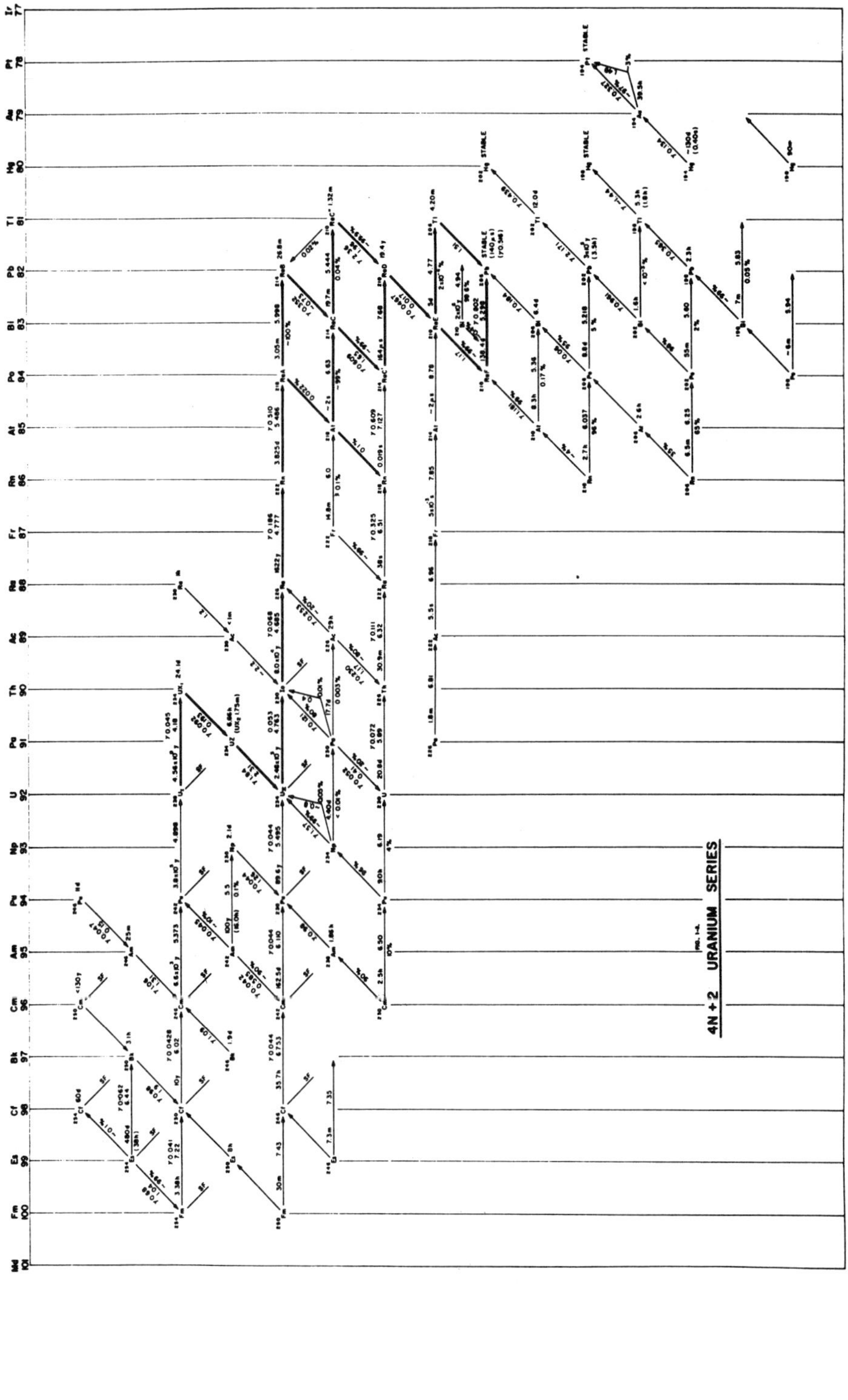

Fig. 1-3.

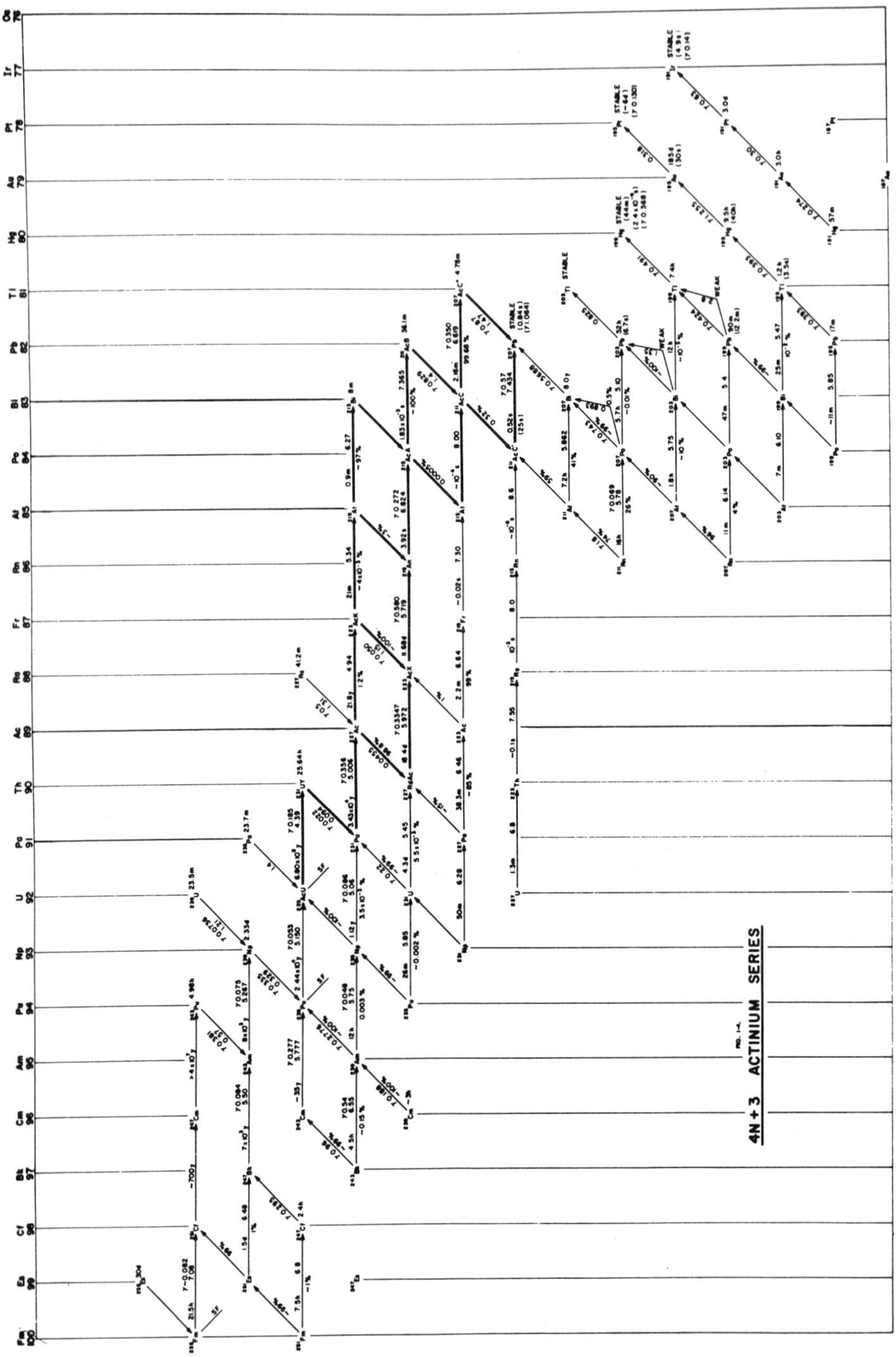

Fig. 1-4.

Figs. 1-1 to 1-4. Conventions used to show the relationships of nuclides in the uranium, thorium, actinium, and neptunium series. (Based on data in References 138, 178, and 194).

1. The relationships are indicated by arrows, where (*a*) Double arrows indicate transitions in naturally occurring radionuclides. (*b*) Single arrows indicate transitions in artificially produced radionuclides. (*c*) The direction of the arrows indicates the type of decay: alpha decay: → beta (negatron) decay: ↙ beta (positron) decay: ↑ orbital electron capture: ↗. (*d*) Spontaneous fission is indicated by a slanting line labeled SF: $\searrow^{SF}$.
2. Radioactive half-lives of the nuclides appear directly to the right of the nuclides (y = year, d = day, h = hour, m = minute, s = second, μs = microsecond).
3. Alpha, beta, and gamma energies (in MeV) are placed directly above appropriate arrows, the gamma energies being preceded by the symbol γ. In each case the energy given corresponds either to the largest energy reported for that transition or to the most abundant energy. (The reader is referred to standard sources of nuclear data for a detailed listing of the energies emitted.) Absence of an energy value indicates that the energy of the transition is not known.
4. Half-lives of metastable states and isomeric transitions are placed directly beneath the other half-lives. The gamma energy, when the decay is to a stable nuclide, is placed in parentheses beneath the half-life.
5. Branching ratios for the different modes of decay are given in percent below the arrows indicating the type of decay.

background radiation. Likewise, with the exception of ^{14}C (5568 yr), the short-lived radionuclides of the abundant elements make a negligible contribution to the human dose from background radiation. Concentrations of radionuclides such as ^{14}C, ^{3}H (12.26 yr), ^{36}Cl (3.08×10^5 yr), and ^{41}Ca (1.1×10^5 yr), are maintained in equilibrium in low concentration at the earth's surface by the (n, p) and (n, γ) reactions of the neutron component of cosmic radiation (about 8 slow neutrons/sq cm/hr at sea level). There are undoubtedly many other naturally occurring radionuclides in the earth's crust arising primarily from capture and fission reactions with cosmic ray neutrons and with the spontaneous fission neutrons of some of the heavy elements such as ^{238}U with 7.0×10^{-3} fission/gm (sec) and ^{232}Th with 4.1×10^{-5} fission/gm (sec). These other radionuclides are, for the most part, relatively short-lived, and their concentration is presumably so low that they have not been detected; their natural occurrence is only of academic interest. Undetectable traces of many radionuclides are produced in the atmosphere and in the earth's crust because of the many types of reactions between the primary and secondary components of cosmic radiation and the elements they encounter.

Table 1-1 gives for illustration the calculated absorbed dose rate (mrad per year) and dose rate equivalent (mrem per year) to critical organs of the body of the standard man (70 kg in weight) from a number of radionuclides naturally occurring within the body. In the case of ^{138}La, ^{50}V, and ^{209}Bi, the concentrations in the body and the resulting doses to the indicated body organs are too low for measurement by ordinary techniques. For ^{226}Ra and

Table 1-1 Radioactivity of the Human Body and Dose to Critical Body Organs from Some of the Naturally Occurring Radionuclides

Radio-nuclides	Half-life (yr)	Type of Decay	% in Natural Element (by wt)	% Natural Element in Body (by wt)	Specific Activity (counts/gm)	Curies in Standard Man	Absorbed Dose Rate[a]
^{40}K	1.27×10^{9}	$\beta^-, \gamma, \epsilon$	0.0119	0.2	7.03×10^{-6}	1.17×10^{-7}	17 TB, 30 Br
^{87}Rb	4.7×10^{10}	β^-	27.85	0.0017	8.73×10^{-8}	2.89×10^{-8}	0.69 TB, 2.8 P
^{226}Ra	1622	α, β^-, γ	~100	1.4×10^{-13}	0.974	9.55×10^{-11}	0.28 TB, 2.8 B (2.8 TB, 28 B)
^{138}La	1.1×10^{11}	$\beta^-, \gamma, \epsilon$	0.089	$<7 \times 10^{-5}$	2.35×10^{-8}	$<1.02 \times 10^{-12}$	$<2 \times 10^{-4}$ TB, $<4 \times 10^{-4}$ B ($<6 \times 10^{-4}$ B)
^{50}V	4×10^{14}	γ, ϵ	0.25	$<1.4 \times 10^{-7}$	1.79×10^{-11}	$<4.39 \times 10^{-18}$	$<1 \times 10^{-9}$ TB, $<5 \times 10^{-9}$ K
^{209}Bi	2×10^{18}	α	100	$<4.3 \times 10^{-7}$	8.54×10^{-16}	$<2.57 \times 10^{-19}$	$<2.1 \times 10^{-10}$ TB, $<1.8 \times 10^{-8}$ K ($<2.1 \times 10^{-9}$ TB, $<1.8 \times 10^{-7}$ K)
^{238}U	4.56×10^{9}	α, γ	99.276	3×10^{-8}	3.29×10^{-7}	6.86×10^{-12}	7.9×10^{-3} TB, 0.12 K (7.9×10^{-2} TB, 1.2 K)

[a] Absorbed dose rate (mrem/yr) in indicated body organ (TB = total body, Br = brain, P = pancreas, B = bone, K = kidney). When the dose rate equivalent (mrem/yr) differs from the absorbed dose, it is given in parentheses. These organs were selected for this table because of the higher dose they receive, that is, they are the critical body organs.

^{238}U, the doses are larger than otherwise because of the accumulation in the body of daughter radionuclides. The same would be true for ^{232}Th, but values are not given in Table 1-1 for this radionuclide because reliable data could not be found for the concentration of this radionuclide in the human body. The dose rate equivalent (mrem per year) is greater than the absorbed dose rate (mrad per year) for those radionuclides that emit α-radiation because the quality factor (commonly referred to as relative biological effectiveness, RBE) for α-radiation is set equal to 10. Also, the dose rate equivalent is greater than the absorbed dose rate for ^{138}La in bone; here, the damage factor n is set equal to 5 for the particulate (β^- and electron-capture, denoted by ϵ) radiation. This damage factor of 5 is applied for particulate radiation emitted by radionuclides deposited in bone except where a radionuclide of Ra is the parent radionuclide; there, $n = 1$.

Table 1-2 summarizes the principal components of natural background dose rate from the internal and external sources to which man is subjected. Values are given of dose rate to gonads, bone, and lungs because of an interest in and possible relation to genetic damage, bone tumors, and lung cancer, respectively. Several points of particular interest in this table are:

1. There is considerable variation in the published estimates of the dose to man, and better data can and should be obtained from more detailed studies.
2. The lung dose (average 200 mrad/yr or 1100 mrem/yr) can vary between individuals by more than an order of magnitude, depending primarily upon the type of construction of the house in which an individual lives.
3. The average dose to the gonads or bone is about 100 mrad/yr. This dose may be less by 10–20% or greater by a factor of 2 or more, depending primarily on the region in which one lives, for instance, the type of soil, elevation, and geomagnetic latitude.

A variety of factors affects both the absorbed dose and the dose equivalent received by man from natural background radiation. Some of these factors are age, sex, eating habits, place of residence, type of dwelling, heating system in the home, the value assigned to the quality factor, the damage factor, and the body organ and type of damage for which dose is considered. For example, Anderson and Langham[28] have shown that the average concentration of potassium (and consequently the absorbed dose from ^{40}K) in the human body is a function of age and sex. If the average concentration of potassium in the body at birth is taken at 100%, it increases to a maximum of about 134% in both sexes at age 8 or 9. Thereafter, in the female, it drops continually and at age 80 reaches about 76%. In the male after age 8 or 9 it drops to a minimum of about 123% at age 12, rises to a second maximum of about 145% at age 16, and drops to 100% at age 80. As another example, Hultqvist[91] and Gabrysh and Davis[67,68] found that in exceptional cases of

Table 1-2 Exposure of Man to Natural Background Radiation[114,56] (mrad/yr)

Type of Exposure		Mean Dose to Gonads (mrad/yr)	Mean Dose to Bone (mrad/yr)	Mean Dose to Lungs (mrad/yr)
Internal, from radionuclides	^{40}K	19[205,106], 20[54,188], 18[38], 18[64], 22[106]	10[54], 11[205], 7[64]	15[64]
	^{226}Ra		3[54], 3.8[205], 6.7[106], 3[64]	0.5[64]
	^{228}Ra		3[54]	
	^{210}Pb		2[54]	
	^{14}C	1[38,188], 0.7[205], 1.5[106], 1.3[64], 1.8[106]	1.6[205], 1.3[64]	
Internal,[91] from $^{220,222}Rn$	^{220}Rn in wooden house			26 av., 45 max., 2.3 min.
	^{220}Rn in brick house			46 av., 210 max., 2.7 min.
	^{220}Rn in concrete house			94 av., 290 max., 3.9 min.
	^{222}Rn in wooden house			19 av., 39 max., 1.3 min.
	^{222}Rn in brick house			58 av., 120 max., 5.7 min.
	^{222}Rn in concrete house			64 av., 140 max., 4.0 min.
External, from ^{238}U and ^{232}Th series and from ^{40}K	Normal regions	47[204 a], (range 28–82)[205 a], 39[102 a]	Approximately the same as for gonads	
	Granite regions in France	162[205]		
	Monazite regions in India	802[205]		
	Monazite regions in Brazil	287[205 a]		
External, from cosmic radiation	Sea level, 0° geo. lat.	23[54], 24[47], 35[205], 28[188], 30[106], 33[106]	Approximately the same as for gonads	
	Sea level, >50° geo. lat.	26[54], 27[47], 41[205], 37[106]		
	10,000 ft, 0° geo. lat.	56[54], 50[47], 89[214], 80[106]		
	10,000 ft, >50° geo. lat.	84[54], 66[47], 128[205], 120[106]		
Total for normal regions near sea level[b]		100 av., 150 max., 70 min.	100 av., 150 max., 70 min. (180), (250), (140)	200 av., 570 max., 70 min. (1100), (4500), (110)

[a] These values were reduced[14] for both sexes by a shielding factor of 0.63.
[b] The values given in parentheses are in units of mrem/yr using an RBE of 10 for α.

poor room ventilation and for certain types of room heating, the ^{220}Rn and ^{222}Rn concentrations in room air may be very high, leading to dose equivalents to the lungs that are more than ten times the common maximum values given in Table 1-2.

There is some uncertainty about the appropriate quality factor QF that should be applied to α-radiation from natural background. However, a QF of 10 was assigned to the α-radiation in obtaining the values of dose rate equivalent given in Tables 1-1 and 1-2. Experimental data on animals exposed at much higher absorbed dose rates seem to indicate that the QF depends in particular on the type of damage under consideration, the dose rate, and the specific ionization of the ionizing particles. In view of the fact that many animal experiments indicate the QF of particulate radiation of high specific activity increases with decreasing dose rate, it is possible that in the case of the low-level background radiation a QF of 10 for α-radiation is too low. It should be noted, also, that in this text a QF of 1.7 is applied to the highly ionizing, low-energy β-radiation emitted by a radionuclide with energy less than 0.03 MeV (maximum). The damage factor n expresses the bone tumor damage of particulate radiation from bone-seeking radionuclides relative to this damage from radium. This factor is reasonably well established experimentally for only a few radionuclides like ^{89}Sr, ^{90}Sr, and ^{239}Pu.

It is extremely important to determine the effect of natural background radiation on man because, presumably, this is the only type of ionizing radiation exposure to which man has established a biological equilibrium. Perhaps for this reason the only reasonably safe level of population exposure from man-made sources of ionizing radiation is an additional exposure of the order of background, or within the limits of common variation of background. To establish a relation between this background dose and biological effect, two studies must be undertaken.

1. Determine the absorbed dose in each body organ from each type of natural background radiation.
2. Conduct epidemiological studies, especially in regions of high background radiation, for example, the monazite sand regions of India and Brazil, to determine whether the natural background radiation has had any perceptible effect on man and his environment.

There is considerable urgency on the part of the health physicist to obtain this information because the permissible exposure levels which he enforces must be safe and reasonable. As time goes on and as man adds to his exposure from other radiation sources, it will become increasingly difficult to obtain meaningful results from such background radiation studies.

The foregoing discussion and the data summarized in Table 1-2 suggest several epidemiological studies that might be undertaken in order to determine the effect of natural radiation on man. External, environmental radiation from the ^{238}U and ^{232}Th series of radionuclides and from ^{40}K contributes about half of the total absorbed dose to most organs of the human body. The Kerala regions of India with a population of about 52,000 and the states of Rio de Janeiro and Espirito Santo of Brazil[205] with a population of about 50,000 provide the most promising areas for environmental studies. There, as a result of the γ-radiation from the thorium-bearing monazite sand, the external environmental absorbed dose ranges from ten to thirty times the world average. Studies in these regions could be made with reasonably adequate control populations. However, suitable medical records have not been maintained and the studies would have to begin without extensive previous medical histories.

Cosmic radiation is the second largest contributor to the natural radiation absorbed dose. It increases by a factor of 2 in going from sea level to 10,000 ft and by about 10% in going from 0 to 50° geomagnetic latitude. This suggests that epidemiological studies of populations living at various elevations might be made. Such studies, however, do not appear to offer much promise because there are too many factors that could introduce serious biases into the data. For example, at a high elevation, sunlight is much brighter and the ultraviolet radiation is far more intense than at lower elevations. At high elevations, air is less dense and the lungs must work harder to obtain the oxygen required. Perhaps most serious of all is the fact that at high elevations the spectrum of the cosmic radiation is quite different from that at sea level. It could well be, for example, that the increase in the primary proton component and in the cosmic showers and bursts (caused primarily by mesons at higher elevations) is of greater significance in producing radiation damage than is simply the change in the total absorbed dose with elevation.

The internal absorbed dose due to $^{220,222}Rn$ which escapes from the concrete, stone, and brick of homes offers an interesting problem for study. The average absorbed dose to the lung can vary by a factor of 100 or more in going from one type of construction to another.[67,68] In a study of this type, it would probably be difficult to obtain adequate controlled population groups. The immediate objective of studies of the dose from radon and its daughter products would be directed primarily to observations of effects on the lungs and other parts of the respiratory system.

Perhaps there is not enough variation in the ^{40}K internal absorbed dose to persons of a given age and sex to warrant a study of this problem for the purpose of establishing a radiation dose–effect relationship. However, a study of the variation of $^{226,228}Ra$ and ^{210}Pb in man's drinking water and food and in his skeleton appears to offer an interesting possibility of obtaining useful dose–effect data on man. No extensive studies have been made of the variation of ^{210}Pb, but many workers have investigated the concentrations of $^{226,228}Ra$ and ^{228}Th in drinking water and in food. Investigators at the Argonne National Laboratory[107] have found that Illinois is a very good area for this study because the concentration of ^{226}Ra varies by a factor of 500 in water from various areas of the state. Auerbach[118] has made a detailed analysis of the average mortality from bone cancer in Illinois during the 11-year period prior to 1950. Data reported by Marinelli[118] indicate a possible significant increase in bone cancer among females living in parts of the state where the ^{226}Ra concentrations in water and in the skeleton are considerably higher than in other areas. Turner et al.[203] have made a study of the ^{226}Ra and ^{228}Th content of common foods and found a wide variation; for example, the concentration of ^{226}Ra varies by a factor of 64 in various cereals, 3400 in nuts, >510 in beverages, >23 in fruits and vegetables, and >310 in meats and fish. Lucas[115] has made a study of the body uptake of ^{226}Ra by prisoners confined at Statesville Penitentiary, Illinois, and by high school boys living in Lockport, Illinois, where the average water content of ^{226}Ra is approximately 100 and 300 times that in the Chicago area, respectively, and made comparisons with respect to uptake from food by persons in the Chicago area. He found the interesting relationship for accretion of radium from lifetime uptake,

$$q_{^{226}\mathrm{Ra}} = 50C_{\mathrm{H_2O}} + 22C_{\mathrm{food}} \qquad (1\text{-}1)$$

in which $q_{^{226}\mathrm{Ra}}$ = body burden of ^{226}Ra(pc), $C_{\mathrm{H_2O}}$ = daily intake of ^{226}Ra in water (pc per day) and C_{food} = daily intake of ^{226}Ra in food (pc per day). Perhaps, a very extensive study of such populations with reference to eating habits and cancer incidence would reveal useful information relative to the effects of background radiation on man.

1-3 HUMAN EXPERIENCE WITH MAN-MADE SOURCES OF IONIZING RADIATION

Except for high radon exposures to some miners, the exposures of man to ionizing radiation continued at a very low level and within the common range of variation of natural background radiation as indicated in Table 1-2 until the discovery of X-rays was first announced by Roentgen on January 4, 1896, and the discovery of radioactivity from uranium was announced by Becquerel in November, 1896. Since then, man has added steadily to his sources of exposure to ionizing radiation. During the past few decades, radiation from high-voltage accelerators, nuclear reactors, and artifically produced radionuclides has made an appreciable and ever-increasing contribution. However, most of this exposure has resulted from diagnostic and therapeutic use of

X-rays and radium products. It was estimated by Laughlin and Pullman[112] in 1957 that in the United States the probable average gonad-absorbed dose from medical sources was about 150 mrad/yr, or 50% more than the average gonad exposure listed in Table 1-2 for natural background radiation.

Early External Exposure

Following the discovery of man-made sources of ionizing radiation, man was not long in learning about both the harmful and useful characteristics of this new source of energy. Grubbe[70] a manufacturer of Crookes tubes in Chicago, Illinois, was using his equipment to study the fluorescence of chemicals even before the public press on January 4, 1896, heralded Roentgen's discovery of X-rays. During January, 1896, he first noticed an erythema, edema, hyperemia, and hyperesthesia on the back of his left hand. Later there was blister formation with skin desquamation and epilation. This was followed by skin cracking, ulcer formation and, finally, development of a scar. His hand was sufficiently painful that he first sought medical aid on January 27, 1896. Realizing from his own experience the destructive power of X-rays, Grubbe on January 29, 1896, treated a patient for carcinoma of the breast with his Crookes tube. This treatment was significant not only because it was one of the earliest (if not the first) therapeutic uses of ionizing radiation, but it is also noteworthy that he acted as one of the first health physicists when he used lead as a shield to protect the rest of the body of the patient. On February 29, 1896, Daniel and Dudley[51] of Vanderbilt University, Nashville, Tennessee, radiographed Dudley's head, and 21 days later they observed skin changes and epilation. Also, in April, 1896, Webster[70] of England repeatedly exposed the elbow joint of a patient to X-ray diagnosis and found that rheumatic pains were relieved by these treatments. On April 13, 1896, Pratt[70] of Chicago began the use of X-rays as a therapeutic agent for the treatment of cancer. In July, 1896, Despeignes[52] of Lyon, France, reported X-ray treatment of a case of carcinoma of the stomach. The patient's general condition was improved and the pain was relieved. In November, 1896, Freund[65] of Vienna began an extended treatment with X-rays of a large, hairy nevus and observed severe skin damage in the patient.

Thus, damage as well as benefits of X-irradiation were observed almost from the beginning of the use of X-rays. Fortunately, because of the low penetrating power of the X-irradiation then available, most of this early damage to man was confined to the skin. Scott[174] of Cleveland reported in March, 1897, on 69 cases of skin damage due to X-rays and in 1902 Codman[46] of Boston reported records of 170 cases of X-ray injury; there were, undoubtedly, many additional, unrecorded X-ray injuries. Even Becquerel[32] learned the need for radiation protection when he carried in his vest pocket a sealed glass tube of radium-bearing barium chloride enclosed in a small pasteboard box. As a result of this exposure, he developed on his chest an erythema which ulcerated and left a scar when it healed. Madame Curie[32] also developed radiation burns on her hands as a result of handling a small tube of radium enclosed in a thin metallic box.

Experiments and practical experiences (only a few of which can be mentioned here) led during the early part of the twentieth century to some information on what to expect when the internal organs of the body are irradiated with the more penetrating X-rays. For example, in 1903 Senn[175] of Chicago reported the treatment of a patient with leukemia. There was improvement in the condition of the patient and the spleen decreased in size. Bergonie and Tribondeau[39] of France in 1904 irradiated the testicles of rats and studied the histological changes. From these studies they formulated the basic law that "immature cells and cells in an active state of division are more sensitive to radiation than are cells which have acquired their adult morphologic and physiological characteristics." In the same year Perthes[161] of Germany, who experimented with ova of *Ascaris megalocephala*, confirmed the hypersensitivity of cells that are in active mitosis. In 1903 Albers-Schonberg[27] of Germany produced sterilization in rabbits and guinea pigs by X-ray exposure. In 1903 Heineke[77,78] of Germany recognized changes in the peripheral blood as a result of X-irradiation of the blood-forming tissues, and he produced leukemia in mice as a result of X-ray exposure. By 1911 Hesse[79] of Germany had collected histories of 94 cases of tumors induced in man by X-rays; 50 of these cases were among radiologists.

Although the first 15 years of experience with ionizing radiation led to a better understanding of the early types of damage that could be expected from excessive radiation exposure, they could not have been expected to serve much warning of what to anticipate from low-level, chronic exposure, or to provide a reliable indication of what to expect many years after an acute exposure. Some types of damage, such as skin cancer, did not make their appearance until 10 to 30 years after the exposure began, and, in some cases, medical men observed malignant skin changes as late as 25 years after they had discontinued the practice of exposing their hands during fluoroscopic examinations. By 1922 Ledoux-Lebard[167] estimated that more than 100 early radiologists had died of occupationally produced cancer. Many radiologists died of radiation-induced leukemia; March[116] found that in the 20-year period prior to 1949 there were nine times as many deaths from leukemia among radiologists as among other physicians.

Several studies have shown that the average life-span of the radiologist is shorter than that of other medical men. For example, Seltser[176,177] found the mean life of radiologists has been shorter than that of medical men who are not radiologists and are thus less likely to be exposed to ionizing radiation.

Table 1-3 indicates that the mean age at death of members of the Radiological Society of North America (radiologists) was less than that of members of the American Academy of Ophthalmologists and Otolaryngologists (ophthalmologists and otolaryngologists) by 4.8 years from 1935 to 1944, by 4 years from 1945 to 1954, and by 2.9 years from 1955 to 1958. Presumably, this decreased life shortening of radiologists during more recent decades is the result of their increasing awareness of the desirability of minimizing exposure.

Table 1-3 Median Age at Death of Members of Medical Societies[176,177]

Society	Specialty	Membership 1935–1958 (person-years)	Median Age at Death 1935–44	1945–54	1955–58
Radiological Society of North America	Radiology	48,895	71.4	72.0	73.5
American College of Physicians	General practice	100,504	73.4	74.8	76.0
American Academy of Ophthalmologists and Otolaryngologists	Ophthalmology Otolaryngology	83,309	76.2	76.0	76.4

Thus, as radiologists have become more and more concerned about the dangers of radiation exposure, they have taken precautionary measures which have paid off in allowing them increasing average life-span as well as minimizing the chance of their deaths from malignancies and other radiation-induced diseases.

Many studies on a number of species of animals have indicated that there is a life shortening of about 10^{-4} life spans per roentgen of low-level, chronic exposure. On this basis, we might estimate that the total dose received by the average radiologist over this period ranged between 400 and 800 R or the annual dose ranged between about 15 and 30 R. In spite of the uncertainty of such extrapolation to man or the average dose received by early radiologists and the variation in energy and in the portion of the body irradiated, this dose estimate has served as one of the few available benchmarks for the establishment of 15 rem/yr as the maximum permissible annual exposure recommended by the National Committee on Radiation Protection (NCRP) and the International Commission on Radiological Protection (ICRP) and now in use as the occupational exposure limit for most organs of the body.

The other long-standing radiation reference standard (especially with reference to internal exposure) has been the maximum permissible body burden of 0.1 μc of ^{226}Ra which corresponds to an average dose to the skeleton of about 30 rem/yr.

Early Internal Exposure

Internal absorbed dose from radium (^{226}Ra—1622 yr), mesothorium (^{228}Ra—5.8 yr), radiothorium (^{228}Th—1.91 yr), and their daughter products has accounted for much suffering and a number of deaths occurring many years after exposure. For example, bronchogenic epithelial carcinomas appeared 10 to 20 years following exposure to radon in the Schneeberg cobalt mines of Saxony and at the Joachimsthal pitchblende mines of Bohemia. These mines contained large concentrations of uranium and other ores such as cobalt, arsenic, and lead as well as silicon. Any one of these might be expected to lead to industrial hazards; however, as pointed out by Sikl[180] the one factor which seemed primarily responsible for the high indicence of lung cancer was the radiation from the daughter products of uranium, that is, ^{226}Ra, ^{222}Rn, ^{218}Po, etc. It had been recognized as early as 1500 that there was a high incidence of lung disease among these miners. In 1879 Herting and Hesse[180] performed the first postmortems on the deceased miners and noted the malignant growth of the lungs. This growth was first shown by Arnstein[180] in 1911 to be a lung carcinoma. Thiele et al.[202] studied 143 miners from the Schneeberg mines over a period of several months. Lung carcinoma was definitely established in 8 cases and pneumoconiosis in 17. In the same period 296 persons from the same district, but not employed in the mines, were examined without finding a single case of lung carcinoma or pneumoconiosis. Peller[162] reported 47 cancer cases including 42 primary pulmonary intrathoracic cancers found from 60 autopsies of 89 miners who died from 1929 to 1938 from among about 400 miners and 100 pensioners of the Joachimsthal mines. The death rate from lung cancer in these mines was about thirty times the normal expectancy.

Beginning at the time of World War I and continuing until about 1930, there were many unfortunate exposures to radium. Some of these were the result of therapeutic injection of radium, the drinking of radium and radiothorium water, and occupational exposures of radium chemists. During this period radium was considered to be a useful therapeutic agent and was administered by physicians in the United States to hundreds of patients. In some cases it was taken as a general tonic and in others it was given as a curative agent for hypertension, anemia, arthritis, and may other human ailments—even for insanity. Just prior to 1931, Schlundt et al.[187] administered ^{226}Ra by injection to 32 patients at the Elgin State Hospital in Illinois. They were given from 70 to 450 μg of ^{226}Ra over a period of two to six months and in 1953 these patients retained from 0.52 to 6.8 μg. This group is of particular importance because (1) it is being studied very carefully by Norris et al.,[154] (2) the initial mixture contained no ^{228}Ra, (3) the initially retained ^{226}Ra is known rather accurately, and (4) it probably provides the most reliable quantitative information on chronic human exposure to radium or any other internally administered radionuclide. These studies led Marinelli[118] to report in 1958, "No case of bone cancer or clinically recognizable pathology due to ^{226}Ra or ^{228}Ra has been reported in individuals carrying burdens less than 0.4 μc." It will be seen under the following discussion of the radium-dial painters that one case under study by Evans et al., has now lowered this level to 0.13 μc of pure ^{226}Ra.[26,59–62] However, in this case the ^{228}Ra content was such that the ^{226}Ra equivalent is estimated to have been 0.62 μc at time of death.

By far the most serious exposures to radium were to young women employed in the radium-dial painting industry. Clark[45] in 1958 stated, "While some were injured, we are reasonably certain that about 2000 luminous-dial painters from the early 1920's are still living, most of them in good health. There are probably several hundred who drank radium water or who had injections of radium salts during the period 1915–1930 who are still very much alive today but, except for a few, we do not know who or where they are. The publicity created by the death of a few was partly responsible for the elimination of the dangers in the use of radium in more recent years." The total number of radium-dial painters and others who took radium by mouth or injection and subsequently died with readily detectable symptoms of radiation damage is not known. The first recorded fatality due to radium in the radium dial industry was in 1925 and was reported by Martland.[117] This was a woman who started work in 1917 at the age of 35 as a dial painter and continued this work until a few months before her death from radiation-induced anemia with leukopenia. Since that time, histories of over fifty such cases have been recorded in the United States. The dial painters ingested radium mainly as a result of the practice of tipping their paintbrushes with their lips. The early deaths were caused primarily by aplastic anemia, tumors of the paranasal sinuses, and osteomyelitis of the jaw and other bone structures, whereas later deaths resulted from osteogenic sarcoma.

In 1941 when an advisory committee of the NCRP[141] established the maximum permissible body burden of radium at 0.1 μg, no serious injury to radium dial workers was known to result with a burden of 10 times this value (or 1 μg). As mentioned above, Evans et al.[26,61] have reported the death of a radium dial worker from radium exposure where the body burden at the time of death was 0.13 μc of pure ^{226}Ra. This woman worked for a clock company from 1924 to 1930, was 16 years of age when employment began, and in 1955 died of osteogenic sarcoma with axillary metastases and generalized carcinomatosis. In 1926 the ratio of ^{228}Ra to ^{226}Ra in her teeth was estimated to be 2.5 so that about 80% of the total dose was delivered by ^{228}Ra. The ^{226}Ra total body burden at death was 0.13 μc (or about 0.62 μc total equivalent of ^{226}Ra). Using the power function data of Norris et al.[154] we may conclude that in this case the bone burden during the early period of occupational exposure was probably the equivalent of an initial single first-day intake retention of about 150 μc of ^{226}Ra. From these data it is evident that a safe body burden of ^{226}Ra should be less than that carried by this woman by perhaps a factor of 10. Thus, if the maximum permissible body burden of ^{226}Ra is to provide an adequate safety factor many years after exposure, it should be no more than about 15 μc on the first day, no more than 0.4 μc 1 year later and no more than 0.08 μc 20 years following a single exposure. These values would be upper limits and perhaps should be reduced still further to provide an adequate safety factor. It should be noted that such a pattern of distribution most certainly would not provide a safety factor of 10 over a 50 year period and, in fact, would not guarantee that there would be no deaths from radium damage throughout this period. Also, it has been shown by J. T. Boyd et al.[16] who studied a group of 62 women that had been employed for 5–7 years during World War II in the watch dial industry, that they now (1965) have about twice as many lymphocyte chromosomal abnormalities as a control group; this is in spite of the fact that these women were subjected to much lower levels of radium exposure than their counterparts just following World War I. Their body contents of ^{226}Ra some 18 years later (1966) as determined by whole-body γ-counting and by radon breath measurements range from undetectable (0.04 μc) to a maximum of 0.56 μc.

Another source of information on the internal dose problem is the large group of persons who were injected with Thorotrast (ThO_2 in suspension) for purposes of diagnostic radiology from about 1928 to 1945. These persons were given 3–15 gm of ^{232}Th (equivalent of 0.4–2 μc of ^{226}Ra) by intravenous injections; 10–25 years after the injections, a number of these persons have developed minor irregularities, such as areas of increased density in some of the bones. Looney[110,111] reports that two patients died of tumors of the liver, 17 and 23 years, respectively, following the administration of Thorotrast. Looney et al.[105] report nine patients in whom malignant tumors have occurred at the site of the Thorotrast injection 6 to 35 years after its administration. Also, they report ten cases of hepatocellular and cholangiocellular tumors 10 to 24 years after injecting Thorotrast, although the absence of a suitable set of controls makes it difficult to attribute the origin of tumors to Thorotrast in all ten cases. Of particular significance is their report of ten cases of hepatic tumors of mesodermal origin 3 to 23 years following the administration of Thorotrast. They point out, "It seems that this rare hepatic tumor is emerging as the predominant neoplasm in the thorotrast patients, as sarcomas of bone have emerged as the predominant neoplasm in patients who received radium medically."

The reader is referred to summary reports by Martland,[117] Evans,[59–62] Silberstein,[179] Looney,[108,109] and many others for more detailed information concerning these radium and thorium exposure cases. The history of early human radiation exposure is of interest to the health physicist not only from a historical point of view but also because of its great value as the principal basis for the selection of levels of maximum permissible exposure.

Recent and Current Sources of Exposures

In addition to natural radiation, early medical use of radium, thorium, X-rays, and the industrial exposure to uranium, thorium, radium, radon, and their daughter products, there are several important sources of human exposure that are of more recent origin. These are currently under intensive study in order to aid in developing a better understanding of the mechanisms of radiation damage and to serve as a guide in the establishment of safe levels of exposure to ionizing radiation. These sources are (1) present-day medical exposures, (2) occupational exposures from accelerators, nuclear reactors, and the radionuclides they produce, (3) environmental contamination from weapons fallout, (4) nuclear weapon exposures at Hiroshima and Nagasaki, (5) radiation accidents, and (6) other sources such as watches, shoe-fitting machines, television, security X-ray inspectoscopes, industrial X-rays, high-voltage rectifiers, exposures in high-flying aircraft, space travel, etc. For completeness, these sources of exposure will be mentioned briefly in this review of the history of radiation damage and protection.

Medical Exposure. Medical exposures to ionizing radiation have increased steadily in frequency and in magnitude of dose during the past three decades. Table 1-4 gives some information relative to the magnitude of the skin dose received from common diagnostic examinations and, for comparison, values of dose are given also for a few other cases, that is, the dose from the treatment of tumors, from shoe-fitting examinations, and at the surface of a watch. It is estimated by a joint study group of the ICRP and the ICRU[205,93,94] that 75–90% of the total dose received by the population-at-large from medical uses of ionizing radiation results from diagnostic uses of X-ray. Perhaps from the point of view of potential damage, the most critical human tissues irradiated are those of the bone marrow, the gonads, and the fetus. Irradiation of bone marrow is of interest primarily in relation to the production of leukemia; gonadal irradiation is of importance because of genetic damage, and fetal irradiation is of concern because it can lead to physical and mental maldevelopment of the child. Perhaps most such defects that stem from exposure to ionizing radiation are never even recognized, and, if they are recognized, the origin is seldom identified.

The United Nations Scientific Committee[205,206] estimates that the mean bone marrow dose is 50–100 mrem/yr from various diagnostic procedures representative of those in such countries as the United Kingdom, the United

Table 1-4 Common X-Ray Exposures to Skin

Type of Exposure	Exposure	Reference
Average chest X-rays at Oak Ridge National Laboratory using full-size X-ray film in direct contact with intensifying screen (1965)—PA exposure	10 mR	72
Average chest X-ray exposure found from survey in England—PA exposure (1954)	160 mR	202
Average chest X-ray in United States (1957)	200 mR	112
Range of measured chest X-ray exposures	10–1000 mR	112, 193, 136, 72
Range of measured dental X-ray exposures	0.02–350 R	72, 162, 136, 112
Average GI tract exposure	22 R	112
Range of measured exposures from fluoroscopic examinations	5–400 R	112, 136, 33
Average routine X-ray examination during first half of pregnancy	46 R	112
Average routine X-ray examination during second half of pregnancy	56 R	112
Treatment of malignant tumors	3000–7000 R	136
Fluoroscopy in shoe-fitting (per 20-second exposure)	7–14 R	136
Exposure at surface of watch (β and γ)	0–50 R/day[a]	

[a] This high dose of 50 R/day was measured by D. M. Davis of Oak Ridge National Laboratory at the surface of the crystal of a wrist watch that had an excessive amount of ^{90}Sr on the dial.

States, and Sweden. The Committee estimates that of this figure, 5–50 mrem/yr are received from fluoroscopic examinations, 17 mrem/yr from GI tract examinations, 10 mrem/yr from mass chest X-ray examinations, and 2 mrem/yr from dental examinations. The Committee points out that for the United States, however, the dose from dental examination is probably 8 mrem/yr instead of 2 mrem/yr, and in countries where fluoroscopy is used extensively the mean bone marrow dose from mass chest X-ray may be much greater than 10 mrem/yr. For example, the mean bone marrow dose arising from mass survey fluoroscopy of the chest was calculated to be 1900 mrem/yr in Austria, 1200 mrem/yr in France, and 1300 mrem/yr in Spain. The genetically significant dose from diagnostic X-ray varies considerably from country to country. For example, the United Nations Scientific Committee[205,206,112] estimates that this annual dose is 50 to 141 mrem/yr in the United States, 58 mrem/yr in France, 14 mrem/yr in the United Kingdom, 10 mrem/yr in Norway, and 6.8 mrem/yr in the Netherlands. The report of the Committee points out the need to reduce these medical exposures as much as possible, consistent with good medical practice. Some of the suggestions for accomplishing this are (1) use shields and X-ray cones of high atomic number to limit the irradiated tissue to a minimum and to include only that tissue under study or under treatment, (2) increase the anode–skin distance to no less than 30 cm in order to protect the skin of the patient, (3) use at least 2 mm of Al filters to remove the low-energy X-rays from the beam, (4) use higher-voltage techniques for diagnostic work, (5) use image intensifiers for fluoroscopic work, and (6) use faster films and faster screens, and dark-adapt the eyes at least 10 minutes before beginning fluoroscopic examinations. Gonad doses could be reduced very decidedly (for example, by a factor of 50–100 for males) by these improved techniques and by the use of better equipment.

Russ made one of the first studies of the general radiation protection problem.[169] Shortly thereafter in 1916, the (British) Röntgen Society unanimously adopted a resolution recognizing hazards from X-ray examinations and recommending a number of precautionary measures.[36] This was followed by their publication of general recommendations* and measures for X-ray and radium protection.[37] Unfortunately, many of these recommendations have attracted very little attention and in some cases have been completely disregarded. A number of radiation protection handbooks have been published by the National Committee† on Radiation Protection and Measurements (NCRP)[142–144] and the International Commission on Radiological Protection (ICRP).[95] These publications should be studied carefully for a more comprehensive understanding of this vital health physics problem. For a quick review of the recommendations of the NCRP and the ICRP concerning radiation protection in the use of medical X-rays, the reader is referred to publications of Taylor,[198,199] who has been chairman of the NCRP since its formation in 1929.

There is considerable variation in the estimates of the present genetically significant dose to the average member of the population from radiation therapy, but in most countries it is considerably less than the diagnostic dose. For example, Clark[44] estimates 12 mrem/yr in the United States, Martin[119,120] estimates 28 mrem/yr in Australia, and Purser and Quist[165] estimate 1–1.5 mrem/yr in Denmark. Table 1-5 summarizes a few typical values of average

* The recommendations[37] are reprinted in Appendix 5, Reference 100.

† In 1964, "Committee" was changed to "Council."

genetic dose from diagnostic and therapeutic use of ionizing radiation. The present exposure of the medical workers is estimated[203] to range from 500 to 5000 mrem/yr. The number of persons in this exposure group is relatively small so that the estimated per capita dose equivalent from this source is less than 1 mrem/yr in most countries.

Skin dose from medical exposure is frequently very large. An inspection of the values of absorbed dose given in Table 1-4 for typical X-ray examinations indicates that some of these values are unnecessarily high. For example, a few organizations deliver a peak dose of 10–20 mrem to the skin per chest X-ray,

Table 1-5 Average Genetic Dose to Each Member of a Population from Diagnostic and Therapeutic Use of Ionizing Radiation

Country (Reference)	Diagnostic (mrem/yr)	Therapeutic (mrem/yr)	Radioisotopes (mrem/yr)
United States[44]	84	12	8
United States[112,206]	137 ± 100	17	0.25–7
Australia[119,120]	159	28	–
Hamburg, Germany[206]	17.7	2.2	0.19
France[206]	58.2	5.6	–
Leiden, Netherlands[206]	6.8	4.1–13.1	–
United Kingdom[206]	14.1	5	0.18
Denmark[165,206]	27.5	1–1.5	–

whereas others are using X-ray equipment and techniques that deliver a skin dose of from 1000 to 5000 mrem per chest X-ray. These latter organizations usually are using antiquated equipment and often attempt to save a few cents on X-ray films by using a 35-mm camera to photograph a fluorescent screen, rather than using a full-size 14 × 17 in. X-ray plate. The dose may be unexcusably high because of the use of a cheap camera with a slow lens. The harm done because of the increased probability of chronic damage to such exposed persons makes this medical practice very wasteful and one that should be corrected.

Unfortunately, some of the states in the United States have no manpower, facilities, equipment, or authority to inspect and correct the various forms of medical X-ray equipment and most of the existing state programs are understaffed and very inadequate. Since about 14% of the United States population is examined each year in the mass chest photofluorogram program, it should be mandatory that the best equipment and techniques be used; it is essential to calibrate this equipment frequently and it is important to seriously evaluate the desirability of continuing chest X-ray programs on a mass scale. Probably it is unwise to include children in mass chest X-ray programs except in areas where there is a high incidence of tuberculosis. Certainly, chest X-rays should be made of persons who have a high risk of tuberculosis, but then only with the best equipment and techniques. Perhaps most members of the population should be given chest X-rays only when tuberculosis tests are positive or when there is an indicated need in individual cases.

A number of surveys of dental X-ray equipment have been made during the past few years. All the extensive surveys made in the United States reveal essentially the same situation: some dentists have removed the filters, decreased the focal distance to the patient's jaw, and are using such insensitive X-ray film or poor film-processing techniques that very high doses are reported. For example, Nolan[153] reported that, in some instances, during a single series of dental X-rays, doses up to 300 R are delivered in the region of the lips, and that some dentists have removed the cone and modified the X-ray tube housing so that the whole body, including the gonads of the patient, is sprayed with X-irradiation. About 400 per 1000 persons in the United States each year receive dental radiograms. Many people are exposed by equipment that has never been calibrated or inspected since it was installed. Much existing dental X-ray equipment needs modification, adjustment, or replacement. Fortunately, since about 1963 the U.S. Public Health Service and some of the state public health agencies are conducting programs to reduce unnecessary dental exposure.

Perhaps the most risky exposure to man is that delivered to the fetus during the early stages of pregancy. Data summarized by Laughlin et al.[112] indicate that in the United States 11.4% of obstetrical patients are given X-ray pelvimetry examinations resulting in a probable average dose of 4 R per examination to the gonads of the fetus. As a consequence an average of about 5 R is probably received *in utero* by the child whose mother undergoes pelvimetry and other associated types of X-ray examinations. The practice of some doctors and certain medical establishments of automatically conducting an X-ray examination of all pregnant women under their care is considered to be an extremely poor demonstration of good health physics principles. Table 1-4 indicates that the skin dose received by the mother during these obstetrical examinations may be quite high and that in exceptional cases the dose equivalent to the fetus may be as high as 30 rem. The Russells[170] point out, from their studies on several species of mammals, that the most critical period in man for the production of gross abnormalities is that from the second to the sixth week of gestation. Exposure at later stages may produce less obvious and possibly more delayed and more numerous effects which, under many circumstances, may be as harmful as gross monstrosities. The Russells have produced serious damage to the embryos of animals at doses as low as 25 R. They point out that even lower doses applied during the critical period may result in serious damage. In view of the fact that early stages of pregnancy often go undetected for some time, the Russells recommended in 1952 that "whenever possible, irradiation involving the uterus in women of childbearing age should be restricted to the two weeks following the last menstrual period, to preclude the possibility of irradiating the fertilized ovum." This recommendation of the Russells is actively subscribed to in Denmark and, as indicated by Hammer-Jacobsen,[75] relatively firm measures are taken there; for example, he suggests that the need for a therapeutic abortion is always indicated in cases where fetal doses are estimated to have exceeded 10 R.

In 1962, the ICRP[96] made a very important and significant recommendation relative to the radiological examination of women of reproductive age as follows:

> "The Commission wishes to call attention to reports of embryonic and fetal sensitivity to ionizing radiation and to emphasize that the possiblity of pregnancy must be taken into account by the attending physician when deciding on radiological examinations that involve the lower abdomen and pelvis of women of reproductive age. The Commission also wishes to point out that the 10-day interval following the onset of menstruation is the only time when it is virtually certain that women of such age are not pregnant. Therefore, it is recommended that all radiological examinations of the lower abdomen and pelvis of women of reproductive age, that are not of importance in connection with the immediate illness of the patient, be limited in time to this period when pregnancy is improbable. The examinations that should be delayed to await the onset of the next menstruation are those that could without detriment be delayed till the conclusion of a pregancy or at least until its latter half."

This has particular significance and may have far-reaching influence in reducing human exposure because, for the first time, the ICRP offers a specific recommendation to members of the medical profession to be very cautious and weigh carefully the need for a radiological examination in view of the high probability of radiation damage to a portion of the population that is particularly susceptible to such damage. In spite of some objections raised by the American College of Radiology (ACR), it is anticipated this ICRP recommendation may be of great influence in significanly reducing radiation damage to the world population. The ACR[25] indicated it approved the motive of the ICRP but disapproved the recommendation. It stated, "Occasions in which this policy becomes important are quite infrequent.... It also believes that the hazards of diagnostic radiation in the hands of a trained radiologist are small in relation to the health benefit gained from properly chosen procedures, properly done." This statement would appear to be open to question in view of Laughlin's estimate above that 11.4% of obstetrical patients are given X-ray pelvimetry examinations with an average fetal dose of 4 R per examination. Also, by their own admission a word of caution should extend to any improperly trained or poorly equipped radiologist and similarly to other medical groups.

Quoting from the ACR,[25] "When untrained people operate unsafe equipment, there is no way of knowing whether or not the examination was justified and how much unnecessary radiation exposure resulted." The fact should not be overlooked that a large number of medical doctors, who not only are not radiologists but also have had little or no training in the use of X-ray equipment, have and use X-ray equipment. Often, this equipment is not in proper operating condition and is frequently used in such a way that the dose to the patient is more than an order of magnitude greater than that which would be delivered by a proper radiographic examination. It should be noted also that in June, 1965, California became the first state in the United States to pass a law requiring that courses in roentgenologic technique and radiation safety be incorporated into the medical school curriculum and that the State Board Examination include questions in these subjects. It would seem that lacking such laws in other states there probably are many in the medical profession who might profit by education and advice in the proper use of X-rays.

It is estimated by Laughlin et al.,[112] that in the United States X-ray examinations—exclusive of dental and mass chest X-rays—are given as follows: 75% by radiologists in hospitals, 13% by general practitioners, 2% by osteopaths, 5% by chiropractors and chiropodists, 1% by pediatricians and orthopedic surgeons, and 4% by other physician specialists. The average genetically effective dose from medical exposure in the United States is estimated by Laughlin et al. to be 4.6 rem/30 yr (154 mrem/yr), of which 39% is from radiography, 33% from fluoroscopic examinations, 15% from obstetrical X-rays, 11% from therapy treatments, and 2% from dental X-rays, photofluorograms, and mass chest X-rays. One encouraging feature of these studies is that during the past few years since the publication of a number of reports, [Refs. 93, 94, 112, 124, 153, 193, 198, 199, 205], there has been considerable effort, especially by some of the radiologists and dentists, to reduce or discontinue unnecessary exposure to the patient and thereby lower materially the major source of both somatic and genetic exposure of the population.

There is considerable uncertainty relative to the average genetically effective dose from medical sources of ionizing radiation in the United States. Plans are under discussion to conduct an extensive national survey and until this is done one can only speculate why the medical exposure in the United States is so much higher than that in other countries such as the United Kingdom and Norway. Some have suggested that the lower limit of 154 ± 100 mrem/yr given by the estimate of Laughlin et al.,[112] is a more likely value, whereas others with equally good evidence believe the higher estimate is closer to the truth. R. H. Chamberlain[42] supported the lower estimate before the Joint Committee on Atomic Energy of Congress on June 6, 1962, in his statement, "It seems reasonable to assume that our own (United States) exposures are probably closer to 25 to 50 millirads per year than the preliminary estimate of 150 milliards." He went on to indicate that because of "... improved X-ray techniques, better selection of patients, faster film emulsions, and intensified images, plus state and local regulation and educational efforts, the value of medical X-ray procedures has increased in recent years while the risk factor was diminishing."

On the other hand, Blatz[34] in 1962 provided equally convincing evidence that the upper limit of dose estimated by Laughlin et al. may be more nearly correct,

"There appears to be no good evidence as to the actual population dose in the United States resulting from the medical and dental use of X-rays. Estimates of X-ray population dose in the United States have been made by either radiologists or radiological physicists. Both groups have access primarily to radiological usage and techniques employed in progressive hospitals and radiologists' offices. New York City registration figures show that only 15% of the X-ray machines in use are in hospitals or radiologists' offices. The records also show that more than half of the units in use are fluoroscopes. Very little is known about fluoroscopic doses but it is our impression that nonradiologists in New York City use fluoroscopes much more than is generally suspected. The inspection of many thousands of X-ray units used for chest radiography shows that more than 50% of the units employ essentially no collimation (beam in excess of 30 inches in diameter), which results in the exposure of the gonads and the eyes to hundreds of times the exposure generally listed as typical for those organs in chest radiography. Two-thirds of the units inspected employ X-ray beams in excess of the liberal 24-inch diameter considered necessary to cover a 14 × 17 inch film (22-inch diagonal). These percentages were found to hold true in hospitals and in radiologists' offices as well as in the offices of nonradiologists. We know very little about the workloads of X-ray equipment in the offices of nonradiologists but expect to have good data within a year or two. A comparison of our inspection results from three and a half years ago when we started, compared to the results gained in the most recent inspections shows little evidence that our attempt at an educational program has resulted in improved techniques. On the other hand, there appears to be some evidence of a decrease in the use of X-rays in certain specialities, particularly pediatrics."

It is of great interest to the health physicist that medical records at the present time are being studied and analyzed in an effort to relate medical exposures to the incidence of various forms of cancer, leukemia, gross abnormalities in children, and genetic changes. Stewart et al.[206,195] made a study of children in England under the age of ten and concluded that prenatal exposure approximately doubles the probability of a child dying of leukemia or cancer before the age of ten. These results have since been confirmed by some investigators in the United States, but other investigators have failed to observe this effect. MacMahon[207,121] studying a population of 73,243 children estimated that the cancer mortality was 40% higher in those who received intrauterine X-ray exposure than in the controls who had not been exposed and that this relationship held for leukemia, neoplasms of the central nervous system, and other neoplasms. The excess cancer mortality in the X-ray group was most marked at ages five to seven years at which time the relative risk was 2. He observed no significant variation with the stage of pregnancy at which the X-ray exposure had been received. The final quantitative conclusion must await additional data. For the present, it is safest to conclude that *in utero* exposure even at relatively low doses is harmful to the child because it increases the probability of the child contracting leukemia and many types of neoplasms and sustaining other forms of damage. Except where the need is imperative, such exposure should be avoided during any stage of pregnancy and especially during the first half.

The X-ray treatment of ankylosing spondylitis seems to offer one of the best bases now available for the evaluation of risk of radiation-induced leukemia. The studies of Court-Brown and Doll,[41] of these patients, indicated 16 to 17 leukemias per 10^4 men/yr following a mean spinal-marrow dose of greater than 1750 R, and 72 cases per 10^4 men/yr following 2250 R. This is to be compared with a natural incidence of leukemia of 0.5 cases per 10^4 men/yr. These and other data prompt these authors to suggest the hypothesis that perhaps there is a linear relation between absorbed dose and leukemia incidence. From this group of men exposed to 1750 R, linear extrapolation into the low-dose region would indicate that there may be about 10^{-6} leukemias/yr/R delivered to the spinal marrow.

In the United States the average medical exposure is probably about the same, if not somewhat greater, than the average background exposure; for example, as estimated by Laughlin et al.,[112] the genetically significant dose-equivalent from medical sources is 4.6 rem/30 yr or 150 mrem/yr. The average medical dose to bone marrow in the United States is not known; it probably is not much different from the gonad dose. In Table 1-2 the average natural background dose to the gonads is given as 100 mrad/yr or 100 mrem/yr, and the corresponding dose to the bone is given as 100 mrad/yr or 180 mrem/yr. Other sources of population exposure to ionizing radiation are much smaller than background and medical exposure.

Exposure from Accelerators and Reactor Operations. The United Nations Scientific Committee reports[206,207] indicate that the genetically significant dose equivalent to the population from all occupational exposure (that is, both research and production) is very small. For example, it was estimated to be 0.2 mrem/yr in 1955 in Austria, 0.3 mrem/yr in 1960 in the Netherlands, and 0.4 mrem/yr in 1959 in the United Kingdom. Perhaps in most countries the occupational exposure is even less. In the United Kingdom the estimated population dose contributed by the atomic energy workers is about 40% of the occupational dose or 0.15 mrem/yr. This includes exposures to radiation (neutrons and γ-rays) from nuclear reactors as well as exposures from associated chemical and engineering operations in the fabrication of fuel elements, separation of the fission products, and transuranic elements, and a wide variety of research operations. The contribution to population exposure from high-voltage accelerators is less than that from nuclear reactors by several orders of magnitude.

In spite of the insignificance of these recent sources of research and industrial exposure in terms of the total dose equivalent to the population-at-large, they are very important to the individuals involved in this work and especially to the health physicists, who expend a great deal of effort to keep these exposures as low as possible and, in any case, below the maximum permissible levels specified by the NCRP and the ICRP. These exposures are also of particular importance because, unlike medical exposures, accurate records are maintained of the accumulated dose of each type of ionizing radiation for each individual employed in these operations. Rather extensive studies of the exposure records are now under way at several of the national laboratories and production operations of the U.S. Atomic Energy Commission. It is to be hoped that useful conclusions can be derived from these studies of many thousands of human exposure records that have accumulated for periods of more than 20 years.

The vast majority of these individual exposure records are far below the maximum permissible values recommended by the ICRP and the NCRP. For example, there have been a number of minor accidents and a few major accidents leading to both external exposure from X-, γ-, β-, and neutron radiation and internal exposure from a variety of radionuclides that emit α-, β-, X-, and γ-radiation. Radionuclides can become distributed in the various organs of the body following intake by ingestion, inhalation, injection into open wounds, and direct absorption through the skin. The human exposures to radiation from accelerators are of particular interest in that they have provided very useful data on cataract production by neutrons. The criticality accidents with fissionable material have been most enlightening because of the data they have furnished on acute exposure to mixed radiation, that is, neutron and γ-radiation. Although the health physicist strives to anticipate and avoid all situations that might lead to a radiation accident, he is equally alert to obtain all possible information for the evaluation of each accident and to determine for various types of radiation exposure the relation between absorbed dose and the resultant radiation damage. Radiation accidents are discussed later in this chapter.

Environmental contamination by radioactive waste from nuclear reactors and their associated operations is an important health physics problem under study. Reasonably economical methods of radioactive waste disposal must be developed within the next few years to provide the necessary and permanent assurance for adequate protection of man and his environment when there is widespread use of and dependence on nuclear energy. Several methods for the long-range and permanent disposal of radioactive waste are under study. Perhaps for many years to come low and intermediate levels of radioactive waste from the chemical processing plants associated with reactor operations can be released into the environment. At present, these wastes are being discharged into seepage basins from which there is slow seepage to nearby rivers or they are diluted with large quantities of water and discharged directly into rivers, lakes, or the ocean. High-level radioactive waste for the most part is being stored in large underground tanks—only postponing the time when a safer and more permanent method of disposal is developed and put into operation. Three of the more promising methods for the disposal of intermediate and high-level waste have been under study at Oak Ridge National Laboratory: (1) liquid injection into deep permeable formations, (2) disposal by hydraulic fracturing, and (3) storage in salt formations [see Refs. 134, 135, 209–211]. Each of the three methods appears to offer certain advantages. However, no one method or none of the three methods would be satisfactory in many localities.

Thus, in planning a large nuclear energy power plant, consideration must be given to siting that will provide methods of adequate disposal of the radioactive wastes from the associated chemical operations. At Oak Ridge National Laboratory the hydraulic fracturing method is proving very satisfactory for the disposal of laboratory waste. By this method the radioactive waste is mixed with cement and clay and pumped under several thousand pounds of hydraulic pressure to a depth of about 1000 ft. Here it passes through an opening that has been cut in the well casing at the proper depth. This fluid under pressure fractures the shale formation and spreads out in uniformly thin layers, about 1 mm thick. This liquid slurry is then permitted to set and harden, and a few weeks or months later another slot is cut a few feet further up in the well casing and another injection of a few tens of thousands of curies of radioactive waste is disposed of. This method has the advantage that the waste is distributed in a thin layer that permits adequate disposal of the heat generated by the fission products. There appears to be no likely mechanism that would return the radioactive material to the ground water or to the surface.

The disposal of radioactive waste in salt formations is actively under study[134,135] in salt mines in the State of Kansas. Here the high-level waste is reduced to a solid and sealed in containers; then, by using especially designed equipment, the containers are placed in slots cut in the floor of the salt mine. This method has the advantage that the fission products may be recoverable later should there be a use for them, but the cost of this method of disposal is probably much greater than that of hydraulic fracturing.

Deep-well injection has not as yet been used as a method of waste disposal, but it may offer many advantages in areas where there are suitable geological formations. This method would have the disadvantage that all the waste would have to be put into solution and all solid radioactive waste would have to be filtered out and disposed of by an auxiliary method of waste disposal. All three methods mentioned have the advantage that presumably suitable geological formations can be found in many parts of the world in which one or more of these methods offers promise of success.

At present the contribution of wastes from reactor operations to the total dose received by the population-at-large is quite negligible. In local environments in the neighborhood of the controlled areas of the large nuclear energy establishments, the radioactive waste is very carefully planned and kept under constant surveillance by the health physicist so that the maximum permissible levels suggested for the environment by the ICRP are not exceeded. It should also be kept in mind that the disposal of radioactive waste from hospitals and medical research establishments presents some of the problems existing in the nuclear energy industry. Although the quantities of waste involved are much smaller, the waste originates in populated areas rather than the isolated locations of the nuclear energy industry. Thus a frequent problem is how much and in what manner may radioactive waste be disposed of in the public sewer. References should be made to the ICRP[97] and NCRP[145–149] publications for suggestions and recommendations relating to this problem.

Environmental Exposure from Fallout. Radioactive fallout from nuclear weapons tests provides a much more effective and extensive mechanism for worldwide radioactive contamination than does radioactive waste disposal from the nuclear energy plants. However, in both cases the radionuclides of principal concern are those which (1) are produced in large concentrations (high fission yield or high neutron cross section for materials exposed during the fission process), (2) have an intermediate half-life (from 1 to 10^4 yr), (3) have a large uptake by the human body, and (4) have a slow turnover (that is, a long biological half-life) in the critical body organs. ^{239}Pu and some of the other transuranic radionuclides meet all four of these requirements for radioactively dangerous materials but because of their great value, their concentration in radioactive waste is usually very low—although often not negligible. In some types of weapons, however, where ^{239}Pu is used as the fissionable material or where ^{238}U is used to provide inertia, the contamination by ^{239}Pu could be relatively high.

The increasing use of the more dangerous radionuclides such as ^{238}Pu (86.4 yr), ^{241}Pu (14 yr), ^{241}Am (458 yr), ^{242}Cm (163 days), ^{210}Po (138.4 days), ^{137}Cs (30 yr), ^{147}Pm (2.6 yr), ^{144}Ce (285 days), and ^{90}Sr (27.7 yr) as sources of energy for small power plants in satellites, space ships, lighthouses, Arctic stations, etc., will in time present serious problems of environmental contamination. For example, the malfunctioning of a United States space rocket[88–90] carrying 17,000 curies of ^{238}Pu and its reentry and incineration in the upper atmosphere over the Indian Ocean in April, 1964, is cause for some concern until more is known about the hazards that may be associated with the inhalation of submicron particles of the more dangerous radionuclides. It can be shown that the dose rate equivalent of these particles when deposited in the respiratory system is thousands of rem per week to the tissue close to them. If, on the other hand, the radioactive decay energy of these particles were dissipated over hundreds of grams rather than over micrograms of tissue, the dose rate equivalent would be negligible. Experimental data relating to this problem are very limited but some animal experiments indicate a high dose in a localized area is more damaging than a small dose to a large mass of tissue but involving the same number of curies. However, there are other experiments that seem to lead to a reverse conclusion.

In certain operations, radionuclides which meet only some of the four requirements above, but not all of them, may under certain circumstances present the principal environmental contamination problems. For example, although ^{131}I (8 days) does not meet the second requirement, it presents perhaps the most serious environmental problem from the standpoint of airborne contamination associated with reactor accidents because of its high fission yield, its gaseous state at ordinary pressure and temperature, its facility to be occluded on dust particles and on exposed surfaces, its large uptake and concentration in the relatively small mass of tissue comprising the thyroid of man and animals, and its efficient turnover from the grass eaten by dairy cows to fresh milk that is largely consumed by young children. By far the radionuclides of greatest concern, with reference to the four characteristics listed above, and of most importance from the standpoint of environmental hazard from both radioactive waste disposal and nuclear weapons fallout, are ^{90}Sr (27.7 yr) and ^{137}Cs (26.6 yr). These two radionuclides rate high in all four of the categories, ^{90}Sr concentrating in the skeleton and ^{137}Cs being distributed throughout the entire body, including the gonads.

There are three classes of fallout from nuclear weapons: (1) local, (2) tropospheric, and (3) stratospheric. The amount of local fallout and its pattern of distribution depend markedly upon the type of weapon, kiloton size, the associated material in the weapon, height of firing, meteorological conditions, rainfall, etc. Fissile materials do not ordinarily contribute appreciably to the fallout material. However, in the case of an abortive firing, ^{239}Pu or ^{235}U may cause the principal local contamination. After a weapon detonation at low elevation, the relatively short-lived fission products present for a few days the principal problems involved in contamination of the local environment for a distance of a few hundred miles. Some of the more important of these radionuclides are ^{89,91}Sr, ^{91,92,93}Y, ^{95,97}Zr, ^{95,97}Nb, ^{99}Mo, ^{103,105}Ru, ^{103,105}Rh, ^{132}Te, ^{131,132,133,135}I, ^{140}Ba, ^{140,141}La, ^{141,143,144}Ce, ^{143,145}Pr, ^{147,149}Nd, ^{149}Pm, and, in some cases, radionuclides such as ^{24}Na, ^{31}Si, ^{45}Ca, ^{59}Fe, ^{58}Co, ^{65}Zn, etc., which may be produced in elements present in the vicinity of the fireball of the weapon. These radionuclides (with half-lives of a few hours to a few months) are characterized by their intense, high-energy β-radiation. They are formed on the larger fallout particles and do not have the noble gases Kr and Xe as parents if the parent half-life is longer than about 10 sec. The dose equivalent of γ-radiation and especially of β-radiation received by persons in the local area following the detonation of a high-kiloton weapon can be very high, as witnessed by the experiences of the Japanese fishermen aboard the *Fukuryu Maru* and the natives and Americans who were exposed in the Marshall Islands.

Table 1-6 Groups Exposed in the Marshall Islands to Fallout Radiation from the Testing of a Hydrogen Weapon on Bikini Atoll in the Spring of 1954

Group and Island (Total No. of Persons)	Distance from Bikini (mi)	Best Estimate of γ-exposure (R)
Marshallese on Rongelap (64)	100	175
Marshallese on Ailinginae (18)	80	69
Marshallese on Utirik (157)	300	14
Americans on Rongerik (28)	150	78

As a result of fallout from the hydrogen weapon test at Bikini Atoll on March 1, 1954, a number of persons in the Marshall Islands were exposed to high doses of radiation. The average γ-dose was measured with instruments, and the results of these measurements as reported by Sondhaus et al.[201] are summarized in Table 1-6. The doses from β- and low-energy γ-radiation were not measured, but Sondhaus estimated that this dose on Rongelap Island was 2000 rep at the level of the dorsum of the feet, 600 rep at hip level, and 300 rep at head level of the Marshallese who were continuously exposed with no shielding. Those who remained indoors and those who wore heavy clothing were provided with significant protection, especially from the β-radiation. The children who went swimming washed a considerable portion of the fallout from their bodies and, as a consequence, fared better than others who did not in the early period cleanse themselves of the fallout material. Dunning[55] has evaluated the biological hazard from this exposure which is due primarily to β-radiation of the surface of the body. Among the 64 Rongelapese receiving an estimated dose of 175 R of γ-radiation, there were 13 with deep lesions, 45 with superficial lesions, 6 with no lesions, and 35 with epilation. Those on Ailinginae receiving 69 rem or less of γ-radiation had mostly superficial lesions.

The residual effects of this exposure to the Marshallese were studied and reported by Conard and Hicking[17,43] in 1965. Nodules were detected 9 to 11 years after fallout exposure in the thyroid glands of 11 of the Rongelap people (in 8 children or adolescents and 3 adults) among the group of 82 who received what was estimated to be the highest exposures. The thyroid glands of six of the cases were treated surgically. Five of the treated cases were children and their nodules were benign, but the nodules in the other case (an adult woman) were malignant with local metastasis. Thyroids of two young adults and a middle-aged man were sufficiently different from normal to be listed as questionable. No nodules were found on thyroids of 75 unexposed children from these islands.

Other possible effects of this radiation exposure that were noted by Conard and Hicking among these 82 exposed Rongelapese were (1) slight retardation of statural growth and bone maturation in exposed children, particularly in boys exposed at less than five years of age, (2) greater incidence of miscarriages, stillbirths, and neonatal deaths to exposed women during the first four years after the incident, but no longer an increased incidence, (3) incomplete recovery of some of the peripheral blood elements, and (4) increased incidence of nevus-like lesions (moles or scars) in areas of previous β-radiation burns of the skin but no evidence of chronic radiation dermatitis or skin cancer was noted. There are no definite indications of effects on birth rate, aging, leukemia, or genotype. There were two deaths from cancer in the exposed group but no clearcut relation to the radiation exposure has been established.

The fishermen on the *Fukuryu Maru* evidenced deep skin lesions and other symptoms of radiation sickness. One of the fishermen died, but there is some question as to whether he died of radiation sickness or hepatitis; in any case, radiation exposure led to his death. It is very likely that if these fisherman had known enough to wash the fallout material from their bodies and from the deck of their fishing vessel, their radiation exposure would have been almost negligible.

The γ-doses from fallout in the communities in the local area surrounding the Nevada test site are given by Dunning.[53] The cumulative exposure dose in the local area (to May, 1958) from all the nuclear tests ranges up to 5 R, for example, 5.25 R at Lincoln Mine, Nevada (population 250–500), 4.28 R at Hurricane, Utah (population 1375), 4.30 R at Bunkerville, Nevada (population 250), etc. Records were not maintained during this period on the ^{131}I fallout levels in the communities immediately downwind from the Nevada test site but there is some evidence that these levels probably were higher by an order of magnitude than the levels measured in the summer of 1962 in Salt Lake City, Utah, where they rose to twice the top of Range III for ^{131}I, a value that had been set as an action point by the Federal Radiation Council. The top of Range III corresponds to human consumption of 1 nc/day of ^{131}I, which is related to a dose rate equivalent of about 7 rem/yr to the thyroid of the average young child.

Fallout material injected into the troposphere (40,000–60,000 ft) returns to the earth over a period of a few months as a result of sedimentation and rainfall, and during this time it is carried around the earth several times. This fallout material includes most of the fission products found in the local fallout and significant ^{90}Sr, $^{90,91}Y$, ^{106}Ru, ^{137}Cs, ^{140}Ba, ^{144}Ce, and ^{147}Pm. Induced radioisotopes are also found in varying quantity, depending on the type of weapon, elevation of the shot and the tracer elements that may have been introduced. Some of these are $^{181,185}W$, $^{57,58,60}Co$, ^{65}Zn, $^{55,59}Fe$, ^{54}Mn, ^{45}Ca, and ^{102}Rh. The gonad and average marrow dose equivalent resulting from fallout material from the troposphere has been very intermittent, reaching a peak shortly after each weapons test. In 1958, the United Nations Scientifiic Committee[205] estimated this dose from external radiation to be about 0.5 mrem/yr. It estimated that the internal dose equivalent to bone marrow in 1958 was about 0.2 mrem/yr and the internal dose equivalent to the thyroid (excluding local areas in the United States within a few hundred miles downwind from the Nevada site) was about 5 mrem/yr.

Perhaps some or all of the radionuclides from a large nuclear weapon explosion above ground are ejected into the stratosphere. However, radionuclides $^{89,90,91}Sr$, $^{90,91}Y$, and $^{135,137}Cs$ that have the relatively long-lived precursors Kr ($T = 9.8$ sec to 3.18 min) and Xe (3.9 min to 9.13 hr) as parents predominate in this fallout material. These radionuclides are stored as atomic or submicron particles in the stratsophere from which they fall out over the entire world. The rate of fallout varies with latitude and has been greatest in the northern hemisphere where most of the tests have been conducted. The mean residence time of material higher than 100 km in the stratosphere is greater than five years while in the lower stratosphere it is less than one year. The United Nations Scientific Committee[207] in 1964 adopted for purposes of calculation an over-all mean residence time of two years for the composite stratospheric fission products. This United Nations Committee estimated the accumulated absorbed dose up to the year 2000 from fallout resulting from weapons tests for the years 1954–1962 will be 76 mrad to the gonads, 163 mrad to bone marrow, and 257 mrad to cells lining the bone surfaces. If there are no further atmospheric tests in this period (1954–2000), the average annual dose over this period will be about 2 mrad/yr to the

Table 1-7 Dose Commitments from Nuclear Explosions

				Dose Commitments as a Result of Weapons Tested in Period 1954 to 1962	
Tissue	Radiation Hazard Principally Concerned	Type of Exposure	Radionuclides	Total Dose to Year 2000 (mrad)	Total Dose to Infinite[a] Time (mrad)
Gonads	Genetic damage	External	Short-lived	21	21
			^{137}Cs	29	29
		Internal	^{137}Cs	13	13
			^{14}C	13	180
			Total	76	243
Cells lining bone surfaces	Bone tumors	External	Short-lived	21	21
			^{137}Cs	29	29
		Internal	^{90}Sr	174	174
			^{137}Cs	13	13
			^{14}C	20	290
			^{89}Sr	0.3	0.3
			Total	257	527
Bone marrow	Leukemia	External	Short-lived	21	21
			^{137}Cs	29	29
		Internal	^{90}Sr	87	87
			^{137}Cs	13	13
			^{14}C	13	180
			^{89}Sr	0.15	0.15
			Total	163	330

[a] By the year 2000 the percent of the eventual radioactive decay will be approximately 100% of short-lived radionuclides, 65% of ^{137}Cs, 63% of ^{90}Sr, and 0.5% of ^{14}C. However, as indicated by a comparison of the numbers in the last two columns the disappearance from man's environment is much more rapid than would be expected from radioactive decay alone because of environmental factors (erosion, dispersion, isolation, etc.).

gonads, 4 mrad/yr to bone marrow, and 6 mrad/yr to cells lining the bone surfaces.

The peak annual absorbed doses during the years just following the weapons tests are an order of magnitude larger than these average dose rates, but even then the dose from fallout has been small compared to the average medical dose in the United States. In a few small population groups, the absorbed dose rate on occasion rose above these values by a factor of 100 or more. Typical of such groups are inhabitants near weapons test sites or persons living in Arctic regions who depend for food almost exclusively on reindeer and caribou meat. With the exception of the dose from ^{14}C, essentially all of the dose from fallout of weapons tested in 1954–1962 will be delivered by the year 2000 as a result of radioactive decay and environmental erosion resulting in isolation of the fission products from man's environment. This United Nations Committee estimates the total absorbed dose (integrated to infinite time) from fallout of weapons tested in 1954 to 1962 to be 243 mrad to the gonads, 330 mrad to the bone marrow, and 527 mrad to cells lining the bone surfaces.

Table 1-7 summarizes data prepared by the United Nations Scientific Committee indicating the distribution of dose commitments from nuclear explosions for weapons tested in the period 1954–1962. The Medical Research Council (United Kingdom) Committee on Protection Against Ionizing Radiations[133] estimates the maximum dose equivalent in 1963 would have been received by children of about one year of age and this might have been 215 mrem to the bone or about one-half this quantity to bone marrow for those exposed under the most adverse conditions. It estimates a dose equivalent of 30 mrem to the bones (or about 15 mrem to bone marrow) in 1963 as the average for this age group in the population-at-large.

With "pure" hydrogen bomb explosions (neglecting the ^{235}U or ^{239}Pu primer and assuming no ^{238}U-bomb casing), there are negligible fission products produced. If only hydrogen bombs had been tested in the period 1954–1962, the dose commitments from fission products would have been far less than indicated in the last two columns of Table 1-7; however, the ^{14}C and ^{3}H would have increased greatly. Table 1-8 summarizes data from Leipunsky[104] comparing the important radionuclides produced by the two types of weapons, each approximately equivalent in explosive power to 10 megatons of TNT. It is to be noted that since the prompt energy released per fission is about ten times that per fusion of the type $^{2}H + {}^{3}H \rightarrow {}^{4}He + {}^{1}n$, the number of fusions is about 10 times the number of fissions in weapons of the same explosive power. These data emphasize that although fission products are lacking in the pure hydrogen bomb explosion, there are about twenty times as many curies of $^{14}C + {}^{3}H$ produced as of $^{90}Sr + {}^{137}Cs + {}^{14}C$ in a fission-type weapon of the same explosive power.

Table 1-8 Reaction Products of a Nuclear Explosion Equivalent to 10 Megatons of TNT

Product	Ordinary Atomic Bomb Explosions	Pure Hydrogen Bomb Explosions
Number of fissions	1.45×10^{27}	—
Assumed total neutrons/fission	2.5	—
Assumed total neutrons/fusion	—	1
Prompt energy/fission (MeV)	178	—
Prompt energy/fusion (MeV)	—	17.6
Number of $^{2}H + {}^{3}H$ reactions	—	1.45×10^{28}
Number of Neutrons	2.2×10^{27}	1.45×10^{28}
Activity of ^{90}Sr (curies)	1.43×10^{6}	—
Activity of ^{137}Cs (curies)	1.57×10^{6}	—
Activity of ^{14}C (curies)	2.34×10^{5}	1.49×10^{6}
Activity of ^{3}H (curies)	—	6.67×10^{7}
Activity total of above (curies)	3.3×10^{6}	6.8×10^{7}

From a comparison of the uptake of these radionuclides in the body and of the maximum permissible body burdens in each case, and with appropriate allowance for external exposures, the genetic hazard of this pure hydrogen bomb may be about ten times that of the ordinary bomb. However, the danger from radiation-induced leukemia and bone tumors may be only $\frac{1}{10}$ to $\frac{1}{100}$ as great, depending to a considerable extent on the availability of the contamination as influenced by the height of detonation and the meteorological conditions of fallout. Therefore, from the standpoint of immediate population damage, the testing of ordinary nuclear weapons is more hazardous than the testing of hydrogen weapons, but with respect to damage to the race as expressed over many generations, hydrogen-weapons tests appear to be considerably more damaging. The introduction of Co, ^{238}U, or other materials into hydrogen weapons could make the immediate as well as the long-range fallout hazards from these tests as great as or greater than those from the testing of ordinary nuclear weapons.

Nuclear Weapon Exposures at Hiroshima and Nagasaki. We have already emphasized the fact that there is no substitute for human data in drawing conclusions relative to the relations between dose equivalent from ionizing radiation, the biological effects, and the resultant damage to man. Some valuable indications or guides can be observed from animal experiments. However, reliable conclusions can be reached only following extensive studies of the effects of ionizing radiation on man. Experiments with various species of animals—including man—have indicated not only a wide variation of radiosensitivity among individuals within a given species but also even wider variations—by a factor of 10 or more—in the average variation from one species to the other. It is for this reason that the importance of studies of human exposure is so great.

Without question the most extensive studies of human exposure to ionizing radiation that have ever been made are now being conducted with reference to the survivors of the nuclear weapons bombing of Hiroshima on August 6, 1945, and Nagasaki on August 9, 1945. Prior to the bombing of these two cities, they had a combined civilian population of 600,000; an unknown number of military personnel and work parties were also in temporary residence in Hiroshima at the time of the bombing. The total number of civilian deaths at the time of the bombing of Hiroshima and for a few weeks following was estimated to be about 60,000, with 70,000 injured survivors. The total number of persons killed in Hiroshima, however, may have exceeded 100,000. In Nagasaki there were approximately 33,000 civilian deaths with 25,000 surviving injured. Table 1-9 indicates the distribution of early radiation symptoms[155] among some of the survivors in Hiroshima and Nagasaki. It is evident from Table 1-9 that the persons closest to the hypocenter and receiving the greatest dose equivalent showed the largest percentages of the various types of radiation damage among the survivors. The situations at Hiroshima and Nagasaki were somewhat similar, as shown in Table 1-9, except that (1) the damage to the survivors close to the hypocenter at Nagasaki was less severe than that in the case of the Hiroshima victims; and (2) at distances greater than 1500 meters at Nagasaki the damage to survivors was, in most cases, greater than that at Hiroshima. These differences reflect differences in the intensity of the explosions, spectral distribution, local shielding, and neutron-to-gamma ratio. The Health Physics Division of the Oak Ridge National Laboratory[24] has had in progress an intensive study of the absorbed dose, shielding, spectral distribution, and angular distribution of the radiation received by each individual survivor.

Table 1-10 summarizes[98] the best absorbed dose estimates obtained from this study by 1965. It is to be observed that the γ-absorbed dose was larger and the neutron absorbed dose was smaller at Nagasaki than at Hiroshima. If the RBE (or quality factor) of the neutron absorbed dose for the various types of damage under study were 1, the dose equivalent at Nagasaki was slightly higher close to the hypocenter and much higher a large distance from the hypocenter than it was at Hiroshima. On the other hand, if the RBE were 10, the dose equivalent at Hiroshima was much higher close to the hypocenter, and less a large distance from the hypocenter, than at Nagasaki. This would argue for a value of RBE greater than 1 for most of the symptoms listed in Table 1-9 but the data are too limited at present for use in the calculation of values of RBE. As the dose data become more refined, they will be compared with the medical information obtained from studies of the

Table 1-9 Frequency of Occurrence of Certain Symptoms in Persons from Hiroshima and Nagasaki Alive 20 or More Days Following Atomic Bombing, as Related to Distance from the Hypocenter (Figures inside the Parentheses Apply Only to Nagasaki and the Others Apply to Hiroshima)

Distance (meters)	No. of People	Epilation %	Purpura %	Oropharyngeal Lesions %	Necrotic Gingivitis %	Diarrhea %	Bloody Diarrhea	Other Hemorrhage
0–1000	749 (789)	60.4 (31.6)	48.9 (21.3)	61.1 (34.9)	10.3 (2.5)	50.1 (42.7)	10.7 (6.8)	38.1 (13.9)
1000–1500	1125 (1882)	30.3 (24.4)	21.4 (17.2)	33.9 (32.3)	3.8 (1.8)	42.3 (39.1)	6.7 (4.6)	17.5 (12.4)
1500–2000	1824 (1034)	8.3 (12.4)	4.3 (8.9)	15.7 (19.3)	0.7 (1.5)	36.5 (34.6)	5.4 (5.4)	6.1 (7.5)
2000–2500	1450 (672)	4.8 (6.1)	1.9 (3.6)	15.9 (15.0)	1.1 (0.6)	35.8 (33.3)	6.6 (3.4)	6.5 (5.4)
2500–3000	700 (644)	2.3 (2.5)	1.9 (1.1)	15.0 (13.0)	0.1 (0.9)	37.1 (30.3)	4.9 (4.3)	5.9 (2.8)
3000–4000	576 (1141)	1.2 (1.4)	1.2 (1.5)	6.8 (10.6)	0 (0.1)	23.3 (23.5)	2.3 (1.8)	3.8 (2.5)
4000–5000	239 (265)	0 (0.4)	1.7 (0.4)	8.8 (7.2)	0 (0)	24.7 (8.7)	0.8 (1.1)	2.1 (1.9)'
0–5000	6663 (6427)	16.6 (14.2)	11.0 (9.8)	22.8 (21.9)	2.2 (1.2)	37.4 (33.3)	6.0 (4.2)	11.3 (7.9)

Table 1-10 Estimates of Absorbed Dose as a Function of Distance from the Hypocenter at Hiroshima and at Nagasaki

	Hiroshima		Nagasaki	
Distance (meters)	Neutron Dose (rad)	Gamma Dose (rad)	Neutron Dose (rad)	Gamma Dose (rad)
0	14,800	10,900	4180	26,400
500	3220	2887	737	7290
1000	192	260	37.1	903
1500	10.1	22.3	1.8	121
2000	0.54	1.9	0.089	17.9

survivors of Hiroshima and Nagasaki. In this way, the Japanese who suffered the first effects of nuclear weapons are contributing to an understanding of the dose–effect relation of ionizing radiation.

The long-range studies of the survivors of Hiroshima and Nagasaki are perhaps of even greater value than the studies of acute symptoms from the standpoint of setting adequate levels of permissible exposure for occupational workers and for the population-at-large. These studies[2–14,23] have indicated an increased incidence in stillbirths among the exposed pregnant women and, within a few years after birth, cases of microcephaly and related mental retardation were observed among the surviving offspring of these women. In addition, studies of children who were *in utero* at the time of the bombing of Nagasaki have revealed more subtle defects such as failure in achieving normal growth, that is, as related to head circumference, height, weight, chest circumference, age at menarche, degree of epiphyseal closure in the wrist, etc. As indicated by the ABCC report, there was a significant increase in leukemia among the exposed Japanese, especially among males and among children under age nine.

The increased leukemia risk apparently started 1.5 to 3 years following radiation exposure and the mean latent period for those exposed at a distance less than 2000 meters from the hypocenter was in the interval between 4 and 8 years following exposure. The latent period before the onset of symptoms of leukemia increased with the distance from the hypocenter of the exposed survivors reaching a peak incidence 15 to 18 years after exposure for those survivors who were exposed at distances of 10,000 or more meters from the hypocenter. The occurrence rates of leukemia per 100,000 population by distance from the hypocenter for Nagasaki and Hiroshima were 72.2 and 130.4 cases within 1000 meters, respectively; 39.9 and 32.5 cases[22] within 1500 meters, respectively, 8.6 and 6.3 cases within 2000 meters, respectively,

4.2 and 4.0 cases beyond 2000 meters, and 2.4 and 2.3 for controls, respectively. These data, so far as they can be evaluated at present, as well as the data of Court-Brown[41] mentioned earlier, support the concept of a linear relation between dose and leukemia incidence at least in the range of higher doses. Also, Hiroshima and Nagasaki survivors exposed within less than 1500 meters of the hypocenter have been found to have a tumor incidence twice that of the control population. Thyroid carcinoma has been found to be significantly more prevalent among the survivors of Hiroshima and Nagasaki who were heavily exposed. Thyroid disorders are strikingly higher among the women than among the men. Some of the Japanese who survived high radiation exposure developed cataracts similar to those developed in the United States among cyclotron workers. Professor Y. Masuda[19] examined, at the Hiroshima Red Cross Hosptial from June, 1953, to March, 1958, 1101 persons exposed to the weapons radiation and found 96 patients or 8.7% with cataracts and 121 or 11% who were highly suspected of having them. The incidence of cataracts was as follows: 10.9% for those exposed within 2000 meters of hypocenter; 4% for those exposed at 2000–3000 meters, and almost none beyond 3000 meters.

It is encouraging that overt lesions visible with the ophthalmoscope have been found (January 1960) in only 154 survivors of the two cities, although plaques on the posterior capsule of the lens have been observed on a large percentage of the survivors who received sufficient exposure to result in almost complete epilation following the bombing. Also, children who were ages 7–10 at the time of the bombing are now found to have a significant loss of visual acuity when compared with the controls. There is not much evidence from the present, very limited observations on the offspring of exposed parents to indicate genetic damage. Changes in sex ratio seem to follow the expected pattern related to exposures but it is too early to evaluate quantitatively the impact of radiation with reference to genetic changes in man. Consistent life shortening that diminishes with the distance from the hypocenter has been observed for the survivors from Hiroshima and Nagasaki, and this increased mortality applies to all causes of death, including tuberculosis, as well as leukemia and other malignant diseases.[18] It is to be hoped that data from individual survivors of bombing in these two Japanese cities, as mentioned above, may shed some light on the appropriate values of RBE that should be applied to neutron exposure for the various types of radiation damage. There is further discussion of these data in later chapters.

Exposures from Accidents Associated with the Nuclear Energy Industry. Finally, one of the most useful sources of radiation exposure data is radiation accidents [Refs. 69, 86, 113, 140, 157, 171–173, 204, and 213]. We have mentioned already the radiation experience (or accidents) with radium and the fact that data concerning them are being studied intensively in order to establish a better reference standard of maximum permissible internal dose. In addition, there have been a few accidents in the various nuclear energy establishments and these cases are being evaluated and followed carefully as a source of useful information. Fortunately, as of March, 1967, so far as can be determined, there have been only eight radiation fatalities associated with critical assemblies and reactor accidents—three at Los Alamos, New Mexico, United States (August 21, 1945, May 21, 1946, and December 30, 1958), three at Idaho Falls, Idaho, United States (January 3, 1961) (there was concurrent lethal blast damage in these three cases), one at Wood River Junction, Rhode Island, United States (July 24, 1964), and one at Vinca, Yugoslavia (October 15, 1958). These fatal accidents together with the other more serious reactor and criticality accidents involving high exposures to individuals from external sources of radiation are summarized in Table 1-11.

The most serious reactor accident in terms of environmental contamination and number of persons exposed to internal sources of radiation was the accident at Windscale, England,[48,158] on October 10, 1957. This accident occurred in No. 1 Pile, a natural uranium-fueled, graphite-moderated, air-cooled reactor used for the production of plutonium. At the time of the accident, the reactor had been shut down for regular maintenance and was being heated to anneal the graphite, a routine procedure to release the stored energy (that is, the Wigner effect). This procedure of bringing the temperature of the graphite to about 200°F above normal had been carried out routinely in a rather uneventful fashion some thirty times previously. Presumably, the nuclear heating was applied too soon following a previous heating, and at too rapid a rate, so there was overheating. This resulted in a fire which burned a number of the uranium cartridges and some of the graphite, affecting altogether approximately 150 channels. Forced air-cooling could not be used to lower the temperature and there was concern about other suggested methods of quenching the fire. Finally, the fire was put out with water some two days after it was first discovered. In the mean-time, fission products had passed up the stack and some of them, especially the ^{131}I gas, had leaked through the filters which were located near the top of the stack. These escaped fission products were deposited over the neighboring countryside. Fortunately, most of the fission products were attached to smoke particles and held up in the stack duct system or deposited on the filters, yet thousands of curies—mostly ^{131}I—fanned out in the smoke plume from the stack, contaminating thousands of acres of grazing and farmland.

It soon became evident that there was no serious problem in connection with the inhalation of these fission products, but that ingestion of milk and possibly other food products such as leafy vegetables, beans, eggs, etc., might result in serious exposures. The exposure potential was especially serious

Table 1-11 Absorbed Dose Received by Persons Who Have Been Exposed in Reactor and Criticality Accidents

Location, Date, and[a] Type of Accident	Estimate of Dose to Total Body[b] Neutron (rad)	Gamma (rad)	Total (rad)	Symptoms[c] and Survival Time	References
Los Alamos, New Mexico, United States (R) June 4, 1945			66.5 (N)	A	73,[d] 74, 83
			66.5 (N)	A	
Los Alamos, New Mexico, United States (SA) August 21, 1945	297	?	297+(N)	(25d)D, E	204,[d] 213, 86, 73, 74
	18	2	20.2 (N)	S	
Los Alamos, New Mexico, United States (SA) May 21, 1946	1230	120	1350 (N)	(9d)D	204,[d] 213, 83
	220	22	242 (N)	D,E	
	120	12	142 (N)	S	
	50	5	55 (N)	A	
	33	3	36 (N)	A	
	25	2	27 (N)	S	
	20	2	22 (N)	A	
Argonne National Laboratory, Illinois, United States (R), June 2, 1952	14.2	145	159 (W)	D	85,[d] 204, 213, 86, 73, 74
	9.6	116	127 (W)	A	
	5.5	55	60.5 (W)	D	
	0.8	10	10.8 (N)	H	
U.S.S.R. (R) <1954			450	D,H	66,[d] 204, 213
			300	D,H,E	
Y-12 at Oak Ridge, Tennessee, United States (LP), June 16, 1958	96	269	365 (N)	D,H,E	81,[d] 204, 213, 160, 86, 50, 103, 31
	89	250	339 (N)	D,E	
	86	241	327 (N)	D,E	
	71	199	270 (N)	D,E	
	62	174	236 (N)	D,E,H	
	18	50.5	68.5 (N)	A	
	18	50.5	68.5 (N)	A	
	6	16.8	22.8 (N)	S	

Vinca, Yugoslavia (R), October 15, 1958	222	214	436	(N)	(32d)D,H,E	87,[d] 204, 213, 86
	227	192	419	(N)	D,H,E	
	217	209	426	(N)	D,H,E	
	225	189	414	(N)	D,H,E	
	165	158	323	(N)	D,H,E	
	112	95	207	(N)	D,E	
Los Alamos, New Mexico, United States (LP), December 30, 1958	900	3500	4400	(N)	(1.4d)D	76,[d] 204, 213, 86, 21
	2.6	118	121	(W)	A	
	1.4	31.5	33	(W)	A	
Idaho Falls, Idaho, United States (R), SL-1 explosion, January, 3, 1961	35,400	?	35,400	(W)	(< second)	163,[d] 181, 164, 82, 58, 182–186
	1860	?	1860+	(W)	(∼ seconds)	
	317	?	317+	(W)	($1\frac{1}{2}$–2 hrs)	
Hanford, Washington United States (LP), April 7, 1962	27	63	90	(W)	?	166[d]
	11	23	34	(W)	?	
	3	13	16	(W)	?	
Wood River Junction, Rhode Island, United States (LP), July 24, 1964	5100	5400	10,500	(W)	(2.05d)D	139[d]
	10–41	50	60–91	(W)	S	
	17–50	47–54	64–104	(N)	S	
	?	3.5	3.5+	(W)	A	
	?	2.5	2.5+	(W)	A	
Mol, Belgium, (R), Mild explosion, December (30), 1965	?	550	550+	(W)	D,E	

[a] (SA) = solid assembly; (R) = reactor; (LP) = liquid process.
[b] (W) = wearing personnel dosimeter at time of accident; (N) = not wearing personnel dosimeter at time of accident.
[c] S = slight symptoms of radiation exposure; D = symptoms of radiation damage; E = epilation; H = hemorrhage; (—d) = days of survival time; A = radiation symptoms absent.
[d] The values from this report were used as the best absorbed dose estimates.

because of the uptake of ^{131}I from the grass by the cattle which passed it into the milk where it became available for uptake by young children. Furthermore, children consume relatively large quantities of milk and the young child (age six months to one year) deposits this ^{131}I in a thyroid of only about one-tenth the mass of the adult thyroid. Awareness of this situation led to the establishment of a restricted area 30 miles long and 6 to 10 miles wide. Extensive monitoring was conducted in this area and thousands of gallons of contaminated milk were destroyed. Hundreds, perhaps thousands, of people received an absorbed dose of a few rad to the thyroids but it is believed that because of the prompt and effective emergency measures, there were no exposures of serious consequence. There was readily detectable population exposure from ^{89}Sr, ^{90}Sr, ^{137}Cs, and a few of the other radionuclides but most of the population dose was from ^{131}I. The highest measured concentration of ^{131}I in a child's thyroid was 0.28 μc which, if maintained for a few weeks, may have delivered a thyroid dose of as much as 16 rad.

Many valuable health physics lessons were learned as a result of this experience. In particular, it was demonstrated that even in what had been considered one of the safest types of reactors, serious accidents can occur—accidents leading to widespread consequences. Furthermore, it was found that preplanning for such emergencies (regardless of how exceedingly unlikely their occurrence) is an important part of all responsible reactor operations. These plans must, among other things, provide for the following: (1) an adequate and well-trained health physics organization that is prepared at all times for almost any type of emergency, (2) well-defined dose limits and action levels at which specific measures such as population evacuation, the use of food substitutes, or the condemnation of foods will be put into force unhesitatingly, (3) a well-developed instrument program that will provide for aerial surveys, the rapid testing of all food supplies, and reliable external and internal monitoring of individuals in the population, (4) a continuing and close arrangement for cooperation between reactor management, the health physics organization, and local, state, and national authorities who can see that proper and prompt action is taken as specified by the health physics organization and the reactor management, and (5) an adequate communications system for contacting the health physicists in the field, the potentially exposed population, the public press, radio, and television.

Space will not permit further discussion of radiation accidents, but the health physicist is encouraged to read the references to accidents given in this chapter because they provide many hints and valuable suggestions which could prove useful in time of radiation emergency. Table 1-12 lists some of the more serious exposures from radioactive materials.

Finally, in the discussion of radiation accidents (their prevention, preplanning, damaging consequences, etc.), it is important to maintain the

Table 1-12 Some of the More Serious Exposures from Radioactive Materials

Location, Type, and Date of Accident	Consequences	Reference
Madison, Wisconsin. Graduate student unknowingly exposed to 200 c source of ^{60}Co. March 29, 1960.	Student became very ill following exposure of 250–300 R. Usual symptoms of suppressed blood count, epilation, etc.	168
Mayaguez, Puerto Rico. Reactor fuel element placed in area where insufficient shielding was provided. July 24, 1962.	Five persons near source received gamma ray absorbed doses of 100, 55, 24, 18, and 18 rad, respectively.	123
Houston, Texas. Employees exposed to ^{192}Ir dust. March 13, 1957.	Slight environmental contamination. Two employees received minor radiation burns. One employee received film-badge dose of 1.7 rad during week.	101 73 74
Argonne National Laboratory, Argonne, Illinois. 50 mg $RaSO_4$ source broken. June 13, 1952.	Approximately 10 mg of Ra spread through 3 wings of building. Several employees inhaled considerable quantities of Ra.	73 74
Eniwetok Proving Ground in South Pacific. Two persons improperly handled fission product samples. May 14,1948.	Two persons received beta burns to hands.	73 74
Los Alamos, New Mexico. Unpacking radioactive materials. September 7, 1948.	One person received beta burns to ankle.	73 74
Los Alamos, New Mexico. Handling radioactive material with torn glove. July 9. 1952.	One person received beta burns to hands.	73 74
Nevada Test Site. Entered exclusion area during tests. March 1, 1955.	One person received 39 R total body dose.	73 74
National Reactor Testing Station (USA). Radioactive particle entered ear. July 27, 1955.	One person had partial loss of hearing.	73 74
Los Alamos, New Mexico. Handling radioactive material with torn glove. April 3, 1956.	One person received beta burns to hand.	73 74
Hanford, Washington. Escape of Pu solution into central room. June 18, 1956.	One person received contaminated skin.	73 74
Rocky Flats, Colorado. Explosion in dry box. June 14, 1957.	Pu lodged in finger, necessitating amputation. 0.4 μg of Pu deposited in facial wound of one man and 0.5 μg of Pu in facial wound of another.	73 74

Table 1-12 (*continued*)

Location, Type, and Date of Accident	Consequences	Reference
University of California Radiation Laboratory, Berkeley, California. Employee held 85 mc of ^{90}Sr in centrifuge tube in hand. June 27, 1957	Employee received blisters on hand.	73 74
Oak Ridge National Laboratory, Oak Ridge, Tennessee. Employee mistakenly entered room containing highly radioactive material. October 4, 1957.	Employee received a total body exposure of 63 rem.	73 74
Oak Ridge National Laboratory, Oak Ridge, Tennessee. Employee entered cell to mix chemicals and check agitator. November 12, 1957.	Employee received a total body exposure of 13 rem.	20

proper perspective and not lose sight of the fact that although ionizing radiation when improperly controlled can be extremely dangerous, it can be and has been used safely (usually for years without a serious accident). Health physics programs in all parts of the world have demonstrated, in fact, that nuclear energy production can be made one of the safest of modern industries. Table 1-13 indicates that there was a wide variation in the safety record of the various nuclear energy research and production operations in 1964 and 1965. Some organizations such as Westinghouse Electric have a remarkably good safety record, whereas others seem to have consistently bad records. Experience has indicated that, for the most part, the wide differences in the safety records of the various organizations listed in Table 1-13 are due to differences in attitude and respect for safety rather than inherent differences in the hazards involved. The three-year average (1961–1963) frequency rate in nuclear energy production activities was 0.93 and the severity rate was 262. The average frequency rate in research activities was 1.72 and the severity rate was 229. As shown in Table 1-14, the nuclear energy industry is as good as or better than most other industries.

Although there have been a few very serious radiation accidents, they have, in fact, been very few in number and perhaps have received excessive publicity. For the most part, large nuclear energy organizations operate for months or, in some instances, for many years without a serious radiation accident. Table 1-15 summarizes the exposure records at Oak Ridge National Laboratory for three typical years. During this period there was only one exposure (in 1963) in excess of the average occupational dose limit of 5 rem/yr and each year over 70% of the employees received essentially no radiation exposure (that is, between 0 and 0.1 rem). The integrated dose records of persons currently employed as of March 1966 are most impressive. Out of a total of 5145 employees, there are only 127 individuals who have accumulated more than 20 rem and only 18 who have accumulated over 50 rem. Some employment records go back to the beginning of the Laboratory in November 1943.

Exposures from Other Sources. Other sources of exposure such as watches, shoe-fitting machines, television, security X-ray inspectroscopes, industrial X-ray machines, high-voltage rectifiers, exposures in high-flying aircraft, etc., may furnish useful data. It is important that health physics efforts are being made to reduce these exposures to the lowest practical limit but as long as such exposures continue, the doses delivered should be determined and the effects on those exposed should be assessed by the health physicist. The high exposure values of watches as shown in Table 1-4 were found in a shipment of watches imported to the United States. The dials of these watches had been painted with an excessive amount of ^{90}Sr. An effort has been made by the company that imported these watches to recall them and replace the dials with safe ones. By May, 1960, about two-thirds of those sold in the United States had been recalled and replaced by the company. Joyet and Miller[99] made a study of the genetic dose received by the Swiss population from luminous dials of watches and clocks and they estimated a 30-yr dose of 100 ± 25 mR. A few years later, Enz et al.[57] confirmed this value with a best estimate of 75 mrem/30 yr as the genetic dose equivalent and 100 mrem/wk as the dose equivalent applied to the wrist of the wristwatch wearer in Switzerland. This dose to the Swiss population is perhaps typical of that to many other populations. It is of interest to note that this gonad dose is about the same as that (integrated to the year 2000) from fallout produced by weapons tested in the period 1954–1962 (see Table 1-7).

A shoe-fitting machine is probably in most cases an improper application of X-rays, since shoes can be fitted on most people just as satisfactorily without the aid of these devices. Sometimes their use has led to unfortunate and uncalled-for overexposure both to the shoe salesman and to the customer. The American Standards Association and some cities and states of the United States have taken positive action to control these devices. For example, as pointed out by Williams,[212] the ASA[29,30] and the New York City Health Department[156] have specified that "the maximum permissible dose per exposure shall not exceed 2 R" and New York City further requires, "There shall not be more than 3 exposures in one day and not more than a total of 12 exposures in 1 year." Unfortunately, in many cases this regulation is enforced only by signs posted in shoe stores to warn the customer. The

Table 1-13 Safety Record in Various Types of Employment in Nuclear Industry

Type of Employment	1964				1965			
	Number of Employees	Number of Disabling Injuries	Frequency Rate[a]	Severity Rate[b]	Number of Employees	Number of Disabling Injuries	Frequency Rate[a]	Severity Rate[b]
Nuclear Energy Research								
Westinghouse Electric, Pittsburgh, Pa.	2827	0	0	0	1505	0	0.00	0
Phillips Petroleum Co., Idaho Falls, Idaho	2145	0	0	0	2132	0	0.00	0
General Electric (KAPL), Schenectady, N.Y.	2025	3	0.24	1	1960	3	0.79	252
Atomics International, Canoga Park, Calif.	3391	6	0.69	863	2230	8	1.43	42
Union Carbide Nucl. Co. (ORNL), Oak Ridge, Tenn.	3921	7	0.99	73	4065	17	2.21	99
Argonne Nat'l Lab., Argonne, Ill. and Idaho Falls, Idaho	5062	10	1.54	83	5165	7	0.71	137
Sandia Corporation, Livermore, Calif.	995	3	0.53	13	1000	2	1.10	139
Sandia Corporation, Albuquerque, N. Mex.	7173	24	1.15	750	7089	11	0.84	158
University of California (Los Alamos Nat'l Lab.), Los Alamos, N. Mex.	3944	10	1.71	34	4062	13	1.79	29
Lawrence Radiation Lab., Berkeley and Livermore, Calif.	8175	36	2.26	38	7882	36	2.29	29
Brookhaven Nat'l Lab., Upton, L.I., N.Y.	3267	17	3.32	48	3462	20	2.71	114

Nuclear Energy Production								
DuPont, Aiken, S.C.	5680	1	0.16	40	4829	3	0.33	255
Dow Chemical, Golden, Colorado	3088	2	0.34	5	2830	1	0.16	4
General Electric, Richland, Washington	7846	8	0.62	53	3426	8	0.79	48
ACF Industries, Albuquerque, N. Mex.	2473	4	0.54	41	2281	5	1.08	85
Bendix, Kansas City, Missouri	8182	16	1.36	146	6158	18	1.30	176
Union Carbide (Y-12 Plant), Oak Ridge, Tennessee	6077	7	0.34	542	5371	4	0.39	605
Union Carbide (K-25 Plant), Oak Ridge, Tennessee	2432	7	0.40	12	2468	8	1.75	222
Union Carbide, Paducah, Kentucky	1113	2	0.85	294	1073	4	1.95	179
National Lead of Ohio, Fernald, Ohio	1719	6	0.01	81	1523	0	0.00	0
Monsanto Research, Miamisburg, Ohio	1624	0	0.00	0	1674	0	0.00	0

[a] Frequency rate is the number of injuries per million man-hours worked.
[b] Severity rate is the number of days lost per million man-hours worked.

Table 1-14 Safety Records in Various Industries

	1964		1965	
	Frequency Rate	Severity Rate	Frequency Rate	Severity Rate
All industries	6.45	693	6.53	695
Automobile	1.86	196	1.72	215
Chemical	3.13	325	3.36	539
Glass	4.14	334	5.14	373
Iron and steel products	9.20	717	9.31	540
Quarry	13.67	2596	12.63	1567
Lumber	17.26	1900	15.89	1843
Mining (underground, except coal)	30.69	4490	29.46	3457

output of these machines, as reported by Williams,[212] ranges from 10 to 116 R to the foot per 20-sec exposure (the exposure time most commonly used). Hemplemann[80] has indicated that there are three types of injury likely from the excessive use of these machines, especially from those with higher output and/or inadequate shielding: (1) interference with normal foot development of the child, (2) acute radiation burns to some feet, resulting in late, permanent damage, and (3) chronic injury of the blood-forming organs of store employees.

Table 1-15 Exposure Records of Oak Ridge National Laboratory Employees

Annual Dose Values				Integrated Dose Values	
Dose Range (rem)	Number of Employees			Dose Range (rem)	Persons Employed at ORNL as of March, 1966
	1963	1964	1965		
0.0–0.1	3681	4213	4453	0.0–0.5	1919
0.1–0.5	1046	695	812	0.5–1.0	875
0.5–1.0	228	238	209	1.0–5.0	1631
1.0–2.0	179	151	125	5.0–10	383
2.0–3.0	26	36	38	10–20	210
3.0–4.0	10	9	10	20–30	64
4.0–5.0	5	1	1	30–50	45
5.0–6.0	1	0	0	50–70	14
≥6.0	0	0	0	70–90	4
				≥90	0
Total	5176	5343	5648	Total	5145

Sources of exposure to ionizing radiation such as home television sets are of concern not because of the likelihood of radiation damage to the individual but because of the large number of persons exposed and the consequent genetic dose to the population and the possibility of a significant increase in leukemia, bone tumors, and other radiation-induced diseases, the incidence of which may relate monotonically to the dose received by various of the body organs. Braestrup and Mooney[35] made a study of this problem and concluded that in the United States the average number of viewing hours is 1000/yr/person. They showed that a typical television set operating at 25 kV and delivering 0.5 mR/hr at 5 cm from the viewing screen would deliver an absorbed dose of 7.5 mrad/yr to the testes and 1.1 mrad/yr to the ovaries for a viewing distance of 200 cm. However, since most sets in the United States operate at voltages less than 25 kV and are provided with a safety glass in front of the screen, they concluded that the average absorbed dose to the gonads is much less than 1 mrad/yr. Both the ICRP[92,95] and the NCRP[150–152] have made similar recommendations, directed to the manufacturers of television sets, in an effort to minimize the radiation hazards associated with their use. The recommendation of the ICRP is, "... the dose-rate at any readily accessible point 5 cm from the surface of any set used in the home or place where the public is likely to be in close proximity shall not exceed 0.5 mR/hr under normal operating conditions (and) all other television equipment such as those used for projection purposes or in closed circuit applications shall be shielded and ... constructed, installed and operated, as to provide adequate protection."

Some organizations have used X-ray machines or so-called inspectoscopes for X-raying employees at exit portals of factories to ascertain that valuable property is not concealed in clothing and stolen. It has been estimated that some organizations may have delivered in this manner more than 50 rad/yr to employees. Perhaps the infrequent use of these machines could be condoned in prisons to locate concealed weapons, but it is doubtful that any employer has the right to X-ray an employee without his knowledge or consent; moreover exposure of this type does not conform to the health physics requirement that all unnecessary exposure should be avoided. It has been suggested frequently that inspectoscopes might be used advantageously to check airplane baggage and passengers so that the possibility of deranged persons shipping or carrying dangerous explosives and other illegal cargo aboard aircraft can be reduced. Perhaps aircraft baggage inspection is a useful and desirable application of these devices, but until they can be made to provide the necessary information with negligible dose per inspection, it is doubtful whether they should be used to inspect passengers without their knowledge and written consent. Here, a negligible dose might be, for example, 5 mR per inspection, allowing for the unlikely event that a passenger might have two inspections per week for the entire year. In this instance, he would just reach the maximum permissible dose limit of 0.5 rem/yr set by the ICRP for members of the population-at-large.

One very common source of present-day overexposure to ionizing radiation is the careless use of industrial radiography and fluoroscopy. Originally, the testing for defects in machines, equipment, structures, etc., was done by X-rays and, at present, this work is augmented with the use of radiation sources such as ^{137}Cs, ^{60}Co, and ^{226}Ra. Sometimes persons doing this work are either poorly trained in the elementary practices of health physics or are using instruments that are not maintained in proper operation. This situation has led to a number of minor radiation accidents and to a few serious exposures. For example, beginning on March 21, 1962, a Mexican family was unknowingly exposed for 3 months and 22 days to a 5-c source of ^{60}Co which presumably had been used in industrial radiography. As reported by the Mexican National Nuclear Energy Commission,[125] this source had been abandoned or was left at a private home in a lead container. Apparently the family did not know the nature of the contents of this container, and if they were provided with any radiation protection instructions, such instructions were certainly completely inadequate. Details of this accidental exposure are given by Rafael Martinez et al.[122] It is not known how the ^{60}Co capsule was removed from the lead container, but sometime during the period between March 21 and April 1 the ten-year-old son of the family happened to find it in the yard. Apparently he carried it in the left front pocket of his trousers. On April 1, it is believed the mother found the capsule and removed it from the pocket of her son's trousers, placing it in a drawer in the kitchen. On April 17, the son's grandmother came to live with the family and at that time noted the blackening of glasses that had been close to the capsule in the drawer. On April 29 the son died and, on July 19, his mother died. In retrospect, it is evident that both son and mother died of acute radiation sickness; the son died following an estimated 24 days of exposure (some of which were in close contact with the source), with absorbed dose estimates ranging between 2940 and 5165 rad. The mother died following an estimated 115 days of exposure, with absorbed dose estimates ranging between 1995 and 2930 rad. On July 22, 1962, the owner of the ^{60}Co source came by the house, claimed the source, and took it away with him, apparently without anyone suspecting the trouble it had caused. Fortunately, on August 13, an alert physician of the Mexican Institute of Social Security suspected for the first time that the common symptoms of members of this family might be due to radiation damage. On August 18, the two-and-a-half-year-old daughter died after receiving an estimated 99 days of exposure, with dose estimates ranging between 1373 and 1872 rad. On August 20, the father and his mother were admitted to the hospital with what appeared to be radiation exposure symptoms. The father was discharged from the hospital on September 6 but was still (1965) under close medical observation. His mother died on October 15, following an estimated 90 days of exposure with absorbed dose estimates ranging between 1518 and 2827 rad. The father had been exposed off and on for 106 days with dose estimates ranging between 994 and 1716 rad.

Another radiation accident somewhat similar in nature occurred in Moscow, USSR.[15] In this case a 19-year-old research worker committed suicide by carrying a sealed ^{137}Cs source of 10 gram-equivalents of radium in the pockets of his trousers and jacket for about 20 hr beginning about 6:00 PM June 8, 1960. Shortly afterwards at 4:40 PM, June 9, he was admitted as a patient in one of the clinical hospitals in Moscow. The estimated mean doses were 3000 rad to the trunk and 2000 rad to the entire body with local doses of up to 165,000 rad in the trunk and 550,000 rad to the left thigh. Four hours after irradiation began, he observed the usual symptoms of malaise, weakness, and nausea; after seven hours he had abdominal pains and liquid stool and vomited several times. During the following days as the swelling in his abdomen increased and as the count of the various formed elements in the blood dropped, his condition grew progressively worse. In spite of intensive medical care, that is, blood transfusions, bone marrow transplants, the use of antibiotics, and finally on the fifteenth day the opening of his abdominal cavity to permit drainage, he died on the eighteenth day in the presence of cardiac-failure phenomena.

All too often persons have been slow in recognizing that any high-voltage discharge tubes are sources of X-rays. One of the best examples of this type of exposure occurred on March 8, 1960, at Lockport, New York.[84,206] Nine men working in the vicinity of an unshielded Klystron microwave tube for about six hours received X-ray exposures estimated as high as 300–500 rad on large portions of their bodies and perhaps 1000–2000 rad to limited body areas. The Klystron, at the time of the exposure, was operating at a reduced voltage of 150 to 175 kV with a filtration of only about 4 mm of copper. The men mistakenly thought there was no radiation hazard, since the high frequency of the radar system was cut off at the time. They were unknowingly exposed and went to their homes as usual after work. That evening several of them became very sick with the early symptoms of radiation sickness, that is, headache, nausea, vomiting, nasal stuffiness, fatigue, irritability, dry mouth, increased thirst, mild chill, erythema blush, etc. When one of the men first reported to a doctor, it was thought he might have the mumps. It was only due to the alertness of one of the doctors consulted that the similarity to the characteristic radiation syndrome was recognized, and all the exposed men were rounded up and provided with early and appropriate medical care. Although it is unfortunate that such radiation accidents have occurred, they nevertheless have served the useful purpose of alerting industry, occupational workers, and the general public to the need for health physics training, guidance, and supervision in a large number of our industries where high-voltage electronic discharge tubes as well as radioisotopes, nuclear reactors, and atom-smashing accelerators are in use. The ICRP[95] takes special note of this problem in its recommendation,

"Such equipment, e.g. electron microscopes, cathode-ray tubes, high voltage electronic rectifiers, transmitting valves, television and image tubes, shall be so constructed, installed and operated as to provide adequate protection. Whenever practicable, such items of equipment shall be shielded and provided with interlocks so as to insure that the places where they are used can be regarded as being outside controlled areas, thus avoiding the need for monitoring procedures and special medical examinations. Particular attention is drawn to the hazards which may arise during the manufacture, testing and repair of all such equipment."

The wide use of radioisotopes has spread the potential of unsuspected radiation exposure to many devices, too numerous to list completely. Some of these are glow buttons, static eliminators, depth gages, radio tubes, calibration sources, "atomic cocktails" for patients, radioisotope-thermocouple power plants, instrument panel illuminators, food irradiators, medical sterilizers, sewage treatment, irradiators of special polymeric materials, etc. Some persons have suggested that the industrial demand and use of radioisotopes may some day eliminate the problems of radioactive waste. This is certainly far from the truth because eventually the remains of these scattered sources in their various applications will require disposal and, in the meantime, many of them will develop leaks or be involved unsuspectedly in fires and explosions. Health physicists everywhere must be on the alert constantly to guard against the careless use of these sources and prevent the spread of radioactive contamination into places of work, homes, and the environment as a result of accidents or human errors with sources of radioisotopes.

Perhaps, in the near future, an important contribution to human exposure from ionizing radiation will be space travel. Already some commercial planes are flying at 30,000–40,000 ft where the cosmic radiation dose rate on some occasions is increased by as much as an order of magnitude; in a few years commercial airliners in supersonic transport will be flying at elevations up to 80,000 ft. The ICRP report prepared by Upton et al.[208] provides some useful information and valuable predictions of the dose that can be expected during such flights. This report points out that the radiation at the proposed flight elevations of the supersonic transports (60,000–80,000 ft) may be grouped into two convenient classes: (1) galactic cosmic radiation (cosmic rays originating from galactic space beyond the solar system) and (2) solar cosmic radiation (associated with solar flares). Both types of cosmic radiation increase with elevation and geomagnetic latitude. The intensity of galactic cosmic radiation is rather constant, decreasing somewhat during high solar flare activity due to the shielding effects of interplanetary magnetic fields that develop. The dose from this cosmic ray component is not expected to exceed 1–2 mrem/hr at 80,000 ft elevation. On the other hand, the solar cosmic radiation is variable and predictable only in a general way. It may build up rapidly and deliver very high doses over a few hours following a solar flare. Usually the dose contribution of this solar cosmic radiation is a small fraction of the dose from the galactic cosmic radiation, but once or twice during the 11-yr solar sunspot cycle, the dose from the solar cosmic radiation may reach peak rates of 1–10 rem/hr at 80,000 ft elevation. Although there is some uncertainty in the dose values and considerable question relative to the appropriate *RBE* or *QF* to be applied to the various components of cosmic radiation (protons, α-particles, neutrons, mesons, electrons, heavy particles, etc.), it is anticipated that if cosmic ray exposure can be avoided during the short and infrequent periods of major solar flares, the exposures to the crew and passengers of these supersonic transport flights can, in general, be maintained within present maximum permissible exposure limits.[132] When man escapes in rockets from the earth's atmosphere, one of the serious problems he must face is the increased intensity of cosmic radiation. The dose rate in some of the belts of trapped charged particles surrounding the earth's geomagnetic equator may reach many rad per hour and, during solar flares, this radiation may reach dose rates of hundreds of rad per hour.

1-4 EARLY RECOMMENDATIONS FOR RADIATION PROTECTION

Man began to recognize many of the hazards of work with ionizing radiation almost immediately following the discovery of X-, γ-, and β-radiation. As pointed out by Taylor,[197] most of the early efforts toward radiation control and health physics were mainly on a voluntary basis and not by force of law, although in some cases the fear of court action as well as moral responsibility motivated some concern for radiation protection. It is unfortunate that during this early period so many radiation workers gave little or no heed to the warnings provided by the overexposures and to the recommendations published by the early pioneers of health physics. Evans,[59–62] Martland,[117] Stone,[199–201] Cantril,[40] Parker,[159] and many others have reviewed in various publications this early history and development of radiation protection. This history indicates that if the studies of Russ[169] and the recommendations of the Röntgen Society[36] had been followed, then much of the suffering and death from overexposure to radium and X-rays during the following 20 years could have been avoided. In Table 1-16, Stone[190–192] summarizes estimates of tolerance dose made by some of the early workers in this field.

One of the greatest handicaps to the professional development of health physics in this early period was the lack of suitable units of dose and instruments for quantitative dosimetry. As Stone[190–192] points out, Roentgen noted that X-rays ionize the air through which they pass, and Perrin[71] as

Table 1-16 Estimation of Tolerance Dose

Date	Author	1/1000 Erythema Dose (Days)	Calculated (R/day)
1902	Rollins	(film blackening)	10
1925	Mutscheller	3	0.2
1925	Sievert	3	0.2
1926	Solomon	0.3	2.0
1927	Dutch Board	15	0.04
1928	Barclay and Cox	3.5	0.17
1928	Kaye	5	0.12
1931	Adv. Comm. (U.S.)	–	0.2
1932	Failla	30	0.1
1932	Stenstrom	3.7	0.16
1936	Adv. Comm. (U.S.)	–	0.1

early as November, 1896, first measured by means of air condensers the loss of electric charge produced by X-ray. However, in the years that followed, this method was almost lost sight of as a means of dosimetry. In 1914, Szilard[71] constructed a dosimeter to measure ionization but it was not until after 1925 when Fricke and Glasser[63] developed the thimble ionization chamber with an "air wall" that this principle was used as the basic method of quantitative dosimetry. Perhaps Rollins was acting considerably before his time when (in 1902) he suggested the use of X-ray film as a means of monitoring exposure. The X-ray films available at the time varied considerably in radiosensitivity but, nevertheless, some means could have been found for their calibration, for example, the use of a radium standard and of calibration control films. This might have provided better dosimetry than was commonly available for 30 to 40 yr thereafter. Rollins underestimated the hazards of ionizing radiation when he suggested that if a photographic plate is not fogged within 7 min, the intensity of the radiation is not harmful. However, the X-ray equipment in his time produced only low-voltage (or very soft) X-rays that caused primarily superficial skin damage, so the dose he recommended for tolerance was not as high comparatively as would otherwise be suggested by the figures given in Table 1-16. Holzknecht[71] in 1902 devised scales for comparison of color changes in fused mixtures of KCl and Na_2CO_3 when exposed to X-rays. Although such studies were the forerunners of chemical dosimetry, the techniques were not advanced to quantitative dosimetry until about 1950. It is interesting to note also that although prior to the close of the nineteenth century Becquerel, Thomson, Rutherford, the Curies, and others studied the α- and β-radiations from uranium and radium products by observing the scintillations produced in various substances, it was not until 50 years later that this technique combined with the photomultiplier came into its own as an important means of radiation dosimetry. The reader is referred to the summaries of Glasser et al.[71] for a quick scan of historic milestones in early radiation dosimetry.

As related by Stone,[199–201] there were no organized efforts to promote methods of radiation protection until those of the Röntgen Society in 1916.[36] These recommendations related primarily to the use of protective shields, restricted working hours, and special medical examinations, including blood tests for workers in X-ray and radium departments, inspections of equipment, etc. They did not consider the important question of tolerance or maximum permissible dose. Even if they had recommended a tolerance dose, the specifications could have been no more than qualitative since it was not until 1928 that a quantitative unit of dose measurement, the roentgen,[132–200] was specified and adopted by the International Commission on Radiological Units (ICRU). Taylor[196,197] recalls that serious efforts to understand and curtail radiation exposure to individuals were not begun until 1920. In July, 1921, the British X-ray and Radium Protection Committee[37] presented its first radiation protection measures. It is noteworthy that at this time this organization did not suggest a value for the tolerance dose. Mutscheller[137] in 1925 presented the first very clear statements relative to the tolerance dose. He gave a tolerance dose of 1/100 of an erythema dose in 30 days. This was based on studies of operators of typical X-ray equipment and was considered "the dose which an operator can, for a prolonged period of time, tolerate without ultimately suffering injury." Many years prior to and since this recommendation of Mutscheller, the reddening of the skin or threshold erythema has been used as an indicator of radiation dose. This concept has served a useful purpose but it is unfortunate that a better defined and more reproducible physical unit was not used many years prior to the adoption by the ICRU of the roentgen which specifically is a unit of X-ray intensity that with the proper corrections and qualifications may be related to the absorbed dose in tissue. The threshold erythema dose depends upon the radiation intensity, the size of the irradiated area, the dose rate, the specific ionization, the region of the body being irradiated, the age and race of the individual, the energy distribution of the radiation, the training of the observer, the complexion of the exposed person (whether fair or dark), as well as the individual biological variations within each characteristic group. Thus, it is no wonder there is difficulty in transcribing early records of erythema dose to roentgen and/or rad units. The threshold erythema dose is not only higher for the higher energy electromagnetic radiation but it is higher for combinations of low- and high-energy radiation. Table 1-17 gives values of the threshold erythema dose as determined at the Memorial Hospital and published by Glasser et al.[71] The roentgen and more recent units of dose will be discussed in Chapter 3.

Table 1-17 Threshold Erythema Doses for Various Qualities of Radiation Ionizing

Energy (keV)	Absorber HVL (mm)	Roentgens for Threshold Erythema Dose (Including Backscatter)
100	1.0 Al	270
140	0.4 Lu	525
200	0.9 Cu	680
700	7.0 Cu	800
1000	3.8 Rb	1000
600–2200	Ra applicators (0.5 Pt)	1060 (av)

Until very recent times, control of both radiation hazards and radiation protection in the United States has depended on an understanding of the problem by the employer and his voluntary willingness to take the initiative and bear the expense to provide radiation protection for the employee. The recommendations of the International Commission on Radiological Protection (ICRP) and the National Committee on Radiation Protection and Measurements (NCRP)* were taken or ignored, depending in large measure on the understanding and integrity of the employer. In many respects, however, the recommendations of the ICRP and NCRP have served the useful purpose of law, without imposing undue restrictions or the rigidity of law, in that when a case of overexposure to ionizing radiation was brought into court and when it was proved that the recommendations of the ICRP or the NCRP were not complied with, then damages were usually collected by the exposed individual Today, however, the situation in the United States has changed, since a number of states have enacted these regulations into law and the Federal Radiation Council (FRC) and other agencies of the Federal government are

* In 1964, the name was changed to the National Council on Radiation Protection.

placing these regulations in the *Federal Register* where they have the force of law. The recommendations of the ICRP and NCRP, and the guides of the FRC, are discussed in detail in Chapter 14.

Several countries such as Sweden, England, and Germany have provided radiation survey and inspection services on request to designated laboratories since about 1925. Since 1941, licensing and inspection of all radiation installations in Sweden have been required by law. In New Zealand, radiation protection regulations requiring licensing and inspection of all radiation sources were established under law in 1951. At present, there is governmental control of radiation sources and there are requirements for radiation protection based on the recommendations of the ICRP in most countries of the world.

1-5 HEALTH PHYSICS—ITS EARLY DEVELOPMENT AND GROWTH

Radiation protection with its many associated activities was first called health physics on the Metallurgical Project at the University of Chicago a few months prior to the operation of the first nuclear reactor on December 2, 1942. The word "health" was used because radiation protection related to the health of the people working with ionizing radiation and the word "physics" was added because the major unique problems were those of physics. Likewise, most of the senior personnel working for the first few years in this new group were physicists. It was recognized then (1942) by some and has since been pointed out by many others that the words "health physics" do not have a unique meaning and do not in themselves convey a correct understanding of the scope and professional activities of the occupation that is called "health physics." However, similarly, neither the early name "natural philosophy" nor the present name "physics" convey a correct understanding of the profession of the physicist.

In 1959, the Health Physics Society adopted official meanings as follows:

"Health Physics is a profession devoted to the protection of man and his environment from unwarranted radiation exposure. A health physicist is a person engaged in the study of the problems and practices of providing radiation protection. He is concerned with an understanding of the mechanisms of radiation damage, with the development and implementation of methods and procedures necessary to evaluate radiation hazards and with providing protection to man and his environment from unwarranted radiation exposure."

In the above definition, the word "radiation" is used to mean "ionizing radiation" or any radiation that directly or indirectly may be capable of producing ionization. Perhaps these definitions are not satisfactory and certainly they do not completely define or even suggest many of the extensive activities included in present-day health physics. Yet in the broad and simple sense, health physics deals with all problems of protection from ionizing radiation and, on this basis, such studies as biophysics, radiobiology, radiation ecology, mechanisms of interaction of radiation with matter, reactor safety, permissible exposure levels, radioactive waste disposal, and radiation surveying and monitoring are all important health physics research and applied programs. Health physics is a branch of science that fills in a wide gap between many of the older sciences at all points where they relate to the protection of man and his environment from ionizing radiation. Some of these related sciences are physics, biology, medicine, chemistry, geology, meteorology, and many of the branches of engineering. Health physics originated only in name in 1942, for prior to that time many men such as W. C. Roentgen, the Curies, A. H. Becquerel, J. J. Thomson, E. Rutherford, P. Villard, B. Szilard, A. H. Compton, H. Fricke, O. Glasser, L. H. Gray, H. S. Martland, A. Mutscheller, L. S. Taylor, G. Failla, R. M. Sievert, R. D. Evans, W. Binks, R. S. Stone, S. T. Cantril, H. M. Parker, and others too numerous to mention devoted a large portion of their time to problems that today could rightly be classified within health physics.

The first group of professional men at the University of Chicago that were called health physicists consisted of E. O. Wollan, *H. M. Parker*, *C. C. Gamertsfelder*, *K. Z. Morgan*, L. A. Pardue, *J. C. Hart*, O. G. Landsverk, and R. R. Coveyou (only those whose names are italicized have continued in the field of health physics). The early organizational growth and development of health physics are described in reports by Stone,[189] Morgan,[126–131] and Cantril and Parker.[49] From Chicago, health physics spread to various other installations and today is a worldwide activity.

Initially, persons designated as health physicists were located only in the large, government-operated, nuclear-energy laboratories. Today they are found also in hospitals, universities, industrial laboratories, military organizations, public health groups, industrial production plants, labor organizations, insurance companies, and international organizations such as the International Atomic Energy Agency and the World Health Organization. From a small beginning in Chicago in 1942, health physics has grown, so that in 1966 about 6000 health physicists are scattered throughout most of the countries of the world. Originally, health physicists were primarily physicists trained by the hard road of experience. Today there are many arrangements and organizations throughout the world for the education and training of health physicists. However, the demand has always exceeded somewhat the supply and we can expect to see an effort to increase the number of health physicists for many years to come.

Originally, health physics was organized as a branch or arm of medicine. Outstanding medical men such as R. S. Stone, S. T. Cantril, and Shields Warren were in charge of these early programs of health physics at a time when it was difficult to anticipate the biological consequences of operating the large reactors and associated chemical plants that were to handle fission products in megacurie lots, quantities millions of times the number of curies of radium available to man in the entire world prior to the atomic age. As the years passed, however, it became evident that if the health physicist did his part properly there would be little and very infrequent need in this industry for the medical specialist. The nuclear energy workers have had a lower incidence rate of tumors, leukemia, cataracts, and other diseases characteristic of radiation damage and a lower accident rate (see Table 1-13) than the average workman in many of the conventional chemistry industries. Radiation accident cases requiring medical care were few and far between (see Tables 1-11 and 1-12) and for the most part the medical departments in the large nuclear energy plants became very fine (but rather standard and conventional), modern, industrial medical departments. One by one, the radiological specialists left the nuclear energy plants and returned to the hospitals where there were many medical cases on which the radiologist or medical doctor could apply his skills in facing the ever-present challenge of human ills and suffering. In the meantime, the need for applied research and engineering activities in health physics continued to increase at a rapid rate. One by one, most of the health physics groups were severed administratively from the industrial medical departments and, in most cases, the better developed and more progressive organizations became full-fledged divisions reporting to top management.

Perhaps the most important landmark of health physics since its original beginning in 1942 was the organization of a Health Physics Society. In June 1955, at Ohio State University, the health physicists were first organized, and in June 1956, at the University of Michigan the constitution and bylaws of a society were adopted and the name "Health Physics Society" was selected. By June 1957, when the Society met at the University of Pittsburgh, it had increased in membership to about 900, including members from many countries. By the Berkeley meeting in 1958, the Society had a going journal with a wide international circulation. By the Gatlinburg (Tennessee) meeting in 1959 a number of Chapters of the Society were organized in various parts of the United States. At the meeting in Boston in 1960, a mechanism was set up by which the American Board of Health Physics began the certification of health physicists. By June 1965, there were 392 certified health physicists and many companies seeking such employees specified in their job descriptions the requirement of certification in health physics. The annual meetings of the Health Physics Society were held in Las Vegas, Nevada (1961),

Chicago, Illinois (1962), New York City (1963), Cincinnati, Ohio (1964), Los Angeles, California (1965), and Houston, Texas (1966). Future meetings are scheduled for Washington (1967), Denver, Colorado (1968), Pittsburgh, Pennsylvania (1969), and Chicago, Illinois (1970). By the time of the meeting in Houston, the Health Physics Society had a membership of 3027. It had 24 Chapters in the United States and, altogether, a membership scattered in 45 different countries. It comprises most of the leaders in the field of radiation protection in the ICRP as well as the leaders in national groups such as the NCRP.

Although the Health Physics Society when organized in Columbus, Ohio, was deliberately called the Health Physics Society to emphasize the desire that it be the international organization, it was realized in 1963 that this initial goal had not been completely accomplished because (1) several large countries such as the USSR and China had no members in it, (2) countries such as Argentina, Brazil, Iraq, Pakistan, Poland, Czechoslovakia, etc., had only one member each, (3) there were several other independent health physics societies in the process of being organized in Europe, and (4) in some parts of the world the Health Physics Society was referred to as the American Health Physics Society. Therefore, the Health Physics Society formed a committee with K. Z. Morgan as its chairman to study this problem. After several meetings of this committee and correspondence with thousands of persons concerned with this problem, it was determined that there existed a strong and almost unanimous desire to form a truly international health physics organization. This led to the calling of a *pro tempore* general assembly in Paris, France, November 30–December 3, 1964, at which there were 45 official delegates representing 15 health physics or radiation protection societies. Altogether, these delegates represented members in over 50 countries. Throughout the meeting there was expressed a genuine spirit of cooperation and a strong desire to form this international association which was called the International Radiation Protection Association (IRPA). A constitution was adopted which states the primary objective of providing a medium where international contacts and cooperation may be obtained among all those engaged in radiation protection work. This organization would assist in every way possible individuals and national societies in their efforts to provide for the protection of man and his environment from the hazards caused by ionizing radiation and thereby facilitate the exploitation of radiation and nuclear energy for the benefit of mankind. At this Paris meeting, provisional officers were elected and it was decided to hold the first general assembly of the IRPA in Rome, Italy, September 5–10, 1966. At the meeting of the Provisional Executive Council of the IRPA in Los Angeles in June, 1965, the following organizations in order of application were voted into membership in the IRPA: the Central European Section of the Health

Physics Society, the Luxembourg Association of Radioprotection, the French Society of Radioprotection, and the Health Physics Society. At the present time (June, 1973) there are affiliated with the IRPA twenty-two organizations as follows:

Asociacion Mexicana de Proteccion Radiologica Y Fisica Medica, A.C. (Mexico)
Association Belge de Radioprotection (Belgium)
Association Luxembourgeoise de Radioprotection (Luxembourg)
Associazione Italiana di Fisica Sanitaria e Protezione contro le Radiazioni (Italy)
British Radiological Protection Association Limited (United Kingdom)
Czechoslovak Society of Nuclear Medicine and Radiation Hygiene (Czechoslovakia)
Eötvös Loránd Physical Society, Health Physics Section (Hungary)
Fachverband für Strahlenschutz (Germany, Switzerland)
Health Physics Society (United States and Canada)
Indian Association for Radiation Protection (India)
Israel Health Physics Society (Israel)
Nederlandse Vereniging voor Stralingshygiëne (Netherlands)
Nippon Hokenbutsuri Kyogikai (Japan)
Nordic Society for Radiation Protection (Denmark, Finland, Iceland, Norway, Sweden)
Osterreichischer Verband für Strahlenschutz (Austria)
Philippine Association for Radiation Protection (Philippines)
Polish Society for Medical Physics, Section of Radiation Protection (Poland)
Radiation Hygiene Section of the Scientific Society of Hygienists of the USSR
Sociedad Argentina de Radioproteccion (Argentina)
Société Française de Radioprotection (France)
South African Association of Physicists in Medicine and Biology,
Subgroup for Health Physicists (South Africa)
Yugoslav Radiological Protection Association (S.F.R. Yugoslavia)

REFERENCES

1. Tomonaga, M., A. Brill, I. Itoga, and R. Heyssel, "Leukemia in Nagasaki Atomic Bomb Survivors," *Atomic Bomb Casualty Commission Rept. 11–59*, Hiroshima and Nagasaki.
2. Heyssel, Robert, A. B. Brill, L. A. Woodbury, E. T. Nishimura, T. Ghose, T. Hoshino, and M. Yamasaki, "Leukemia in Hiroshima Atomic Bomb Survivors," *Atomic Bomb Casualty Commission Rept. 02–59*, Hiroshima and Nagasaki, 1959.
3. Hollingsworth, J. W., "Delayed Radiation Effects in Survivors of the Atomic Bombings," *Atomic Bomb Casualty Commission Rept. 01–60*, Hiroshima and Nagasaki.
4. Sutow, W. W. and E. West, "Studies on Nagasaki (Japan) Children Exposed in Utero to the Atomic Bomb," *Atomic Bomb Casualty Commission Rept. 31–59*, Hiroshima and Nagasaki.
5. Kato, H., and S. Ueda, "Influence of Concomitant Variables Upon Mortality Rate Comparison," *Atomic Bomb Casualty Commission Rept. 08–63*, Hiroshima and Nagasaki.
6. Anderson, R. E., T. Yamamoto, A. Yamada, and D. W. Will, *Arch. Pathol.* **78,** 618 (1964).
7. Socolow, E. L., A. Hashizume, S. Neriishi, and R. Niitani, *New Engl. J. Med.* **268,** 406 (1963).
8. Angevine, D. M., and S. Jablon, *Ann. N.Y. Acad. Sci.* **114,** 823 (1964).
9. Doll, R., *Gerontol. Clin.* **4,** 211 (1962).
10. Anderson, R. T., *Arch. Pathol.* **79,** 1 (1965).
11. "Irradiation In Utero," *JAMA* **192,** 410 (1965).
12. Inada, G. J., and C. W. Hall, *Am. J. Ophtalmol.* **59,** 216 (1965).
13. Burrow, G. N., H. B. Hamilton, and Zdevck Hrubec, *Yale J. Biol. Med.* **36,** 430 (1964).

14. Burrow, G. N., H. B. Hamilton, and Zdevck Hrubec, *JAMA* **192,** 97 (1965).

15. Alekseyeva, O. G. et al., "A Case of Acute Radiation Sickness in Man," TT63-24437, Clearing House for Federal Scientific and Technical Information, U.S. Dep. Comm. Springfield, Virginia (Code 410.20) (1963).

16. Boyd, J. T., W. M. Court-Brown, J. Vennart, and G. E. Woodcock, *Brit. Med. J.* **1,** 377 (1966).

17. "Fundamental Nuclear Energy Research," p. 114, U.S. At. Energy Comm. (1965).

18. Johnson, K. G., and A. Ciocco, "Human Radiation Effects," *Atomic Bomb Casualty Commission Tech. Rept. 10–65* (1965).

19. Masuda, Y., *Rept. 70th General Meeting Japanese Ophthalmological Society*, Sanyo Press, Japan, 1966.

20. *Oak Ridge Nat. Lab. Rept.* ORNL CF-57-11-91, Oak Ridge, Tennessee (1957).

21. Shipman, T. L., C. C. Lushbaugh, D. F. Peterson, W. H. Langham, P. S. Harris, and J. N. P. Laurence, *J. Occupational Med., Special Suppl.* **3,** (1961).

22. *Leukemia Reaches Final Stage*, Nishinhon Press, Japan, 1966.

23. Hollingsworth, D. R., H. B. Hamilton, H. Tamagaki, and G. W. Beebe, *Medicine* **42,** 47 (1963).

24. Ritchie, R. H., and G. S. Hurst, "Penetration of Weapons Radiation: Application to the Hiroshima-Nagasaki Studies," *Atomic Bomb Casualty Commission Rept. 26–59*, Hiroshima and Nagasaki, 1959.

25. "On X-Ray and Pregnancy—ACR Differs with International Body," *Am. Coll. Radiol. Bull.* **20,** No. 5 (1964).

26. "Radium and Mesothorium Poisoning and Dosimetry and Instrumentation Techniques in Applied Radioactivity," *U.S. At. Energy Comm. Rept. AECU*-3504, Oak Ridge, Tennessee, 1957.

27. Albers-Schonberg, H., *Muench. Med. Wochsch.* **50,** 1895 (1903).

28. Anderson, E. C., and W. H. Langham, *Science* **130,** 713 (1959).

29. *Safety Code for the Industrial Use of X-Rays*, American Standards Association, New York, (1946).

30. "Shoe Fitting Fluoroscopes," *Am. Col. Radiol. Monthly Newsletter* **6,** 1 (1950).

31. Andrews, G. A., B. W. Sitterson, A. L. Kretchmar, and M. Brucer, *Health Phys.* **2,** 134 (1959).

32. Becquerel, H., and P. Curie, *Compt. Rend.* **132,** 1289 (1901).

33. Bell, A. L. L., *Radiology* **40,** 139 (1943).

34. Private Communication to K. Z. Morgan from Hanson Blatz, Director of Office of Radiation Control, New York, August 20, 1962.

35. Braestrup, C. B., and R. T. Mooney, *Science* **130,** 1071 (1959).

36. *J. Röntgen Soc.* **12,** 67 (1916).

37. *J. Röntgen Soc.* **17,** 100 (1921).

38. Burch, P. R. J., and F. W. Spiers, *Science* **120,** 719 (1954).

39. Bergonie, J., and L. Tribondeau, *Compt. Rend.* **47,** 400 (1904).

40. Cantril, S. T., *Industrial Medicine on the Plutonium Project* **20,** 36, McGraw-Hill, New York, 1951.

41. Court-Brown, W. M., and R. Doll, *Leukaemia and Aplastic Anaemia in Patients Irradiated for Ankylosing Spondylitis*, H.M. Stationery Office, London, 1957.

42. Chamberlain, R. H., *Am. Coll. Radiol. Bull.* **18,** No. 6, 1 (1962).

43. Conard, R. A., and A. Hicking, *JAMA* **192,** 457 (1965).

44. Clark, S. H., *Bull. Atomic Scientists* **12,** 14 (1956).

45. Clark, S. D., *JAMA* **168,** 761 (1958).

46. Codman, E. A., *Philad. Med. J.* **9,** 438 (1902).

47. Compton, A. H., *Phys. Rev.* **43,** 387 (1933).

48. Committee of Inquiry, *Accident at Windscale No. 1 Pile on 10th October, 1957*, H.M. Stationery Office, London, 1957.

49. Cantril S. T., and H. M. Parker, *The Status of Health and Protection at Hanford Engineer Works*, CH-3570 (1946).

50. Callihan, A. D., and J. T. Thomas, *Health Phys.* **1,** 363 (1959).

51. Daniel, J., *Science* **3,** 562 (1896).

52. Despeignes, V., *Lyon Med.* **82,** 428 and 503 (1896).

53. Dunning, G. M., *Health Phys.* **1,** 255 (1958).

54. Dudley, R. A., "Natural and Artificial Radiation Background of Man," in Low Level Radiation, A. M. Brues, Ed., *Publ. No. 59, AAAS*, Washington, 1959.

55. Dunning, G. M., *Health Phys.* **1,** 379 (1959).

56. Eisenbud, Merril, *Environmental Radioactivity*, McGraw-Hill, New York, 1963.

57. Enz, H., R. LeCoultre, and P. Lerch, *Nucl. Sci. Abst.* **18,** 474 (1964).

58. "Earlier Utility Survey Showed Little Concern on Siting, SL-1," *Nucleonics News* **19,** No. 5 (1961).

59. Evans, R. D., *Am. J. Public Health* **23,** 1017 (1933).

60. *Mass. Inst. Technol. Ann. Prog. Rept.*, MIT-952-2 (1965).

61. *Mass. Inst. Technol. Ann. Prog. Rept.* MIT-952-1 (1964).

62. *Mass. Inst. Technol. Ann. Prog. Rept.* NYO-9505 (1963).

63. Fricke, H., and O. Glasser, *Am. J. Roentgenol. Radium Therapy Nucl. Med.* **13,** 453 (1925).

64. Ford, M. R., and K. Z. Morgan, Unpublished calculations using data from *ICRP Publ. 2*, Report of Committee II on Permissible Dose for Internal Radiation, Pergamon Press, London, 1959.

65. Freund, L., *Elements of General Radio-Therapy for Practitioners*, Rebman Company, New York, 1904.

66. Guskova, A. K., and G. D. Baisogolov, *Peaceful Uses of Atomic Energy* **11,** 35, United Nations, New York, 1956.

67. Gabrysh, A. F., and F. J. Davis *Civil Eng.* **839,** 89 (1954).

68. Gabrysh, A. F., and F. J. Davis, *Nucleonics* **13,** 50 (1955).

69. Gerstner, H. B., *US Armed Forces Med. J.* **19,** 313 (1958).

70. Grubbe, E. H., *Radiology* **21,** 156 (1933).

71. Glasser, O., E. H. Quimby, L. S. Taylor, J. E. Weatherwax, and R. H. Morgan, *Physical Foundations of Radiology*, Hoeber, New York, 1961.

72. Gupton, E. D., unpublished measurements, Health Physics Division, Oak Ridge National Laboratory, Oak Ridge, Tennessee (1965).

73. *A Summary of Accidents and Incidents Involving Radiation in Atomic Energy Activities, June 1945 Through December 1955*, TID 5360, USAEC Techn. Inform. Serv. Extension, Oak Ridge, Tennessee, 1956.

74. *A Summary of Industrial Accidents in USAEC Facilities*, TID D-5360, Suppl. 2, USAEC Techn. Inform. Serv. Extension, Oak Ridge, Tennessee, 1959.

75. Hammer-Jacobsen, E., *Danish Med. Bull.* **6,** 113 (1959).

76. Harris, P. S., *Health Phys.* **5,** 37 (1961).

77. Heineke H., *Muench. Med. Wochsch.* No. 48, 2089 (1903).

78. Heineke, H., *Muench. Med. Wochsch.* No. 18, 784 (1904).

79. Hesse, Otto, *Fortschr. Gebiete Roentgenstrahlen* **17,** 82 (1911).

80. Hempelmann, L. H., *New Engl. J. Med.* **241,** 335 (1949).

81. Hurst, G. S., R. H. Ritchie, and L. C. Emerson, *Health Phys.* **2,** 121 (1959).

82. Horan, J. R., and W. P. Gammill, *Health Phys.* **9,** 177 (1963).

83. Hoffman, J. G., and L. H. Hempelmann, *Am. J. Roentgenol. Radium Therapy Nucl. Med.* **77,** 144 (1957).

84. Hansen, C. L., J. W. Howland, M. Ingram, and H. Mermagen, *US Armed Forces Med. Serv. Digest* **12,** 2 (1961).

85. Hasterlik, R. J.,and L. D. Marinelli, *Peaceful Uses of Atomic Energy* **11,** 25, United Nations, New York, 1956.

86. *Radiation Accidents: Dosimetric Aspects of Neutron and Gamma-Ray Exposures,* ORNL-2748, Part A, Oak Ridge Natl. Lab., Oak Ridge, Tennessee, 1959.

87. Hurst, G. S., R. H. Ritchie, F. W. Sanders, P. W. Reinhardt, J. A. Auxier, E. B. Wagner, A. D. Callihan, and K. Z. Morgan, *Health Phys.* **5,** 179 (1961).

88. *Health and Safety Laboratory Fallout Program Quarterly Summary Rept.* HASL-149, USAEC N.Y. Operations Office (Oct. 1964).

89. *Health and Safety Laboratory Fallout Program Quarterly Summary Rept.* HASL-155, USAEC N.Y. Operations Office (Jan. 1965).

90. *Health and Safety Laboratory Fallout Program Quarterly Summary Rept.* HASL-158, USAEC N.Y. Operations Office (April 1965).

91. Hultqvist, B., *Studies on Naturally Occurring Ionizing Radiations*, Almqvist and Wiksell, Uppsala, Sweden (1956).

92. *ICRP Suppl. 6*, Brit. Inst. Radiology, London, 1955.

93. ICRP and ICRU, *Phys. Med. Biol.* **2,** 107 (1957).

94. ICRP and ICRU, United Nations Rept. A/AC.82/G/R. 117, New York (1957).

95. *ICRP Publ. 3*, Pergamon Press, London (1960).

96. *ICRP Publ. 6*, Pergamon Press, London (1964).

97. *ICRP Publ. 5*, Pergamon Press, London (1964).

98. Auxier, J. A., H. H. Hubbell, and J. S. Cheka, Health Physics Division, Oak Ridge Natl. Lab., Oak Ridge, Tennessee, unpublished dose data on Hiroshima and Nagasaki (1965).

99. Joyet, G., and M. Miller, *Experientia* **16,** 342 (1960).

100. Kaye, G. W. C., *X-rays*, Longmans, Green, London, 1929.

101. "M. W. Kellogg Co. Radiation Incident on March 13, 1957," *Nucleonics* **15,** 42 (Dec. 1957).

102. Lowder, W. M., H. L. Beck, and W. J. Condon, *Nature* **202,** 745 (1964).

103. McLendon, J. D., *Health Phys.* **2,** 21 (1959).

104. Leipunsky, O. I., *Congressional Rec.* **3,** 2423 (1959).

105. Looney, W. B., J. B. Hursch, M. Colodzin, and L. T. Stedman, *Acta Union Intern. Contra Cancrum* **16,** 435 (1960).

106. Libby, W. F., *Science* **122,** 57 (1955).

107. Lucas, H. F., and F. H. Ilcewicz, *J. Am. Water Works Assoc.* **50,** 1523 (1958).

108. Looney, W. B., *J. Bone Joint Surg.* **37A,** 1169 (1955).

109. Looney, W. B., *J. Bone Joint Surg.* **38A,** 175, 392 (1956).

110. Looney, W. B., *Am. J. Roentgenol. Radium Therapy Nucl. Med.* **83,** 163 (1960).

111. Looney, W. B., TID-20,008, Dept. of Commerce, Washington 1964.

112. Laughlin, J. S., and I. Pullman, *The Biological Effects of Atomic Radiation, Section III,* Natl. Acad. Sci., Washington, 1957.

113. *Radiation Accidents and Emergencies in Medicine, Research and Industry*, Charles C Thomas, Springfield, Illinois, 1965, L. L. Lanzl, J. H. Pingel, and J. H. Rust, eds.

114. *Background Radiation*, NYO-4712, USAEC N.Y. Operations Office, 1956.

115. Lucas, H. F., Jr., *Argonne Natl. Lab. Radiol. Phys. Div. Semiannual Rept.*, ANL-6297, Chicago, Illinois, 1961, p. 55.

116. March, H. C., *Am. J. Med. Sci.* **220,** 282 (1950).

117. Martland, H. S. *Collection of Reprints on Radium Poisoning, 1925–1939*, AEC Tech. Inform. Serv. Extension, Oak Ridge, Tennessee, 1951.

118. Marinelli, L. D., *Am. J. Roentgenol. Radium Therapy Nucl. Med.* **80,** 729 (1958).

119. Martin, J. H., *Med. J. Australia*, p. 79 (July 1958).

120. Martin, J. H., *Proc. Coll. Radiol. Australia* **1,** 103 (1957).

121. MacMahon, B., *J. Natl. Cancer Inst.* **28,** 1173 (1962).

122. Martinez, R., G. Cassab, G. Ganem, E. Guttman, M. Lieberman, L. Vater, M. Linarea, and H. Rodriguez, *Rev. Med. Inst. Mex. Seguro Social* **3,** 14, Suppl. 1 (1964).

123. McCandless, J. B., *JAMA* **192,** 185 (1965).

124. Medical Research Council, *The Hazards to Man of Nuclear and Allied Radiations*, H.M. Stationery Office, London, 1956.

125. *First Report About a Radiation Accident*, Mexican National Nuclear Energy Commission (translation by Leo Goodman, Secretary, Atomic Energy Tech. Comm. Industrial Union Dept., AFL–ClO), 1962.

126. Morgan, K. Z., *Natl. Safety News* **53,** 25 (1946).

127. Morgan, K. Z., *Sci. Monthly* **63,** 93 (1946).

128. Morgan, K. Z., *Health Phys.* **1,** 1 (1958).

129. Morgan, K. Z., *Eng. J.* (*Montreal*) **31,** 154 (1948).

130. Morgan, K. Z., *Med. Physics.* **2,** 766 (1948).

131. Morgan, K. Z., *Sci. Am.*, p. 5 (Jan., 1948).

132. Morgan, K. Z., *Science* **139,** 565 (1963).

133. Medical Research Council, *The Exposure of the Population from Fallout*, H.M. Stationery Office, London, 1964.

134. *Health Phys. Div. Ann. Progr. Rept.*, ORNL-3697, Oak Ridge Natl. Lab., Oak Ridge, Tennessee, 1964.

135. *Health Phys. Div. Ann. Progr. Rept.*, ORNL-3849, Oak Ridge Natl. Lab., Oak Ridge, Tennessee, 1965.

136. Moeller, D. W., J. G. Terrill, and S. C. Ingraham, Public Health Rept **68,** 57 (1953).

137. Mutscheller, A. M., *Am. J. Roentenol. Radium Therapy Nucl. Med.* **13,** 65 (1925).

138. *Nucl. Data Sheets*, Natl. Acad. Sci.-Natl. Res. Council, Washington (sheets through August, 1960, were consulted).

139. *The Nuclear Aspects of the Accidental Criticality at Wood River Junction, Rhode Island, July 24, 1964*, United Nuclear Corp., New Haven, Connecticut, 1964.

140. *Nuclear Incident at United Nuclear Corporation, Wood River Junction, Rhode Island, July 24, 1964*, United Nuclear Corp., New Haven, Connecticut, 1964.

141. *Safe Handling of Radioactive Luminous Compounds, Natl. Bur. Std.* (*U.S.*) *Handbook* **27,** (1941).

142. *Medical X-ray Protection Up To Two Million Volts, Natl. Bur. Std.* (*U.S.*) *Handbook* **42,** (1949).

143. *X-Ray Protection, Natl. Bur. Std.* (*U.S.*) *Handbook* **60,** (1955).

144. *Medical X-Ray Protection Up To Three Million Volts, Natl. Bur. Std.* (*U.S*) *Handbook* **76,** (1961).

145. *Safe Handling of Radioactive Isotopes, Natl. Bur. Std.* (*U.S.*) *Handbook* **42,** (1949).

146. *Control and Removal of Radioactive Contamination in Laboratories, Natl. Bur. Std.* (*U.S.*) *Handbook* **48,** (1951).

147. *Recommendations for Waste Disposal of Phosphorus-32 and Iodine-131 for Medical Users, Natl. Bur. Std.* (*U.S.*) *Handbook* **49,** (1951).

148. *Recommendations for the Disposal of Carbon-14 Wastes, Natl. Bur. Std.* (*U.S.*) *Handbook* **53,** (1953).

149. *Radioactive Waste Disposal in the Ocean, Natl. Bur. Std.* (*U.S.*) *Handbook* **58,** (1954).

150. National Committee on Radiation Protection, *Am. J. Roentgenol., Radium Therapy Nucl. Med.* **84,** 152 (1960).

151. National Committee on Radiation Protection, *Radiology* **75,** 122 (1960).

152. National Committee on Radiation Protection, *Health Phys.* **5,** 1 (1961).

153. Nolan, W. E., and H. W. Patterson, *Radiology* **61,** 625 (1953).

154. Norris, W. P., T. W. Speckman, and P. F. Gustafson, *Am. J. Roentgenal. Radium Therapy Nucl. Med.* **73,** 785 (1955).

155. Neel, J. V., and W. J. Schull, "The Effect of Exposure to the Atomic Bombs on Pregnancy Termination in Hiroshima and Nagaski," *Natl. Res. Council Publ. No. 461*, Washington, 1956.

156. New York City Department of Health, Amendment to Sanitary Code, Section 107a, filed January 22, 1948.

157. *The Acute Radiation Syndrome*, ORINS-25, Oak Ridge Inst. Nucl. Studies, Oak Ridge, Tennessee, 1959.

158. "Incident at Windscale," *Atomic Indus. Forum* **4,** 10 (1957).

159. Parker, H. M., *Advan. Biol. Med. Phys.* **1,** 223 (1948).

160. *Accidental Radiation Excursion at the Y-12 Plant*, Y-1234, Union Carbide Nuclear Company, Oak Ridge, Tennessee, 1958.

161. Perthes, G., *Fortschr. Gebiete Roentgenstrahlen* **8,** 12 (1904).

162. Peller, S., *Human Biol.* **11,** 130 (1939).

163. Peterson, D. F., *Health Phys.* **9,** 231 (1963).

164. "AEC's Pittman Reports on the SL-1 Accident," *Nucleonics* **19,** No. 3 (1961).

165. Purser, P. R., and C. F. Quist, *Acta Radiol.* **48,** 267 (1957).

166. Roesch, W. C., C. C. Gamertsfelder, H. V. Larson, J. M. Nielsen, and E. C. Watson, *Health Phys.* **9,** 757 (1963).

167. Rolleston, Humphrey, *Quart. J. Med.* **23,** 101 (1930).

168. Rossi, E. C., A. A. Thorngate, and F. C. Larson, *J. Lab. Clin. Med.* **59,** 655 (1962).

169. Russ, Sydney, J. Röntgen Soc. **12,** 38 (1916).

170. Russell, E. B., and W. L. Russell, *Radiology* **58,** 369 (1952).

171. Savic, P. P., *Bull. Inst. Nucl. Sci.*, "*Boris Kidrich*" (*Belgrade*) **9,** 1 (1959).

172. Saenger, E. L., *Am. J. Roentgenol, Radium Therapy Nucl. Med.* **84,** 715 (1960).

173. *Medical Aspects of Radiation Accidents*, U.S. Gov. Printing Office, Washington 1963.

174. Scott, N. S., *Am. X-Ray J.* **1,** 57 (1897).

175. Senn, N., *Med. Record* (*N.Y.*) **64,** 281 (1903).

176. Seltser, Raymond, "Continuing Studies of Mortality of American Radiologists," presented at the annual meeting of the APHA, Kansas City, Mo. (1963).

177. Seltser, Raymond, *JAMA* **190,** 90 (1964).

178. Strominger, D., J. M. Hollander, and G. T. Seaborg, *Rev. Mod. Phys.* **30,** (1958).

179. Silberstein, H. E., *Radium Poisoning*, AEC Document M-1695 (1945).

180. Sikl, H., *Acta Unio Intern. Contra Cancrum* **6,** (1950).

181. "Reactor Accident at Idaho Falls," *Nucl. Power* (Feb. 1961).

182. "Interim Report on SL-1 Accident," *Nucl. Power* (March 1961).

183. "SL-1 Films Show Core Damage," *Nucl. Power* (April 1961).

184. Photograph showing "A View Looking Straight Down on Top of the SL-1 Reactor Involved in the January 3 Accident," *Nucl. Power* (May 1961).

185. "SL-1 Reactor to be Decontaminated," *Nucl. Power* (June 1961).

186. "New SL-1 Accident Theory," *Nucl. Power* (Dec. 1961).

187. Schlundt, H., J. T. Nerancy, and J. P. Morris, *Am. J. Roentgenol., Radium Therapy Nucl. Med.* **30,** 515 (1933).

188. Spiers, F. W., *Brit. J. Radiol.* **29,** 409 (1956).

189. Stone, R. S., *Proc. Am. Phil. Soc.* **90,** 11 (1946).

190. Stone, R. S., *Western J. Surg. Obstet. Gynecol.* **54,** 153, 201 (1946).

191. *Protection in Diagnostic Radiology*, Rutgers University Press, New Brunswick, New Jersey, 1959, p. 70.

192. Stone, R. S., *Radiology* **58,** 639 (1952).

193. Stanford, R. W., and J. Vance, *Brit. J. Radiol.* **28,** 266 (1955).
194. Sullivan, W. H., *Trilinear Chart of Nuclides*, USAEC (1957) (revisions through August 1960 were consulted).
195. Stewart. A., J. Webb, D. Giles, and D. Hewitt, *Lancet* **271,** 447 (1956).
196. Taylor, L. S., *J. Wash. Acad. Sci.* **46,** 69 (1956).
197. Taylor, L. S., *Proc. Intern. Conf. Peaceful Uses At. Energy Geneva, 1955* **13,** 15, United Nations, New York, 1956.
198. Taylor, L. S., *Southern Med. J.* **49,** 826 (1956).
199. Taylor, L. S., *Am. J. Roentgenol., Radium Therapy Nucl. Med.* **78,** 983 (1957).
200. Taylor, L. S., *Health Phys.* **1,** 306 (1958).
201. *Some Effects of Ionizing Radiation*, AEC Report TID-5358, Washington 1956.
202. Thiele, H., Rostoski, Saupe, and Schmorl, *Muench. Med. Wochschr.* **1,** 24 (1924).
203. Turner, R. C., J. M. Radley, and W. V. Mayneord, *Health Phys.* **1,** 268 (1958).
204. Thoma, G. E., and N. Wald, *J. Occupational Med.* **1,** 420 (1959).
205. *Report of the United Nations Scientific Committee on the Effects of Atomic Radiation*, Suppl. No. 17 (A/3838), New York, 1958.
206. *Report of the United Nations Scientific Committee on the Effects of Atomic Radiation*, Suppl. No. 16 (A/5216), New York, 1962.
207. *Report of the United Nations Scientific Committee on the Effects of Atomic Radiation*, Suppl. 14 (A/5814), New York, 1964.
208. Upton, A. C., *Health Phys.* **12,** 209 (1966).
209. Boegly, W., *Health Phys.* **12,** 417 (1966).
210. deLaguna, W., and E. G. Struxness, "Waste Disposal by Hydraulic Fracturing," presented at the Health Physics Society meeting, Cincinnati, Ohio, 1964.
211. Shaikh, M. U., D. G. Jacobs, and F. L. Parker, *A Study of the Movement of Radionuclides Through Saturated Porous Media*, ORNL-TM-1681 Oak Ridge Natl. Lab., Oak Ridge, Tennessee, 1966.
212. Williams, C. R., *New Engl. J. Med.* **241,** 333 (1949).
213. *Radiation Accidents: Medical Aspects of Neutron and Gamma-Ray Exposures*, ORNL-2748 (Part B), Oak Ridge Natl. Lab., Oak Ridge, Tennessee, 1961.

PROBLEMS

1. Assuming that a reliable method could be developed to account for the observed mutations in man and animals, what studies could be made to account for the 2 to 10% of mutations in man that are attributable to natural background radiation?

2. List factors other than ionizing radiation that contribute to the natural mutations in man.

3. What fraction of the total mutations in man are considered to be of the so-called visible or readily detectable type? Of what consequence are non-visible types?

4. If ionizing radiation has brought about beneficial mutations that are important in the evolution of man, why is not low-level exposure to the individual considered beneficial?

5. Why should we at low dose levels be more concerned about the number of body cells that have been exposed to and affected by ionizing radiation than with those that have been killed by it?

6. The four series of radioactive elements of high atomic number have atomic masses given by the relation $(4n + b)$. (*a*) Why are there no series for which $b = 5$ or greater? (*b*) Why are there no series of the type $(5n + b)$? (*c*) Why is the $(4n + 1)$ series missing in nature? (*d*) Why is the $(4n + 3)$ series called the actinium series? (*e*) Why are there no values of n greater than 59 for any of the naturally occurring radionuclides? (*f*) Why are there no values of n less than 51 for any of the naturally occurring radionuclides? (*g*) What is the largest value of the series $(4n + b)$ that has been produced artificially?

7. With so many short-lived radionuclides in nature, why was not the natural background on the earth much higher 10^9 yr ago than it is today?

8. Give two types of nuclear reactions that can be carried out with sources of high-energy particles that produce only other radionuclides within the same series, $(4n + b)$. Give four such reactions that produce only radionuclides in other series.

9. Why is the natural concentration of ^{239}Pu in the earth very low? Name two likely nuclear reactions that would account for the natural background of ^{239}Pu prior to the time of man-made sources of radiation.

10. What are the principal factors that account for the high rates of dose equivalent in the bone from natural deposits of ^{226}Ra?

11. Why is there such a wide variation in the dose equivalent in the lungs of man from natural background radiation?

12. Why are studies of human exposure to natural background radiation of particular importance to the health physicist?

13. Why are studies of radium in man and animals of particular interest to the health physicist?

14. Why should man retain about twice as much of the radium that he takes in from water as that which he takes in from food?

15. How was it possible for people in the United States to be experimenting with X-rays and to be damaged from their use only a few days after the announcement of the discovery of X-ray by Roentgen?

16. Why weren't more persons seriously injured in the early and rather reckless use of X-rays?

17. Basically, why would we expect *in utero* children to be much more radiosensitive than adults? Why might very old persons be more radiosensitive than middle-aged persons?

18. Why were the early experiences of man with ionizing radiation an insufficient source of information upon which to establish safe, reliable, and reasonable levels of exposure for the occupational worker and for members of the population-at-large?

19. How can the life-shortening data of Seltser et al., combined with information on the slope of the curve of "life-shortening as a function of dose," be used to conclude that the early effective dose of radiologists may have been of the order of 15 to 30 R/yr? How might this relatively low dose be explained in view of the claim of some investigators that the early radiologists received much higher doses?

20. Why were so many young radium-dial painters seriously injured before action was taken to curb these exposures?

21. What would be the level of maximum permissible body burden of ^{226}Ra that would assure no consequent bone tumors in man?

22. What type of neoplasm in man appears to be characteristic of chronic damage from high levels of exposure of ThO_2?

23. What are the present-day estimates of gonadal exposure to medical sources of the average man in the United States? What fraction is this of his total exposure and of his exposure to man-made sources?

24. Why was it a much simpler task to reduce human exposure to shoe-fitting machines than to unnecessary medical exposure?

25. Why does it appear that medical exposure in the United States may be much higher than in most other countries?

26. Chest X-ray skin doses commonly range from 10 to 500 mR. List ten factors that might be expected to account for unnecessarily high exposures of this type.

27. From studies of the biological effects of ionizing radiation on the offspring of exposed pregnant mice, what types of damage might be expected as a result of large exposures to pregnant women?

28. What recommendation to prevent fetal damage does the ICRP give with reference to radiological examinations of women in the childbearing age?

29. About what is the ratio of dose of the usual fluoroscopic examination to the usual radiographic examination of the same part of the body?

30. Discuss the advantages and disadvantages from the standpoint of reducing unnecessary human exposure resulting from using high-voltage rather than low-voltage X-rays.

31. What are some of the arguments that could be given opposing the practice of some medical men in the United States who routinely prescribe X-ray examinations of their patients when they are considered to be or are known to be pregnant?

32. List ten things that can be done by the dentist which will reduce dental exposure to the patient in some cases by one to two orders of magnitude.

33. Have medical doctors in the United States who X-ray their patients been required in their period of training to take courses in radiation physics or X-rays and to pass examinations on these subjects?

34. Explain why, in general, most of the measures that can be taken to reduce unnecessary medical diagnostic exposure can be expected to result in more radiographic medical information.

35. If there is a linear relationship between radiation-induced leukemia and dose equivalent (not a proved relationship) and if the average bone marrow dose in the United States is 150 mrem per year per person, how many leukemias are produced per year under equilibrium conditions in a country of 2×10^8 persons?

36. Why is the health physicist particularly interested in a detailed study of cases of human exposure to ionizing radiation? Why study a variety of human exposure cases? Why not base conclusions of effects of exposure to man on animal studies?

37. Why not dump high-level, liquid, radioactive waste directly into the ocean from chemical processing plants located on the coasts of the United States?

38. Name four characteristics of radionuclides which, in general, tend to classify them among the radionuclides more dangerous to man.

39. Within certain limits, α- or β-radiation may be considered to be attenuated exponentially. Show that if a small radioactive particle is deposited in the body (particle so small that the attenuation of α- or β-radiation within the particle is negligible), the dose equivalent rate at a distance r (cm) from the particle in tissue (of density ρ) is given by the equation

$$R = \frac{2.9 \times 10^{-5} BSW(QF)e^{-\mu_t r}}{\rho r^2} \qquad \text{rem/wk}$$

in which B is the specific activity in microcuries per particle, μ_t is the attenuation coefficient of the α- or β-radiation in tissue (cm^{-1}), QF is the quality factor, W is the energy required to produce an ion pair (*ev*/*ip*), and S is the specific ionization (ip/cm).

40. Show that the average dose equivalent rate for α- or β-radiation in a sphere of radius r (cm) about the particle in tissue is given by the equation

$$R = \frac{8.6 \times 10^{-5} BSW(QF)(1 - e^{-\mu_t r})}{\rho r^3 \mu_t} \qquad \text{rem/wk}$$

41. When $r \geq$ the range of the α- or β-radiation in tissue and the average effective energy is ξ (MeV), show that the average dose equivalent rate in a sphere of radius r (cm) about the particle is given by the equation

$$R = \frac{85 B\xi(QF)}{\rho r^3} \qquad \text{rem/wk}$$

42. If the dose equivalent rate from a fallout mixture originating from a nuclear detonation at time after the detonation $t = t_1$ is found to be R_1 (rem/hr at 1 meter above ground), derive the equation in terms of R_1, t_1, and t_2 for the integrated dose equivalent at this same point for the exposure period $(t_2 - t_1)$. Reduce equation to its simplest form. Assume the Wigner-Way relationship, that is, $R \sim t^{-1.2}$.

43. Why is ^{131}I often the radionuclide of principal concern in estimating hazards that may be associated with a major accident of a nuclear reactor?

44. What simple procedure could have been used by the inhabitants of the Marshall Islands or the fishermen on the *Fukuryu Maru* to reduce the dose from weapons fallout to negligible proportions?

45. What might you expect to be the type of radiation damage that could make its appearance in the populations in certain parts of Nevada and Utah that received large fallout doses from weapons tests prior to May, 1958?

46. Why would you expect the longer-lived radionuclides that are daughters of relatively long-lived noble gases produced in the fission process to be the predominant radioactive material found in the stratosphere from surface detonations of fission-type weapons?

47. How does the average medical dose in the United States compare with the average dose from weapons fallout?

48. Why is the ratio of the gonad dose from weapons fallout integrated to infinity to the dose integrated to the year 2000 greater for the thermonuclear weapon than for the fission-type weapon?

49. Why is the number of fusions about ten times the number of fissions for weapons of the same explosive power?

50. For weapons of the same explosive power, what is the approximate ratio of curies of $^{14}C + ^{3}H$ to curies of $^{90}Sr + ^{137}Cs + ^{14}C$ for thermonuclear and fission-type weapons?

51. Although studies of the dose-effect relationships of the survivors of nuclear weapons explosions over Hiroshima and Nagasaki are of extreme value in setting maximum permissible dose values for occupational exposure, there are some serious weaknesses in such extrapolations. Name the most important weakness in these extrapolations.

52. Discuss two of the principal types of chronic radiation damage that have been observed in the survivors of the nuclear weapons explosions over Hiroshima and Nagasaki.

53. Explain what is meant by the Wigner Effect in graphite.

54. List five of the most important health physics lessons learned from the Windscale reactor accident in 1957.

55. Assuming (for this purpose) that the recorded values of absorbed dose as listed in Table 1-11 for the various serious reactor and criticality accidents are correct, list five reasons why the medical syndrome for the June 16, 1958, Oak Ridge accident was so much less severe than that of the Vinca accident of October 15, 1958.

56. Why might it be advisable to excise tissue or even to amputate a finger when accidents with ^{239}Pu or other of the more dangerous radionuclides result in a large local deposition of the radionuclide in the body?

57. Give a general discussion indicating the safety of the nuclear energy industry in comparison with other modern industries. Discuss the record of radiation safety in the United States.

58. List three types of probable injury that may result from excessive use of shoe-fitting machines.

59. Name ten common uses of radioisotopes.

60. What is meant by the term "threshold erythema dose"? Give eight reasons why this was an unreliable unit. About what was the magnitude of this unit?

61. Give a brief review of the history of health physics.

CHAPTER 2

The Passage of Heavy Charged Particles, Gamma Rays, and X-Rays through Matter

J. Neufeld, L. C. Emerson, F. J. Davis, and J. E. Turner

This Chapter treats the interaction of radiation and matter as a problem in physics. In particular, a quantitative description is given for the passage of heavy charged particles, γ- and X-rays through matter. The treatment of these subjects differs from the later treatments of the tissue dose from neutrons (Chapter 4), β-particles (Chapter 8), and X- and γ-rays (Chapter 9) where dosimetric aspects of the interactions with tissue are emphasized.

A. HEAVY CHARGED PARTICLES

2-1 INTRODUCTION

A charged particle passing through matter loses energy as a result of electromagnetic interactions with the atoms and molecules in the surrounding medium. The character of these interactions and the mechanism of the energy loss depend on the charge and velocity of the particle and on the character of the medium. It will be assumed that the particle is "heavy," that is, it has a mass considerably exceeding the electron mass. Heavy particles include, for example, protons, α-particles, mesons, and atomic and molecular ions; electrons and positrons are excluded.

Because a heavy charged particle loses a relatively small fraction of its energy in a single collision with an atomic electron, any phenomena associated with its slowing-down can be described by formulas that are continuous functions of its energy.

2-2 ELASTIC AND INELASTIC COLLISIONS

The collisions of a moving charged particle with the atoms and molecules in the medium may be elastic or inelastic. The effect of an elastic collision is to produce a recoil atom having an energy which is a considerable fraction of the energy of the incident particle, and the probability of such a collision is appreciable only when the velocity of the particle is considerably lower than the electronic velocity v_0 in the ground state of hydrogen, which is given by

$$v_0 = \frac{e^2}{\hbar} = 2.183 \times 10^8 \text{ cm/sec} \tag{2-1}$$

where e is the electronic charge and $\hbar$ is Planck's constant divided by 2π. For velocities higher than v_0, the elastic collisions are negligible and the main mechanism for the energy dissipation is due to inelastic collisions that result in ionization and excitation of the surrounding atoms.

Radiation protection and dosimetry are concerned primarily with inelastic collisions, and the following discussion therefore is concerned primarily with these.

2-3 CHARGE FLUCTUATION

In analyzing the slowing-down process of an ion passing through a material, account must be taken of the effect of the medium on the charge of the ion. At speeds $v \leqslant v_0$, a moving ion continually captures and loses orbital electrons, so that the charge of the ion fluctuates. At a given instant the ion can exist in various "charge states" with definite probabilities. Let $P(i)$ represent the probability of the ion having charge i. The average charge of an ion $\bar{i}$ and the average square charge $\overline{i^2}$ are then given by

$$\bar{i} = \sum_i iP(i) \tag{2-2}$$

and

$$\overline{i^2} = \sum_i i^2P(i) \tag{2-3}$$

where the summation extends over all possible charge states of the ion.

Some of the effects of charge fluctuation are illustrated in Figs. 2-1, 2-2, and 2-3, which are taken from Neufeld and Snyder (1961). Figures 2-1 and 2-2 show graphs based on experimental data which represent the relationship between the average charge $\bar{i}$ and the velocity v of a nitrogen or oxygen ion passing through matter. A relationship between the average square charge

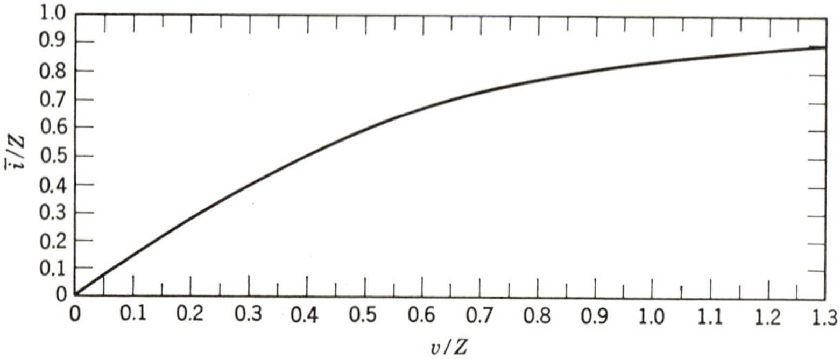

Fig. 2-1. Average charge i of nitrogen and oxygen ions as a function of velocity v/Z (Z = atomic number and v is in atomic units, $e^2/\hbar = 2.183 \times 10^8$ cm/sec) (Neufeld and Snyder, 1961).

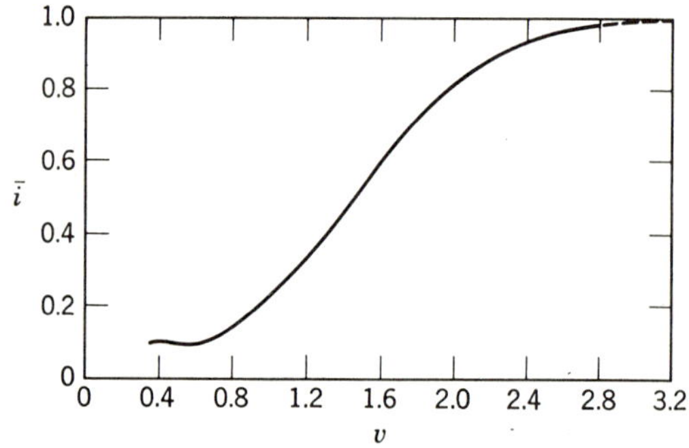

Fig. 2-2. Average charge $\bar{i}$ of a hydrogen ion as a function of velocity v (in atomic units, $e^2/\hbar = 2.183 \times 10^8$ cm/sec) (Neufeld and Snyder, 1961).

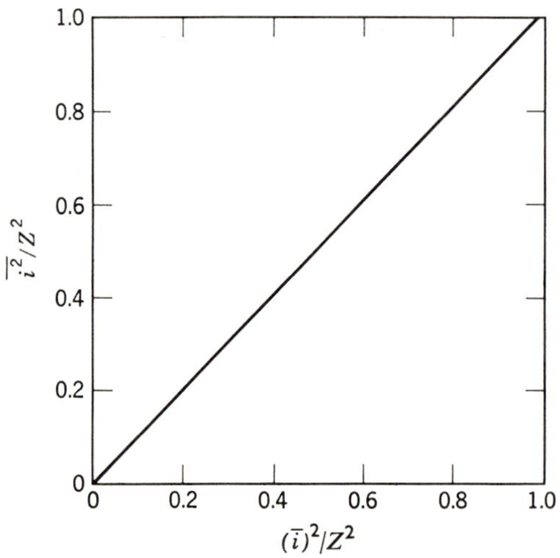

Fig. 2-3. Average square charge $\overline{i^2}$ as a function of the square of the average charge $\bar{i}$ of nitrogen and oxygen ions (Neufeld and Snyder, 1961).

$\overline{i^2}$ and an average charge of a nitrogen or hydrogen ion based on experimental data is shown in Fig. 2-3. Note that $\overline{i^2}/i^2$ is essentially equal to unity.

The rate of energy loss of any charged particle passing through a material is proportional to the square of its electric charge. Therefore, when the charge fluctuates, the average rate of energy loss is proportional to the average square of the particle charge. The "effective charge" of a moving ion, designated as Z_1e, is defined as

$$Z_1 e = (\overline{i^2})^{1/2} \tag{2-4}$$

Since $\overline{i^2} \cong (\bar{i})^2$, the effective charge of an ion is for all practical purposes equal to its average charge, that is, $Z_1e \approx \bar{i}$. At speeds $v \geqslant v_0$, an ion is completely stripped of its orbital electrons and the effective charge is equal to its nuclear charge.

2-4 CLASSICAL CALCULATION OF THE ENERGY LOSS

General

The charge fluctuations described contribute in part to the slowing down process of the moving ion. This contribution becomes larger when the velocity of the ion decreases, and for sufficiently low velocities ($v \ll v_0$), the energy loss due to charge fluctuations may represent a substantial fraction of the total energy loss of the ion.

In the following calculations the effect of charge fluctuations on the energy loss is not taken into account. To simplify the process it is assumed that the ion has constant charge Z_1e and the energy loss is attributed mainly to an idealized mechanism in which the ion does not undergo any charge exchange during the collisions with the surrounding atoms and molecules.

Formulation of the Problem

There are essentially two methods of calculating the linear rate of energy loss, or stopping power, $-dE/dx$, of a charged particle passing through a medium. One of these, based on classical considerations and due to Bohr (1913, 1915), is presented in this section. The quantum mechanical method was developed mainly by Bethe (1930, 1933) and is discussed in Section 2-5. According to Bohr (1948), the range of validity of the classical picture requires the validity of the inequality

$$\kappa = \frac{2Z_1e^2}{\hbar v} \gg 1 \tag{2-5}$$

where v is the velocity of the ion. The quantum mechanical treatment requires that

$$\kappa = \frac{2Z_1e^2}{\hbar v} \ll 1 \tag{2-6}$$

The calculation of stopping power is based on simplified assumptions concerning the structure of the material in which the ion moves. The medium is represented as an assembly of free electrons at rest and distributed uniformly in space; the charged particle is moving swiftly ($v \gg v_0$), so that the electrons do not move appreciably during a collision. Under these conditions only

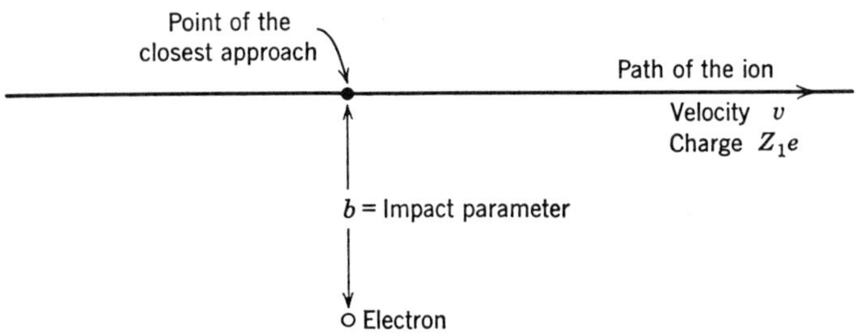

Fig. 2-4. Schematic track of an ion used in calculations.

small momentum transfers from the ion to the electrons occur, and since the ion has a relatively large mass, its trajectory is substantially unaffected by these momentum transfers.

The collision of the moving ion with an electron is represented schematically in Fig. 2-4. The ion has a velocity v and passes the electron at an "impact parameter" b, which is the distance of closest approach of the ion to the electron. We regard the time $t = 0$ as the instant of the closest approach.

The electric field of the ion at the position of the electron has a component perpendicular to the particle track, designated as $E_\perp$, and a component parallel to the particle track, designated as $E_\parallel$. The magnitudes of the components are given as functions of the time by

$$E_\perp = \frac{Z_1eb}{(v^2t^2 + b^2)^{3/2}} \tag{2-7}$$

$$E_\parallel = \frac{Z_1evt}{(v^2t^2 + b^2)^{3/2}} \tag{2-8}$$

Since the electron remains stationary during the collision and since the parallel force at time t is cancelled by the force at the time $-t$, that is, $E_\parallel(t) = -E_\parallel(-t)$, the electron receives no net momentum in the direction of motion of the ion. The total momentum change of the ion from the collision with the electron is due only to the $E_\perp$ component of the field. This momentum change, designated as $p_\perp$, is given by the time integral of the force $eE_\perp$:

$$p_\perp = \int_{-\infty}^{\infty} eE_\perp \, dt = 2\,\frac{Z_1e^2b}{v^3}\int_0^{\infty}\frac{dt}{(t^2 + b^2/v^2)^{3/2}} = \frac{2Z_1e^2}{bv} \tag{2-9}$$

The amount of energy lost by the ion in the collision is equal to the amount of energy gained by the electron from the passage of the ion. Therefore the energy lost by the ion is given by

$$\Delta E_b = \frac{p_\perp{}^2}{2m} = \frac{2Z_1{}^2e^4}{mv^2b^2} \tag{2-10}$$

where ΔE_b is the change in energy of the ion and m is the mass of the electron.

If there are n electrons per cubic centimeter, then in a cylindrical section lying between impact parameters b and $(b + db)$ and having length dx, there are $2\pi nb\, db\, dx$ electrons. Hence the total energy change of the ion in moving a distance dx is given by

$$dE = \int_{b_{\min}}^{b_{\max}} 2\pi bn(-\Delta E_b)\, db\, dx \tag{2-11}$$

where $b_{\min}$ and $b_{\max}$ represent the "minimum" and "maximum" impact parameters, which are discussed further below. Substituting Eq. (2-10) in Eq. (2-11), and performing the integration over db, we obtain for the stopping power, or linear rate of energy loss of the ion,

$$-\frac{dE}{dx} = \frac{4\pi Z_1{}^2e^4}{mv^2}\, n \ln \frac{b_{\max}}{b_{\min}} \tag{2-12}$$

Stopping power can be expressed in ergs per centimeter or in other units having the dimensions of energy/length.

Minimum Impact Parameter

Equation (2-10) demonstrates that the energy transfer $-\Delta E_b$ is inversely proportional to the square of the impact parameter so that close collisions involve very large energy transfers. In order to apply our approximate calculations to determine $b_{\min}$, we equate the maximum possible energy transfer to the expression (2-10) in which we set $b = b_{\min}$.

Since the velocity of the ion is considerably higher than that of the electron, it was assumed that the electron remained stationary during the collision. Following the collision, however, the electron acquires a velocity v_2 and the velocity of the ion decreases from v to v_1. The conservation of energy for the collision can be expressed by writing

$$\tfrac{1}{2}Mv^2 = \tfrac{1}{2}Mv_1{}^2 + \tfrac{1}{2}mv_2{}^2 \tag{2-13}$$

where M is the mass of the ion. For the conservation of momentum we have

$$M\mathbf{v} = M\mathbf{v}_1 + m\mathbf{v}_2 \tag{2-14}$$

The maximum possible momentum transfer corresponds to a "head-on" collision in which the velocity vectors in Eq. (2-14) lie in the same direction. Replacing the vectors by their magnitudes in Eq. (2-14) and eliminating v_1 from Eqs. (2-13) and (2-14) we obtain for the maximum momentum transfer

$$(mv_2)_{\max} = \frac{2mM}{m + M}\, v \tag{2-15}$$

Since we have assumed $m/M \ll 1$, we obtain, approximately,

$$(mv_2)_{\max} = 2mv \tag{2-16}$$

The maximum energy an electron can acquire as a result of a collision with an ion is

$$(\Delta E)_{\max} = \frac{mv_2^{\,2}}{2} = 2mv^2 \tag{2-17}$$

which is the maximum value of the energy lost $-\Delta E_b$ by the ion. Using this expression in Eq. (2-10), we find that the impact parameter corresponding to the maximum energy transfer is

$$b_{\min} = \frac{Z_1 e^2}{mv^2} \tag{2-18}$$

Maximum Impact Parameter

For large impact parameters the duration of the collision becomes comparable with the orbital periods of atomic electrons, and the electrons can no longer be treated as if they were free. The effect of the passage of the ion on a bound electron depends on the relative values of the collision time τ defined as

$$\tau = \frac{b}{v} \tag{2-19}$$

and the period of oscillation $T = 2\pi/\omega$ (ω is the orbital frequency of the electron). The net transfer of momentum to an electron is most effective when $\tau \ll T$. For interaction times which are comparable to or larger than this period, the probability of a quantum transition in the atom, with an accompanying energy loss by the ion, is negligible. Thus, in order for energy to be exchanged, we have

$$\frac{b}{v} < \frac{1}{\omega} \tag{2-20}$$

Therefore, the maximum value of the impact parameter is given by

$$b_{\max} = \frac{v}{\omega} \tag{2-21}$$

Stopping Power

Substituting Eqs. (2-18) and (2-21) into (2-12), we obtain for the stopping power

$$-\frac{dE}{dx} = \frac{4\pi Z_1{}^2 e^4 n}{mv^2} \ln \frac{mv^3}{\omega Z_1 e^2} \tag{2-22}$$

We should now take into account that the medium is not comprised of oscillating electrons. It contains atoms and molecules which are highly complicated dynamic systems. Let Z be the number of electrons in each atom and N be the number of atoms per cubic centimeter in the substance traversed by the ion. Then $n = NZ$. If we assume that an atom is sufficiently distant from the moving ion, then the field exerted by the ion does not vary greatly within the space occupied by the atom. In such cases the interaction is of "dipole type" and the atom responds to the electric field of the ion as an assembly of independent harmonic oscillators of frequencies ω_n and oscillator strengths f_n where $\sum f_n = Z$. We shall introduce the concept of "mean frequency" ω_m which is related to the oscillator frequencies and oscillator strengths as follows:

$$Z \ln \omega_m = \sum f_n \ln \omega_n \tag{2-23}$$

Consequently, the energy loss of a moving ion can be expressed as

$$-\frac{dE}{dx} = \frac{4\pi Z_1{}^2 e^4}{mv^2} NZ \ln \frac{1.123 mv^3}{\omega_m Z_1 e^2} \tag{2-24}$$

The coefficient 1.123 in the logarithmic expression given above was obtained by Bohr on the basis of calculations which are more detailed than those given here.

Bohr's formula can be applicable when the inequality (2-5) is fulfilled. This is equivalent to a relationship

$$v \ll 2Z_1 v_0 \tag{2-25}$$

which gives the upper bound for the quantity v. It is also necessary that the velocity of the ion be large when compared to the orbital velocity μ_s of the K electrons in the atoms of the surrounding media, that is,

$$\eta_s = \frac{v}{\mu_s} > 1 \tag{2-26}$$

An additional requirement, which was formulated by Bohr for the validity of the expression (2-24), is that

$$\frac{\kappa}{\eta_s} < 1 \tag{2-27}$$

Alternative Expression for Stopping Power

There is an alternative expression (Bohr, 1948) for the energy loss of a moving ion when Eq. (2-27) is replaced by the inequality

$$\frac{\kappa}{\eta_s} > 1 \tag{2-28}$$

This alternative expression is as follows:

$$\begin{aligned} -\frac{dE}{dx} &= \frac{2\pi Z_1{}^2 e^4 NZ}{mv^2} \ln \frac{\eta_s^3}{\kappa} \\ &= \frac{2\pi Z_1{}^2 e^4}{mv^2} NZ \ln \frac{2^{1/2} m^{3/2} v^4}{\hbar^{1/2} \omega_m^{3/2} Z_1 e^2} \end{aligned} \tag{2-29}$$

A different expression for the energy loss as described by Eq. (2-29) was obtained by Neufeld (1954).

The classical formulations of Bohr are particularly applicable in determining energy loss of heavy ions having a large effective charge. As an example, consider a fission fragment moving in hydrogen. In the beginning of the range of the fission fragment, $Z_{\text{eff}} \sim 20$ and $v \sim 1.4 \times 10^9$ cm/sec. Consequently, $\kappa \sim 6$ and $\eta_s \sim 6$ and the stopping power should be determined by means of the Bohr formula (2-24). However, close to the end of the range $Z_{\text{eff}} \sim 2$ and v is somewhat above 2.18×10^8 cm/sec. This gives $\kappa \sim 4$ and $\eta_s \sim 2$, and consequently, the Bohr formula (2-29) is applicable.

2-5 QUANTUM MECHANICAL CALCULATIONS

Stopping Power

When the incident particle has an effective charge Z_1e and velocity v such that the inequality (2-6) holds, the orbital (classical) consideration on which Bohr's formulation is based breaks down. Classically there is no difficulty in defining the position and momentum of an electron at the same time. However, quantum mechanically this is a violation of the uncertainty principle. Taking Δx as the uncertainty in position of the electron in the x direction and Δp_x as the uncertainty in the x component of momentum of the electron, the Heisenberg uncertainty relation states that $\Delta x \, \Delta p_x \geqslant \hbar$. If we try to construct wave packets to give meaning to a classical trajectory, we can see that the path can be defined only within the uncertainty principle. Because of this uncertainty the concept of an impact parameter is no longer applicable. Other quantum effects are based on discreteness of possible energy transfers resulting from the interaction of an incident particle with the surrounding atoms.

It is convenient in the quantum mechanical treatment of the problem to differentiate between two types of collisions between the swiftly moving heavy particle and the atomic electrons: hard and soft collisions. In hard collisions the energy Q imparted by the moving particle is relatively large and satisfies the inequality $Q > H$. The quantity H should be large when compared to the binding energy of the electron and sufficiently small so that the effective collision parameter is large when compared to the atomic dimensions. In practical cases H is of the order of 10^4–10^5 eV. In soft collisions the energy imparted by the moving particle extends from $Q = H$ down to the minimum possible energy transfer, which is of the order of the excitation energy of an orbital electron in an atom in the surrounding medium.

According to Bethe, the energy losses at a velocity $\beta = v/c$ (c = velocity of light) due to soft collisions can be expressed as

$$-\left(\frac{dE}{dx}\right)_{Q<H} = \frac{2\pi Z_1{}^2 e^4}{mv^2} n\left[\ln \frac{2mv^2 H}{I^2(1-\beta^2)} - \beta^2\right] \tag{2-30}$$

where I is mean excitation energy of atoms in the stopping medium and is defined as

$$Z \ln I = \sum f_{nk} \ln I_{nk} \tag{2-31}$$

where f_{nk} are the oscillator strengths corresponding to the transitions between energy levels n and k, and I_{nk} are the corresponding energies of transition.

The mean excitation energy I can be calculated accurately for light elements. For heavy elements, I can be approximated by the relationship

$$I = I_0 Z \tag{2-32}$$

which was derived by Bloch (1933a,b) based on the Thomas-Fermi statistical model of the atom; I_0 is a constant determined by experiment and approximately equal to Rydberg's constant.

In the framework of the preceding theory, the quantity I is dependent on the stopping substance and substantially independent of the velocity of the ionizing particle. In practice, I is determined empirically from the measurements of energy losses and ranges of charged particles in various substances. A summary of values of I is given in Fig. 2-5 (Turner, 1964).

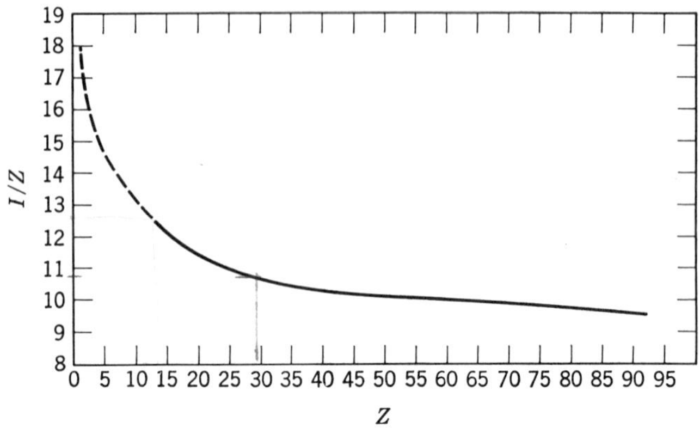

Fig. 2-5. Values of I/Z in electron volts for the chemical elements. The dashed line for $Z < 13$ (aluminum) represents an average of values of I/Z that depend on different chemical states of the element (based on Turner, 1964).

The energy losses due to hard collisions can be expressed in the form

$$-\left(\frac{dE}{dx}\right)_{Q>H} = \frac{2\pi Z_1{}^2 e^4}{mv^2}\, n\left(\ln \frac{Q_{\max}}{H} - \beta^2\right) \tag{2-33}$$

where $Q_{\max}$ represents the largest possible energy transfer from the incident particle to an initially stationary electron. Using relativistic considerations and applying a simplifying assumption that the kinetic energy of the heavy particle is small when compared to its rest mass, we obtain

$$Q_{\max} = \frac{2mv^2}{1 - \beta^2} \tag{2-34}$$

Combining Eqs. (2-30), (2-33), and (2-34), we obtain for the total stopping power of a heavy particle

$$-\frac{dE}{dx} = \frac{4\pi Z_1{}^2 e^4}{mv^2}\, n\left[\ln \frac{2mv^2}{I} - \ln(1 - \beta^2) - \beta^2\right] \tag{2-35}$$

For a medium composed of a number of elements i, the quantity $n \ln I$ can be replaced by the weighted average $\Sigma_i n_i \ln I_i$ (Bragg rule).*

Shell Corrections

The formula (2-35) is based upon the assumption that the speed of the ion is much greater than the orbital speeds of the electrons in the atoms of the

* Compare, for example, the numerical results for soft tissue (Tables 2-1 and 2-2).

stopping medium. Since the maximum possible speed of the ion is c, the velocity of light, this criterion cannot always be satisfied. Inner-shell electrons in heavy elements such as lead, for example, move with speeds comparable with c. The stopping power of K- and L-shell electrons has been calculated separately and correction terms to be added to Eq. (2-35) have been estimated (Walske, 1952 and 1956). Such corrections for all atomic shells are needed for the heavy elements, and shell correction curves for a number of elements throughout the periodic system have been developed for a wide range of ion energies (Bichsel, 1963; Fano, 1963; Fano and Turner, 1964).

Density Effect

In calculating the energy losses of moving charged particles, the effects due to the polarization of the medium were not considered. These effects are negligible when the energy of the particle is in the nonrelativistic range. The energy loss is given accurately by the formulas derived above. For relativistic particles the screening due to the polarization of the medium becomes important, especially for dense media. Consequently, the interaction between the particle and the surrounding atoms is weakened, and this decreases the energy losses. This phenomenon was noted by Swann and calculated by Fermi (1940) on the basis of an idealized atomic model represented by a single oscillator. More detailed models were considered by Wick (1942), Halpern and Hall (1948), and, particularly, by Sternheimer (1956, 1959), who tabulated the appropriate correction terms for a number of stopping substances.

Ranges

The average track length of an ion is the integral of the stopping power, $-dE/dx$. The analytic expression (2-35) is not valid at low velocities. We can, however, express the range $R(E)$ of an ion of energy E as an integral of Eq. (2-35) down to some cutoff energy E_0, at which the theory is still valid, and an experimentally measured residual range $R_0(E_0)$:

$$R(E) = R_0(E_0) + \int_{E_0}^{E} \left(-\frac{dE}{dx} \right)^{-1} dE \qquad \text{(2-36)}$$

This expression describes the range on the assumption that the ion slows down in a continuous manner, that is, that the quantity $(-dE/dx)^{-1}$ is a continuous function of energy. Since charged particles are slowed down by means of a discrete number of energy losses, the range (2-36) represents a statistical average. The phenomenon of pathlength fluctuation is called straggling.

Table 2-1 Composition and Mean Excitation Energies Used for Soft Tissue

Element	Z_i	N_i (atoms/cm^3)	I_i (eV)
Hydrogen	1	5.98×10^{22}	18
Carbon	6	9.03×10^{21}	78
Nitrogen	7	1.29×10^{21}	85
Oxygen	8	2.45×10^{22}	98

Table 2-2 Stopping Power and Ranges of Protons in Soft Tissue

Energy (MeV)	Range (cm)	$-dE/dx$ (MeV/cm)	Energy (MeV)	Range (cm)	$-dE/dx$ (MeV/cm)	Energy (MeV)	Range (cm)	$-dE/dx$ (MeV/cm)
400	84.060	2.965	240	36.165	3.921	80	5.278	8.451
395	82.379	2.983	235	34.897	3.972	75	4.701	8.881
390	80.707	3.001	230	33.647	4.024	70	4.153	9.369
385	79.047	3.020	225	32.413	4.079	65	3.634	9.926
380	77.397	3.040	220	31.196	4.137	60	3.145	10.569
375	75.757	3.060	215	29.996	4.197	55	2.688	11.320
370	74.129	3.081	210	28.813	4.260	50	2.262	12.211
365	72.511	3.102	205	27.648	4.326	45	1.869	13.283
360	70.905	3.124	200	26.502	4.395	40	1.510	14.602
355	69.310	3.146	195	25.373	4.468	35	1.185	16.265
350	67.726	3.169	190	24.264	4.544	30	0.895	18.433
345	66.155	3.193	185	23.173	4.624	25	0.643	21.385
340	64.595	3.217	180	22.101	4.709	20	0.428	25.662
335	63.047	3.243	175	21.049	4.798	19	0.390	26.761
330	61.511	3.269	170	20.017	4.892	18	0.353	27.971
325	59.987	3.295	165	19.005	4.992	17	0.318	29.309
320	58.476	3.323	160	18.014	5.098	16	0.285	30.799
315	56.978	3.352	155	17.044	5.210	15	0.254	32.468
310	55.493	3.381	150	16.095	5.329	14	0.224	34.351
305	54.021	3.411	145	15.167	5.457	13	0.195	36.494
300	52.562	3.443	140	14.262	5.593	12	0.169	38.956
295	51.116	3.475	135	13.380	5.738	11	0.144	41.819
290	49.684	3.509	130	12.520	5.894	10	0.121	45.190
285	48.266	3.544	125	11.683	6.062	9	0.100	49.226
280	46.862	3.580	120	10.870	6.243	8	0.080	54.153
275	45.473	3.617	115	10.082	6.440	7	0.063	60.315
270	44.098	3.656	110	9.318	6.653	6	0.047	68.268
265	42.737	3.696	105	8.579	6.885	5	0.034	78.971
260	41.392	3.737	100	7.865	7.139	4	0.022	94.241
255	40.062	3.781	95	7.178	7.418	3	0.012	118.037
250	38.747	3.825	90	6.517	7.727	2	0.005	161.090
245	37.448	3.872	85	5.884	8.069	1	0.000	268.045

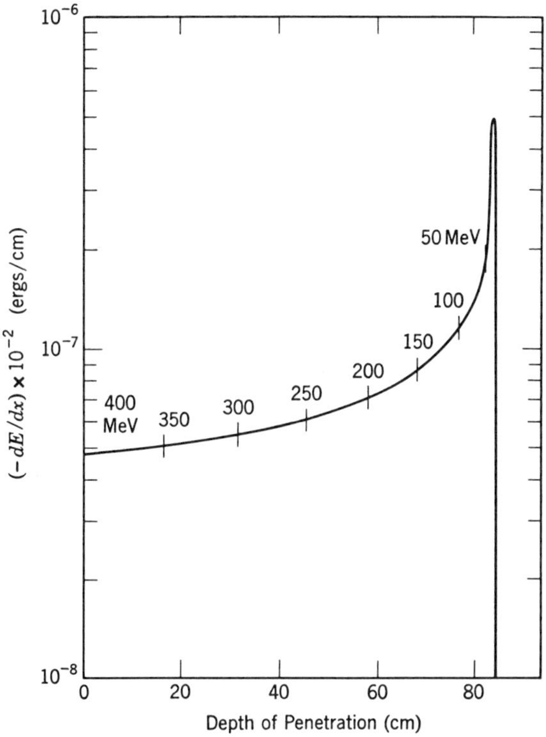

Fig. 2-6. Stopping power of 400 MeV protons as a function of depth of penetration in soft tissue. The depths at which the proton energy is reduced in steps of 50 MeV are shown. The value at the peak is an average at that depth due to straggling and other factors.

Numerical Results for Soft Tissue

As an example of practical importance in health physics, we give numerical values for protons having energies as high as 400 MeV in soft tissue. For this purpose we write Eq. (2-35) in the form ($Z_1 = 1$)

$$-\frac{dE}{dx} = \frac{4\pi e^4}{mv^2}\Sigma_i N_i Z_i\left[\ln\frac{2mv^2}{I_i} - \ln(1-\beta^2) - \beta^2\right] \tag{2-37}$$

Here $n \ln I$ has been replaced, in accordance with the Bragg rule, by

$$n \ln I \rightarrow \Sigma_i N_i Z_i \ln I_i \tag{2-38}$$

where N_i is the number of atoms per unit volume with atomic number Z_i.

To a good approximation, soft tissue may be represented by the composition and mean excitation energies shown in Table 2-1. The resulting values of the stopping power for protons are given in Table 2-2. Equation (2-36) was used to calculate ranges. Formula (2-37) is adequate for protons in tissue in this energy range without significant shell or density-effect corrections. The ranges were obtained by choosing $E_0 = 10$ MeV and using for $R_0(E_0)$ the measured proton range in water (0.121 cm) at this energy. Below 10 MeV the theoretical values in Table 2-2 thus fit smoothly onto measured values for protons in water, which approximates soft tissue in its electronic composition. A plot of the stopping power, expressed in other units, is shown in Fig. 2-6.

A complete theory of radiation dose from proton penetration in soft tissue must include, in addition to the energy losses by electronic interaction treated here, the interactions of protons with atomic nuclei (Turner et al., 1964). At 400 MeV, 25 to 30% of all protons incident on soft tissue will generate secondary nucleons in a nuclear collision before they penetrate to a depth of 30 cm.

B. GAMMA AND X-RAYS

2-6 FUNDAMENTAL PROCESSES

Electromagnetic radiation interacts with matter through processes which are quite different from those encountered by charged particles. In general, heavy charged particles lose energy to individual atomic electrons through a series of electromagnetic interactions, each of which results in only a relatively small energy transfer. Thus, the energy loss occurs in a large number of small steps with the result that the charged particle is gradually slowed down. On the other hand, individual photons interact with matter by processes which can remove the entire photon energy in a single event. This event may be absorption, in which the photon completely disappears, or scattering, in which the direction of the photon, and perhaps its energy, is altered. The continuous slowing down of a charged particle passing through a particular material leads to a specific range in that material. In contrast with this, the single-step absorption and scattering of a beam of electromagnetic radiation lead to an exponential decrease in intensity with distance. Thus, there is no specific value for the range of electromagnetic radiation.

Four distinct types of interactions are possible (Fano, 1953); these are with: (1) atomic electrons, (2) nucleons, (3) the electric field surrounding charged particles, and (4) the meson field surrounding nucleons. Two types of scattering processes occur: elastic or coherent scattering, in which the internal energy of the interacting system remains unchanged, and inelastic or incoherent, in which there is an internal energy change. The probabilities for various events depend only on the energy of the photon regardless of its source. Physically, γ- and X-ray photons are identical, the term γ-ray referring to a photon of nuclear origin.

2-7 SCATTERING OF ELECTROMAGNETIC RADIATION

Were it not for scattering processes, calculations of the penetration of electromagnetic radiation through matter would be simple. Scattering produces secondary photons having a continuous distribution of energy and direction. A complete description of the penetration process must take into account this secondary source of photons distributed along the tracks of the primary photons.

Classical Description of Scattering

In this description of the scattering process, the interaction considered is that between the electric field of the incident photon and a single unbound electron (Compton and Allison, 1935). The free electron vibrates under the influence of the electric field with the frequency of the incident wave. The secondary radiation from the vibrating electron, which behaves as a Hertzian oscillator, has the same frequency as the incident radiation. In effect, the incident photon is elastically scattered in this process.

According to classical electromagnetic theory, the magnitude of the electric field at a position **r** due to an electron of charge e and mass m_0 having an acceleration **a** is given by

$$E_s = \frac{ae}{rc^2} \sin \theta \tag{2-39}$$

where θ is the angle between **a** and **r** and c is the speed of light. The magnitude of the acceleration may be expressed in terms of the force on the electron eE_0 by

$$a = \frac{eE_0}{m_0} \tag{2-40}$$

where E_0 is the magnitude of the electric field of the incident photon. Equation (2-39) can now be written

$$E_s = \frac{e^2E_0}{rm_0c^2} \sin \theta \tag{2-41}$$

Since the ratio of the scattered and incident intensity, I_s/I_0, is equal to $(E_s/E_0)^2$, we may write

$$I_s = I_0 \frac{E_s^{\,2}}{E_0^{\,2}} = I_0 \frac{e^4}{r^2m_0^{\,2}c^4} \sin^2 \theta \tag{2-42}$$

The angular distribution of the scattered intensity can be obtained by calculating the two components of the incident intensity I.

One component $I_{\parallel}$ is associated with the amplitude lying in the plane formed by the incident beam and the scattered beam and the other $I_{\perp}$ perpendicular to this. Since for unpolarized incident radiation the electric vector is randomly oriented in the plane normal to the direction of the incident photons, it follows that

$$I_{\parallel} = I_{\perp} = \tfrac{1}{2}I_0 \tag{2-43}$$

The scattered intensity may be written in terms of these components with the help of Eq. (2-42); that is,

$$I_{s\parallel} = I_{\parallel} \frac{e^4}{r^2 {m_0}^2 c^4} \sin^2 \theta_{\parallel}$$

$$I_{s\perp} = I_{\perp} \frac{e^4}{r^2 {m_0}^2 c^4} \sin^2 \theta_{\perp} \tag{2-44}$$

where $\theta_{\parallel}$ and $\theta_{\perp}$ are the angles between the direction of $\mathbf{r}$ and $\mathbf{E}_{\parallel}$ and $\mathbf{E}_{\perp}$. Angle $\theta_{\perp} = \pi/2$, by definition and $\theta_{\parallel}$ is the complement of the angle between the incident and scattered photon beams.

Finally, from the relation $I_s = I_{s\parallel} + I_{s\perp}$ we have

$$I_s = \tfrac{1}{2}I_0 \frac{e^4}{r^2 {m_0}^2 c^4} \cos^2 \phi + \tfrac{1}{2}I_0 \frac{e^4}{r^2 {m_0}^2 c^4}$$

$$= I_0 \frac{e^4}{2r^2 {m_0}^2 c^4} (1 + \cos^2 \phi) \tag{2-45}$$

where ϕ is the scattering angle, that is, the angle between the directions of the incident and scattered photons. In terms of the classical electron radius $r_0 \equiv (e^2/m_0c^2)$ this expression may be written as

$$I_s = I_0 \frac{{r_0}^2}{2r^2} (1 + \cos^2 \phi) \tag{2-46}$$

The quantity $r_0^2(1 + \cos^2 \phi)/2$ is the differential scattering cross section $d\sigma/d\Omega$ for the number of photons scattered into a unit solid angle about ϕ per scattering electron. It is also called the Thomson differential cross section (Thomson, 1906). The variation of the cross section (which is energy-independent) with angle is shown in Fig. 2-7.

This classical result, although not rigorous, nevertheless provides a physical picture of the scattering process. As will be seen below, the Thomson formula agrees with the quantum mechanical result when the photon energy is negligible compared with the rest energy of the electron.

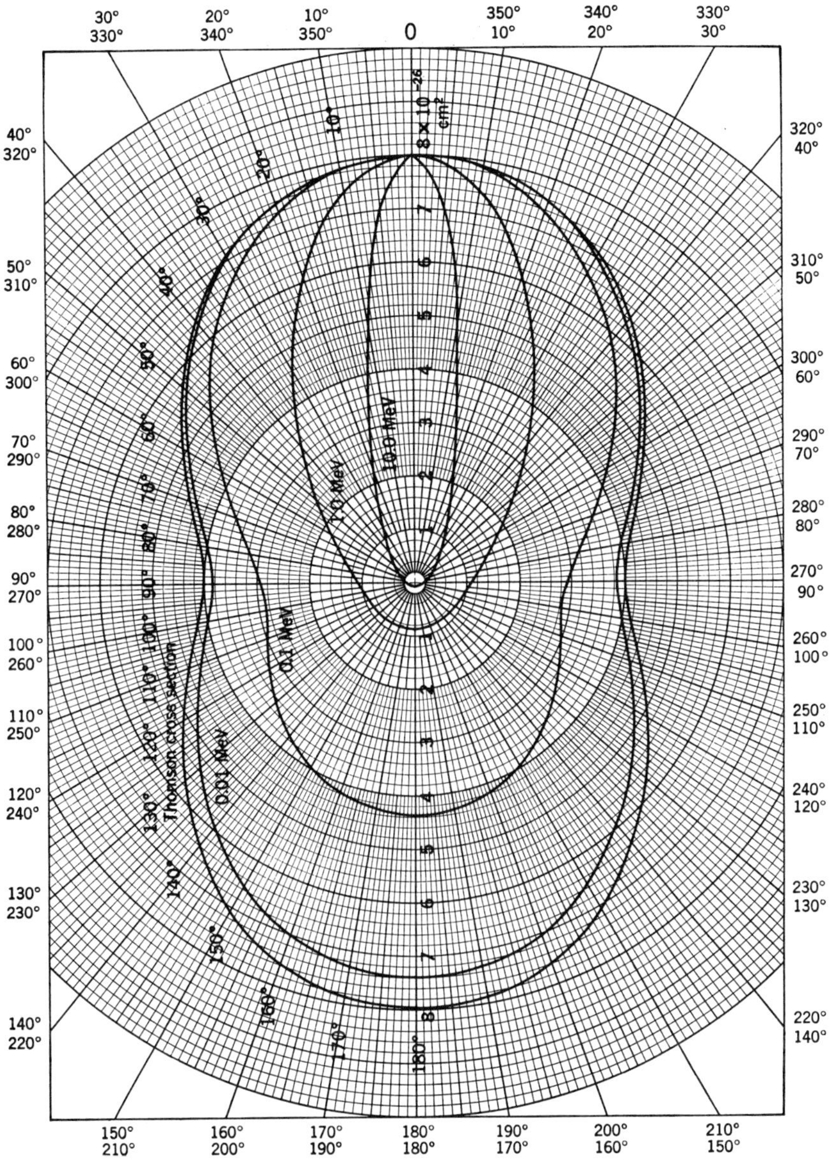

Fig. 2-7. Differential cross section $d\sigma/d\Omega$ for photon scattering.

Quantum Mechanical Description of the Scattering

Following the introduction of quantum mechanics in the mid-1920's many investigators again looked at the general problem of scattering. Their results, although not in complete agreement with the then available experimental data, offered a great improvement over the classical result. The problem was not completely resolved until 1929 when Klein and Nishina (1929) treated the problem by using the newly developed relativistic electron theory of Dirac. Their result, the so-called Klein-Nishina formula, to a very good approximation describes scattering by loosely bound electrons. Corrections to the formula seem to amount to only about 1%.

The differential cross section is not strongly dependent on the state of polarization of the incident photons. However, except for scattered photons, which are partially polarized, we generally deal with unpolarized radiation and no serious errors result from neglecting this effect.

The Klein-Nishina differential scattering cross section for unpolarized photons is a function of both the energy of the incident photon and the scattering angle and is given by

$$\frac{d\sigma}{d\Omega} = f(E, \phi)$$

$$= \frac{r_0^2}{2}\left[\frac{1}{1+\alpha(1-\cos\phi)}\right]^2\left[1+\cos^2\phi+\frac{\alpha^2(1-\cos\phi)^2}{1+\alpha(1-\cos\phi)}\right] \tag{2-47}$$

where $\alpha = h\nu/m_0c^2$ is the ratio of the photon energy to the rest energy of the electron. This represents the cross section per electron for scattering photons of initial energy $h\nu$ into a solid angle $d\Omega$ about the scattering angle ϕ. For low energies $h\nu \ll m_0c^2$, that is, $\alpha \to 0$, and the Klein-Nishina cross section approaches the classical Thomson value.

The differential cross section for the electromagnetic energy scattered, $d\sigma_s/d\Omega$, can be obtained from $d\sigma/d\Omega$ by multiplying by the fraction of the original energy carried off by the scattered photon. As is shown in Section 2-8, this fraction is given by $1/[1+\alpha(1-\cos\phi)]$; therefore

$$\frac{d\sigma_s}{d\Omega} = \frac{1}{1+\alpha(1-\cos\phi)}\cdot\frac{d\sigma}{d\Omega}$$

$$= \frac{r_0^2}{2}\left[\frac{1}{1+\alpha(1-\cos\phi)}\right]^3\left[1+\cos^2\phi+\frac{\alpha^2(1-\cos\phi)^2}{1+\alpha(1-\cos\phi)}\right] \tag{2-48}$$

It is important that the distinction between these two cross sections be understood.

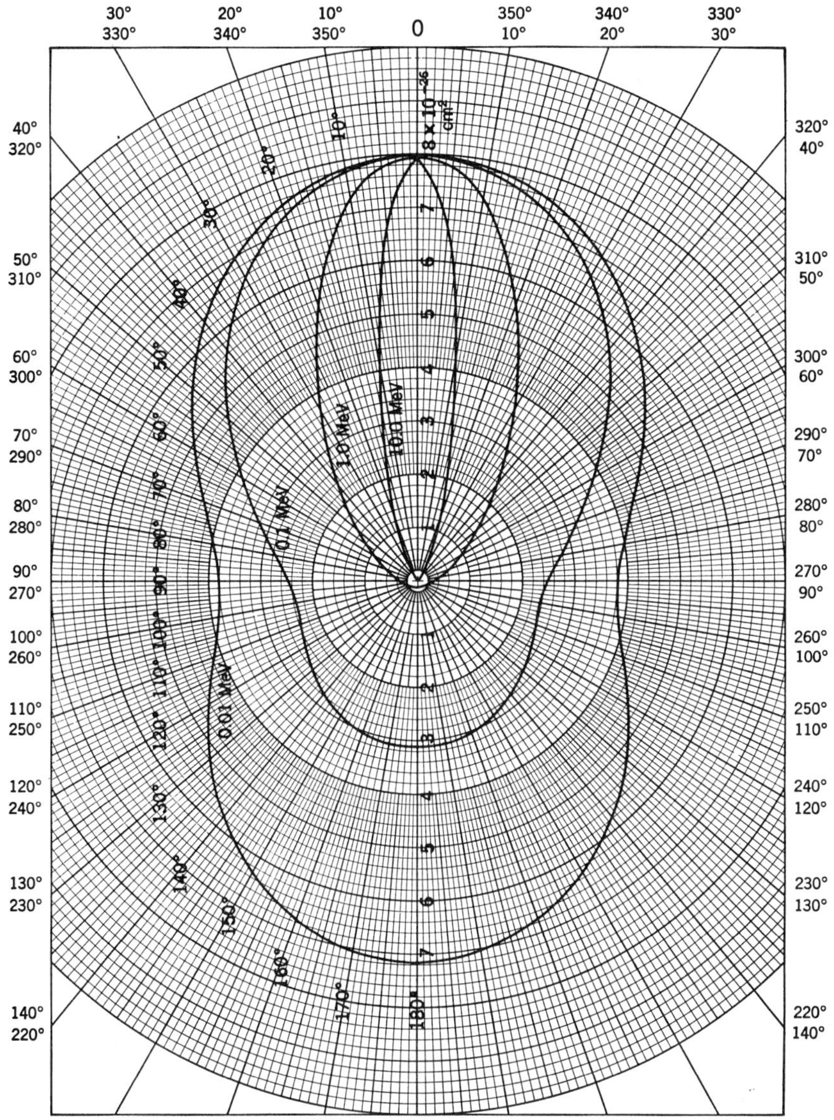

Fig. 2-8. Differential cross section $d\sigma_s/d\Omega$ for photon energy scattering.

The angular dependences of the differential cross sections are shown in Figs. 2-7 and 2-8. The very strong peaking of both scattered photons and scattered energy in the forward direction at high energies is clearly evident. This phenomenon is due to relativistic effects and is characteristic of many physical processes involving electromagnetic radiation.

In practice, it is necessary to correct the theoretical cross sections for binding effects in the case of small-angle scattering and low initial energies (Grodstein, 1957).

Other Scattering Processes

There are numerous other processes by which electromagnetic radiation can be scattered; these account, however, for a very small proportion of the total scattering. The most predominant is Rayleigh scattering, which is a process involving the bound lattice electrons. This type of scattering is coherent and, because the recoil involves the entire atom, there is essentially no energy lost by the photon; hence the process is elastic. Except for high-energy photons or high-Z materials the scattering angles are small. Thomson scattering by the nucleus is also a possible process, but its probability is small. Nuclear resonance scattering and Delbrück scattering, involving the nuclear potential, are at just the limit of detectability.

2-8 ABSORPTION OF ELECTROMAGNETIC RADIATION

Unlike scattering, in which one process predominates, three processes contribute significantly to the absorption of energy. These are the photoelectric effect, the Compton effect, and pair production.

The Photoelectric Effect

An incident photon can interact with the atomic system and lose its entire energy to one of the more tightly bound electrons. The electron is ejected from the atom with a kinetic energy given by the Einstein relation

$$T = h\nu - \phi_B \tag{2-49}$$

where ϕ_B is the binding energy of the electron. The presence of the nucleus is required to conserve momentum, although the energy transfer to the nucleus is negligible. The probability of this interaction is highest for those electrons most tightly bound. About 80% of the interactions involve the K-shell electrons. The process is of the threshold type in that the photon energy must exceed the shell binding energy for the interaction to take place.

The methods of calculating the cross section for the photoelectric effect are well established, but because of the complexity of the calculations, it has been necessary to resort to various approximations. Unfortunately, these approximations hold only for particular energy regions, and so no single equation for the cross section is valid for all energies.

A nonrelativistic quantum mechanical treatment utilizing the Born approximation yields a result which, although not quantitatively correct, does exhibit the main characteristics of the cross section. This result, taken from Heitler (1954), gives the cross section for the K-shell interaction

$$\tau_k = \frac{32\sqrt{2}}{3}\pi r_0^2\left(\frac{m_0c^2}{h\nu}\right)^{7/2} Z^5\left(\frac{e^2}{\hbar c}\right)^4 \qquad (2\text{-}50)$$

where Z is the atomic number of the absorbing atom and $e^2/\hbar c = \frac{1}{137}$ is the fine-structure constant. This relation is approximately correct only when $h\nu/m_0c^2 \ll 1$, $Z/137^2 \ll 1$, and in the region far from the K-shell binding energy. However, the strong dependence on Z and $h\nu$, which is the most characteristic feature of the photoelectric cross section, is evident. The dependence on energy is so pronounced below about 10 keV that the photoelectric effect accounts for most of the energy absorption in all materials. For a high-Z material such as lead ($Z = 82$), the photoelectric effect is the dominant absorption process up to energies of about 900 keV.

A discussion of the formulas for the cross section in the various energy regions along with their approximations is given by Grodstein (1957). This reference also contains a useful table of the cross sections for many materials for energies between 10 keV and 100 MeV. The dependence of the cross section on energy and atomic number is seen in Fig. 2-9. These curves, drawn from the data tabulated in the Grodstein reference, show the photoelectric cross section plotted versus energy for materials with atomic numbers between 13 and 82.

Photoelectric absorption, leaving an inner-shell vacancy, is always accompanied by the emission of characteristic X-radiation or Auger electrons. These X-rays have a wide range of energies. In a high-Z material the energies may exceed 100 keV, whereas in a low-Z material they are of the order 10–100 eV.

The angular distribution of the ejected photoelectrons has been discussed by Davisson and Evans (1952), who point out that the distribution is neither unidirectional nor isotropic. Theoretical calculations of the angular distribution have been carried out by Fischer (1931) for the nonrelativistic case and by Sauter (1931) for the relativistic region. These authors show that the probability of emission along the path of the incident photon is zero at all energies and that, in the low-energy limit, the distribution follows a

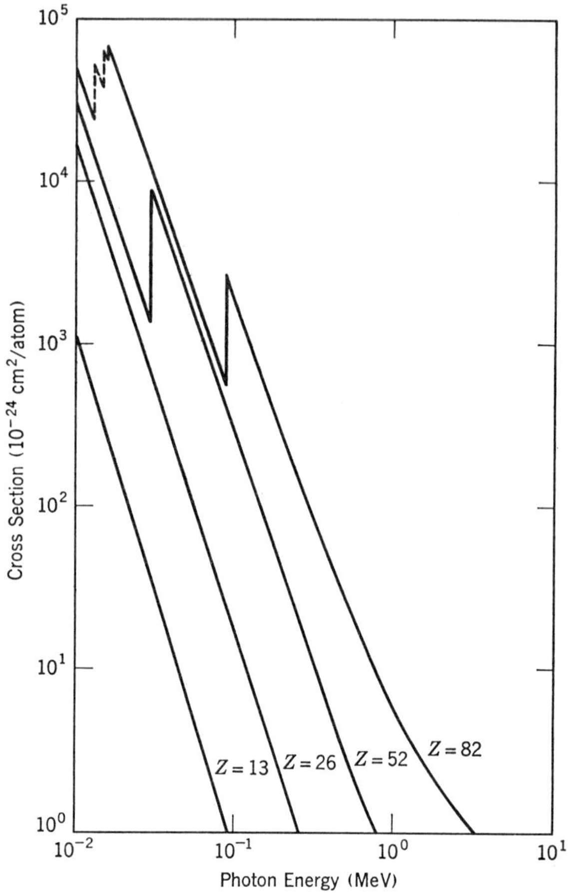

Fig. 2-9. Cross sections for the photoelectric effect.

sine-square law. A characteristic peaking of the distribution in the forward direction, similar to that shown in Figs. 2-7 and 2-8, occurs as the photon energy increases.

The Compton Effect

In the intermediate energy region (~0.5–~5 MeV) the dominant photon interaction, known as the Compton effect, is incoherent scattering from electrons. A. H. Compton (1923) showed that the energy absorbed by the electron and the directions taken by the scattered photon and recoil electron were simply related.

The process is illustrated in Fig. 2-10 where the energy and momentum of both photon and electron are shown. It is necessary to use the relativistic forms for the electron mass m and kinetic energy T, that is,

$$m = m_0(1 - \beta^2)^{-1/2} \tag{2-51}$$

and

$$T = (m - m_0)c^2 = m_0c^2[(1 - \beta^2)^{-1/2} - 1] \tag{2-52}$$

where β is the ratio of the speed of the electron to that of light v/c. For incident photons of energy $h\nu$ and scattered photons of energy $h\nu'$ we have

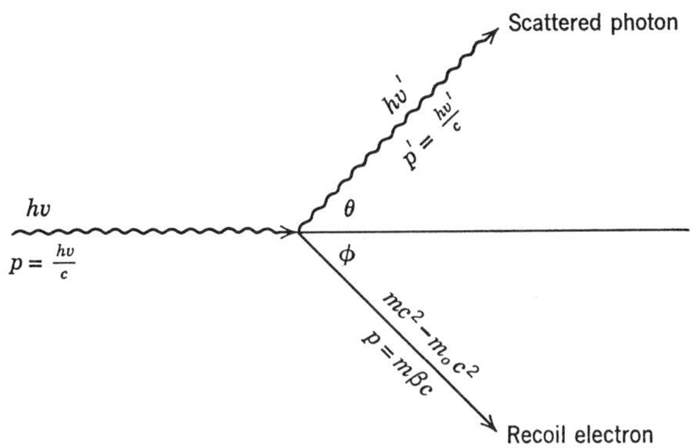

Fig. 2-10. Schematic representation for the Compton process.

from the conservation of energy

$$h\nu = h\nu' + T \tag{2-53}$$

For the conservation of momentum in the directions parallel and perpendicular to the direction of the incident photon we may write

$$\frac{h\nu}{c} = \frac{h\nu'}{c}\cos\theta + mv\cos\phi \tag{2-54}$$

and

$$0 = \frac{h\nu'}{c}\sin\theta - mv\sin\phi \tag{2-55}$$

The solution to Eqs. (2-53)–(2-55) may be expressed in a number of ways which describe the energy and angle of the scattered photon. The energy of the scattered photon is given by

$$h\nu' = \frac{m_0c^2}{\dfrac{m_0c^2}{h\nu} + 1 - \cos\theta} \tag{2-56}$$

The energy of the recoil electron is

$$T = \frac{(h\nu)^2(1 - \cos\theta)}{m_0c^2 + h\nu(1 - \cos\theta)} \tag{2-57}$$

The angles θ and ϕ are related by

$$\cot\phi = -\left(1 + \frac{h\nu}{m_0c^2}\right)\tan\frac{\theta}{2} \tag{2-58}$$

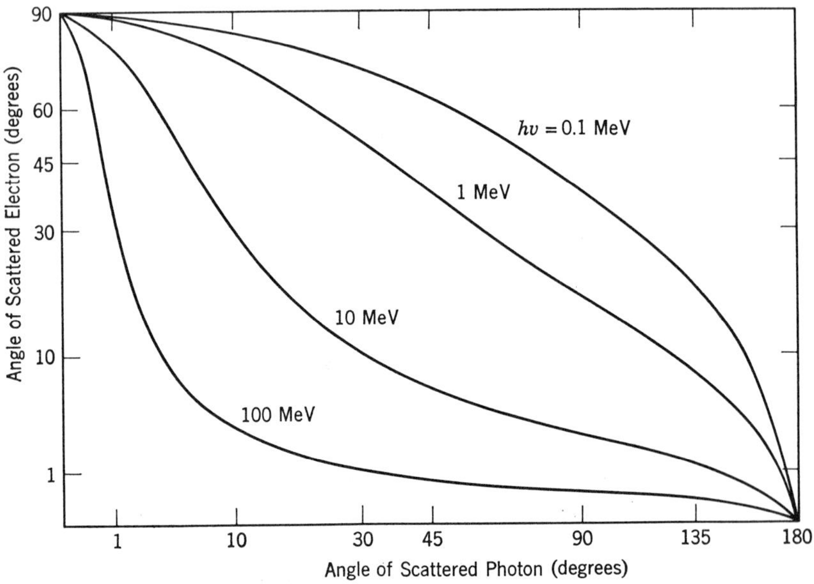

Fig. 2-11. Angular dependence of the Compton scattered electrons.

Several interesting observations can be made from these relations. From Eq. (2-56) it is seen that $h\nu' \sim h\nu$ in the limit of small-angle scatter ($\theta \to 0$) and in the low-energy limit ($h\nu \ll m_0c^2$). Note that, in the low-energy limit, the Compton effect is identical with the Thomson scattering discussed in Section 2-7. In the high-energy limit Eq. (2-56) reduces to

$$h\nu' = \frac{m_0c^2}{1 - \cos\theta} \tag{2-59}$$

showing that the scattered photon energy depends only on the scattering angle and not on the original energy.

The relation between θ and ϕ, shown in Fig. 2-11 for several values of $h\nu$, indicates that the electron distribution becomes peaked in the forward direction at high values of the incident photon energy. The angular dependence of the relative energy of the recoil electron and scattered photon is shown in Fig. 2-12. At small angles most of the energy is retained by the photon whereas at large angles the opposite is true.

The total cross section for Compton scattering is obtained by integrating

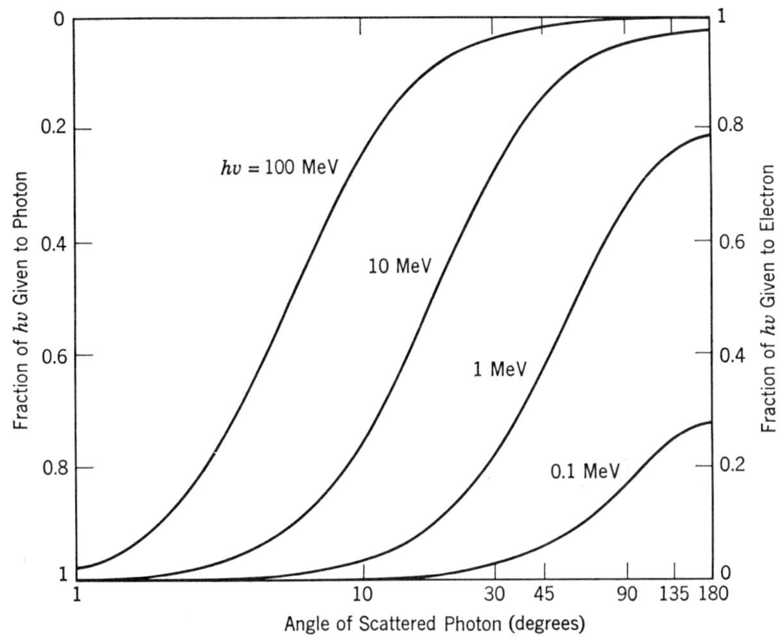

Fig. 2-12. Angular dependence of the fractional energy carried off by the scattered photon and electron.

the Klein-Nishina scattering formula, Eq. (2-47), over all solid angles, that is,

$$\sigma = \int \frac{d\sigma}{d\Omega}\, d\Omega \qquad (2\text{-}60)$$

We find that

$$\sigma = \frac{2\pi e^4}{m_0{}^2c^4}\left\{\frac{1+\alpha}{\alpha^2}\left[\frac{2(1+\alpha)}{1+2\alpha} - \frac{1}{\alpha}\ln(1+2\alpha)\right] + \frac{1}{2\alpha}\ln(1+2\alpha) - \frac{1+3\alpha}{(1+2\alpha)^2}\right\} \qquad (2\text{-}61)$$

where $\alpha = h\nu/m_0c^2$ is the same as it is in Eq. (2-47). Physically, σ represents the probability per electron per square centimeter that a photon entering a surface will undergo a Compton interaction.

The scattering cross section, σ_s, is obtained by integrating Eq. (2-48) over all solid angles, that is,

$$\sigma_s = \int \frac{d\sigma_s}{d\Omega}\, d\Omega$$
$$= \frac{2\pi e^4}{m_0{}^2 c^4}\left[\frac{4\alpha^2}{3(1+2\alpha)^3} - \frac{(1+\alpha)}{\alpha^2(1+2\alpha)^2}(1+2\alpha-2\alpha^2) + \frac{1}{2\alpha^3}\ln(1+2\alpha)\right] \tag{2-62}$$

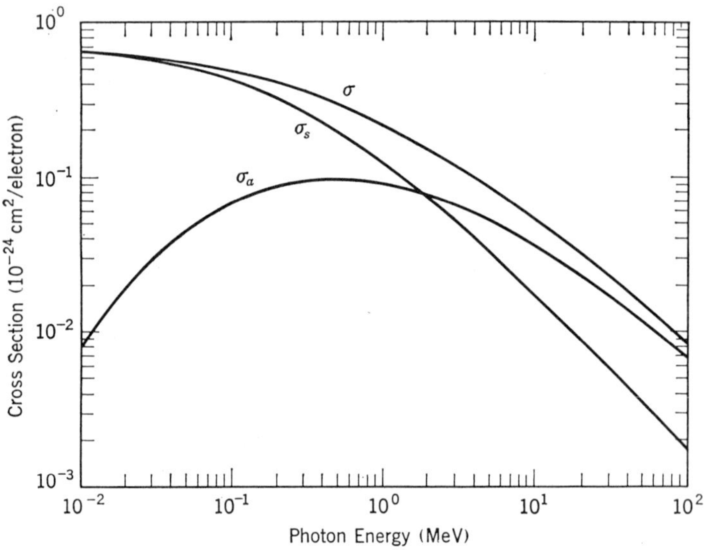

Fig. 2-13. Cross sections for the Compton process.

The energy absorption cross section, σ_a, is defined by

$$\sigma_a = \sigma - \sigma_s \tag{2-63}$$

Because energy is conserved, the recoil energy of the electron is equal to the difference in energies of the incident and scattered photons. Since the electron range is small, all of its energy is dissipated in the vicinity of the interaction. The variation with energy of the cross section in Eq. (2-63) is shown in Fig. 2-13. Expressions for the cross sections can be simplified for very large or very small photon energies. The following expansions are given by Lea (1955) and Bethe and Ashkin (1953). For $\alpha \ll 1$:

$$\begin{aligned}\sigma &\cong \tfrac{8}{3}\pi r_0{}^2(1 - 2\alpha + 5.2\alpha^2 - \cdots)\\ \sigma_s &\cong \tfrac{8}{3}\pi r_0{}^2(1 - 3\alpha + 9.4\alpha^2 - \cdots)\\ \sigma_a &\cong \tfrac{8}{3}\pi r_0{}^2(\alpha - 4.2\alpha^2 + \cdots)\end{aligned} \tag{2-64}$$

For $\alpha \gg 1$

$$\sigma \cong \pi r_0^2 \frac{1}{\alpha}(\ln 2\alpha + \tfrac{1}{2}) \tag{2-65}$$

In the low-energy limit, σ approaches the classical limit $\frac{8}{3}\pi r_0^2$. This is the Thomson cross section which may be obtained directly by integration of Eq. (2-46).

Pair Production

The third important process by which electromagnetic radiation is absorbed by matter is pair production. A photon, interacting with the electric field surrounding a charged particle, can disappear in the creation of an electron-positron pair. The photon energy must be at least twice the electron rest mass energy (or 1.02 MeV) in order for the interaction to be energetically possible. In the field of a nucleus,

$$h\nu = T_+ + T_- + 2m_0c^2 \tag{2-66}$$

where T_+ and T_- are the kinetic energies of the positron and electron. In this case the excess energy above $2m_0c^2$ is distributed between the electron-positron pair, a negligible amount going to the nucleus. When pair production takes place in the field of an electron the threshold is $4m_0c^2$, and the electron may absorb an appreciable fraction of the excess energy (Perrin, 1933).

The basis for most of the pair-production cross sections reported in the literature has been the calculation based on the Born approximation carried out by Bethe and Heitler (1934). Various corrections, depending on the energy region of interest, are applied to these results (Grodstein, 1957). When screening by atomic electrons is neglected, the cross section in the energy region $m_0c^2 \ll h\nu \ll 137m_0c^2Z^{-1/3}$ is given approximately by

$$\kappa = \frac{r_0^2}{137} Z^2(\tfrac{28}{9} \ln 2\alpha - \tfrac{218}{27}) \tag{2-67}$$

The strong dependence on atomic number Z and the weak dependence on energy α are evident. At very high energies the log term must be modified, with the result that the cross section becomes even less dependent on energy. Values of the pair-production cross section as a function of energy are shown in Fig. 2-14.

Calculation of the angular distribution of the created pair is difficult at low energies. At high energies the distribution peaks characteristically in

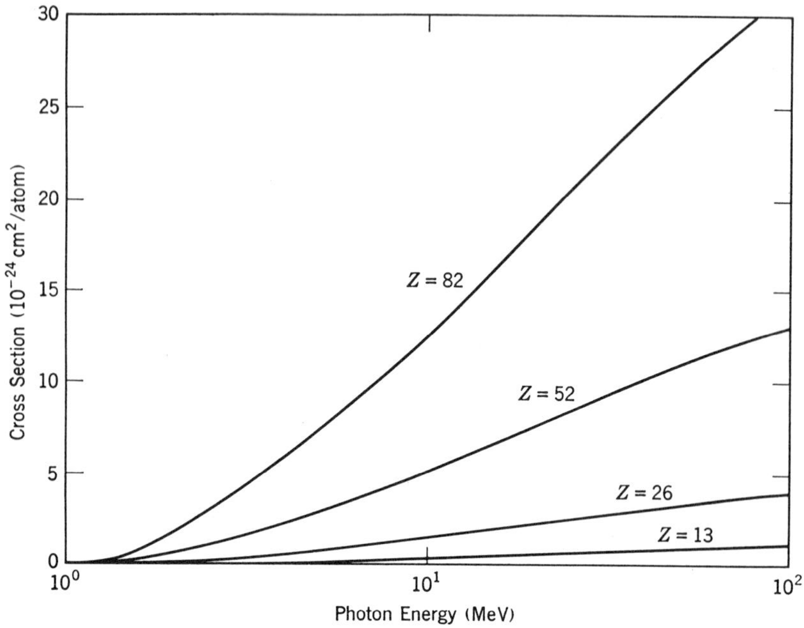

Fig. 2-14. Cross sections for the pair-production process.

the forward direction. The *average* angle between the direction of the incident photon and that taken by either one of the pair is given by

$$\theta_0 = \frac{m_0c^2}{T} \tag{2-68}$$

where T is the electron energy (Bethe and Ashkin, 1953).

Cross sections for pair production in the field of an electron are typically one or more orders of magnitude smaller than the values shown in Fig. 2-14.

Other Absorption Processes

Other absorptive processes occur with small probabilities. The most important of these is the nuclear photoeffect. The cross section increases rapidly with energy above a threshold, passes through a maximum, and then decreases with energy. Depending on Z the position of the maximum is generally between 10 and 20 MeV. Below 10 MeV the cross section is entirely negligible. The photoneutron cross sections for several materials have been tabulated by Grodstein (1957).

2-9 THE ATTENUATION LAW

The Attenuation Coefficient

Consider a beam of photons of intensity I_0 incident upon a plane of absorbing material as diagrammed in Fig. 2-15. At a depth x within the material this number has been reduced to I because of interactions with the absorber.

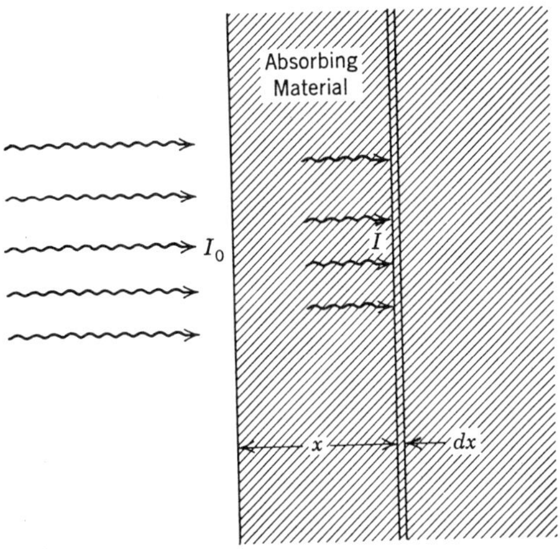

Fig. 2-15. Attenuation of a photon beam passing through a medium.

Within an incremental distance dx at x there is a further reduction in I by an amount dI. The probability for interaction within dx is dI/I; that for interaction per unit distance is $(dI/I) \cdot (1/dx)$. This latter probability has been given the name "linear attenuation coefficient" and is generally designated by the symbol μ with dimensions of reciprocal centimeters (cm^{-1}). In terms of μ, we may write for the decrease in intensity within dx

$$-dI = \mu I \, dx \tag{2-69}$$

This simple differential equation, when solved by using the boundary condition that $I = I_0$ at $x = 0$, yields the familiar Lambert's law of exponential attenuation

$$I = I_0 e^{-\mu x} \tag{2-70}$$

The attenuation coefficient μ is related to the total microscopic cross section σ_t by $\mu = N\sigma_t$ where N is the number of targets per unit volume. For this reason μ is sometimes referred to as the macroscopic cross section. If we

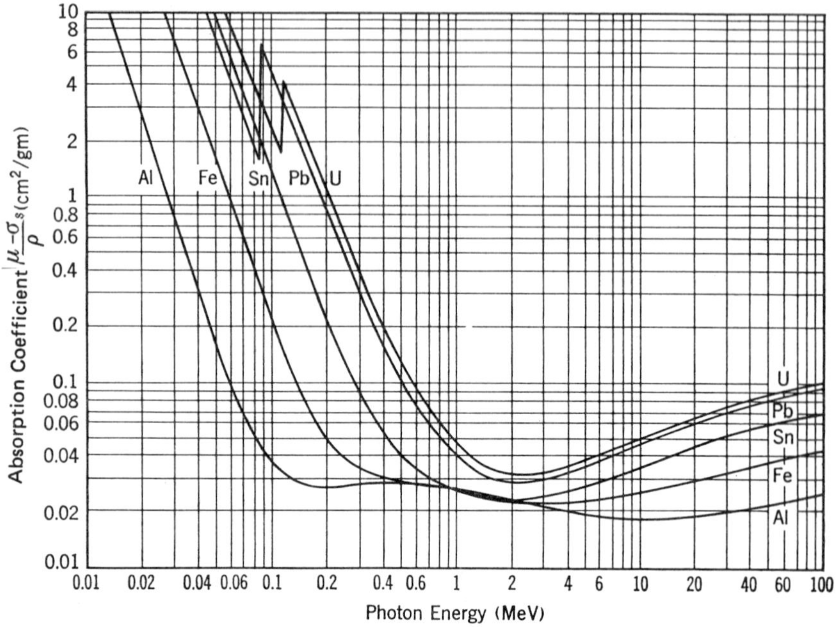

Fig. 2-16. Mass-absorption coefficients for various elements.

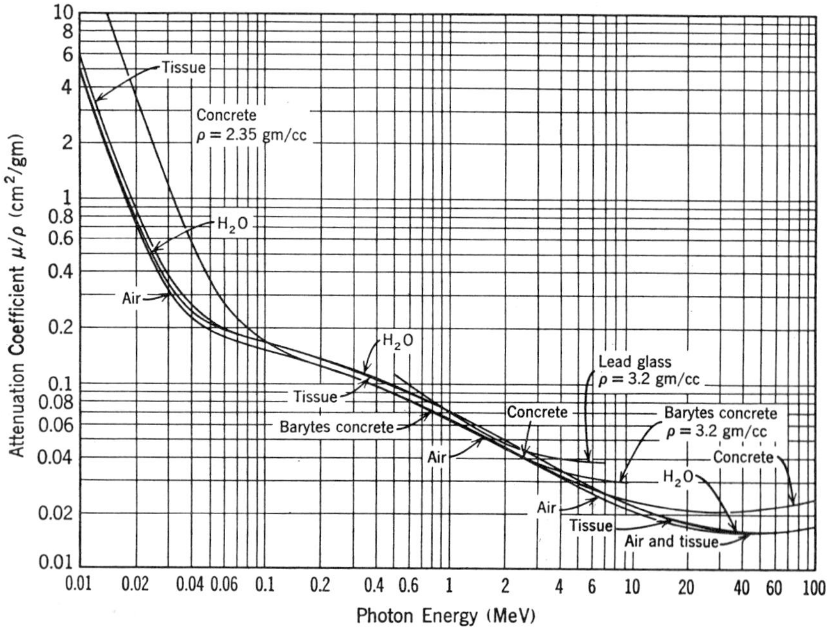

Fig. 2-17. Mass-attenuation coefficients for various materials.

neglect the minor interaction processes, the linear attentuation coefficient is given by

$$\mu = N[(\tau_K + \tau_L + \cdots) + Z\sigma + \kappa] \tag{2-71}$$

where N is the number of atoms per unit volume. Another commonly used quantity is the mass attenuation coefficient, which has the units square centimeters per gram (cm²/gm). It is obtained by dividing the linear attenuation coefficient by the density ρ. Mass-attenuation coefficients for a number of materials are shown in Figs. 2-17 and 2-18.

Absorption of Electromagnetic Energy

The absorption of energy, which is the physical quantity used in specifying radiation dose, is described in terms of an absorption coefficient. The absorption coefficient is obtained from the attenuation coefficient by subtracting the portion associated with photon scattering. Since the only important scattering process is Compton scattering, the linear absorption coefficient is given by $\mu - \sigma_s$ and the mass-absorption coefficient by $(\mu - \sigma_s)/\rho$. Values of the latter quantity are shown in Figs. 2-16 and 2-19.

In using $\mu - \sigma_s$ to describe energy absorption in a medium, we exclude the part of the original photon energy that escapes from the site of interaction.

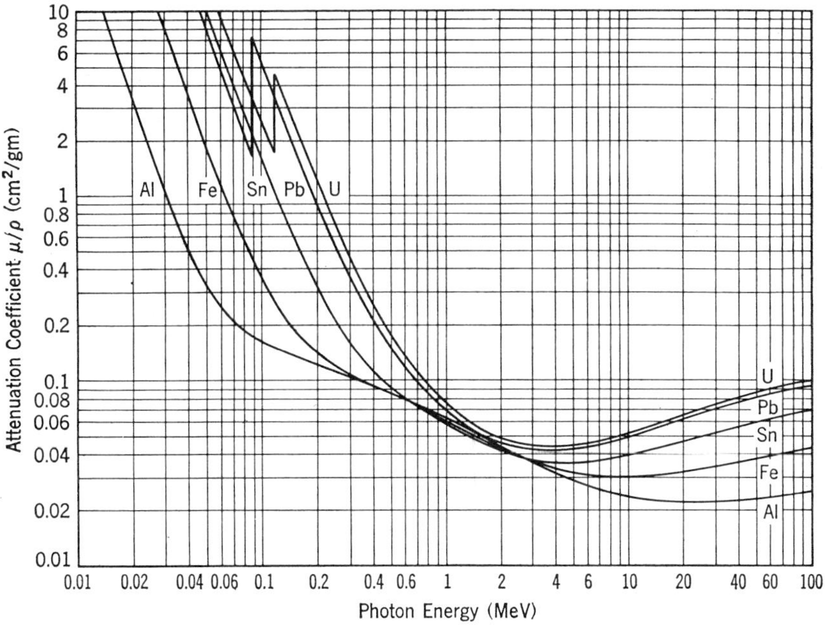

Fig. 2-18. Mass-attenuation coefficients for various elements.

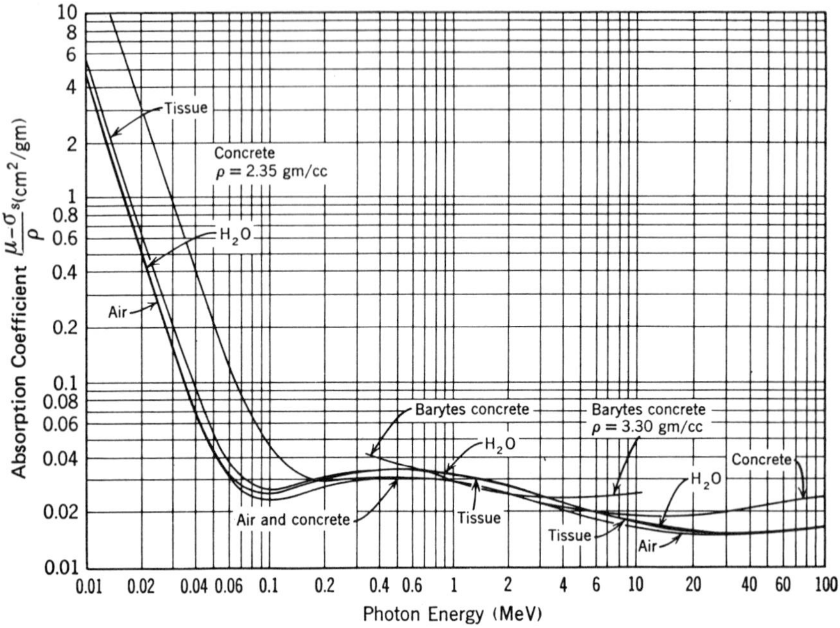

Fig. 2-19. Mass-absorption coefficients for various materials.

In the photoelectric effect, the electron binding energy is often released in the form of an X-ray. Usually, the X-ray has low energy and is absorbed nearby. In pair production, however, photons of the accompanying annihilation radiation have an energy of 510 keV and generally travel some distance before being absorbed. As a rule, neglect of this escape of energy is warranted because the contribution of the pair production to the over-all energy-absorption coefficient is small.

Absorption at Very Low Energies

More and more research in health physics is being carried out in the very low-energy region between a few electron volts and a few hundred electron volts. It is in this region that the final increments of the original photon energy are imparted to matter. Figure 2-20 shows the absorption coefficient for aluminum over the energy region from 1 eV to 1 MeV (Philipp and Ehrenreich, 1964). This curve is a composite made up of experimental data from a number of sources.

The sharp breaks at 1560 eV and 87 eV are the K and $L_{2,3}$ X-ray edges, respectively. The missing L_1 edge has been reported at 113 eV (Fohlman et al., 1965). For photons whose energy is insufficient to cause L-shell

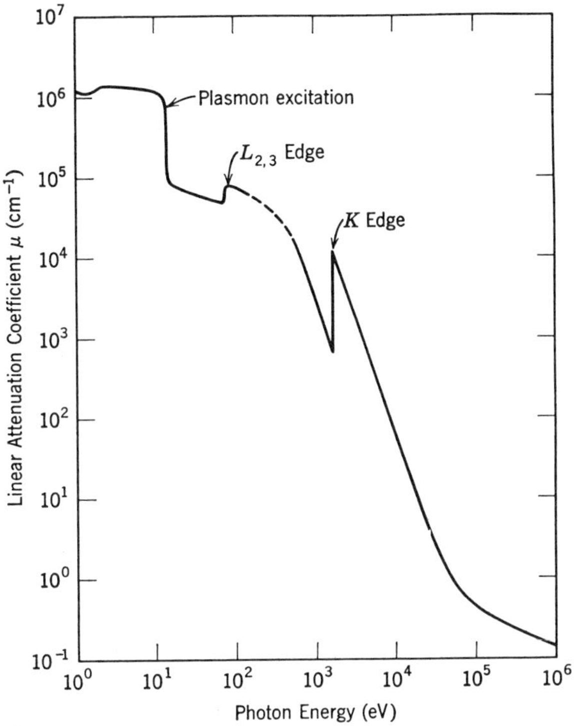

Fig. 2-20. Linear attenuation coefficients for aluminum between 1 eV and 1 MeV.

ionization, the most important absorption mechanisms are interband excitation of valence electrons, a process which is not expected to introduce much structure into the absorption curve. Interpretation of this portion of the curve is difficult and requires a knowledge of the band structure. The sharp change in absorption at 15 eV is associated with the excitation of the volume plasmon (Blackstock et al., 1955).

2-10 BUILDUP OF SCATTERED RADIATION

Calculations of the photon intensity at some depth in an absorbing medium by the use of Eq. (2-70) generally give results which are too low, due to neglect of the contribution of secondary radiation. This secondary radiation consists mostly of Compton scattered photons with the addition of X-rays resulting from photoelectric interactions, annihilation radiation from the pair-production process, and bremsstrahlung from the slowing down of energetic electrons within the absorber. This increase in intensity, which is called buildup, depends in a complicated way on photon energy, the characteristics of the absorber, and the geometrical arrangement. In terms of a buildup factor B, the equation for the intensity can be written

$$I = BI_0e^{-\mu x} \tag{2-72}$$

Neglect of buildup can lead to sizable errors in calculations of radiation dose. At low photon energies and with moderate depth, the buildup factors generally range between 1 and 10, although values in excess of these are not uncommon in special situations.

Buildup factors are defined for the number flux, energy flux, and radiation dose. Unless more sophisticated calculations are desired, however, the differences among these are usually neglected. Exact calculations of the buildup factor are extremely complicated and necessitate the use of digital computers. In practice, other than for simple geometry, various approximations to the buildup factor are used. These are discussed in some detail in Chapter 9. The most comprehensive tables of the buildup factors have been given by Goldstein and Wilkins (1954); excerpts of these tables have been published (Rockwell, 1956) for many commonly encountered cases.

REFERENCES

Bethe, H., *Ann. Phys.* **5,** 325 (1930).

Bethe, H., *Handbuch der Physik* Vol. 24, Part I, p. 491 (1935).

Bethe, H. A., and J. Ashkin, "Passage of Radiations through Matter," in *Experimental Nuclear Physics*, (E. Segri, ed.), John Wiley and Sons, New York, 1953.

Bethe, H. A., and W. Heitler, *Proc. Roy. Soc.* (*London*) **A146,** 83 (1934).

Bichsel, H., in *American Institute of Physics Handbook*, McGraw-Hill, New York, 1963, 2nd Ed., pp. 8–20 ff.

Blackstock, A. W., R. H. Ritchie, and R. D. Birkhoff, *Phys. Rev.* **100,** 1078 (1955).

Bloch, F., *Ann. Phys.* **16,** 285 (1933a).

Bloch, F., *Z. Phys.* **81,** 363 (1955b).

Bohr, N., *Phil. Mag.* **25,** 10 (1913).

Bohr, N., *Phil. Mag.* **30,** 581 (1915).

Bohr, N., *Kgl. Danske Videnskab. Selskab, Mat.-fys, Medd.* **18,** No. 8 (1948).

Compton, A. H., *Phys. Rev.* **21,** 483 (1923).

Compton, A. H., and S. K. Allison, *X-rays in Theory and Experiment*, Van Nostrand, Princeton, New Jersey, (1935).

Davisson, C. M., and R. D. Evans, *Rev. Mod. Phys.* **24,** 2, 79 (1952).

Fahlman, A., K. Hamrin, R. Nordberg, C. Nordling, and K. Siegbahn, *Phys. Rev. Letters* **14,** 5, 127 (1965).

Fano, U., *Nucleonics* **11,** 8, 8 (1953).

Fano, U., *Ann. Rev. Nucl. Sci.* **13,** 1 (1963).

Fano, U., and J. E. Turner, in *Studies in Penetration of Charged Particles in Matter, Nat. Acad. Sci. Nat. Res. Council Rept.* **1133** (1964), pp. 49-67.

Fermi, E., *Phys. Rev.* **57,** 485 (1940).

Fischer, J., *Ann. Physik.* **8,** 821 (1931).

Garber, F. W., "Low Energy Electron and Proton Transmission in Aluminum and Silver," *Oak Ridge Natl. Lab. Rept.*, ORNL-TM-1153 (1965).
Goldstein, H., and J. E. Wilkins, "Calculating of the Penetration of Gamma Rays," Report NYO-3075, *U.S. Dept. Comm., Office Tech. Serv. Rept.* Washington, D.C. (1954).
Grodstein, G. W., *Natl. Bur. Std. (U.S.) Circ.* **583** (1957).
Halpern, O., and H. Hall, *Phys. Rev.* **73,** 477 (1948).
Heitler, W., *The Quantum Theory of Radiation*, Oxford University Press, London, 1954, 3rd Ed.
Klein, O., and Y. Nishina, *Z. Phys.* **52,** 853 (1929).
Lea, D. E., *Actions of Radiation on Living Cells*, Cambridge University Press, London, 1955.
Neufeld, J., *Phys. Rev.* **95,** 1128 (1954).
Neufeld, J., and W. S. Snyder, in *Selected Topics in Radiation Dosimetry*, Intern. Atomic Energy Agency, Vienna (1961), p. 35.
Perrin, F., *Compt. Rend.* **197,** 1100 (1933).
Philipp, H. R., and H. Ehrenreich, *J. Appl. Phys.* **35,** 5, 1416 (1964).
Rockwell, T., III, ed., *Reactor Shielding Design Manual*, Van Nostrand, Princeton, New Jersey, 1956.
Sauter, F., *Ann. Physik* **11,** 454 (1931).
Sternheimer, R. M., *Phys. Rev.* **103,** 511 (1956).
Sternheimer, R. M., *Phys. Rev.* **115,** 137 (1959).
Thomson, J. J., *Conduction of Electricity through Gases*, Cambridge University Press, London, 1906.
Turner, J. E., in *Studies in Penetration of Charged Particles in Matter, Nat. Acad. Sci. Nat. Res. Council Rept.* **1133** (1964), p. 99.
Turner, J. E., C. D. Zerby, R. L. Woodyard, H. A. Wright, W. E. Kinney, W. S. Snyder, and J. Neufeld, *Health Phys.* **10,** 783 (1964).
Walske, M. C., *Phys. Rev.* **88,** 1283 (1952).
Walske, M. C., *Phys. Rev.* **101,** 940 (1956).
Wick, G. C., *Nuovo Cimento* **1,** 302 (1942).

PROBLEMS

1. Which of the criteria (2-5) and (2-6) is satisfied for a 10 MeV proton?

2. Verify Eqs. (2-7) and (2-8).

3. Show that the stopping power as given by Eq. (2-12) has the dimensions ergs per centimeter.

4. Verify Eq. (2-15).

5. The maximum energy loss, Eq. (2-17), can be obtained by viewing the ion-electron collision in the rest frame of the ion. The head-on collision of the electron with the ion then resembles the elastic collision of the electron with a rigid wall. Show that the maximum momentum change in this case is 2 mv, and also verify Eq. (2-18).

6. Calculate the stopping power of a 200 MeV proton in aluminum.

7. A 10 MeV proton has the same stopping power as a deuteron of what energy? At what energy does an α-particle have this stopping power?

8. Why is Eq. (2-32) better for heavier elements than for light elements?

9. Write Eq. (2-35) in nonrelativistic form.

10. Explain by qualitative arguments why the inclusion of shell corrections would be expected to result in a decrease in the value of the stopping power.

11. What is the physical origin of straggling?

12. What is the density of the tissue whose composition is given in Table 2-1?

13. Calculate the stopping power of a 300 MeV proton in tissue using the data in Table 2-1 and compare with the value given in Table 2-2.

14. Assuming the mass-absorption coefficient for 1 MeV γ-rays in lead to be 0.0684 cm^2/gm and the density of lead to be 11.34 gm/cm^3, calculate the linear, atomic, and electronic absorption coefficients.

15. Compute the fraction of the dose penetrating a 2-inch slab of lead, assuming the dose buildup factor for 1 MeV γ-rays in lead fits the equation

$$B(\mu x) = 1 + 0.35\mu x - 0.007(\mu x)^2.$$

16. Derive Eqs. (2-56) through (2-58).

17. Derive Eqs. (2-45) through (2-48).

18. What is the predominant angle of emission of photoelectrons produced by 1.0 MeV γ-rays?

19. Compute the energy of a backscattered γ-ray produced by a 2.0 MeV primary photon.

20. What is the maximum energy of a backscattered γ-ray? A γ-ray scattered at right angles?

21. Show that the Compton wavelength h/m_0c is equal to the wavelength of a photon with an energy equal to the rest energy of an electron.

22. Show that for a Compton recoil electron ejected at an angle of 30° the greatest fraction of the original photon energy is transferred to the electron when the original photon energy is equal to twice the rest energy of the electron.

23. Aluminum, $Z = 13$, $A = 27$, has a density of 2.7 gm/cm^3. For 0.10 MeV γ-rays, the energy-absorption coefficient (μ_a) is 0.101 cm^{-1} and the mass attenuation coefficient (μ/ρ) is 0.160 cm^2/gm. If the average fractional Compton scattered energy

$$\left\langle \frac{h\nu'}{d\Omega} \right\rangle_{\text{Ave}} = \frac{1}{h\nu} \cdot \frac{\displaystyle\int h\nu \frac{d\sigma}{d\Omega}\, d\Omega}{\displaystyle\int \frac{d\sigma}{d\Omega}\, d\Omega}$$

has the value of 0.861 at 0.10 MeV, what is the microscopic photoelectric cross section (square centimeters per atom) for Al at this energy? The energy-absorption cross section given was calculated by assuming that the photoelectric process is purely absorptive.

24. Obtain the expression for the Compton scattering cross section σ_s by carrying out the integration of $d\sigma_s/d\Omega$, given by Eq. (2-48), over all solid angles.

25. Obtain the classical Thomson cross section, $(8/3)\pi r_0^2$, by the integration of Eq. (2-46) and show that it agrees with the quantum mechanical result in the low-energy limit.

26. A point isotropic photon source lies in an infinite homogeneous medium characterized by an attenuation coefficient μ, a Compton scattering cross section σ_s, and therefore an energy-absorption coefficient of $\mu - \sigma_s$. Assuming only that energy is conserved, derive an expression for the coefficient β in the linear buildup relation, $B = 1 + \beta\mu r$, in terms of μ and σ_s. The buildup factor may be defined as follows: If the point source emits ϕ photons of energy E, the photon energy passing through an infinitely thin spherical shell of radius r will be $B\phi E e^{-\mu r}$.

CHAPTER 3

Radiation Quantities and Units

M. F. Fair

3-1 INTRODUCTION

Many units of radiation have been used, but only those that have a practical application in health physics will be discussed here. Ideally, these should be agreed upon by international groups that are to use them. The International Commission on Radiological Units and Measurements (ICRU) is the principal organization responsible for selecting and defining radiation quantities and units. Therefore we begin with *ICRU Report* 10*a* which sets forth the official definitions of quantities and units used in the field of radiation physics.

3-2 OFFICIAL QUANTITIES AND UNITS—*ICRU REPORT 10a*

The following pages are reprinted from *Report* 10*a*, "Radiation Quantities and Units," of the International Commission on Radiological Units and Measurements (Ref. 1).

* * * * *

1. INTRODUCTION

There has recently been much discussion of the fundamental concepts and quantities employed in radiation dosimetry. This has arisen partly from the rapid increase in the number of individuals using these concepts in the expanding field of nuclear science and technology, partly because of the need for extending the concepts so that they would be of use at higher photon energies and for particulate as well as for photon radiation, but chiefly because of certain obscurities in the existing formulation of the quantities and units themselves.

The roentgen, for example, was originally defined to provide the best quantitative measure of exposure to medium energy X radiation which the measuring techniques of that day (1928) permitted. The choice of air as a standard substance was not only convenient but also appropriate for a physical quantity which was to be correlated with the biological effect of X rays, since the effective atomic number of air is not very different from that of tissue. Thus a given biological response could be reproduced approximately by an equal exposure in roentgens for X-ray energies available at that time. Since 1928 the definition of the roentgen has been changed several times, and this has reflected some feeling of dissatisfaction with the clarity of the concept.

The most serious source of confusion was the failure to define adequately the radiation quantity of which the roentgen was said to be the unit.[1] As a consequence of this omission the roentgen had gradually acquired a double role. The use of this name for the unit had become recognized as a way of specifying not only the magnitude but also the nature of the quantity measured. This practice conflicts with the general usage in physics, which permits, within the same field, the use of a particular unit for all quantities having the same dimensions.

Even before this, the need for accurate dosimetry of neutrons and of charged particles from accelerators or from radionuclides had compelled the International Commission on Radiological Units and Measurements (ICRU) to extend the number of concepts. It was also desired to introduce a new quantity which could be more directly correlated with the local biological and chemical effects of radiation. This quantity, *absorbed dose*, has a generality and simplicity which greatly facilitated its acceptance, and in a very few years it has become widely used in every branch of radiation dosimetry.

The introduction of absorbed dose into the medical and biological field was further assisted by defining a special unit—the *rad*. One rad is approximately equal to the absorbed dose delivered when soft tissue is exposed to one roentgen of medium voltage X radiation. Thus in many situations of interest to medical radiology, but not in all, the numbers of roentgens and rads associated with a particular medical or biological effect are approximately equal and experience with the earlier unit could be readily transferred to the new one. Although the *rad* is merely a convenient multiple of the fundamental unit, erg/g, it has already acquired, at least in some circles, the

[1] Fränz, H. and Hübner, W. "Concepts and Measurement of Dose," *Proceedings of Second International Conference on the Peaceful Uses of Atomic Energy, Geneva 1958*, P/971 **21**, 101, United Nations, Geneva (1958).

additional connotation that the only quantity which can be measured in rads is absorbed dose. On the other hand, the rad has been used by some authors as a unit for a quantity called by them *first collision dose;* this practice is deprecated by the Commission.

Being aware of the need for preventing the emergence of different interpretations of the same quantity, or the introduction of undesirable, unrelated quantities or units in this or similar fields of measurement, the ICRU set up, during its meeting in Geneva in September 1958, an *Ad Hoc* Committee. The task of this committee was to review the fundamental concepts, quantities, and units which are required in radiation dosimetry and to recommend a system of concepts and a set of definitions which would be, as far as possible, internally consistent and of sufficient generality to cover present requirements and such future requirements as can be foreseen. The committee was instructed to pay more attention to consistency and rigor than to the historical development of the subject and was authorized to reject any existing quantities or units which seemed to hinder a consistent and unified formulation of the concepts.

Bertrand Russell[2] in commenting on the use and abuse of the concept of infinitesimals by mathematicians, remarks: "But mathematicians did not at first pay heed to (these) warnings. They went ahead and developed their science, and it is well that they should have done so. It is a peculiar fact about the genesis and growth of new disciplines that too much rigor too early imposed stifles the imagination and stultifies invention. A certain freedom from the strictures of sustained formality tends to promote the development of a subject in its early stages, even if this means the risk of a certain amount of error. Nonetheless, there comes a time in the development of any field when standards of rigor have to be tightened."

The purpose of the present reexamination of the concepts to be employed in radiation dosimetry was primarily "to tighten standards of rigor." If, in the process, some increased formality is required in the definitions in order to eliminate any foreseeable ambiguities, this must be accepted.

2. GENERAL CONSIDERATIONS

The development of the more unified presentations of quantities and units which is here proposed was stimulated and greatly assisted by mathematical models of the dosimetric field which had been proposed by some members of the committee in an effort to clarify the concepts. It appeared, however, that the essential features of the mathematical models had been incorporated

[2] Russell, B., *Wisdom of the West*, p. 280 (Doubleday and Co., Inc., New York, 1959).

into the definitions and hence the need for their exposition in this report largely disappeared. The mathematical approach is published elsewhere.[3]

As far as possible, the definitions of the various fundamental quantities given here conform to a common pattern. Complex quantities are defined in terms of the simpler quantities of which they are comprised.

The passage to a "macroscopic limit" which has to be used in defining point quantities in other fields of physics can be adapted to radiation quantities and a special discussion of this is included in the section headed "limiting procedures."

The general pattern adopted is to give a short definition and to indicate the precise meaning of any special phrase or term used by means of an explanatory note following the definition. There has been no attempt to make the list of quantities which are defined here comprehensive. Rather, the Commission has striven to clarify the fundamental dosimetric quantities and a few others (such as activity) which were specifically referred to it for discussion.

It is recognized that certain terms for which definitions are proposed here are of interest in other fields of science and that they are already variously defined elsewhere. The precise wording of the definition and even the name and symbol given to any such quantity, may at some future date require alteration if discussions with representatives of the other interested groups of scientists should lead to agreement on a common definition or symbol. Although the definitions presented here represent some degree of compromise, they are believed to meet the requirements in the field of radiation dosimetry.

3. QUANTITIES, UNITS, AND THEIR NAMES

The Commission is of the opinion that the definition of concepts and quantities is a fundamental matter and that the choice of units is of less importance. Ambiguity can best be avoided if the defined quantity which is being measured is specified. Nevertheless, the special units do exist in this as in many other fields. For example, the hertz is restricted, by established convention, to the measurement of vibrational frequency, and the curie, in the present recommendations, to the measurement of the activity of a quantity of a nuclide. One does not measure activity in hertz nor frequency in curies although these quantities have the same dimensions.

It was necessary to decide whether or not to extend the use of the special dosimetric units to other more recently defined quantities having the same

[3] Rossi, H. H., and Roesch, W. C., "Field Equations in Dosimetry, *Radiation Res.* **16,** 783 (1962).

dimensions, to retain the existing restriction on their use to one quantity each or to abandon the special units altogether. The Commission considers that the addition of further special units in the field of radiation dosimetry is undesirable, but continues to recognize the existing special units. It sees no objection, however, to the expression of any defined quantity in the appropriate units of a coherent physical system. Thus, to express absorbed dose in ergs per gram or joules per kilogram, exposure in coulombs per kilogram or activity in reciprocal seconds, are entirely acceptable alternatives to the use of the special units which, for historical reasons, are usually associated with these quantities.

The ICRU recommends that the use of each special unit be restricted to one quantity as follows:

The rad—solely for absorbed dose,
The roentgen—solely for exposure,
The curie—solely for activity.

It recommends further that those who prefer to express quantities such as absorbed dose and kerma (see below) in the same units should use units of an internationally agreed coherent system.

Several new names are proposed in the present report. When the absorbed dose concept was adopted in 1953, the Commission recognized the need for a term to distinguish it from the quantity of which the roentgen is the unit. In 1956 the Commission proposed the term *exposure* for this latter quantity. To meet objections by the ICRP, a compromise term, "exposure dose" was agreed upon.[4] While this term has come into some use since then, it has never been considered as completely satisfactory. In the meantime, the basic cause of the ICRP objection has largely disappeared since most legal codes use either the units rad or rem.

Since in this report the whole system of radiological quantities and units has come under critical review, it seemed appropriate to reconsider the 1956 decision. Numerous names were examined as a replacement for exposure dose, but there were serious objections to any which included the word dose. There appeared to be a minimum of objection to the name *exposure* and hence this term has been adopted by the Commission with the hope that the question has been permanently settled. It involves a minimum change from the older name exposure dose. Furthermore, the elimination of the term "dose" accomplishes the long-felt desire of the Commission to retain the term dose for one quantity only—the absorbed dose.

The term "RBE dose" has in past publications of the Commission not been included in the list of definitions but was merely presented as a "recognized symbol." In its 1959 report the Commission also expressed misgivings over the utilization of the same term, "RBE," in both radiobiology and radiation

[4] For details see ICRU, 1956 Report, *NBS Handbook* **62,** p. 2 (1957).

protection. It now recommends that the term RBE be used in radiobiology only and that another name be used for the linear-energy-transfer-dependent factor by which absorbed doses are to be multiplied to obtain for purposes of radiation protection a quantity that expresses on a common scale for all ionizing radiations the irradiation incurred by exposed persons. The name recommended for this factor is the *quality factor*, (QF). Provisions for other factors are also made. Thus a *distribution factor*, (DF), may be used to express the modification of biological effect due to non-uniform distribution of internally deposited isotopes. The product of absorbed dose and modifying factors is termed the *dose equivalent*, (DE). As a result of discussions between ICRU and ICRP the following formulation has been agreed upon:

The Dose Equivalent

1. For protection purposes it is useful to define a quantity which will be termed the "dose equivalent", (DE).

2. (DE) is defined as the product of absorbed dose, D, quality factor, (QF), dose distribution factor, (DF), and other necessary modifying factors.

$$(DE) = D(QF)(DF) \cdots$$

3. The unit of dose equivalent is the "rem." The dose equivalent is numerically equal to the dose in rads multiplied by the appropriate modifying factors.

Although this statement does not cover a number of theoretical aspects (in particular the physical dimensions of some of the quantities) it fulfills the immediate requirement for an unequivocal specification of a scale that may be used for numerical expression in radiation protection.

Another new name is that for the quantity which represents the kinetic energy transferred to charged particles by the uncharged particles per unit mass of the irradiated medium. This is the same as one of the common interpretations of a concept "first collision dose," that has proved to be of great value in the dosimetry of fast neutrons. The concept is also closely related to the energy equivalent of exposure in an X-ray beam. The name proposed, *kerma*, is based on the initials of *k*inetic *e*nergy *r*eleased in *ma*terial.

Still another new name is the *energy fluence* which is here attached to the quantity in the 1953 ICRU report called *quantity of radiation*. The latter term was dropped in the 1956 ICRU report but the concept—time integral of intensity—remains a useful one and the proposed term appears to be acceptable in other languages as well as English. A related quantity *particle fluence*, which is equivalent to the quantity *nvt* used in neutron physics, is included to round out the system of radiation quantities.

The quantity for which the curie is the unit was referred to the committee for a name and definition. Hitherto the curie has been defined as a *quantity of the radioactive nuclide* such that 3.7×10^{10} disintegrations per second occur in it. However, it has never been specified what was meant by quantity of a nuclide, whether it be a number, mass, volume, etc. Meanwhile the custom has grown of identifying the number of curies of a radionuclide with its transformation rate. Because of the vagueness of the original concept, because of the custom of identifying curies with transformation rate and because it appeared not to interfere with any other use of the curie, the Commission recommends that the term *activity* be used for the transformation rate, and that the curie be made its unit. It is recognized that the definition of the curie is of interest to other bodies in addition to the ICRU, but by this report we recommend that steps be taken to redefine it as $3.7 \times 10^{10}\, s^{-1}$, i.e., as a unit of activity and not of quantity of a nuclide.

It is also recommended that the term *specific gamma ray constant* be used instead of *specific gamma ray emission* for the quotient of the exposure rate at a given distance by the activity. The former term focuses attention on the *constancy* of this quotient for a given nuclide rather than the *emission* of the source.

4. DETAILED CONSIDERATIONS

A. Limiting Procedures

Except in the case of a uniform distribution of sources throughout a large region, radiation fields are in general non-uniform in space. They may also be variable in time. Many of the quantities defined in this report have to be specified as functions of space or time, and in principle they must therefore be determined for sufficiently small regions of space or intervals of time by some limiting procedure. There are conceptual difficulties in taking such limits for quantities which depend upon the discrete interactions between radiations and atoms. Similar difficulties arise with other macroscopic physical quantities such as density or temperature and they must be overcome by means of an appropriate averaging procedure.

To illustrate this procedure we may consider the measurement of the macroscopic quantity "absorbed dose" in a non-uniform radiation field. In measuring this dose the quotient of energy by mass must be taken in an elementary volume in the medium which on the one hand is so small that a further reduction in its size would not appreciably change the measured value of the quotient energy by mass and on the other hand is still large enough to contain many interactions and be traversed by many particles.[5] If it is

[5] In interpreting radiation effects the macroscopic concept of absorbed dose may not be sufficient. Whenever the statistical fluctuations around the mean value are important, additional parameters describing the distribution of absorbed energy on a microscopic scale are necessary.

impossible to find a mass such that both these conditions are met, the dose cannot be established directly in a single measurement. It can only be deduced from multiple measurements that involve extrapolation or averaging procedures. Similar considerations apply to some of the other concepts defined below. The symbol Δ precedes the symbols for quantities that may be concerned in such averaging procedures.

In the measurement of certain material constants such as stopping power, absorption coefficient, etc., the limiting procedure can be specified more rigorously. Such constants can be determined for a given material with any desired accuracy without difficulties from statistical fluctuations. In these cases the formulae quoted in the definitions are presented as differential quotients.

B. Spectral Distribution and Mean Values

In practice many of the quantities defined below to characterize a radiation field and its interaction with matter are used for radiations having a complex energy spectrum. An important general concept in this connection is the *spectral concentration* of one quantity with respect to another. The spectral concentration is the ordinate of the distribution function of the first quantity with respect to the second. The independent quantity need not always be energy or frequency; one can speak of the spectral concentration of flux density with respect to quantum energy or of the absorbed dose with respect to linear energy transfer. The interaction constants (such as μ, S, and W) referred to in this report are often mean values taken over the appropriate spectral distributions of the corresponding quantities.

C. Units

For any of the quantities defined below the appropriate unit of an internationally agreed coherent system can be used. In addition certain special units are reserved for special quantities:

the rad for absorbed dose,
the roentgen for exposure,
the curie for activity.

D. Definitions

(1) *Directly ionizing particles* are charged particles (electrons, protons, α-particles, etc.) having sufficient kinetic energy to produce ionization by collision.

(2) *Indirectly ionizing particles* are uncharged particles (neutrons, photons, etc.) which can liberate directly ionizing particles or can initiate a nuclear transformation.

(3) *Ionizing radiation* is any radiation consisting of directly or indirectly ionizing particles or a mixture of both.

(4) The *energy imparted* by ionizing radiation to the matter in a volume is the difference between the sum of the energies of all the directly and indirectly ionizing particles which have entered the volume and the sum of the energies of all those which have left it, minus the energy equivalent of any increase in rest mass that took place in nuclear or elementary particle reactions within the volume.

Notes. (a) The above definition is intended to be exactly equivalent to the previous meanings given by the ICRU to "energy retained by matter and made locally available" or "energy which appears as ionization, excitation, or changes of chemical bond energies". The present formulation specifies what energy is to be included without requiring a lengthy, and possibly incomplete, catalogue of the different types of energy transfer.

(b) Ultimately, most of the energy imparted will be degraded and appear as heat. Some of it, however, may appear as a change in interatomic bond energies. Moreover, during the degradation process the energy will diffuse and the distribution of heat produced may be different from the distribution of imparted energy. For these reasons the energy imparted cannot always be equated with the heat produced.

(c) The quantity *energy imparted to matter* in a given volume is identical with the quantity often called *integral absorbed dose* in that volume.

(5) The *absorbed dose* (D) is the quotient of ΔE_D by Δm, where ΔE_D is the energy imparted by ionizing radiation to the matter in a volume element, Δm is the mass of the matter in that volume element and Δ has the meaning indicated in Section 4.A.

$$D = \Delta E_D / \Delta m$$

The special unit of absorbed dose is the *rad.*

$$1 \text{ rad} = 100 \text{ erg/gm} = 1/100(\text{J/kg})$$

Note. J is the abbreviation for Joule

(6) The *absorbed dose rate* is the quotient of ΔD by Δt, where ΔD is the increment in absorbed dose in time Δt and Δ has the meaning indicated in Section 4.A.

$$\text{Absorbed dose rate} = \Delta D / \Delta t$$

A special unit of absorbed dose rate is any quotient of the rad by a suitable unit of time (rad/d, rad/min, rad/h, etc.).

(7) The *particle fluence*[6] or *fluence* (Φ) of particles is the quotient of ΔN by Δa, where ΔN is the number of particles which enter a sphere[7] of cross-sectional area Δa and Δ has the meaning indicated in Section 4.A.

$$\Phi = \Delta N/\Delta a$$

(8) The *particle flux density* or *flux density* (φ) of particles is the quotient of $\Delta\Phi$ by Δt where $\Delta\Phi$ is the particle fluence in time Δt and Δ has the meaning indicated in Section 4.A.

$$\varphi = \Delta\Phi/\Delta t$$

Note. This quantity may also be referred to as particle fluence rate.

(9) The *energy fluence* (F) of particles is the quotient of ΔE_F by Δa, where ΔE_F is the sum of the energies, exclusive of rest energies, of all the particles which enter a sphere[8] of cross-sectional area Δa and Δ has the meaning indicated in Section 4.A.

$$F = \Delta E_F/\Delta a$$

(10) The *energy flux density* or *intensity* (I) is the quotient of ΔF by Δt where ΔF is the energy fluence in the time Δt and Δ has the meaning indicated in Section 4.A.

$$I = \Delta F/\Delta t$$

Note. This quantity may also be referred to as energy fluence rate.

(11) The *kerma*[9] (K) is the quotient of ΔE_K by Δm, where ΔE_K is the sum of the initial kinetic energies of all the charged particles liberated by indirectly ionizing particles in a volume element of the specified material, Δm is the mass of the matter in that volume element and Δ has the meaning indicated in Section 4.A.

$$K = \Delta E_K/\Delta m$$

Notes. (a) Since ΔE_K is the sum of the initial kinetic energies of the charged particles liberated by the indirectly ionizing particles, it includes not only the kinetic energy these charged particles expend in collisions but also the energy they radiate in bremsstrahlung. The energy of any charged particles is also included when these are produced in secondary processes occurring within the volume element. Thus the energy of Auger electrons is part of ΔE_K.

[6] This quantity is the same as the quantity, *nvt*, commonly used in neutron physics.

[7] This quantity is sometimes defined with reference to a plane of area Δa, instead of a sphere of cross-sectional area Δa. The plane quantity is less useful for the present purposes and it will not be defined. The two quantities are equal for a unidirectional beam of particles perpendicularly incident upon the plane area.

[8] See footnote 7.

[9] Various other methods of specifying a radiation field have been used, e.g., for a neutron source the "first collision dose" in a standard material at a specified point (see Introduction).

(b) In actual measurements Δm should be so small that its introduction does not appreciably disturb the radiation field. This is particularly necessary if the medium for which kerma is determined is different from the ambient medium; if the disturbance is appreciable an appropriate correction must be applied.

(c) It may often be convenient to refer to a value of kerma or of kerma rate for a specified material in free space or at a point inside a different material. In such a case the value will be that which would be obtained if a small quantity of the specified material were placed at the point of interest. It is, however, permissible to make a statement such as: "The kerma for air at the point P inside a water phantom is · · ·" recognizing that this is a shorthand version of the fuller description given above.

(d) A fundamental physical description of a radiation field is the intensity (energy flux density) at all relevant points. For the purpose of dosimetry, however, it may be convenient to describe the field of indirectly ionizing particles in terms of the kerma rate for a specified material. A suitable material would be air for electromagnetic radiation of moderate energies, tissue for all radiations in medicine or biology, or any relevant material for studies of radiation effects.

Kerma can also be a useful quantity in dosimetry when charged particle equilibrium exists at the position and in the material of interest, and bremsstrahlung losses are negligible. It is then equal to the absorbed dose at that point. In beams of X or gamma rays or neutrons, whose energies are moderately high, transient charged-particle equilibrium can occur; in this condition the kerma is just slightly less than the absorbed dose. At very high energies the difference becomes appreciable. In general, if the range of directly ionizing particles becomes comparable with the mean free path of the indirectly ionizing particles, no equilibrium will exist.

(12) The *kerma rate* is the quotient of ΔK by Δt, where ΔK is the increment in kerma in time Δt and Δ has the meaning indicated in Section 4.A.

(13) The *exposure* (X) is the quotient of ΔQ by Δm, where ΔQ is the sum of the electrical charges on all the ions of one sign produced in air when all the electrons (negatrons and positrons), liberated by photons in a volume element of air whose mass is Δm, are completely stopped in air and Δ has the meaning indicated in Section 4.A.

$$X = \Delta Q/\Delta m$$

The special unit of exposure is the roentgen (R).[10]

$$1R = 2.58 \times 10^{-4}\mathrm{C/kg}$$

[10] This unit is numerically identical with the old one defined as 1 e.s.u. of charge per 0.001293 gram of air. C is the abreviation for coulomb.

Notes. (a) The words "charges on all the ions of one sign" should be interpreted in the mathematically absolute sense.

(b) The ionization arising from the absorption of bremsstrahlung emitted by the secondary electrons is not to be included in ΔQ. Except for this small difference, significant only at high energies, the exposure as defined above is the ionization equivalent of the kerma in air.

(c) With present techniques it is difficult to measure exposure when the photon energies involved lie above a few MeV or below a few keV.

(d) As in the case of kerma [4.D(11), note (c)], it may often be convenient to refer to a value of exposure or of exposure rate in free space or at a point inside a material different from air. In such a case the value will be that which would be determined for a small quantity of air placed at the point of interest. It is, however, permissible to make a statement such as: "The exposure at the point P inside a water phantom is . . . ."

(14) The *exposure rate* is the quotient of ΔX by Δt, where ΔX is the increment in exposure in time Δt and Δ has the meaning indicated in section 4.A,

$$\text{Exposure rate} = \Delta X/\Delta t$$

A special unit of exposure rate is any quotient of the roentgen by a suitable unit of time (R/s, R/min, R/h, etc.).

(15) The *mass attenuation coefficient* (μ/ρ) of a material for indirectly ionizing particles is the quotient of dN by the product of ρ, N, and dl where N is the number of particles incident normally upon a layer of thickness dl and density ρ, and dN is the number of particles that experience interactions in this layer.

$$\frac{\mu}{\rho} = \frac{1}{\rho N}\frac{dN}{dl}$$

Notes. (a) The term "interactions" refers to processes whereby the energy or direction of the indirectly ionizing particles is altered.

(b) For X or gamma radiations

$$\frac{\mu}{\rho} = \frac{\tau}{\rho} + \frac{\sigma}{\rho} + \frac{\sigma_{\text{coh}}}{\rho} + \frac{\kappa}{\rho}$$

where τ/ρ is the mass photoelectric attenuation coefficient, σ/ρ is the total Compton mass attenuation coefficient, σ_{coh}/ρ is the mass attenuation coefficient for coherent scattering, and κ/ρ is the pair-production mass attenuation coefficient.

(16) The *mass energy transfer coefficient* (μ_K/ρ) of a material for indirectly ionizing particles is the quotient of dE_K by the product of E, ρ, and dl where E is the sum of the energies (excluding rest energies) of the indirectly ionizing

particles incident normally upon a layer of thickness dl and density ρ, and dE_K is the sum of the kinetic energies of all the charged particles liberated in this layer.

$$\frac{\mu_K}{\rho} = \frac{1}{E\rho}\frac{dE_K}{dl}$$

Notes. (a) The relation between fluence and kerma may be written as

$$K = F(\mu_K/\rho)$$

(b) For X- or gamma rays of energy $h\nu$

$$\frac{\mu_K}{\rho} = \frac{\tau_a}{\rho} + \frac{\sigma_a}{\rho} + \frac{\kappa}{\rho}$$

where

$$\frac{\tau_a}{\rho} = \frac{\tau}{\rho}\left(1 - \frac{\delta}{h\nu}\right)$$

(τ/ρ = the photoelectric mass attenuation coefficient, δ = average energy emitted as fluorescent radiation per photon absorbed.)
and

$$\frac{\sigma_e}{\rho} = \frac{\sigma}{\rho}\frac{E_e}{h\nu}$$

(σ/ρ = total Compton mass attenuation coefficient, E_e = average energy of the Compton electrons per scattered photon.)
and

$$\frac{\kappa_a}{\rho} = \frac{\kappa}{\rho}\left(1 - \frac{2mc^2}{h\nu}\right)$$

(κ/ρ = mass attenuation coefficient for pair production, mc^2 = rest energy of the electron.)

(17) The *mass energy-absorption coefficient* (μ_{en}/ρ) of a material for indirectly ionizing particles is $\mu_K/\rho(1 - G)$ where G is the proportion of the energy of secondary charged particles that is lost to bremsstrahlung in the material.

Notes. (a) When the material is air, μ_{en}/ρ is proportional to the quotient of exposure by fluence.

(b) μ_K/ρ and μ_{en}/ρ do not differ appreciably unless the kinetic energies of the secondary particles are comparable with or larger than their rest energy.

(18) The *mass stopping power* (S/ρ) of a material for charged particles is the quotient of dE_s by the product of dl and ρ, where dE_s is the average

energy lost by a charged particle of specified energy in traversing a path length dl, and ρ is the density of the medium.

$$\frac{S}{\rho} = \frac{1}{\rho}\frac{dE_s}{dl}$$

Note. dE_s denotes energy lost due to ionization, electronic excitation, and radiation. For some purposes it is desirable to consider stopping power with the exclusion of bremsstrahlung losses. In this case S/ρ must be multiplied by an appropriate factor that is less than unity.

(19) The *linear energy transfer* (L) of charged particles in a medium is the quotient of dE_L by dl where dE_L is the average energy locally imparted to the medium by a charged particle of specified energy in traversing a distance of dl.

$$L = dE_L/dl$$

Notes. (a) The term "locally imparted" may refer either to a maximum distance from the track or to a maximum value of discrete energy loss by the particle beyond which losses are no longer considered as local. In either case the limits chosen should be specified.

(b) The concept of linear energy transfer is different from that of stopping power. The former refers to energy imparted within a limited volume, the latter to loss of energy regardless of where this energy is absorbed.

(20) The *average energy* (W) *expended in a gas per ion pair formed* is the quotient of E by N_w, where N_w is the average number of ion pairs formed when a charged particle of initial energy E is completely stopped by the gas.

$$W = E/N_w$$

Notes. (a) The ions arising from the absorption of bremsstrahlung emitted by the charged particles are not to be counted in N_w.

(b) In certain cases it may be necessary to consider the variation in W along the path of the particle, and a differential concept is then required, but is not specifically defined here.

(21) A *nuclide* is a species of atom having specified numbers of neutrons and protons in its nucleus.

(22) The *activity* (A) of a quantity of a radioactive nuclide is the quotient of ΔN by Δt where ΔN is the number of nuclear transformations which occur in this quantity in time Δt and Δ has the meaning indicated in Section 4.A.

$$A = \Delta N/\Delta t$$

The special unit of activity is the curie (c).

$$1\ \text{c} = 3.7 \times 10^{10}\ \text{s}^{-1}\ \text{(exactly)}$$

Table 3-1 Table of Quantities and Units

No.	Name	Symbol	Dimensions[a]	Units: MKSA	Units: cgs	Units: Special
4	Energy imparted (integral absorbed dose)	—	E	J	erg	gm rad
5	Absorbed dose	D	EM^{-1}	J kg^{-1}	erg gm^{-1}	rad
6	Absorbed dose rate	—	$EM^{-1}T^{-1}$	J kg^{-1} s^{-1}	erg gm^{-1} s^{-1}	rad s^{-1} etc.
7	Particle fluence or fluence	Φ	L^{-2}	m^{-2}	cm^{-2}	
8	Particle flux density	ϕ	$L^{-2}T^{-1}$	m^{-2} s^{-1}	cm^{-2} s^{-1}	
9	Energy fluence	F	EL^{-2}	J m^{-2}	erg cm^{-2}	
10	Energy flux density or intensity	I	$EL^{-2}T^{-1}$	J m^{-2} s^{-1}	erg cm^{-2} s^{-1}	
11	Kerma	K	EM^{-1}	J kg^{-1}	erg gm^{-1}	
12	Kerma rate	—	$EM^{-1}T^{-1}$	J kg^{-1} s^{-1}	erg gm^{-1} s^{-1}	
13	Exposure	X	QM^{-1}	C kg^{-1}	esu gm^{-1}	R (roentgen)
14	Exposure rate	—	$QM^{-1}T^{-1}$	C kg^{-1} s^{-1}	esu gm^{-1} s^{-1}	Rs^{-1} etc.
15	Mass attenuation coefficient	$\frac{\mu}{\rho}$	L^2M^{-1}	m^2 kg^{-1}	cm^2 gm^{-1}	
16	Mass energy transfer coefficient	$\frac{\mu_K}{\rho}$	L^2M^{-1}	m^2 kg^{-1}	cm^2 gm^{-1}	
17	Mass energy absorption coefficient	$\frac{\mu_{en}}{\rho}$	L^2M^{-1}	m^2 kg^{-1}	cm^2 gm^{-1}	
18	Mass stopping power	$\frac{S}{\rho}$	EL^2M^{-1}	J m^2 kg^{-1}	erg cm^2 gm^{-1}	
19	Linear energy transfer	L	EL^{-1}	J m^{-1}	erg cm^{-1}	keV(μm)$^{-1}$
20	Average energy per ion pair	W	E	J	erg	eV
22	Activity	A	T^{-1}	s^{-1}	s^{-1}	c (curie)
23	Specific gamma-ray constant	Γ	QL^2M^{-1}	cm^2 kg^{-1}	esu cm^2 gm^{-1}	Rm^2 h^{-1} c^{-1} etc
	Dose equivalent	DE	—	—	—	rem

[a] It was desired to present only one set of dimensions for each quantity, a set that would be suitable in both the MKSA and electrostatic-cgs systems. To do this it was necessary to use a dimension Q, for the electrical charge, that is not a fundamental dimension in either system. In the MKSA system (fundamental dimensions M, L, T, I) Q represents the product IT; in the electrostatic-cgs system (M, L, T) it represents $M^{1/2}L^{3/2}T^{-1}$.

Note. In accordance with the former definition of the curie as a unit of quantity of a radioactive nuclide, it was customary and correct to say: "Y curies of P–32 were administered . . . ." It is still permissible to make such statements rather than use the longer form which is now correct: "A quantity of P–32 was administered whose activity was Y curies."

(23) The *specific gamma ray constant* (Γ) of a gamma-emitting nuclide is the quotient of $l^2(\Delta X/\Delta t)$ by A, where $\Delta X/\Delta t$ is the exposure rate at a distance l from a point source of this nuclide having an activity A and Δ has the meaning indicated in Section 4.A.

$$\Gamma = \frac{l^2 \Delta X}{A\,\Delta t}$$

Special units of specific gamma ray constant are $\mathrm{R}m^2h^{-1}\mathrm{c}^{-1}$ or any convenient multiple of this.

Note. It is assumed that the attenuation in the source and along l is negligible. However, in the case of radium the value of Γ is determined for a filter thickness of 0.5 mm of platinum and in this case the special units are $\mathrm{R}m^2h^{-1}\mathrm{gm}^{-1}$ or any convenient multiple of this.

* * * * *

Recommended values of the quality factor (QF) of radiation of different linear energy transfer (LET) applicable to exposure to radiation from external sources are shown in Table 3-2, which is taken from Ref. 2. Present knowledge of the biological effectiveness of radiation of different LET does not warrant fine distinctions. Therefore, ranges rather than individual figures are given in the table. For any range of LET it is safer to use the higher of the two values of QF given for that range, but a value obtained by linear interpolation is acceptable.

Table 3-2 LET-QF Relationship, Heavy Ionizing Particles
(X-rays, Electrons, and Positrons of Any LET: QF = 1)

Average Linear Energy Transfer (LET) in Water (keV per micron)	QF
3.5 or less	1
3.5 to 7.0	1–2
7.0 to 23	2–5
23 to 53	5–10
53 to 175	10–20

3-3 RELATIONSHIP BETWEEN γ-FLUX DENSITY AND EXPOSURE RATE

In the study of the interaction of γ-rays with air, it is very useful to relate energy fluence incident on air to the exposure as well as to relate γ-flux density to exposure rate. Using the definitions given in Section 3-2, we proceed as follows.

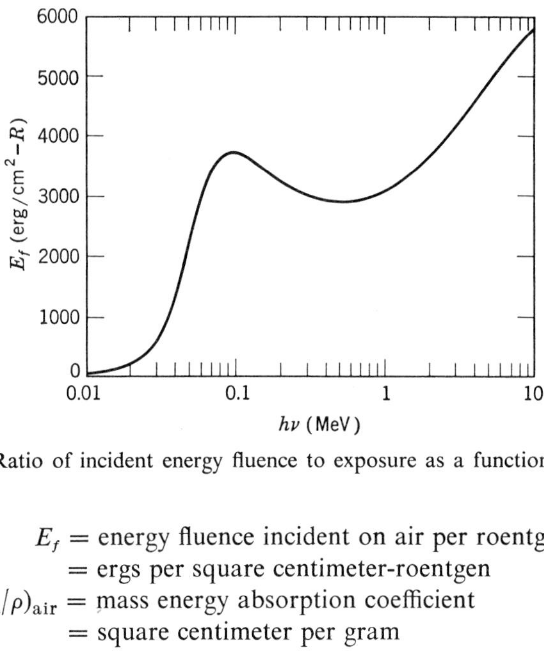

Fig. 3-1. Ratio of incident energy fluence to exposure as a function of photon energy.

Let $\quad E_f$ = energy fluence incident on air per roentgen
= ergs per square centimeter-roentgen

and $\quad (\mu_{en}/\rho)_{air}$ = mass energy absorption coefficient
= square centimeter per gram

Then clearly

$$(E_f)(\mu_{en}/\rho) = 87.7 \text{ ergs/gm-R}$$

and therefore

$$E_f = \frac{87.7}{(\mu_{en}/\rho)_{air}} \quad \frac{\text{ergs}}{\text{cm}^2\text{-R}} \tag{3-1}$$

Figure 3-1 shows a plot of E_f versus $h\nu$, where $h\nu$ gives the energy of the γ-ray.

From this relationship between incident energy fluence and exposure we can easily obtain the more useful relationship between γ-flux density and exposure rate.

The energy fluence (E_f) is expressed in terms of γ-fluence Φ thus:

$$E_f = 1.6 \times 10^{-6} \times \Phi E \tag{3-2}$$

$$\frac{\text{ergs}}{\text{cm}^2\text{-R}} = \frac{\text{ergs}}{\text{MeV}} \times \frac{\text{photons}}{\text{cm}^2\text{-R}} \times \frac{\text{MeV}}{\text{photon}}$$

Introducing this into Eq. (3-1), we find:

$$\Phi = \frac{5.48 \times 10^7}{E(\mu_{en}/\rho)_{\text{air}}} \ \frac{\text{photons}}{\text{cm}^2\text{-R}} \tag{3-3}$$

It is more convenient to express this as a correspondence so that one can state

$$1\,\frac{m\text{R}}{\text{hr}} \approx \frac{0.0196}{E(\mu_{en})_{\text{air}}} \ \frac{\text{photons}}{\text{cm}^2\text{-sec}} \tag{3-4}$$

EXAMPLE. For 2-MeV photons one finds $(\mu_{en})_{\text{air}} = 3 \times 10^{-5}/\text{cm}$. Therefore

$$1\,\frac{\text{mR}}{\text{hr}} \approx 330 \ \frac{\text{photons}}{\text{cm}^2\text{-sec}}$$

Figure 3-2 is a plot showing the γ-flux density which is equivalent to 1 R/hr as a function of γ-energy. With a monochromatic beam of photons the

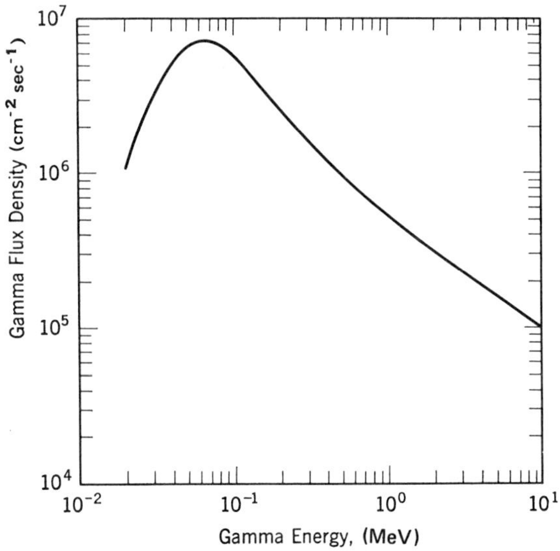

Fig. 3-2. Gamma flux density equivalent to 1 R/hr as a function of γ-energy.

absorbed energy in units of ergs per gram was

$$E_f\left(\frac{\mu_{en}}{\rho}\right)_{\text{air}} = 1.6 \times 10^{-6}\Phi E\left(\frac{\mu_{en}}{\rho}\right)_{\text{air}}$$

With a heterogeneous beam of photons the absorbed energy in units of ergs per gram would be

$$\int_0^{Em} 1.6 \times 10^{-6} E\Phi(E)\left(\frac{\mu_{en}}{\rho}\right)_{\text{air}} dE \tag{3-5}$$

where $\phi(E)$ is the number of photons per square centimeter per unit energy interval. It must be noted that the mass energy absorption coefficient is also a function of the energy.

3-4 RELATIONSHIP BETWEEN EXPOSURE AND ABSORBED DOSE

The absorbed energy in air at a point where the exposure is 1 R is 87.7 ergs/gm. The absorbed dose at a point in air where the exposure is R roentgens is given by:

$$D_{\text{air}} = 0.877R \quad \text{rad} \tag{3-6}$$

The number of rad per roentgen in air is 0.877. In practice one can measure the exposure and calculate the absorbed dose using Eq. (3-6).

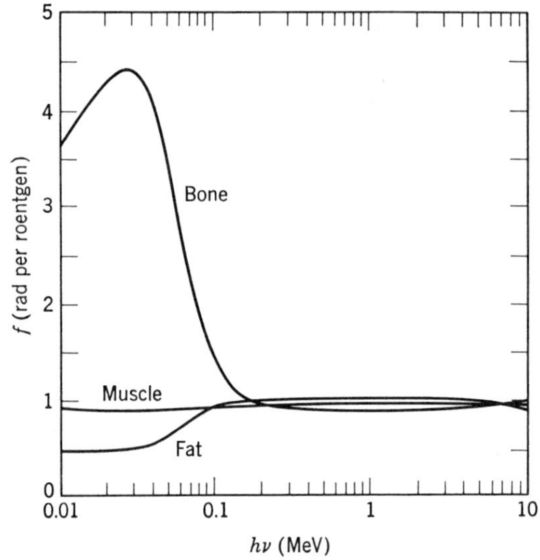

Fig. 3-3. Absorbed dose in rad per roentgen f as a function of photon energy.

When the medium involved is something other than air, and if electronic equilibrium exists, it can easily be shown that Eq. (3-6) must be multiplied by the ratio of the mass-absorption coefficient of the medium to that for air. Thus

$$D_m = 0.877R \frac{(\mu_{en}/\rho)_m}{(\mu_{en}/\rho)_{\text{air}}} = fR \quad \text{rad} \tag{3-7}$$

The number of rad per roentgen in the medium is labeled f. Figure 3-3 shows f as a function of γ-energy for such absorbers as bone, muscle, and fat.

If electronic equilibrium does not exist one can still calculate the absorbed dose by using the Bragg-Gray principle (see Chapter 4). It is useful and instructive to examine Eq. (3-7) under conditions of low-, intermediate-, and high-energy γ-rays.

Low Energy (Less than 0.1 MeV)

As described in Section 2-8, the photoelectric effect predominates. Therefore we can set $\mu_{en} \cong \tau$. Furthermore, it is well known that $\tau \propto Z^4$, where Z is the atomic number of the absorber. Therefore

$$D_m \propto (Z_m/Z_{\text{air}})^4.$$

The absorbed dose in rad varies as the ratio of the atomic numbers of the absorber and air raised to the fourth power. This accounts for the large variation at low energy shown in Fig. 3-3.

Intermediate Energy (between 0.1 MeV and 1 MeV)

The Compton effect (Section 2-8) predominates. Therefore we can set $\mu_{en} \cong \sigma_a$. It is easily shown that since $\sigma_a \propto Z/A$ the absorbed dose varies thus:

$$D_m \propto \frac{(Z/A)_m}{(Z/A)_{\text{air}}}$$

Figure 3-3 shows that in this region the variation of absorbed dose with energy is quite small.

High Energy (Greater than 1 MeV)

The pair-production process (Section 2-8) predominates in this region. Therefore $\mu_{en} \cong \kappa$. The pair-production process varies as the square of the atomic number of the absorber. Therefore it is seen that

$$D_m \propto \left(\frac{Z_m}{Z_{\text{air}}}\right)^2$$

The variation of absorbed dose with energy in this region would be greater than in the Compton region but less than in the photoelectric region.

In dealing with a continuous spectrum such as an X-ray spectrum rather than a monochromatic spectrum which has been assumed up to now, a slight modification must be made in the definition of f. Let

$$f(E) = \text{rad per roentgen at energy } E$$

and

$$R(E) = \text{roentgens per unit energy interval}$$
$$= \text{photon spectral distribution in terms of roentgens}$$

Then

$$D = \int_0^{E_{\max}} R(E) \cdot f(E) \cdot dE$$

and

$$R = \int_0^{E_{\max}} R(E)\, dE$$

Therefore $\bar{f} = D/R$, the average rad per roentgen over the entire photon spectrum.

3-5 APPLICATIONS OF BASIC CONCEPTS

To apply the basic concepts of exposure and absorbed dose to several practical problems, it is first necessary to define a few terms.

R_a = exposure in roentgens at a point p in air
= air dose at p

R_s = exposure in roentgens at point p if the surface of a body or phantom is at the point
= surface dose at p

R_d = exposure in roentgens at some point within the body or within the phantom, measured by using an R-meter

R_s/R_a = backscatter factor

$(R_d/R_s) \times 100$ = percent depth dose

EXAMPLE 1. Given: An X-ray beam of 250 KVP with a field area of 100 sq cm and HVL = 1 mm Cu. Assume $\bar{f} = 0.95$ (soft tissue), $\bar{f} = 1.76$ (bone), and the backscatter factor = 1.36. Given that $R_a = 1000$ R, then $D_a = 0.877$ R = 877 rad. At the air–tissue interface:

$R_s = 1000 \times 1.36 = 1360$ R
$D_s = 0.877 \times 1360 = 1193$ rad in air
$D_s = 0.95 \times 1360 = 1290$ rad in tissue

At the 80% isodose level:

$$D_t = 0.80 \times 0.95 \times 1360 = 1030 \text{ rad in tissue}$$
$$D_b = 0.80 \times 1.76 \times 1360 = 1920 \text{ rad in bone}$$

This example illustrates the way in which we can obtain the absorbed dose from the exposure. In summary: (1) Measure the exposure in air at the point where the surface of the tissue is to be placed. (2) Find the exposure at the depth of interest using isodose curves. (3) Calculate the absorbed dose using Eq. (3-7).

EXAMPLE 2. Given: The depth dose distribution along the axis of an X-ray beam of energy 500 KVP, as shown in Fig. 3-4. At a depth of 5 cm,

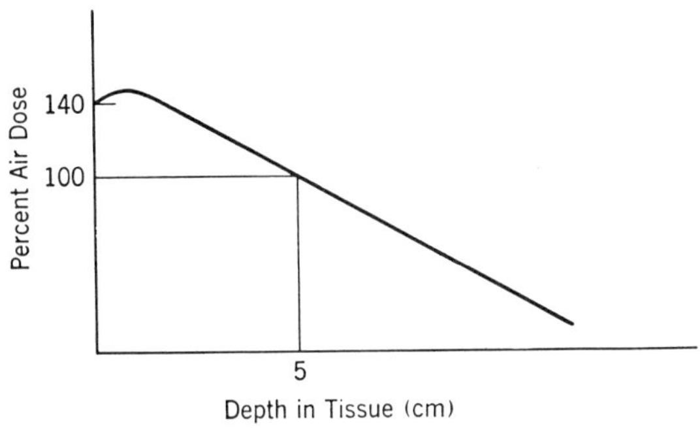

Fig. 3-4. See Example 2.

the depth at which one finds the blood-forming organs, the exposure is identical to that in free air at the point where the surface of the body lies. Also, since the energy falls in the Compton region of the spectrum, $f \cong 1$. Therefore, since $D_t = fR$, $|D_t| = |R|$.

The exposure in roentgens in free air at a point is equal in magnitude to the absorbed dose in rad to the blood-forming organs when the surface of the body is placed at the same point. The health physicist makes much use of this fact in radiation surveys.

REFERENCES

1. "Radiation Quantities and Units," *ICRU Rept. 10a*, published as *Natl. Bur. Std.* (*U.S.*) *Handbook* **84** (1962).

2. "Report of the RBE Committee to the International Commissions on Radiological Protection and on Radiological Units and Measurements, *Health Phys.* **9**, 357 (1963).

PROBLEMS

1. Precise measurements with an ionization chamber show that a single 5.30 MeV α-particle from ^{239}Pu produces 1.47×10^5 ion pairs in standard air. What is the average energy required to produce a single ion pair?

2. A patient is given a local therapeutic exposure of γ-radiation of 100R. If the mass absorption coefficient for air corresponding to the radiation is 0.10 cm^2/gm, calculate the amount of energy per square centimeter that passed into the patient.

3. What is the allowable working time in a radiation field consisting of 25 mrad/hr γ-rays, 15 mrad/hr fast neutrons, and 50 mrad/hr thermal neutrons? Assume a maximum permissible dose equivalent of 100 mrem.

4. What is the mass in grams of 1 c of 5.2-yr ^{60}Co?

5. Find the activity in curies of 0.1 gm of ^{131}Ba. The half-life is 12 days.

6. The total mass-absorption coefficient of lead for 300 keV γ-rays is 0.42 cm^2/gm. Find the linear attenuation coefficient.

7. Assume that hard γ-rays are absorbed in air by the Compton effect. If the linear absorption coefficient for Compton absorption for 2 MeV photons is 3×10^{-5} per centimeter of standard air, calculate the energy absorbed (MeV/cm^3-sec) by a cubic centimeter of air upon which is incident a γ-ray flux density of 10^6 photons/cm^2-sec. What is this in terms of milliroentgens per second?

8. A γ-ray beam falls on a thin-window Geiger counter. The counter has an efficiency of 1.5% for this 2 MeV γ-ray. The flat-plate area of the counter is 5 cm^2. How many counts per minute would you expect to correspond to 1 mR/hr?

9. A standard thimble chamber is exposed to a radiation field and collects a charge of 3×10^{-11} coul in 10 sec. If the volume of the chamber is 1.2 cc, what was the value of the exposure rate in roentgens per second?

10. A substance contains 30% C, 40% N, and 30% O, by mass. Find the effective atomic number at low energy (photoelectric region) and at intermediate energy (Compton region).

11. A person has been occupationally exposed from age 18 to age 24. He has received a total accumulated dose equivalent of 9 rem. How much may he receive in the next year? How many years could this person be exposed at the rate of 3 rem/13 weeks?

12. A person at age 30 has received the amount of radiation permitted according to $D = 5(N - 18)$. In the following 13-week period he receives 2.5 rem. How much may he receive in the next 13-week period?

13. A technician who is handling radium stands behind a protective shield so that only his hands and forearms are exposed. He receives 0.5 rem at each handling. During a 13-week period, how many times would he be permitted to carry out this procedure?

CHAPTER 4

The Physical Basis of Radiation Dosimetry

R. H. Ritchie

4-1 INTRODUCTION

Implied in the term radiation dosimetry is the central problem in the field. One is given an environment consisting of corpuscular radiations distributed in an unknown fashion with respect to quantity, nature, direction, and energy. The problem of dosimetry is that of relating physical measurements of this environment to doses which are important from the standpoint of radiation physics and radiobiology.

We shall consider two approaches to radiation dosimetry. The first gives primary attention to the radiation field itself; in concept, measurements of the energy-angle distribution of particles in the field could be carried out to obtain a complete description of the radiation environment in a given locale. This information could be employed, for example, to compute the detailed distribution of radiation dose within a human placed there. The second characterizes the field primarily in terms of its interaction with matter located in the region of interest. This approach is of great practical importance and has formed the basis for almost all of radiation dosimetry.

The distinction between the two approaches is somewhat arbitrary in that we know the radiation field only through its interaction with matter. The point of the second approach is to avoid a complete characterization of the field by taking advantage of the fact that the doses of interest are usually integral ones, requiring sums over the various spectral components of the field, weighted by more or less well-known importance factors at the different energies. Thus, instead of obtaining a complete description of the energy distribution of the field merely to perform a sum over this distribution to find the dose, we try to utilize methods by which the sums are in a sense automatically performed.

The term *dose* is biologically oriented and originated during the early medical use of X-rays. It is now defined in terms of physically measurable quantities. The ideal unit of dose from the standpoint of radiation protection and biophysics would be one which would produce the same biological effect independent of the kind and energy of the radiation. Such an ideal is probably unattainable because of the extreme complexity of radiation-induced damage in living systems. In practice, we employ a physical dose unit which gives nearly the same biological effect independent of the energy of a given kind of radiation.

A reasonable choice for dose is the energy absorbed by (or alternatively imparted to) a unit mass of matter. Hence a unit like the rad occupies a useful position in modern dosimetry, even though more sophisticated measures of radiation fields, such as linear energy transfer (LET), have assumed important positions in recent years. Definitions of various units of radiation dose were given in Chapter 3.

In this chapter the physical concepts upon which the second approach to dosimetry is based will be discussed. A well-known method for relating a physical measurement in a radiation field to a dose in that field is embodied in the Bragg-Gray principle. This principle is important historically and in the practical design of many γ-ray and neutron dosimeters. It has been considered in great detail by various authors since its original formulation and a discussion follows in the framework of a more general principle of radiation dosimetry.[13]

4-2 VARIATION OF DOSE WITH ENERGY

The first-collision dose to a given material and its variation with the energy of the impinging corpuscular radiation are important concepts in radiation dosimetry. In the physical situation implied by the term "first collision" a beam of ionizing radiation is incident on a mass so small that the attenuation of the beam in the material is small. The first-collision dose is usually expressed in terms of the energy imparted to a unit mass of the material per unit time and per unit flux of the incident beam.

The central importance of this quantity is due (in addition to its simplicity) to the fact that many radiation measuring instruments have an energy response approximating the first-collision dose–energy relationship. Such an instrument may be used to measure the first-collision dose in a free radiation field or, equally important, it may be employed in the form of a small probe to measure the dose inside a large mass of material placed in the field. In the latter situation, it gives a physical measure of the dose to a small element of mass inside the larger one in a region where the dose may have a large spatial gradient and a magnitude quite different from the free field value.

Because of the importance of the first-collision dose to radiation dosimetry, numerical data follow for certain important materials. A more detailed discussion of tissue dose is given in Chapters 3 and 8.

First-Collision Dose from Gamma Radiation

The three important ways in which γ-rays interact with matter in the range of photon energies from 10 keV to 10 MeV are: (1) the photoelectric effect, (2) Compton scattering, and (3) pair production. These and other basic photon processes of less importance to dosimetry are discussed in Chapter 2.*

In the photoelectric process, the ejected electron acquires kinetic energy equal to the photon energy minus the binding energy of the electron in its initial state in the atom. With Compton scattering, the struck electron may assume different kinetic energies depending upon the angle of scatter of the photon. At energies above twice the electron rest energy ($2mc^2$), where pair production is possible, the photon energy in excess of this value is taken up as kinetic energy by the electron-positron pair.† For purposes of calculating first-collision dose, we assume that the energy of secondary ionizing radiations, for example, electrons and positrons, liberated in these processes is dissipated locally.‡ This is a good approximation in the energy range to be considered here, where we may choose a mass of material to be small compared with the attenuation length of the primary γ-rays and large compared with the range of the electrons.

Defining $D^{(\gamma)}(E)$ to be the first-collision dose from γ-rays of energy E in units of rad/(photon-cm^{-2}), we may write for an arbitrary material composition,

$$D^{(\gamma)}(E) = 1.602 \times 10^{-8} \sum_i N_i\{\tau_i(E)\epsilon_i^{pe}(E) + \sigma_i(E)\epsilon^c(E) + k_i(E)\epsilon^{pp}(E)\} \tag{4-1}$$

where N_i = number of atoms of the ith element per gram of the material,
$\tau_i(E)$ = photoelectric cross section of the ith element (in square centimeters per atom)

* See also the brief summary in Ref. 6 for a description of these phenomena as they bear on the subject of radiation shielding.

† The energy of annihilation photons is not included in the definition of first collision dose, since their attenuation in matter occurs in a distance comparable with the attenuation length of the primary photons.

‡ That is, in a distance which is small when compared with the mean range of the primary radiation.

$\sigma_i(E)$ = Compton scattering cross section of the ith element (in square centimeters per atom)

$k_i(E)$ = pair-production cross section of the ith element (in square centimeters per atom)

$\epsilon_i^{pe}(E)$ = average kinetic energy acquired by electrons ejected from atoms of the ith element (in MeV); $\epsilon_i^{pe} = E - \bar{E}_B{}^i$, where $\bar{E}_B{}^i$ is the average binding energy of electrons in the ith atom

$\epsilon^c(E)$ = average kinetic energy transferred to electrons in the Compton scattering process (in MeV); $(\epsilon^c E) = E\left\{1 - \frac{3}{8}\left[\frac{\ln(1 + 2E_0)}{E_0{}^3} + \frac{2(2E_0{}^3 - 3E_0{}^2 - 1)}{E_0{}^2(1 + 2E_0)^2} + \frac{8E_0{}^2}{3(1 + 2E_0)^3}\right]\right\}$

where $E_0 = (E/0.511 \text{ MeV})$

$\epsilon^{pp}(E)$ = average kinetic energy absorbed by the electron positron pair (in MeV); $\epsilon^{pp} = E - 1.022$.

The summation extends over all types of element i in the target. The factor 1.602×10^{-8} converts MeV per gram to rad. Both ϵ^c and ϵ^{pp} have been taken to be independent of the kind of atom, because in the range of energies where these processes are important the photon energy is much larger than the electronic binding energy.* This is not necessarily true for the photoelectric process, where for some elements both K- and L-shell photoelectric processes may be important. In this case an average binding energy $\bar{E}_B{}^i$ must be employed.

Figure 4-1 shows the relation between first collision dose in rad/(photon-cm^{-2}) and the photon energy for soft tissue, bone, air, and carbon. An approximate expression for the first collision tissue dose–energy relationship is

$$D^{(\gamma)}(E) = 3.7 \times 10^{-9}E \quad \text{rad/(photon-cm}^{-2}) \tag{4-2}$$

where E is the photon energy in MeV. This equation is valid to $\pm 30\%$ in the energy range $0.05 \text{ MeV} < E < 7 \text{ MeV}$. The ratios of the rad doses in bone, air, and carbon to that in tissue at various energies are given in Fig. 4-2.

The salient aspects of these curves are easily understood. At low energies the dominant interaction between photons and atoms results in photoelectric ejection of electrons. The cross section for this process increases very rapidly as the photon energy decreases, and varies with Z, the atomic number of the atom, approximately as Z^n at a given photon energy. Here n is roughly 4 or 5. Thus in the energy range below ~0.03 MeV the first

* At these energies the molecular state of the various atoms in the material is not important, since these binding energies are of the order of a few volts.

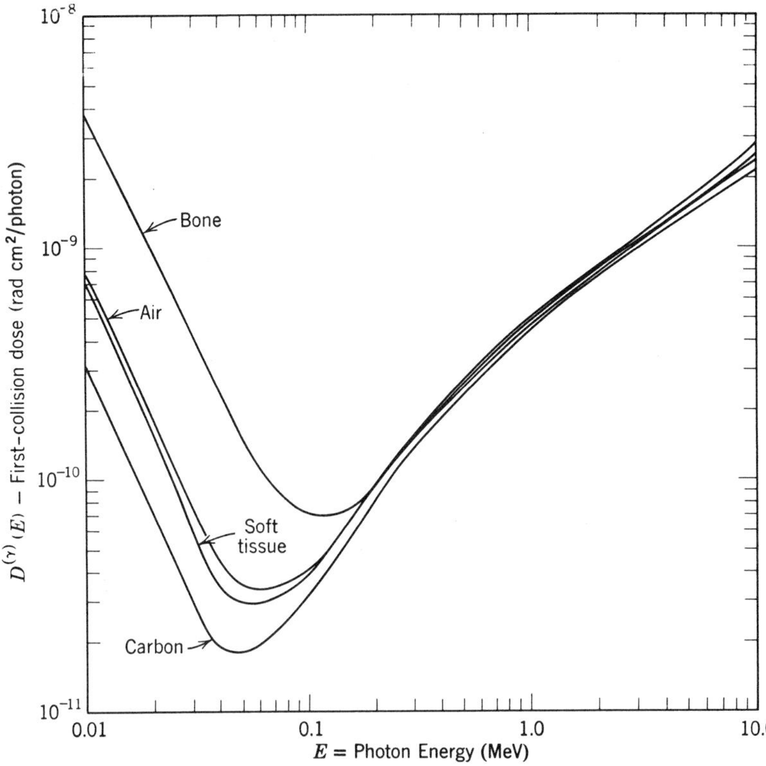

Fig. 4-1. Variation with γ-ray energy of the first-collision γ-dose to bone, air, soft tissue, and carbon.

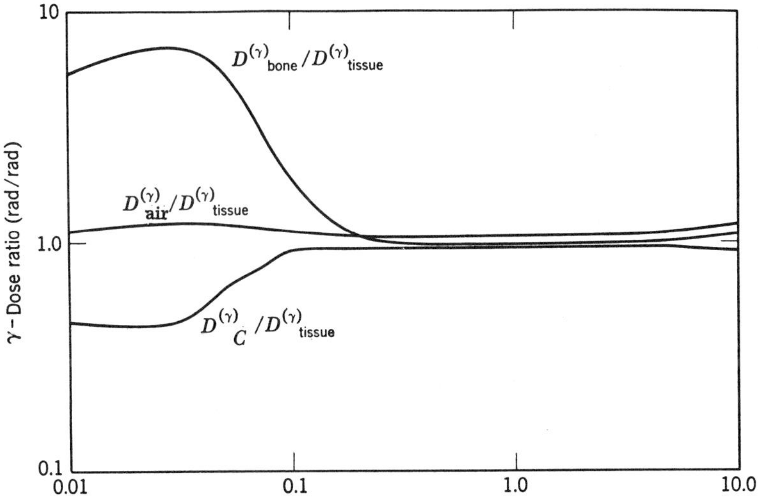

Fig. 4-2. First-collision γ-dose ratios for several substances as a function of γ-ray energy.

collision dose to bone is an order of magnitude larger than that to soft tissue because of the presence of relatively high-Z atoms in the former. At photon energies in the region 0.02–1 MeV the dominant interaction is Compton scattering. Since the cross section for this process is proportional to the total number of electrons in a given atom and since the electronic density in most materials is approximately constant, we find only minor differences in dose in this energy range. Similar considerations hold for higher energies where pair production becomes important.

First Collision Dose for Neutrons

The first-collision neutron dose $D^{(n)}$ imparted to a particular substance when irradiated by neutrons of energy E may be written

$$D^{(n)}(E) = 1.602 \times 10^{-8} \sum_i \sum_j N_i \sigma_{ij}(E) \bar{\epsilon}_{ij}(E) \tag{4-3}$$

where N_i = number of nuclei of type i per gram of substance

σ_{ij} = cross section of the ith kind of nucleus for the reaction in which ionizing particles of type j are produced

$\bar{\epsilon}_{ij}$ = average kinetic energy acquired by the jth kind of ionizing secondary ejected in an interaction between a neutron and the ith type of nucleus

The averaging process used to obtain $\bar{\epsilon}_{ij}$ is understood to be carried out over all possible energies which the jth kind of particle may acquire. In the simple case of elastic scattering which is isotropic in the center-of-mass system of the neutron and the struck nucleus, we have

$$\bar{\epsilon}_{ij}(E) = \frac{2mM_i}{(m + M_i)^2} E \tag{4-4}$$

where m and M_i are the masses of the neutron and the ith kind of nucleus, respectively. This kind of elastic encounter obtains for the scatter of neutrons on hydrogen nuclei in the range $0 < E < 14$ MeV.* For nuclei heavier than hydrogen, anisotropic center-of-mass scattering sets in at considerably lower energies. In general, the larger M_i becomes the more limited is the range of energies in which Eq. (4-4) is valid. However, in substances rich in hydrogen, such as soft tissue, the first-collision dose may be computed with reasonable accuracy up to ~10 MeV using Eq. (4-4) to compute the energy imparted in elastic encounters with all nuclei. This is so because collisions of neutrons with hydrogen nuclei are responsible for most of the energy absorbed in this case (see Fig. 4-3).

* When E becomes as small as a few eV, the chemical binding of hydrogen nuclei in molecules becomes important and this relation is no longer valid.

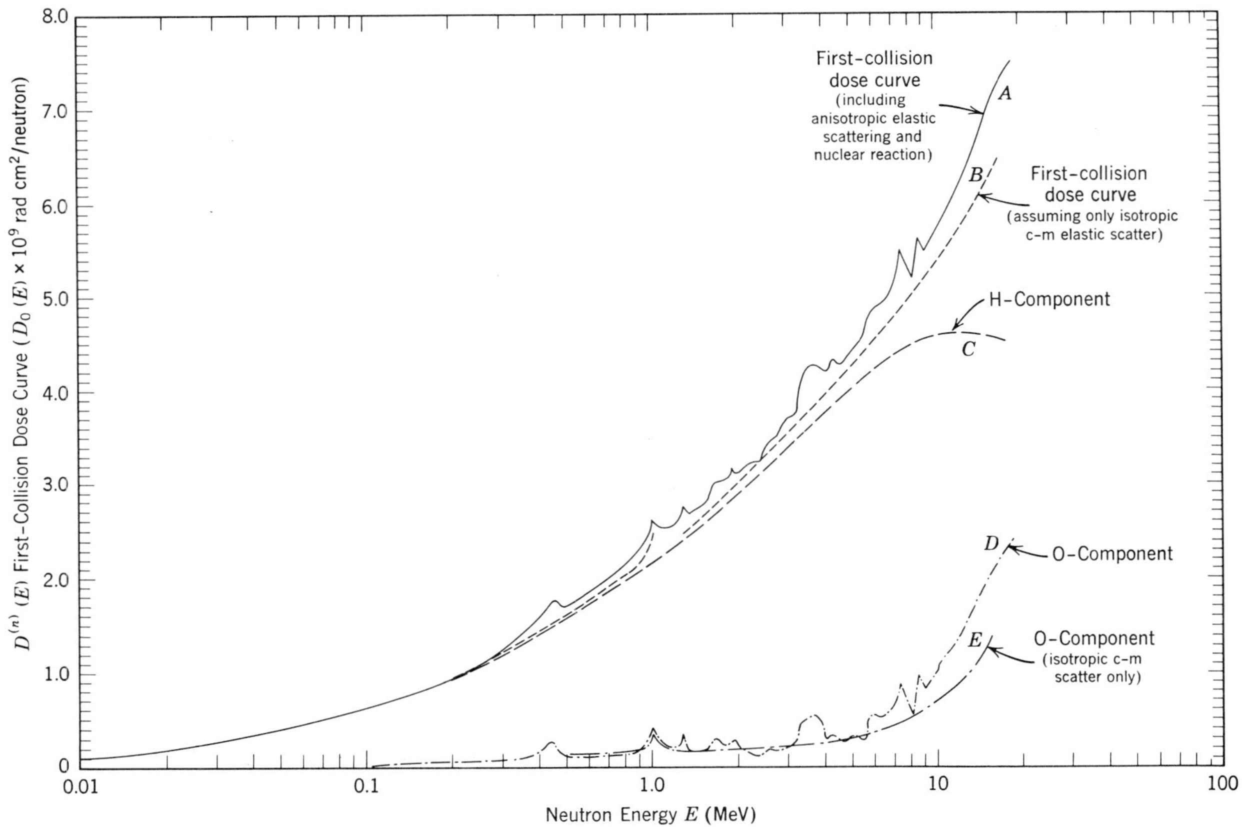

Fig. 4-3. First-collision neutron dose to soft tissue as a function of neutron energy. Curve A is calculated with consideration of anisotropic elastic scattering and inelastic scattering. Calculation of curve B assumed isotropic center-of-mass scattering and neglected inelastic processes. Also shown are various components of the total dose. Curve C shows the dose to hydrogen in soft tissue. Curve D gives the dose to oxygen including forward elastic and inelastic processes while Curve E shows the dose to oxygen calculated on the isotropic center-of-mass model.

If elastic scattering is anisotropic in the center-of-mass system (sometimes called "forward" elastic scattering), we may write

$$\bar{\epsilon}_{ij}(E) = \frac{2mM_i}{(m + M_i)^2} [1 - f_1^{(i)}(E)]E \tag{4-5}$$

where the differential elastic scattering cross section of the ith nucleus in center-of-mass coordinates is expanded as follows:

$$\sigma_i^e(E, \theta) = \frac{\sigma_{i0}^e(E)}{4\pi} \sum_{l=0}^{\infty} (2l + 1) f_l^{(i)}(E) P_l(\cos\theta) \tag{4-6}$$

In this formula $\sigma_{i0}^e(E)$ is the total elastic scattering cross section of the ith nucleus ($f_0^{(i)} = 1$), $P_l(x)$ is the lth Legendre polynomial, and θ is the scattering angle.

If the neutrons induce a nuclear reaction of the ith kind of nucleus in which a charged particle is emitted with reaction energy Q_{ij}, then

$$\bar{\epsilon}_{ij}(E) - (E + Q_{ij}) \operatorname{St}(E + Q_{ij}) \tag{4-7}$$

where Q_{ij} may be either positive or negative, depending upon whether the reaction is exoergic or endoergic, and St (x) is the unit step function.*

For purposes of computing first-collision dose, γ-rays emitted subsequent to inelastic neutron scattering on nuclei are not considered, since these are not absorbed locally.

Figure 4-3 shows the first-collision dose to soft tissue as a function of neutron energy. Curve A was calculated by accounting for isotropic center-of-mass scattering on oxygen and nitrogen, together with the reactions $O^{16}(n, \alpha)C^{13}$ and $O^{16}(n, p)N^{16}$, using data available in report form [Refs. 14, 18, 23, and 175]. Curve B was computed by neglecting both anisotropic scatter and nuclear reaction products. There are only small differences between these curves, even at the highest energy considered. There is a partial compensation between the effect of forward elastic scatter, which usually tends to make the energy absorbed less than it is when scatter is isotropic, and the effect of nuclear reactions. With the exception of the $^{14}N(n, p)^{14}C$ process, the important nuclear reactions in soft tissue are endoergic and have thresholds in the few-MeV region, where forward elastic scatter also begins to be important.

The doses to the hydrogen and oxygen components in soft tissue are shown in Curves C and D, respectively. Curve E shows the oxygen component computed by assuming only isotropic elastic scatter to occur. One concludes that the first-collision dose to matter containing an appreciable atom-number fraction of hydrogen may be computed with good accuracy by considering only elastic isotropic scatter.

* St $(x) = 0$ for $x < 0$, and St $(x) = 1$ for $x > 0$.

Figure 4-4 gives the first-collision dose to CH_2, H_2O, bone, and carbon. Isotropic center-of-mass scattering was assumed for the first three of these substances; anisotropic scattering data were used for carbon. Figure 4-5 shows the ratio of CH_2 dose to soft-tissue dose as a function of energy. Both doses are first-collision values computed by assuming isotropic center-of-mass scattering for all elements. This ratio varies in an irregular way with energy, but the variations are rather narrow in width and, when averaged over a reasonably smooth spectrum of energies, may be considered constant at $\sim$1.45 to a good approximation for many purposes. CH_2 is important since it is used in the Hurst absolute fast neutron dosimeter described in Chapter 6.

4-3 BRAGG-GRAY PRINCIPLE

Physical measurements in a radiation field must be related to quantities of importance in dosimetry. A method for obtaining the dose to the walls of a small gas-filled cavity from ionization measured in the cavity is contained in the Bragg-Gray principle. This principle has assumed great practical importance in the design of γ-ray dosimeters. It is discussed only briefly here because it is considered as a special case of a more general dosimetry principle below.

The use of cavity-ionization chambers for the measurement of X- and γ-ray dose was put on a firm basis by the careful work of Gray,[7,8] who also considered the cavity chamber for the measurement of fast-neutron dose.[9] He showed that for a cavity chamber, the "Bragg-Gray" relation may be written

$$D^{(\gamma)} = \frac{WJS}{100} \tag{4-8}$$

where $D^{(\gamma)}$ = dose to the wall in rad

W = energy in ergs required to produce an ion pair in the gas contained in the cavity

J = number of ion pairs formed per gram of the gas

S = mass stopping power of the wall material relative to that of the gas for the ionizing particles associated with the primary radiation.

Gray concluded from a detailed examination of data then available that W was a constant for electrons in air irrespective of their energy. The value he gave was 32.5 eV/ion pair, whereas more recent data indicate that a value somewhat less than 34 eV is probably more nearly correct. The independence of electron energy seems valid for electrons of initial energies above $\sim$0.5

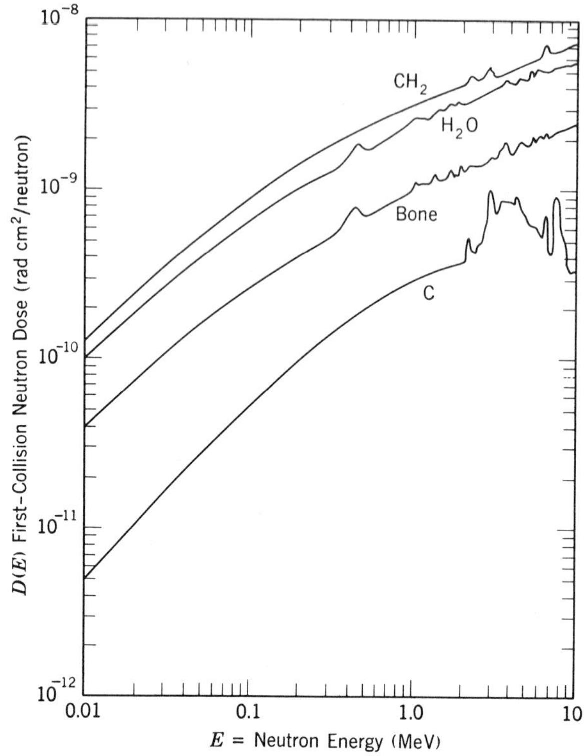

Fig. 4-4. First-collision neutron dose to CH_2, H_2O, bone, and carbon as a function of neutron energy. Isotropic center-of-mass scatter was assumed in calculating all of these data.

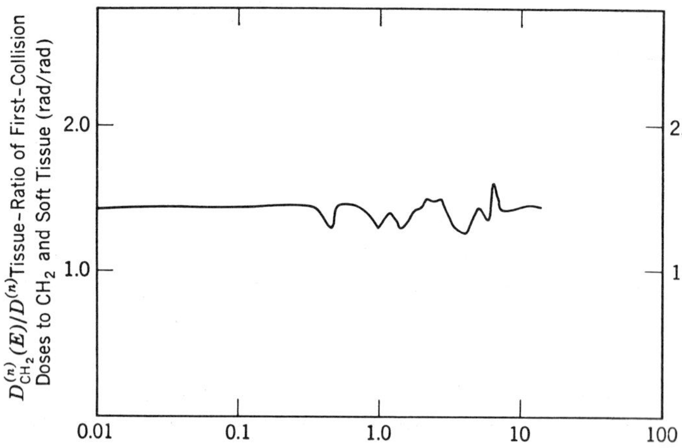

Fig. 4-5. The ratio of first-collision neutron dose in CH_2 to that in soft tissue as a function of neutron energy.

*k*eV, and W for both electrons and α particles may vary rather markedly from one gas to another. A more complete discussion of W is given in Section 5-2.

The original treatment of Gray assumed that S was the ratio of stopping power in the wall material relative to that in the gas and that S was not dependent upon energy. Laurence[16] later made a natural extension of this treatment, taking S to be averaged over the spectrum of ionizing particles that cross the cavity. Other workers have considered the production of energetic δ-rays in the cavity to obtain still another approximation to S.* It is found that where the gas and the cavity wall are reasonably close in atomic number, the use of an average S gives an excellent approximation for most purposes.

For the Bragg-Gray relation to hold, it is necessary that (1) the dimensions of the cavity be small compared with the range of the secondaries in the cavity and (2) charged-particle equilibrium shall obtain, that is, that the wall thickness be greater than the range of the most energetic secondaries in the wall and small compared with the attenuation length of the primary radiation in the wall. If the cavity gas and wall material have the same atomic composition, the first requirement may be relaxed. In this case, neglecting the effect of chemical binding upon stopping power, which is known to be small as long as the velocity of the secondaries is not too low, we may write $S = \rho$, where ρ is the ratio of the density of the wall material to that of the gas.

4-4 A GENERAL PRINCIPLE OF RADIATION DOSIMETRY

Consider a radiation field consisting of a particular kind of corpuscular radiation distributed in energy and angle according to the function $N(E, \hat{\Omega})$, where $N(E, \hat{\Omega})$ is the number of particles per unit energy at energy E and per unit solid angle in the direction of the unit vector $\hat{\Omega}$. The interaction of this field with a medium can be characterized in a number of ways. Among these are (1) the locally absorbed dose in an infinitesimal volume at an arbitrary point in the medium, (2) the distribution of absorbed dose with stopping power (the LET distribution), (3) special cases of (1), such as the absorbed dose at the point of maximum dose in the medium or the dose at some critical point in the medium, (4) the energy imparted (as contrasted to the energy absorbed) per gram in an infinitesimal volume element of the medium, for example, the first-collision dose, (5) the dose significant for an important biological effect to a living medium, or (6) the biological dose for a specified medium obtained from an assumed relation between RBE and LET (the rem dose from particles of various specific ionization). For future

* Useful reviews of these developments and other aspects of cavity chamber theory are given in Reference 19.

reference a particular dose–energy relationship will be designated by $D_{jk}(E)$ where j refers to the medium and k refers to the type of dose or dose distribution. The "dose," $D_{jk}(E)$, may be any unique function of radiation energy, E; in the more general sense it is not restricted to quantities having dimensions of energy/mass.

In principle, any of the doses may be calculated using the $N(E, \hat{\Omega})$ distributions for each kind of radiation incident on the medium. In practice, however, such calculations may be quite difficult to perform and, moreover, the input data may not be known with sufficient accuracy. Experimental methods, on the other hand, may be devised such that the lack of accurate data for the function $N(E, \hat{\Omega})$ does not invalidate the results. Specifically, dosimeters may be designed so that their response at various energies follows the dose–energy relationship for a particular one of the dose quantities over the interesting range of energies. Also the response to radiations other than the one desired may be made negligibly small by careful design.

Consider the interaction of a given kind of radiation with a test system. Suppose that a beam of this radiation having energy E and unit flux impinges on the system. There will then be generated in the test system a certain distribution of interaction products. These products may be recoil nuclei, recoil electrons, radioactive nuclei, etc. Let $n_{l,\beta}^{(\alpha)}(\epsilon_l, E)\, d\epsilon_l$ be the number of these products of type l which dissipate an amount of energy in the range of energies between ϵ_l and $(\epsilon_l + d\epsilon_l)$ in the αth subsystem of the test system. The quantity β is understood to symbolize a set of parameters $\beta_1, \beta_2, \ldots \beta_n$ which characterizes the configuration of the test system, that is, its dimensions, nuclear composition, the physical distribution of subsystems, and any other quantities required to specify it completely.

Suppose that we wish to measure the dose (or dose distribution) $D_{jk}(E)$ in this same beam of radiation. We may obtain a measure of the dose $D_{jk}(E)$ by utilizing a general test system which is unrelated, in a direct sense, to the jth material or the kth physical effect. In order to show how this may be done, we define a response $R_i^{(\alpha)}(E)$ of the αth subsystem in terms of an operator $O_i^{(\alpha)}(\epsilon_l)$ which acts on the distribution $n_{l,\beta}^{(\alpha)}(\epsilon_l, E)$ as follows:

$$R_i^{(\alpha)}(E) = \sum_l O_i^{(\alpha)}(\epsilon_l) n_{l,\beta}^{(\alpha)}(\epsilon_l, E) \tag{4-9}$$

(The functional dependence of $R_i^{(\alpha)}(E)$ upon the configuration parameters β is not shown explicitly.) This generalized response leads to considerable flexibility in practice, since for each operator O_i a different response is at our disposal. To illustrate the nature of the O_i, consider the following expression:

$$O_i^{(\alpha)} \equiv \int_0^\infty d\epsilon_l \epsilon_l \tag{4-10}$$

Then the single-valued response $R_i^{(\alpha)}(E)$ is

$$R_i^{(\alpha)}(E) = \sum \int_0^\infty d\epsilon_l \epsilon_l n_{l,\beta}^{(\alpha)}(\epsilon_l, E) \equiv \epsilon_T^{(\alpha)}(E) \tag{4-11}$$

where $\epsilon_T^{(\alpha)}(E)$ is the total energy absorbed in the subsystem. This will be recognized as the conventional quantity measured in Bragg-Gray dosimetry since ionization-chamber methods, calorimetric methods, and chemical methods all give a measure of the total absorbed energy. In contrast, if we use pulse-counting methods in which the energy-distribution function $n_{l\beta}^{(\alpha)}(\epsilon_l, E)$ is available, we may employ different operators $O_i^{(\alpha)}$, each of which leads to a new response $R_i^{(\alpha)}(E)$. Several examples of the use of the energy-loss operator to generate various response functions follow.

A generalization of dosimetry may now be stated as follows.

> Let operators $O_i^{(\alpha)}$ act upon the energy-loss distribution function $n_{l,\beta}^{(\alpha)}(\epsilon_l, E)$ generated in a test system having a suitable configuration $\beta_1, \beta_2, \ldots, \beta_n$. A given dose function $D_{jk}(E)$ is measurable with error less than δ in an energy range $E_1 \leq E \leq E_2$, provided there exist operators $O_i^{(\alpha)}$ and geometrical parameters $\beta_1, \beta_2, \ldots, \beta_n$ such that the inequality
>
> $$\left| \sum_{l,\alpha} O_i^{(\alpha)}(\epsilon_l) n_{\beta,l}^{(\alpha)}(\epsilon_l, E) - D_{jk}(E) \right| \leq \delta \tag{4-12}$$
>
> is satisfied for all energies E in the specified range.

In carrying out the matching process, we may choose an operator O_i and find a configuration $\beta_1, \beta_2, \ldots, \beta_n$ such that the relation (4-12) is satisfied, or alternatively we may fix the β and find an appropriate operator $O_i^{(\alpha)}$. The generality of the concept expressed by Eq. (4-12) may be emphasized by noting three important points: (1) any type of dose for any kind of matter, D_{jk}, may be considered, (2) the energy operator, $O_i(\epsilon)$, may be chosen arbitrarily, and (3) the geometrical parameter of the test system, β, may be less restricted, depending on the choice of energy operator $O_i(\epsilon)$.

Summary of Methods Based on First Moment-of-Energy Distribution

Important special cases for $R_i(E)$ result when the operator uses a moment of order m of the energy ϵ, that is

$$O_i^{(\alpha)} = \int_B^\infty d\epsilon_l \epsilon_l{}^m \tag{4-13}$$

where B is a lower limit to the energy loss. Many of the now-existing methods of dosimetry may be classified as the special cases where $m = 1$ and $m = 0$. These two cases are considered in detail.

In Eq. (4-13), if $m = 1$, then $R_i(E) = \epsilon_T(E)$; calorimetric, chemical, and all total-ionization methods may be classified in this category.

Calorimetric Method. Radiation energy absorbed in the test medium is largely degraded to heat, and hence the change in temperature of the medium may be used to determine the amount of energy absorbed. The temperature change is given by

$$\Delta T = \left(\frac{1}{c}\right) \sum_l \int_0^\infty \epsilon_l n_{\beta l}(\epsilon_l, E) F_l(\epsilon_l)\, d\epsilon_l \qquad (4\text{-}14)$$

where $\epsilon_l n_{\beta l}(\epsilon_l, E)\, d\epsilon_l$ is the amount of energy absorbed in a unit mass of the medium characterized by a geometrical parameter β by charged particles of type l having energy between ϵ_l and $(\epsilon_l + d\epsilon_l)$; $F_l(\epsilon_l)$ is the fraction of the charged-particle energy which is degraded to heat; and c is the thermal capacity of the substance. To a good approximation $F_l(\epsilon_l)$ is a constant near unity. Thus,

$$\Delta T = \frac{\epsilon_T}{c} \qquad (4\text{-}15)$$

where ϵ_T is the total amount of energy absorbed in the calorimetric system.

Chemical Method. When energy is absorbed in many chemical systems,* the chemical composition is altered to such a degree that dose measurements may be readily made in terms of these changes. If Y indicates the yield of some observable chemical change, then

$$Y = \sum_l \int_0^\infty \epsilon_l n_{\beta l}(\epsilon_l, E) G_l(\epsilon_l)\, d\epsilon_l \qquad (4\text{-}16)$$

where $G_l(\epsilon_l)$ indicates the yield per unit of energy absorbed by particles of type l at energy ϵ_l. In some cases the value of G is essentially independent of particle type and ϵ_l; hence,

$$Y = G\epsilon_T \qquad (4\text{-}17)$$

where ϵ_T is the total amount of energy absorbed per gram in the chemical system.

Ionization Method. The most sensitive and most widely used method of dosimetry is based on the ionization of gases. In this case the number of ion pairs I corresponding to $\epsilon > B$ is given by

$$I = \sum_l \int_B^\infty \frac{\epsilon_l n_{\beta l}(\epsilon_l, E)\, d\epsilon_l}{W_l(\epsilon_l)} \qquad (4\text{-}18)$$

where $W_l(\epsilon_l)$ is the energy required for a particle of type l at energy ϵ to produce a pair of ions. The fact that W for many gases is almost independent

* The term chemical system is used in the general sense to include photographic film, silver-activated glass, etc.

of l and ϵ_l over wide ranges of ϵ_l is well known. Hence if B approaches 0,

$$I = \frac{\epsilon_T}{W} \tag{4-19}$$

where ϵ_T is the total amount of energy absorbed in the gas.

We have set the operator O_i and the value of B for the preceding methods of dosimetry,* and we must now comment on the geometrical arrangement required to satisfy the relation (4-12). When we desire to measure dose to type (1), that is, absorbed dose at an arbitrary point within an irradiated medium, the dimensions of the test material (that is, the detecting element of the calorimeter, chemical system, or ionization chamber) must be small compared to the attenuation length of the primary radiation.

If, in addition, the detector dimensions are small compared to the range of the charged particles, it is only necessary that the materials of the detector be such that the ratio of the stopping power in the materials of the detector to that of the irradiated medium is independent of the energy of the charged particle. Often it is not practical to make the detector dimensions small compared to the range of charged particles. In this case we must add to the above requirement on stopping power the more stringent requirement that the ratio of the first-collision dose in the materials of the detector to that in the irradiated medium be independent of the energy of the primary radiation. Thus, when absorbed energy is used as the response of the test system, the choice of materials which may be used in the detector is limited, depending on the composition of the irradiated medium.

Another difficulty arises from the fact that detectors that are large enough to establish charged-particle equilibrium are often too large to serve as satisfactory radiation probes. Hence these devices are used in air and measure absorbed-dose quantities that are characteristic of their geometry. Efforts are sometimes made to interpret from these readings the first-collision dose. This interpretation requires calculations which take into account buildup and attenuation of the primary radiation as a function of primary radiation energy.

It may be well to elaborate on the preceding remarks as they apply to the ionization method. The various theories of cavity ionization† all deal with rather detailed consideration of an ideal situation, that is, the relationship between (1) the ionization in a cavity which is small compared to the range of charged-particle secondaries and (2) the energy absorbed per gram of the enclosing medium. In practice we cannot always create a small cavity at the

* When I is measured with ionization chambers, B approaches 0. However, in the counting method using the Bragg-Gray principle described in Reference 15, a finite value of B is used so that fast neutrons can be measured in the presence of γ-radiation.

† Compare, in particular, References 7, 8, 9, 16, and 19.

point of interest as required by the idealization described. In the practical case this cavity is fixed within a detector and hence is surrounded with the material of the walls of the detecting device. Usually the thickness of the detector walls is made greater than the range of the charged particles so that its readings in air are approximately the first-collision dose. However, such methods can be accurate only over a limited range of radiation energies. For example, the equilibrium dose for γ-rays is meaningful only when $E_\gamma < 3$ MeV and for fast neutrons only at energies less than 10 to 20 MeV. It will be seen in following sections that these restrictions are not inherent in the more general approach expressed by the relation (4-12).

Summary of Methods Based on Zero Moment-of-Energy Distribution

If in Eq. (4-13), and hence in Eq. (4-9), we let $m = 0$, we find that $R_i(E) = n_T(B, E)$ where $n_T(B, E)$ is the total number of energy losses greater than the energy B. This response function, because of the technical simplification it introduces, has been used extensively. It has been shown[12] that the first-collision tissue dose for neutrons in the energy range from 0.3 to 14 MeV may be obtained by simply counting the number of recoil proton pulses produced in a proportional counter. In this case the number of counts above a bias level B, required to make the instrument insensitive to γ-radiation, is governed by three sources of recoil protons so arranged that the count rate is directly proportional to the first-collision dose in tissue.

Dennis and Loosemore[3] have developed a proportional counter that is nondirectional in response. In this case the counts versus energy-response curve follows the recommendations of the ICRP (as given in *Brit. J. Radiol* Suppl. No. 6, 1955); thus a direct indication of maximum permissible dose is obtained.

Similarly, Cheka[1] has shown that a nuclear track emulsion can be used with suitable proton radiators such that the number of track counts is proportional to the maximum absorbed dose in a man, over a range of neutron energy.

Skjöldebrand[21] has developed a spherical detector whose count rate vs. energy response is adjusted to the multiple-collision neutron dose curve of Snyder and Neufeld.[22] Thus this instrument indicates the maximum dose that a man would receive if located in the radiation field.

DePangher and Roesch[2,4] have shown that a paraffin moderator and BF_3 counter can be arranged so that count rate is approximately proportional to first-collision tissue neutron dose rate over a range of neutron energies from 0.1 to 5.1 MeV. The arrangement is similar to the Hanson-McKibben long counter,[11] but irradiation is done with neutrons normally incident to the axis of the cylinder.

All of these examples deal with methods of measuring various types of fast-neutron doses. Another example illustrates the fact that count-rate methods may be applied to γ-dosimetry. Wagner and Hurst[24] have shown that the γ-ray exposure dose in roentgens can be measured over a wide range of energy with a small halogen-filled Geiger-Müller counter if it is suitably shielded with lead and tin. Because of the discharge mechanism involved, the only response function which can be used here is the number of counts in the detector.

A final example illustrates the situation where the response corresponding to $m = 0$ is applied to more than one subsystem, that is, $\alpha > 1$. Consider neutron threshold detectors[10,20] where it has been shown that the neutron dose can be obtained by determining the number of neutron interactions in various detectors. In Eq. (4-12) let

$$\sum_{l,\alpha} O_i^{(\alpha)}(\epsilon_l) n_{\beta l}^{(\alpha)}(\epsilon_l, E) = \sum_{\alpha} a^{(\alpha)} \int_0^{\infty} n^{(\alpha)}(\epsilon_l, E)\, d\epsilon_l = \sum_{\alpha} a^{(\alpha)} n_T^{(\alpha)}(E) \qquad \text{(4-20)}$$

where $n_T^{(\alpha)}(E)$ is the total number of neutron interactions in subsystem α leading to a measurable product (for example, activation or fission product), and $a^{(\alpha)}$ is an arbitrary coefficient for each subsystem α. From the references cited,[10,20] it may be seen that Eq. (4-12) is satisfied over a range of neutron energies.

The Use of the Second Moment

The instruments utilizing the second moment of a charged particle distribution have not yet been made. However, the conceptual design of such an instrument has been considered.[13] The procedure involves first selecting the operator O_i^{α} for the case $m = 2$ in Eq. (4-13). The response function is then

$$R_i(E) = \int_B^{\infty} \epsilon^2 n(\epsilon, E)\, d\epsilon \qquad \text{(4-21)}$$

We may first consider the problem of finding a geometrical arrangement β which yields a response with the second-moment operator that is closely proportional to (for example) the first-collision neutron dose to soft tissue. Calculations have shown that a response of this nature may be achieved, using a simple plane radiator and detector system irradiated normally by a beam of fast neutrons. Although the geometrical arrangement considered is rather special in that it is computed only for monodirectional neutrons, it is clear that a similar design could be carried out to obtain a detector having isotropic response to neutrons.

The problem of finding an electronic system which operates on the secondary ionization distribution according to Eq. (4-12) could be solved by using a binary scalar system with a particular discriminator voltage biasing arrangement patterned after an earlier device.[5]

The earlier device was designed to yield the first moment of a distribution of pulses fed into the system and achieved this by means of a simple but ingenious scheme in which the nth stage in the binary scaling circuit is biased so that it is sensitive to pulses larger than nV, where V is the bias on the first stage. The second moment may be achieved simply by making the bias on the nth stage equal to $\sqrt{n} \cdot V$.

Use of the Generalized Operator

It is interesting to consider a more general operator than any of those discussed above. The theoretical design of an instrument using such a generalized response has been carried out elsewhere[13] to illustrate the operator concept.

This design consists of two parts. First, we determine $n(\epsilon, E)$, the distribution of energy losses ϵ in a particular test system for a range of primary energies E. This distribution may be computed, or if this is not possible it could be measured experimentally. In this case, $n(\epsilon, E)$ was calculated for protons losing energy in a plane-parallel system containing hydrogen and irradiated normally by fast neutrons. Second, we design an electronic system which operates upon $n(\epsilon, E)$ to give a particular dose $D_{jk}(E)$. In the example considered, two different $D_{jk}(E)$ functions were attained by choosing properly a set of parameters in the electronic system.

The electronic system utilizes the principle of binary addition and subtraction, partly because this is simple to perform with existing instrumentation. The physical processes occurring in the system may be represented mathematically by the following expression for the system response as a function of E:

$$R(E) = A\sum_{p=1}^{N} R_p(E) = A\sum_{p=1}^{N}(-1)^{m_p}2^{r_p}\int_{E_1(p)}^{E_2(p)} n(\epsilon, E)\, d\epsilon \qquad (4\text{-}22)$$

where $R_p(E)$ is the response of the pth channel to neutrons of energy E and where A is a constant. This channel records recoil energies between $E_1(p)$ and $E_2(p)$. The number of recoils losing energy in this range is weighted by the factor $(-1)^{m_p}2^{r_p}$. The weight factor 2^{r_p} may be easily achieved by using a simple binary computational unit; r_p is zero or a positive integer, whereas m_p is either 1 or 2 and symbolizes subtraction and addition, respectively, performed by another electronic component.

Since there are five parameters which may be chosen here, r_p, m_p, $E_1(p)$, $E_2(p)$, and N, there is a rather large number of possible arrangements. One expects that the conditions of Eq. (4-12) may be satisfied for most dose functions over a wide range of energies. It is found that for the plane-parallel hydrogeneous radiation-detector system we may simulate either the first-collision tissue dose or the maximum permissible rem dose with good accuracy by a proper choice of the adjustable parameters.

REFERENCES

1. Cheka, J. S., *Nucleonics* **12,** No. 6. 40 (1954).
2. DePangher, J., *Double Moderator, Neutron Detector*, Hanford Laboratories, Rept. HW-54584, Richland, Washington (1957).
3. Dennis, J. A., and W. R. Loosemore, *A Fast Neutron Counter for Dosimetry*, Atomic Energy Research Establishment Report EL/R 2149, Harwell (1957).
4. DePangher, J., and W. C. Roesch, *Phys. Rev.* **100,** 1793 (1955).
5. Glass, F. M., and G. S. Hurst, *Rev. Sci. Instr.* **23,** 67 (1952).
6. Goldstein, H., *Fundamental Aspects of Reactor Shielding*, Addison-Wesley, Reading, Massachusetts, 1959.
7. Gray, L. H., *Proc. Roy. Soc.* (*London*) **A122,** 647 (1929).
8. Gray, L. H., *Proc. Roy. Soc.* (*London*) **A156,** 578 (1936).
9. Gray, L. H., *Proc. Cambridge Phil. Soc.* **40,** 72 (1944).
10. Hurst, G. S., J. A. Harter, P. N. Hensley, W. A. Mills, M. Slater, and P. W. Reinhardt, *Rev. Sci. Instr.* **27,** 153 (1956).
11. Hanson, A. O., and J. L. McKibben, *Phys. Rev.* **72,** 673 (1947).
12. Hurst, G. S., R. H. Ritchie, and R. N. Wilson, *Rev. Sci. Instr.* **22,** 981 (1951).
13. Hurst, G. S., and R. H. Ritchie, *Health Phys.* **8,** 117 (1962).
14. Hughes, D. J., and R. B. Schwartz, *Neutron Cross Sections*, Brookhaven Natl. Lab. Rept. BNL-325, 1958, 2nd ed.
15. Hurst, G. S., *Brit. J. Radiol.* **27,** 353 (1954).
16. Laurence, G. C., *Can. J. Res.* **A15,** 67 (1937).
17. Lustig, H., H. Goldstein, and M. H. Kalos, *Neutron Cross Sections of Nitrogen*, Nuclear Development Corporation of America Rept. NDA-86-1, White Plains, New York, 1957.
18. Lustig, H., *Differential Elastic Neutron Scattering Cross Section of Oxygen*, Nuclear Development Corporation of America, Rept. NDA-2111-3(A), White Plains, New York, 1959.
19. "Stopping Powers for Use With Cavity Chambers," *Natl. Bur. Std.* (*U.S.*) *Handbook* **79,** (1961).
20. Reinhardt, P. W., and F. J. Davis, *Health Phys.* **1,** 169 (1958).
21. Skjöldebrand, R., *J. Nucl. Energy* **1,** 299 (1955).
22. Snyder, W. S., and J. Neufeld, *Brit. J. Radiol.* **28,** 342 (1955).
23. Troubetzkoy, E. S., *Fast Neutron Cross Sections of Iron Silicon, Aluminum and Oxygen*, Nuclear Development Corporation of America Rept. NDA-2111-3(C), White Plains, New York, 1959.
24. Wagner, E. B., and G. S. Hurst, *Health Phys.* **5,** 20 (1961).

PROBLEMS

1. What is the absorbed dose rate (rad/hr) when a small mass of tissue is uniformly exposed to a beam of 1 MeV photons with a flux density of 10^7 photons/cm^2 sec?

2. Prove that the average fractional energy loss in the isotropic elastic scattering of neutrons on nuclei of mass number A is $f = 2A/(A + 1)^2$ where A is in units of the neutron mass. Assume that the probability of a neutron of energy E being scattered into dE' at E' is proportional to dE' when $(1 - \alpha)E \leq E' \leq E$, and is zero otherwise, where $\alpha = 4A/(A + 1)^2$.

3. Calculate the neutron response of a carbon wall–"carbon gas" ionization chamber for the neutron energies 1, 5, and 20 MeV. Assume that neutron scattering on carbon is isotropic in the center-of-mass system and neglect nonelastic processes ($\sigma_{\text{elastic}} = \sigma_{\text{total}}$). Use cross sections from Ref. 14.

4. For an exposure of 50 R, calculate the number of ion pairs produced and the energy absorbed in 1 gm of air. Use $W = 34$ eV/ion pair.

CHAPTER 5

Detection and Measurement of Ionization

L. W. Cochran

5-1 INTRODUCTION

It has long been known that radiation which has sufficient energy to ionize matter can damage living tissue. In fact, the earliest commonly used unit of radiation exposure dose, the roentgen, was defined in terms of the ionization produced in air. More recently, the trend has been to specify instead the absorbed dose measured in terms of the quantity of energy deposited by the radiation per unit of mass of irradiated material at the site of interest. Many of the practical devices employed in the determination of dose or dose rate depend upon an evaluation of the ionization produced in the detector by the radiation in question. In such instruments as ionization chambers and count-rate meters which employ gas-filled detectors we are concerned with the quantity of ionization produced by the radiation and with the subsequent history of the electrons and gaseous ions produced for information leading to the dose or dose rate being measured. Ionization in gases and the behavior of the products of ionization in typical detectors are the subjects of primary concern in this chapter.

5-2 CREATION OF IONS

Basic Ideas of Excitation and Ionization

When an ionizing particle traverses a gas, part of its energy is lost in ionizing the gas molecules and part is lost in processes other than ionization. Hence the number of ion pairs (electrons and positive ions) produced will always be less than the ratio of the energy lost by the ionizing particle and the ionization potential V_I of the gas. From the point of view of ion yield, a portion of the energy is wasted since the ion yield is the measurable quantity. The situation is simplified somewhat if attention is restricted for the present to ionizing particle velocities which are large compared to the Bohr orbit velocity $v_0 = e^2/\hbar$ where the symbols have the same meaning as in Chapter 2. With this restriction energy losses occur predominantly by direct impact between the particle and single orbital electrons of the gas (Section 2-2) and may be generally classified into three types.

1. Excitation collisions in which the molecule is raised to a higher energy state but in which ionization does not occur.
2. Light ionizing collisions in which the energy lost is of the order of the ionization potential V_I of the gas.
3. Violent collisions in which energetic secondary electrons, called δ-rays, are produced.

Types 1 and 2 are most important from the standpoint of the total energy lost by the primary particle. Violent collisions are relatively rare and contribute little to the final total ionization; furthermore, the energetic secondary electrons dissipate their energy ultimately in collisions of the types 1 and 2. Since the relative numbers of excitation and light ionization collisions are practically independent of the nature and velocity of the primary particle, a constant amount of wasted energy (on the average) is associated with the formation of an ion pair. Because of this constancy, it has been found convenient to define the parameter W, the average energy lost per ion pair produced, as

$$W = \frac{E}{N_{\mathrm{i}}} \tag{5-1}$$

where E is the energy lost by the particle and N_{i} is the number of ion pairs formed. From the above arguments, it follows that W will have a characteristic value for each gas but will be nearly independent of the nature and velocity of the primary particles for velocities appreciably greater than v_0.

An interesting analysis of the division of energy losses has been made by Platzman for energetic electrons in helium.[23] The energy-balance equation may be written

$$E = N_{\mathrm{i}}\bar{E}_{\mathrm{i}} + N_{\mathrm{ex}}\bar{E}_{\mathrm{ex}} + N_{\mathrm{i}}\bar{\epsilon} \tag{5-2}$$

where E is the total energy lost by the primary particle, N_{i} is the ultimate number of singly charged atomic ions produced at an average energy expenditure $\bar{E}_{\mathrm{i}}$, N_{ex} is the number of excited atoms produced at an average energy expenditure $\bar{E}_{\mathrm{ex}}$ and $\bar{\epsilon}$ is the average kinetic energy of the subexcitation electrons (electrons having energies less than the lowest excitation level in helium). From Eqs. (5-1) and (5-2)

$$W = \frac{E}{N_{\mathrm{i}}} = \bar{E}_{\mathrm{i}} + \bar{E}_{\mathrm{ex}}\frac{N_{\mathrm{ex}}}{N_{\mathrm{i}}} + \bar{\epsilon} \tag{5-3}$$

and therefore

$$\frac{W}{V_I} = \frac{\bar{E}_i}{V_I} + \left(\frac{\bar{E}_{ex}}{V_I}\right)\left(\frac{N_{ex}}{N_i}\right) + \frac{\bar{\epsilon}}{V_I} \tag{5-4}$$

The experimental value of (W/V_I) is 1.70 and all of the quantities on the right-hand side of Eq. (5-4) have been evaluated from information that is independent of the absolute measurement of W. Hence the completeness of the energy balance can be verified, the numbers corresponding to these algebraic quantities being

$$1.70 \approx 1.06 + (0.85)(0.40) + 0.30 \tag{5-5}$$

The quantity 1.06 exceeds unity because of the energy wasted in producing excited ions and multiply charged ions. The quantity 0.85 is the average excitation energy relative to V_I; it can be estimated successfully because in helium, as in all the noble gases, all excitation levels lie fairly close to the ionization limit. The quantity 0.40 is the ratio of the total number of excited atoms to the total number of singly ionized atoms produced, and is equal to the ratio of corresponding average cross sections. The quantity 0.30 is the mean energy relative to V_I per ion pair, left in the subexcitation electrons and ultimately dissipated as heat.

At the other extreme, particles of large mass (greater than that of the α-particle), traveling at velocities which are large when compared with orbital-electron velocities, are rapidly stripped of their orbital electrons and the resulting bare nucleus, like any charged particle, loses energy by collisions with the electrons of the gaseous atoms. As the velocity of the nucleus decreases and becomes comparable with orbital-electron velocities, the processes of electron capture and loss become important, and the base nucleus becomes a moving ion whose charge fluctuates statistically. Ultimately, the moving ion will reach thermal energy and become neutral. Theoretical aspects of this slowing-down process have been reviewed and summarized by Northcliff.[24]

W-Values and Their Application to Dosimetry

Historically, the early indication of a direct proportionality between energy lost by an ionizing particle and the ionization produced in a gas, together with the general application of gas-filled detectors to dosimetry measurements, gave importance to accurate experimantal determinations of W for various gases and different ionizing particles. Calculations of W are complex and have been made only for the simplest gases. A detailed calculation[8] of W for atomic hydrogen, using theoretical cross sections gave a value of 36 eV/ion pair compared with the measured value of 36.3 eV/ion pair for molecular hydrogen. Another calculation[22] yielded a value of 43 eV/ion pair for helium, in good agreement with experiment. Calculations for more complex gases have not yet been made.

The experimental determination of W for electrons has been the subject of investigations for more than forty years. Various investigators have used β-rays, photons, and accelerated electron beams from heated filaments as sources ranging in energy from a few electron volts to more than 20 MeV. Measurements have been made in various types of ionization chambers, in proportional counters, in cloud chambers, and in calorimeters. Although air has been the most frequently studied gas, all of the more common gases have been investigated by one technique or another. Much of the earlier work suffered from questionable gas purity; yet, considering the variety of techniques employed, the results are in rather good agreement. Two groups of experiments will be discussed here: (1) experiments utilizing low-energy electrons and (2) experiments utilizing radioactive α-particle sources.

W for Electrons

The question of the variation of W with particle energy is of primary concern and the proportional counter technique is a powerful method for

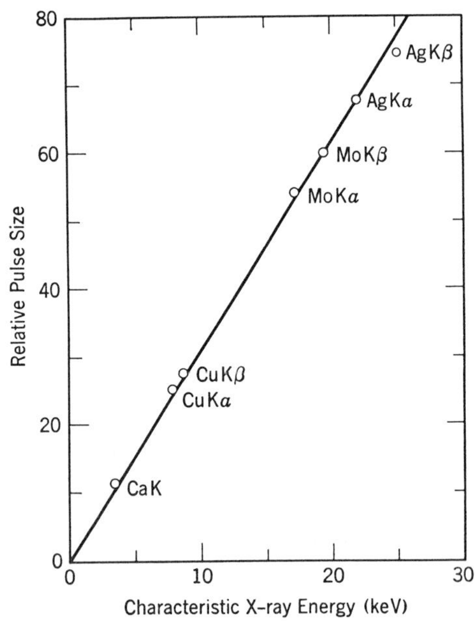

Fig. 5-1. Relationship between electron energy and ionization in nitrogen; the photoelectrons were obtained by bombarding a nitrogen-filled proportional counter with characteristic X-rays of the elements indicated.[7]

studying this question. Characteristic X-rays bombarding a proportional counter will be absorbed by the counter gas, resulting in the liberation of monoenergetic photoelectrons which, in turn, produce definite ionization pulses in the counter. From the distribution in pulse sizes the mean ionization produced by the photoelectrons can be inferred. This method was used by Curran, Angus, and Cockcroft[7] to show the direct proportionality between energy and ionization in argon in the range of 10 to 40 keV and in nitrogen in the range of 5 to 25 keV; their results for nitrogen are shown in Fig. 5-1. Further investigations have extended the energy range down to 240 eV and up to more than 1 MeV with the result that the constancy of W with electron energy has been established to an experimental accuracy of about 2%.

The purity of the gas under study is of paramount importance in the determination of absolute values of W for different gases. Even trace amounts of impurities in the noble gases cause marked changes in the ionization efficiency and result in low W-values. Because of the attention to gas purity, the results of Jesse and Sadauskis[20] listed in the column headed W_β in Table 5-1 are probably the most reliable data presently available for electrons in the several gases studied.

Table 5-1 Values of the Ionization Potential V_I and W-Values for α- and β-Particles for Several Gases.

Note the variation of the ratios W_β/V_I and W_α/W_β

Gas	V_I (eV)	W_β (eV/ion pair)	W_α (eV/ion pair)	W_β/V_I	W_α/W_β
He	24.58	42.3	42.7	1.72	1.01
Ne	21.56	36.6	36.8	1.70	1.01
Ar	15.75	26.4	26.4	1.68	1.00
Kr	14.00	24.2	24.1	1.73	1.00
Xe	12.13	22.0	21.9	1.81	1.00
H_2	15.42	36.3	36.3	2.35	1.00
N_2	15.5	35.0	36.6	2.26	1.05
O_2	13.55	30.9	32.5	2.28	1.05
Air	—	34.0	35.5	—	1.04
CO_2	14.4	32.9	36.0	2.28	1.09
CH_4	13.10	27.3	29.2	2.08	1.07
C_2H_6	11.6	24.8	26.6	2.14	1.07
C_2H_4	10.8	26.2	28.0	2.43	1.07
C_2H_2	11.4	25.9	27.5	2.27	1.06
C_4H_{10}	10.34	—	26.4	2.55	—
C_3H_8	11.3	—	26.3	2.32	—
BF_3	17.0	—	36.0	2.12	—

W for α-Particles

The use of α-particles from radioactive sources offers several advantages for the measurement of W-values. Many α-sources are monoenergetic, with energies.that are precisely known. The ranges in gases at convenient working pressures permit the use of ionization chambers of reasonable dimensions while still completely stopping the α-particles in the active volume of the

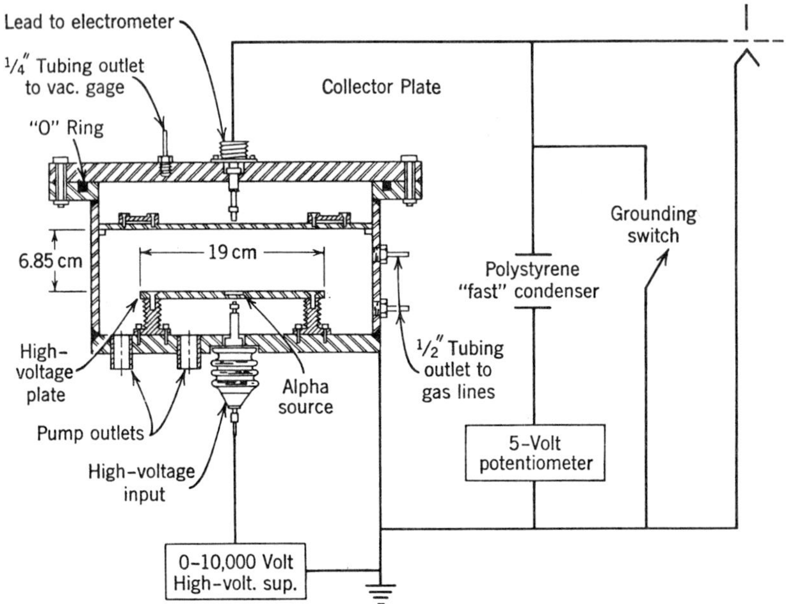

Fig. 5-2. Ionization chamber and electrical circuit for measuring W for α-particles.[3]

chamber. Some results are listed in the column headed W_α in Table 5-1. The arrangement used by Bortner and Hurst,[3] shown in Fig. 5-2, illustrates the straightforward nature of one of these measurements. Under the action of the uniform electric field in the chamber, the free electrons move toward the positively charged plate and the positive ions move toward the collector plate, thus constituting an electric current through the gas. This current is measured by the rate-of-drift method with a vacuum tube electrometer used as a null indicator. The value of W is given as

$$W = \frac{eNE}{C} \cdot \frac{\Delta V}{\Delta t} \tag{5-6}$$

where N is the number of counts per minute of the α-source having 2π geometry, E is the energy of the α particle in electron volts, C is the capacity of the condenser in farads, e is the charge of the electron in coulombs, and $\Delta V/\Delta t$ is the measured rate of change in potential of the condenser in volts per minute.

Although there appear to be certain systematic variations in the ratios of W/V_I and W_α/W_β in Table 5-1, no rigorous explanation has accounted for them. Because of nuclear interactions and the mechanism of energy loss at low velocities, one might expect that W_α will be somewhat larger than W_β

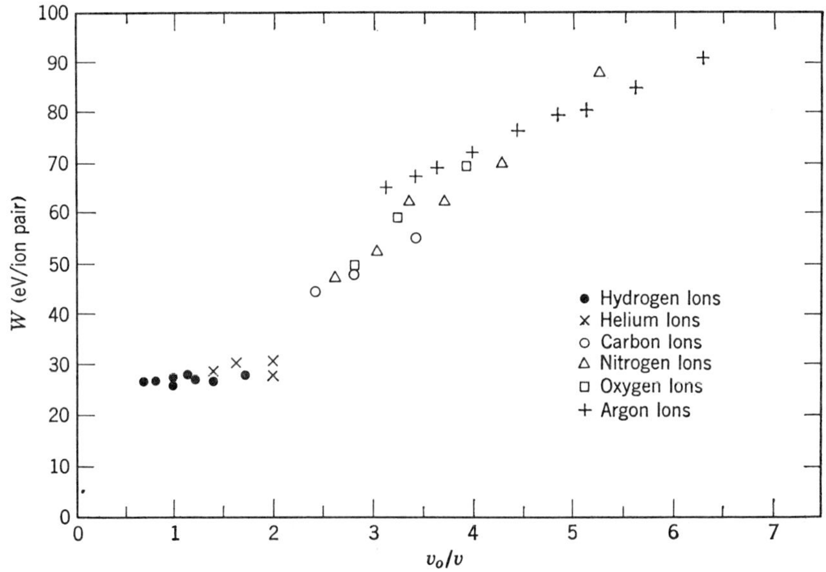

Fig. 5-3. Variation of W with the ratio of the electron velocity in the first Bohr orbit, $v_0 = e^2/\hbar$, and the velocity of the ion in argon.[25]

in a given gas and it is observed to be about 5% greater in some diatomic gases and about 7% greater in some more complex gases; however, the ration W_α/W_β is about unity for hydrogen and the noble gases.

Experiments with protons and deuterons of energies comparable to that of α-particles have yielded results in good agreement with the α-particle results. However, for heavier particles, it is established that the ionization efficiency decreases with the energy of the particle and Fig. 5-3 gives the results for measurements[25] with heavy ions ranging in mass from hydrogen to argon and varying in energy from 8.5 to 100 keV. Here W is plotted against the ratio $v_0 = e^2/\hbar$, and the velocity v of the ion. The increase in W for ratios greater than unity is appreciable and nearly independent of the mass of the ion.

The decrease in ionization efficiency is more marked if the ionizing particle is a fission fragment or a recoil particle from α-decay. Because of their relatively large masses, these particles will reach orbital-electron velocities while still possessing quite large energies and, in accordance with the qualitative arguments given above, they will lose energy thereafter by nuclear collisions and capture and loss of electrons. Plots of total ionization versus energy for fission fragments appear to be linear but do not extrapolate to zero energy for zero ionization. This has led to an expression for the energy of a fission fragment of the form

$$E = WJ + E_0 \tag{5-7}$$

where W is the value of W for α-particles, J is the total observed ionization, and E_0, called the "ionization defect," is about 5.7 and 6.7 MeV for the light and heavy fragments, respectively. The use of the ionization defect concept is somewhat unfortunate since it implies that ionization ceases abruptly at the particle energies noted, whereas a more valid picture is one of increasing W with decreasing energy down to near-zero energy. It is certain that charged particles traveling through a gas are capable of producing ionization at energies much lower than 5 MeV. Experiments utilizing 100 to 200 keV recoil particles from α-decay have given W-values 1.8 to 4.5 times the W-values found for α-particles in the same gases.

Inpurities

It was mentioned previously that early measurements of W were sometimes inaccurate due to impurities in the gas. This is particularly true of the noble gases and the measurements in helium best illustrate the large changes in ionization efficiency that result from even trace amounts of other gases. Although early measurements gave values of W of about 30 eV/ion pair for helium, theoretical calculations indicated that the value should be greater than 40 eV/ion pair. Using very carefully purified helium gas, Jesse and Sadauskis[19] and Bortner and Hurst[2] obtained experimental values of 42.7 and 46.0 eV/ion pair, respectively, and by the deliberate addition of small amounts of contaminant gases showed that the low values obtained earlier were due to impurities. The helium atom has a metastable state at an energy of 19.7 eV above the ground state. If the charged particle passing through the gas interacts to excite this state, that energy absorbed will be wasted, so far as the production of ion pairs is concerned, in pure helium. However, if impurity molecules with an ionization potential less than 19.7 eV are present, the long lifetime of the metastable state provides an opportunity for interaction by collision resulting in ionization of the impurity molecules and a resultant lowered value of W. This effect has been observed in the other noble gases and it is interesting that various contaminant gases give lower W-values in argon even though they may have ionization potentials greater than the energy of the lowest metastable state at 11.5 eV.

Mixtures

It is sometimes necessary to employ a mixture of two gases to achieve a desired operating characteristic of an ionization chamber or proportional counter. Here the interpretation of data may require a knowledge of the effective W for the mixture. It has been found[26] that an effective value W_{ij} can be calculated for mixtures of several pairs of molecular gases using the empirical equation

Table 5-2 Constants for Calculation of W_{ij} from Eq. (5-8) for Certain Binary Mixtures of Gases

Gas	f_i or f_j	Gas	f_i or f_j
H_2	0.5	CH_4	1.8
N_2	1.0	C_2H_2	3.3
O_2	1.3	C_2H_4	3.4
CO_2	1.8	C_2H_6	3.5

$$W_{ij} = (W_i - W_j)Z_{ij} + W_j \tag{5-8}$$

where

$$Z_{ij} = \frac{P_i}{P_i + \left(\dfrac{f_j}{f_i}\right)P_j}$$

Here W_i and W_j are the W-values of the pure constituent gases present in the mixture at partial pressures P_i and P_j, respectively. The quantities f_i and f_j are empirical constants tabulated in Table 5-2 for various gases.

5-3 DIRECT-CURRENT MEASUREMENT OF IONIZATION

Recombination and Electron Capture

The electrons and positive ions formed in a gas during ionization may be partially or totally lost to observation by recombining with each other. If the gas is electronegative or is a mixture containing electronegative components, recombination may be preceded by electron capture to form negative ions. One classification[21] of the mechanisms by which recombination occurs is based on the possible schemes by which the excess energy of the system may be disposed of. Another classification is based on the past history of the ions.

1. Preferential recombination between the positive ion and the negative ion (or electron) from which it was originally separated.
2. Columnar recombination between positive ions and negative ions (or electrons) distributed along the track of the ionizing particle.
3. Volume recombination between positive ions and negative ions (or electrons) from the tracks of different particles.

Preferential recombination in nonelectronegative gases results when the liberated electron is subjected to a collision very close to its point of liberation and is scattered to the neighborhood of the parent ion. In electronegative gases the electron may be captured very close to its point of liberation to form a negative ion which will then combine with the parent positive ion. Except in electronegative gases at high pressures, preferential recombination is quite improbable. Columnar recombination, on the other hand, is a common process and becomes particularly important in an ionization chamber containing electronegative gases. Volume recombination may be important in d-c chambers operated in high-intensity radiation fields.

The mechanism of electron capture in collision with gas molecules to form negative ions is of greater importance in the operation of a pulse-ionization chamber than for d-c instruments. The probability that an electron will undergo capture to form a negative ion is expressed by the attachment cross section σ_a or the attachment probability h, defined as the probability that the electron will be captured by a neutral gas molecule in a single collision. The concept of attachment probability is most useful in the discussion of electron swarms and thus generally refers to the average behavior of the electrons in the swarm. It is related to the average attachment cross section by the equation

$$\bar{\sigma}_a = h\bar{\sigma}_t \tag{5-9}$$

where $\bar{\sigma}_t$ is the average total cross section for electron scattering.

Consider now an electron drifting in the direction of a uniform electric field toward the collector plate of an ionization chamber with a drift velocity w. It will have a mean agitation velocity u, different from the drift velocity w, and, writing λ for the mean free path per unit pressure for collisions with electronegative gas molecules, will experience uP/λ collisions per second with electronegative gas molecules present at a partial pressure P. Then the probability of attachment while traveling a distance dz in the direction of the field will be $(huP/\lambda w)\,dz$ and the fractional number of electrons lost by attachment during the drift of the swarm will be

$$\frac{dn}{n} = -\frac{huP}{\lambda w}\,dz \tag{5-10}$$

For uniform field h, u, λ, and w will not be functions of z, and integration of Eq. (5-10) yields

$$n = n_0 e^{-\frac{huPz}{\lambda w}} \tag{5-11}$$

for the number of electrons not undergoing attachment over a drift distance z, with n_0 electrons present in the swarm at $z = 0$. The attachment probability is measured as a function of E/P, the ratio of electric field intensity to gas pressure, and the variation of h with average electron energy is deduced from the results. Values of h as a function of electron energy for several electronegative gases have been listed by Healey and Reed[15]; the maximum value for most such gases is about 10^{-5} to 10^{-3} per collision.

Although oxygen is the electronegative contaminant most commonly encountered in radiation dosimetry, other troublesome impurities are H_2O NH_3, HCl, SiF_4, and the halogens. Many laboratory gases contain H_2O and oxygen as impurities, whereas HCl is contained in some solder fluxes commonly used in the construction of chambers and counters; SiF_4 is formed in the reaction of BF_3 and water vapor in the presence of glass. Attachment in pure oxygen and in mixtures of oxygen and other gases has been investigated extensively. For electron energies above 1 eV the attachment probability increases to quite large values through the mechanism of dissociative capture of the electron

$$O_2 + e^- \rightarrow O^- + O$$

This is of little concern in the operation of ionization counters since the value of E/P at which these energies are reached is greater than that generally used. However, this would be a factor in the operation of counters where higher values of E/P are employed and the need for gas purity with respect to the elimination of oxygen is obvious. The attachment probability has a minimum at an electron energy of 1 eV and increases at lower energies. The mechanisms of capture at energies well below 1 eV are quite complex, particularly in mixtures of oxygen and other gases.

Electrometer Response and Sensitivity

Information about the total amount of ionization produced in an ionization chamber can be obtained by DC measurements of the current resulting from the motion of the electrons and ions through the chamber gas under the action of the applied electric field. These small currents, typically from 10^{-14} to 10^{-8} amp, are most conveniently measured by electrometers of the vacuum tube, vibrating capacitor, or quartz fiber types. The two methods of current measurement are known as the IR-drop method and the rate-of-drift method.

In Fig. 5-4, C includes the capacitance of the chamber, the connectors, and of any capacitor added to provide a more convenient operating range. The shorting switch S_2 is used to protect the electrometer from surges due to the connection of the battery to the circuit and to discharge of C at the beginning of a measurement. When the applied potential is sufficient to collect the

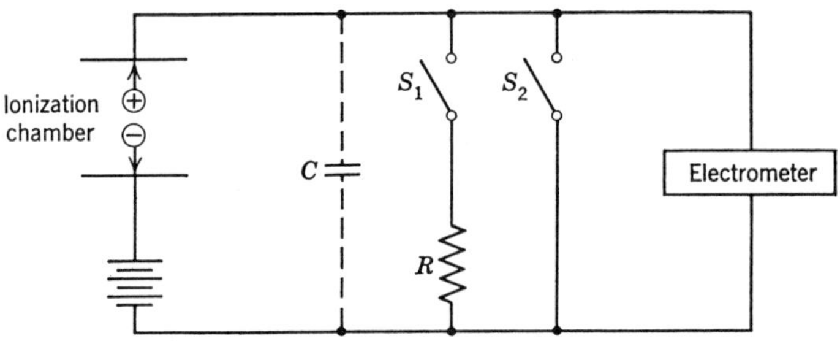

Fig. 5-4. Ionization chamber connections to illustrate *IR* and rate-of-drift methods of measuring current.

electrons and ions in the chamber, the magnitude of the resulting current I is proportional to the rate of ion production and, hence, to the intensity of of the radiation field. When S_2 is open and S_1 closed, this current flows to the RC network resulting in a potential across the electrometer given by

$$V(t) = IR(1 - e^{-t/RC}) \tag{5-12}$$

where t is the time measured from the moment of switching. The limiting, steady-rate value of $V(t)$ is

$$V = IR \tag{5-13}$$

If R is known, a measurement of V yields the value of the current I. If both S_1 and S_2 are opened at the start of a measurement, the flow of the constant current I into C results in a change of potential across the electrometer with time; this is given by

$$\frac{\Delta V}{\Delta t} = \frac{I}{C} \tag{5-14}$$

In this rate-of-drift method the current is determined from the measured value of $\Delta V/\Delta t$ and a knowledge of the value of C. A convenient variation of the circuit in which ΔV is determined from an auxiliary potentiometer with the electrometer used as a null indicator is shown in Fig. 5-2.

Most investigators prefer the rate-of-drift method because of the difficulty in making accurate determinations of the high values of R and the relatively long response time resulting from the use of high resistances in the measurements of small currents. This latter objection is overcome by the use of feedback circuits in most electrometers.

In many d-c measurements involving ionization chambers it is important only that the pressure of the chamber gas and the applied electric field remain constant in order that the measured current be proportional to the intensity of the radiation field. However, in laboratory investigations it is frequently necessary to know that the measured current accurately represents the total collection of all ions produced in the chamber. In nonelectronegative gases at relatively low pressures, this condition of saturation can be assured by increasing the applied field until the measured current becomes constant. However, in electronegative gases and in mixtures containing electronegative gases, and particularly at higher pressures, it is doubtful that true saturation is ever achieved due to capture and recombination. Under these circumstances the theory of Jaffe[18] suggests plotting the reciprocal of the measured current against the reciprocal of the applied field; the $1/I$ intercept of the resulting straight line, extrapolated to $1/E = 0$, yields the value of the saturation current. Plots of $1/I$ versus $1/E$ for different gas pressures should extrapolate to the same value of $1/I$.

In recent years improvements in the design of vacuum tube electrometers have made these the most convenient and satisfactory instruments for the measurement of currents in the range of 10^{-6} to 10^{-13} amp. Below 10^{-13} amp the currents and the fluctuations in the currents that flow into or out of the grid of the input tube, even when no signal is present, limit their usefulness. Quartz fiber and vibrating capacitor electrometers may have quite high sensitivities and are preferred for the measurement of currents less than 10^{-13} amp. These types have the added advantages of exceptional stability and freedom from drift, but suffer in comparison with the vacuum tube electrometers in most other respects. A more detailed review of electrometers has been given by Fairstein.[10]

Chamber Design and Construction

Ionization chambers are of so many types, depending on the intended application, that only the most general design and construction features will be considered here. The simple parallel-plate chamber shown in Fig. 5-2 will be used to illustrate some of these features. This chamber was designed for the measurement of W-values, using an α-particle ionization source.

The body of the chamber and the electrodes are made of stainless steel* with Heliarc welded joints. The electrodes are circular metal plates and the collector electrode is surrounded with a guard-ring as is customary to minimize edge effects. Neoprene and silicone O-rings are used as gaskets to seal the chamber cover and connector to the collector electrode. Fluorothene insulators support the collector electrodes and Teflon is used as an insulator in both the high-voltage connector assembly and in the connector to the collector plate. Both are excellent insulators and fluorothene, because of its better mechanical properties, is preferable in applications requiring machining. The high voltage is connected through a modified porcelain standoff insulator; Kovar glass and Kovar porcelain lead-throughs are frequently used. A smooth finish on all surfaces is desirable and is essential on the surfaces which will be at high voltages. Before assembly, all surfaces are thoroughly cleaned with dilute acids or other suitable agents, carefully rinsed with distilled water and acetone or alcohol, and then thoroughly dried.

Commercial laboratory gases have been improved in purity sufficiently in recent years to make the problem of gas purification less serious than it was earlier. In chambers operated at high pressures it is still important to ensure the removal of electronegative contaminants such as oxygen; the removal of molecular gases from a chamber-filling of the noble gases is essential for reliable operation. Electronegative contaminants can be effectively removed from most gases by circulation through calcium turnings heated to 300°C or uranium turnings heated to 200°C.† Helium is purified by passage through a charcoal trap cooled to liquid-nitrogen temperatures and molecular gases, as well as other noble gases, can be effectively removed from argon by multistage distillation using liquid-nitrogen cooled traps.

5-4 PULSE IONIZATION CHAMBERS

Whereas d-c measurements utilizing ionization chambers give information only about the total ionization produced in the chamber, pulse measurements permit the determination of the disintegration rate and energy spectrum of the radiation source as well. However, proper interpretation of the results obtained from such measurements requires some understanding of the behavior of electrons and ions in gases under the actions of applied electric fields and of the mechanism of pulse formation and amplification.

* Brass and aluminum are frequently used in the construction of these components.

† Special care must be exercised if hydrogen is circulated through heated uranium turnings because of the large quantities of heat liberated in the reaction. The use of uranium purifiers is reviewed in Chapter 1 of Reference 10.

Electrons and Ions in Gases

The passage of an ionizing particle through a gas results in the formation of free electrons, having energies ranging from zero to 4 or 5 keV, and of positive ions. Within a time of the order of 10^{-7} to 10^{-6} sec after the passage of the particle the free electrons, because of collisions with the gas molecules, will be reduced in energy to thermal values. In the absence of an applied electric field, the electrons will have a distribution in velocities given by the Maxwellian distribution function. If $f(u)\,du$ represents the fractional number of electrons with speeds between u and $(u + du)$ then

$$f(u)\,du = \left(\frac{m}{2\pi kT}\right)^{3/2} e^{-(mu^2/2kT)} 4\pi u^2\,du \tag{5-15}$$

where m is the mass of the electron, k is the Boltzmann constant, and T is the absolute temperature of the gas.

If there is an applied electric field E, the electrons will drift through the gas with a characteristic drift velocity w which in general is a function of E, the gas pressure P, and the nature of the gas. This drift velocity along the direction of the field should be distinguished from the random velocity of agitation u upon which it is superimposed. For moderate E/P values of a few volts per centimeter per millimeter Hg, w will be of the order of 1 cm/μsec. whereas u will be of the order of 100 times as great. While drifting through the gas in this manner, the electrons gain energy from the field between collisions with the gas molecules and lose energy in collisions, very quickly attaining an equilibrium condition characterized by an average energy of agitation denoted by the constant k_1 times the gas-kinetic energy $(\frac{3}{2})kT$. In this process the distribution in velocity will change and several equations have been proposed to describe the distribution; that due to Druyvesteyn[9] may be written

$$f(u)\,du = \frac{4u^2}{\Gamma(\frac{3}{4})\beta^3} e^{-v^4/\beta^4}\,du \tag{5-16}$$

where

$$\beta = \left[\frac{3kT\Gamma(\frac{3}{4})}{m\Gamma(\frac{5}{4})}\right]^{1/2}$$

and Γ is the gamma function. A comparison of the relative distributions in velocity for the Maxwellian and Druyvesteyn functions is given in Fig. 5-5.

Diffusion of Electrons

Electrons released as a result of the interaction of a moving charged particle with the gas of an ionization chamber or counter will not drift precisely

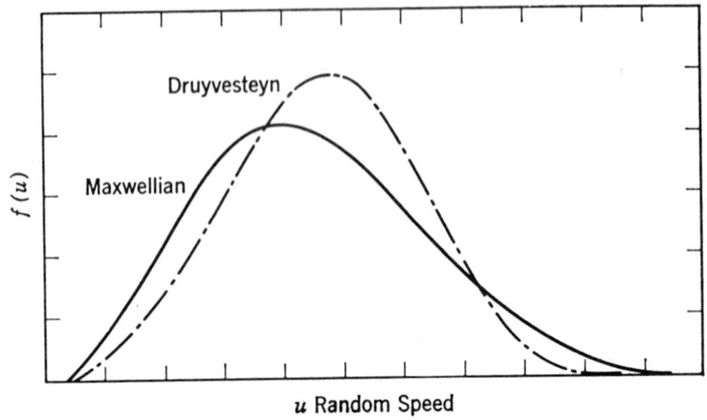

Fig. 5-5. Maxwellian and Druyvesteyn velocity distributions for electron swarms having the same average electron energy.

along the lines of force of the electric field. Instead, because of collisions with the gas molecules, they will tend to spread out by diffusion as they move toward the collecting electrode. In the absence of an external field and under conditions of uniform total pressure and small density variation, the diffusion of electrons across a surface in a gas is proportional to the space rate of change of the electron density at right angles to the surface. The constant of proportionality is the diffusion coefficient K. The expression for the electron transport per unit area per unit time may be written (Fick's law)

$$\mathbf{G} = n\mathbf{v}_t = -K \operatorname{grad} n \tag{5-17}$$

where $\mathbf{v}_t$ is the transport velocity and $n(x, y, z, t)$ is the number of electrons per unit volume in the gas. In one dimension Eq. (5-17) may be written

$$G(z) = nv_{tz} = -K \frac{\partial n}{\partial z} \tag{5-18}$$

Consider the transport of electrons $G(z)$ entering the region between two planes at z and $(z + dz)$ and the transport out of the region,

$$G(z + dz) = G(z) + \frac{\partial G}{\partial z} dz$$

The net inflow of electrons is

$$G(z) - G(z + dz) = -\left(\frac{\partial G}{\partial z}\right) dz = \frac{-\partial}{\partial z}\left(-K \frac{\partial n}{\partial z}\right) dz \tag{5-19}$$

With no sources or sinks in the region, the equation of continuity yields

$$\nabla \cdot \mathbf{G} = -\frac{\partial n}{\partial t} \tag{5-20}$$

Therefore

$$\frac{\partial n}{\partial t} = \frac{\partial}{\partial z}\left(K \frac{\partial n}{\partial z}\right) \tag{5-21}$$

and if K is independent of z

$$\frac{\partial n}{\partial t} = K \frac{\partial^2 n}{\partial z^2} \tag{5-22}$$

If now a uniform electric field is applied in the z direction, the number of electrons crossing unit area per unit time becomes

$$n\mathbf{v}_t = -K \operatorname{grad} n + n\mathbf{w} \tag{5-23}$$

where $\mathbf{w}$ is the mean drift velocity in the direction of the field. Equation (5-23) may be rewritten using the fact that the electron partial pressure is $P_e = \frac{1}{3}nm\overline{u^2}$, where $\overline{u^2}$ is the mean square velocity of the electron. Assuming $\overline{u^2}$ constant in the steady state throughout the gas, we then obtain from Eq. (5-23) the relation

$$\frac{\partial P_e}{\partial z} = \frac{w}{K} P_e - \frac{P_e v_t}{K} = \begin{matrix}\text{momentum transferred to unit volume of gas}\\ \text{molecules per second by the electrons}\end{matrix} \tag{5-24}$$

Since, in the absence of an electric field, $-(P_e v_t/K)$ is the momentum transferred to a unit volume in unit time, wP_e/K represents the momentum transferred to a unit volume in unit time due to the field, and is also neE. There, $neE = wP_e/K$, or

$$\frac{w}{K} = \frac{neE}{P_e} \tag{5-25}$$

Applying the equation of continuity (5-20) to Eq. (5-23) gives

$$\frac{1}{K}\frac{\partial n}{\partial t} = \frac{1}{K}\frac{\partial G}{\partial z} = \frac{\partial}{\partial z}\left(-\frac{\partial n}{\partial z} + n\frac{w}{K}\right)$$

or

$$\frac{1}{K}\frac{\partial n}{\partial t} = -\frac{\partial^2 n}{\partial z^2} + \frac{w}{K}\frac{\partial n}{\partial z} \tag{5-26}$$

if w/K is constant through the gas in the steady state. But in the steady state $\partial n/\partial t = 0$, and so Eq. (5-26) becomes

$$\frac{\partial^2 n}{\partial z^2} = \frac{w}{K}\frac{\partial n}{\partial z} \tag{5-27}$$

The basic diffusion equation may now be written

$$\nabla^2 n = \frac{w}{K}\frac{\partial n}{\partial z} \tag{5-28}$$

Equations (5-28) has been solved for the particular boundary conditions imposed by the configuration of the apparatus used in the experimental investigation of electron diffusion. From the experimental determination of w/K we can obtain the average energy of agitation of the electrons. The energy factor k_1 is given by

$$k_1 = \frac{\frac{1}{2}m\overline{u^2}}{\frac{1}{2}M\overline{\Omega^2}} = \frac{\text{kinetic energy of agitation of electrons}}{\text{kinetic energy of agitation of molecules}}$$

where M is the mass and $\overline{\Omega^2}$ the mean square velocity of the molecules, and atmospheric pressure may be written $P_0 = \frac{1}{3}NM\overline{\Omega^2}$ where N is the number of molecules per unit volume at atmospheric pressure. It follows that Eq. (5-25) can be written

$$\frac{w}{K} = \frac{neE}{P_e} = \frac{neE}{\frac{1}{3}nm\overline{u^2}} = \frac{NeE}{k_1 P_0}$$

or

$$\frac{w}{KP} = \frac{Ne}{P_0}\frac{1}{k_1}\frac{E}{P} \qquad \text{(5-29)}$$

where P is the pressure at which the experiment is performed. The energy thus obtained is uncertain to perhaps 15% because the energy distribution of the electrons in the swarm is not known. Healy and Reed[15] have tabulated the results of such measurements and other parameters, inferred from such data combined with independent measurements of w, for many gases as a function of E/P. At a given value of E/P, electrons have much greater agitation energies in the noble gases than in the molecular gases; for example, at $E/P = 1$, $k_1 = 285$ for argon and $k_1 = 25$ for nitrogen. This difference in behavior is due to the onset of inelastic collisions in the molecular gases at low electron energies, leading to the excitation of low-lying rotational and vibrational states in the molecule. The noble gases, on the other hand, have no low-energy states and, since little energy is lost in elastic collisions due to the disparity of masses, the average electron energy reaches quite high values at relatively low values of E/P.

Ion Mobility

As a result of the early study of the motion of ions in gases it was determined that the drift velocity in a given gas at a given pressure was proportional to the applied electric field. This relationship is usually written

$$w = 760\,\mu\,\frac{E}{P} \qquad \text{(5-30)}$$

where w is the drift velocity of the ions in centimeters per second, E is the electric field in volts per centimeter, P is the gas pressure in millimeters of Hg, and μ is the mobility constant expressed in (centimeters/second per volt/centimeter). For a given ion of a specified charge in a given gas the mobility is nearly constant over extended ranges of E/P as indicated in Table 5-3.

Table 5-3 Mobility of Positive Ions in Gases at Standard Atmospheric Pressures

Gas	Ion	E/P (volt/cm/mm Hg)	μ (cm/sec/volt/cm)	E/P (volt/cm/mm Hg)	μ (cm/sec/volt/cm)
He	He^+	2	10.3	16	9.5
	He_2^+	2	20.2	12	19.9
Ne	Ne^+	0	4.2	18	3.3
	Ne_2^+	3	6.7	18	6.2
Ar	Ar^+	0	1.6	32	1.4
	Ar_2^+	8	2.6	32	2.4
Kr	Kr^+	14	0.9	55	0.7
	Kr_2^+	12	1.2	55	1.1
Xe	Xe^+	10	0.6	90	0.4
	Xe_2^+	20	0.8	80	0.7
O_2	O_2^+	40	1.8	200	1.1
N_2	$N_2^+N_4^+$	40	2.7	200	0.8
	Li^+	0–55	4.2	70	5.2
	NH_3^+	0–25	3.3	35	3.8
	Na^+	0–17	3.0	25	3.3
	K^+	0–10	2.7	23	3.0
	Rb^+	0–6	2.3	60	3.3
	Cs^+	0	2.2	28	2.3

In the noble gases the mobility decreases slowly, and nearly linearly, with increasing E/P and is greater for the heavier ions in a given gas. In nitrogen, the mobility decreases with increased mass of the ion and appears to be quite constant over the lower range of E/P for a given ion. For the pressures and values of E/P commonly used in detecting instruments utilizing gaseous ionization, it is usually sufficient to consider the ion-mobility constant, leading to a calculated drift velocity proportional to the electric field. At very high values of E/P, μ cannot be considered constant. Since the constancy of μ depends on the assumption that the average energy of the ion is sensibly $\frac{3}{2}\,kT$, this means that the energy acquired from the field between collisions is small compared with kT. At high values of E/P this is not true and the ion velocity appears to approach

$$w = \text{constant} \left(\frac{E}{P}\right)^{1/2} \tag{5-31}$$

Electron Mobility

From previous discussions of the gain in energy of an electron from the electric field between collisions and the losses of energy in collisions, we would not expect to be able to define a true mobility constant for electrons as was done for ions; also, the experimental relationship between w and E/P is usually more complex. An approximate relationship between the drift velocity w in the direction of the field and the larger agitation velocity u can be derived from the following argument. Between collisions the electron, traveling with the agitation velocity u, undergoes an acceleration Ee/m due to the electric field. Since a time $L/(uP)$ elapses between collisions, where L is the mean free path of the electron at unit gas pressure, it follows that the electron has an average velocity $\frac{1}{2}(Ee/m)(L/uP)$ in the direction of the field. If this velocity is identified as the drift velocity, one can write

$$w = \frac{EeL}{2umP} \tag{5-32}$$

It is interesting to observe that for a fixed value of L and of E/P, the drift velocity w is inversely proportional to the agitation velocity u.

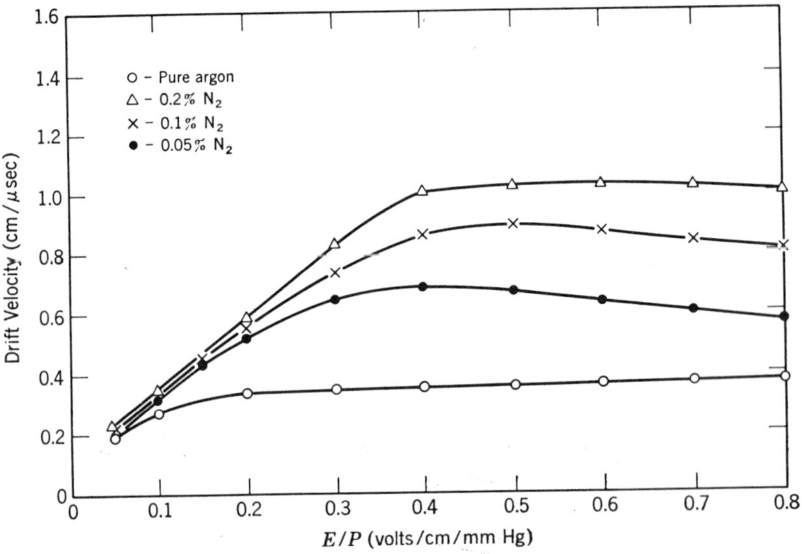

Fig. 5-6. Electron drift velocity in argon-nitrogen mixtures.[4]

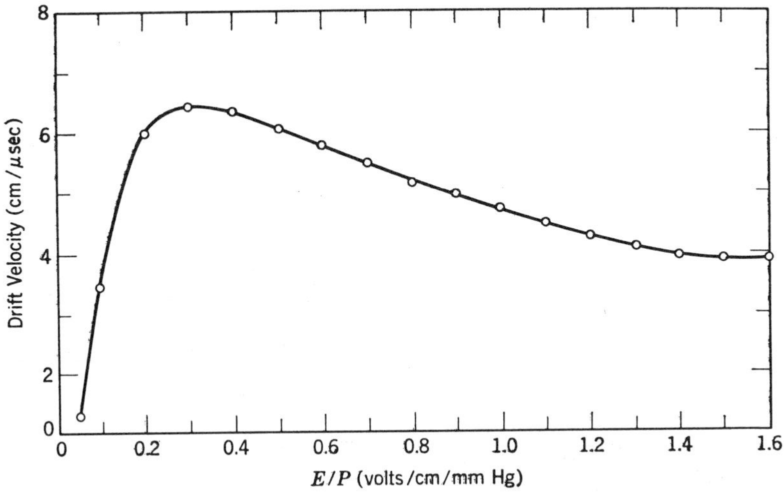

Fig. 5-7. Electron drift velocity in a mixture of 90% argon and 10% methane.[4]

Experimental measurements of w as a function of E/P are shown in Figs. 5-6 through 5-10. In Fig. 5-6 the increase in w of electrons in argon due to the addition of even a few tenths of one percent of nitrogen is clearly demonstrated. This result is in qualitative agreement with the mechanism of reduction of agitation energy, and consequent increase in drift velocity according to Eq. (5-32), caused by inelastic collisions with the nitrogen molecules. Figure 5-7 gives the results for a 90% argon–10% methane mixture commonly used as a filling gas for counters. Figures 5-8 and 5-9 give the results for methane, for carbon dioxide, and for several mixtures of

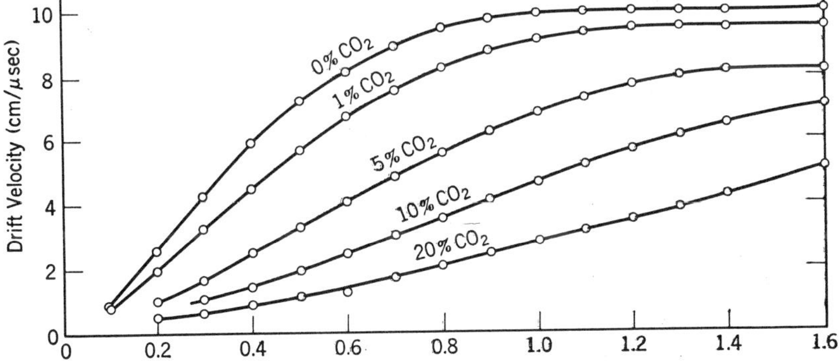

Fig. 5-8. Electron drift velocity in methane–carbon dioxide mixtures.[4]

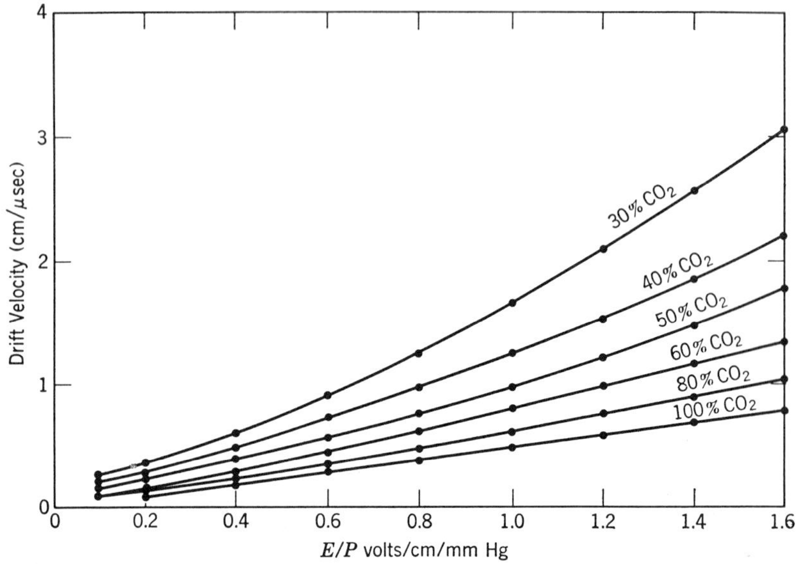

Fig. 5-9. Electron drift velocity in methane–carbon dioxide mixtures.[4]

the two gases. These are of practical interest in that a wide range of drift velocities is possible in the various mixtures. Figure 5-10 gives the results for ethylene (C_2H_4) and for cyclopropane (C_3H_6), two gases having the same relative composition as polyethylene $(C_2H_4)_n$. These gases have been used with polyethylene to make the Bragg-Gray type ionization cavities used extensively in fast-neutron dosimetry.

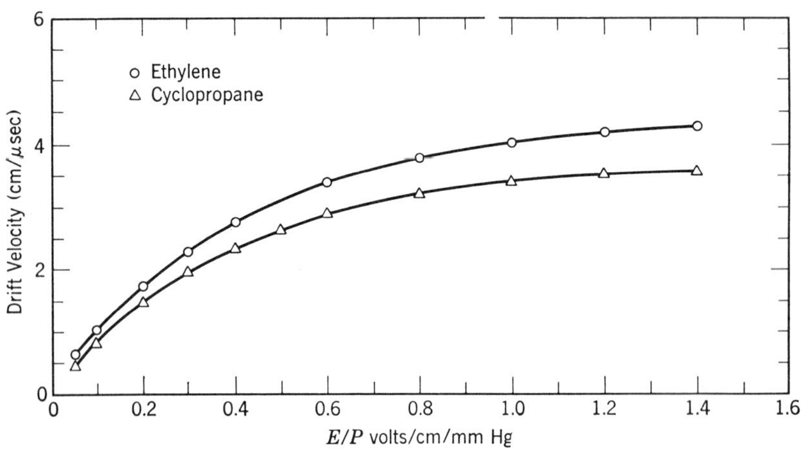

Fig. 5-10. Electron drift velocity in ethylene and in cyclopropane.[4]

In some gases the variation of w with E/P can be expressed to sufficient accuracy by an empirical expression for a range of values of E/P. Examples are given in Table 5-4 and the variations from the kinetic theory expression given in Eq. (5-30) should be noted.

Pulse Shape in a Parallel-Plate Chamber

Although the pulse profile will be developed here only for a parallel-plate chamber, it will be derived in a general manner readily applicable to any chamber possessing axial symmetry (for instance, cylindrical or spherical

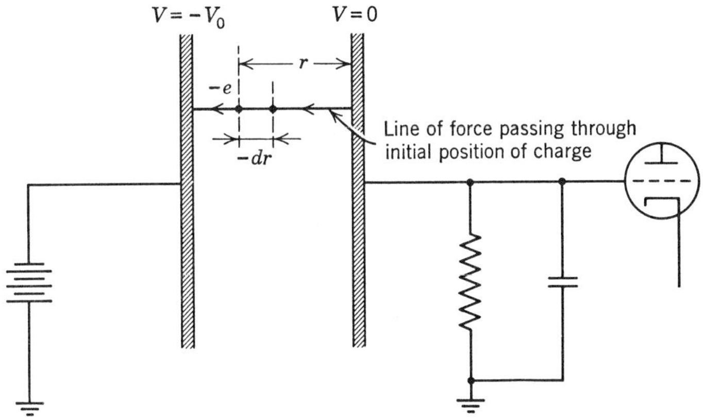

Fig. 5-11. Generalized schematic for development of pulse profile in ionization chamber.

chambers), following the method of Franzen and Cochran.[13] With a potential V_0 between the electrodes, the electric field at any point can be specified by a single coordinate r measured along the lines of force, as indicated in Fig. 5-11. It is assumed that the time constant of the electronic circuitry connected to the collector electrode is long compared to the electron collection time, so that during the collection time the electrodes are effectively insulated from each other. Conservation of energy requires that the work done by the electric field E in displacing a charge $-e$ from r to $(r - dr)$ be equal to the change in the electrostatic energy stored in the capacitance C of the chamber. Thus

$$E(-e)(-dr) = d\left(\frac{Q^2}{2C}\right) \approx V_0\, dQ \qquad (5\text{-}33)$$

where Q is the charge stored in the capacitance C by virtue of the potential difference V_0, since $Q \gg e$. The current flowing in the external network connected to the chamber is

$$i_e = \frac{dQ}{dt} = \frac{Ee}{V_0}\left(\frac{dr}{dt}\right) = -\frac{Ee}{V_0} w \tag{5-34}$$

where w is the drift velocity of the charge $-e$. Since the passage of the charged particle results in the liberation of equal numbers of electrons and positive ions, the total current i would be the sum of all the contributions due to the motions of all the charges. Since the drift velocity of the electrons

Table 5-4 Variation of Electron Drift Velocity with E/P for Several Gases

Gas	Range of E/P (volt/cm/mm Hg)	w (cm/μsec)	Reference
N_2	0.1–2.0	$0.447\,(E/P) + 0.268$	4
CO_2	0.2–2.0	$0.50\,(E/P)$	4
CO_2	0.5–7.0	$0.41\,(E/P)^{1.5}$	15
H_2	0.05–2.0	$1.0\,(E/P)^{0.5}$	6
H_2	2–20	$0.3\,(E/P) + 0.8$	6
BF_3	5–25	$1.0\,(E/P)$	1
Air	1–10	$0.42\,(E/P) + 0.9$	17

is of the order of 1000 times that of the heavy ions, this current will consist initially of a large component arising from the electron motion and a much smaller component arising from the ion motion. After collection of the electrons, this smaller component continues until complete ion collection is realized. Since the change in charge dQ of the chamber capacity C is

$$dQ = i\,dt \tag{5-35}$$

and the signal $V(t)$ appearing as a voltage change at the collector electrode is

$$V(t) = \frac{Q(t)}{C} \tag{5-36}$$

it follows that

$$V(t) = \frac{1}{C}\int_0^t i(t)\,dt \tag{5-37}$$

In practice, the amplifiers used with pulse ionization chambers respond only to the fast part of the signal, that caused by the motion of the electrons. Since, as shown in Table 5-4, it is possible to express the drift velocity w as a rather simple function of E/P in some instances and, for chambers with axial symmetry, E can be expressed as a function of the coordinate r, it is possible to write analytical expressions for the signal generated after the passage of a particle through the chamber. For this purpose we write

$$w = \mu\left(\frac{E}{P}\right)^n \tag{5-38}$$

where μ and n are appropriate empirical constants. This expression is valid for several gases if values of E/P are chosen sufficiently large that the constant terms in the empirical equations of Table 5-4 can be neglected. For the variation of E with r we write

$$E(r) = kr^{-m} \tag{5-39}$$

where for the plane-parallel chamber, $k = V_0/d$ and $m = 0$, with d the separation of the plates; for a cylindrical chamber, $m = 1$ and $k = [V_0/\log (b/a)]$ where b and a are the radii of the outer and inner cylinders, respectively; for the spherical chamber $m = 2$ and $k = V_0[ab/(b - a)]$, where b and a are the radii of the outer and inner spheres. Combining Eqs. (5-38) and (5-39) gives

$$w = \mu(k/P)^n r^{-mn} \tag{5-40}$$

The time required for the electron to move from the initial position r_0 to the position r will be

$$\begin{aligned} t &= \int_{r_0}^{r} \left(\frac{1}{w}\right) dr \\ &= \frac{1}{\mu}\left(\frac{P}{k}\right)^n \frac{1}{mn + 1}\,(r_0^{mn+1} - r^{mn+1}) \end{aligned} \tag{5-41}$$

Since the collection time T is the time required for the electron to move from the position r_0 to $r = a$, this becomes

$$T = \frac{1}{\mu}\left(\frac{P}{k}\right)^n \frac{1}{mn + 1}\,(r_0^{mn+1} - a^{mn+1}) \tag{5-42}$$

Combining Eqs. (5-41) and (5-42) gives

$$r(t) = r_0\left\{1 - \left[1 - \left(\frac{a}{r_0}\right)^{mn+1}\right]\frac{t}{T}\right\}^{1/(mn+1)} \tag{5-43}$$

for the position of the electron as a function of time.

In general, the current is given by

$$i(t) = \frac{-Ee}{V_0}\,w = \frac{e}{V_0}\,kr^{-m}\,\frac{dr}{dt} \tag{5-44}$$

and the signal at the collector is

$$V(t) = \frac{1}{C}\int_0^t i(t)\,dt$$

For the plane-parallel chamber, $m = 0$, $k = V_0/d$ and $a = 0$ so that

$$i(t) = \frac{e}{d}\frac{dr}{dt}$$

and

$$r(t) = r_0\left(1 - \frac{t}{T}\right)$$

which gives

$$V(t) = \frac{-er_0}{Cd}\frac{t}{T} \tag{5-45}$$

Thus the signal changes linearly with time, reaching a saturation value of $-er_0/(Cd)$ at $t = T$. This pulse profile is that due to a single electron; for a track formed parallel to the plates of a parallel-plate chamber at a distance r_0 from the collector electrode and containing N free electrons, the saturation value would be

$$V(T) = -N\frac{er_0}{Cd} \tag{5-46}$$

Pulse Profile With Electron Capture

If the chamber contains electronegative gas components, the signal will be changed by the effective loss of electrons from the swarm due to capture and the subsequent slow drift velocity of the negative ions. If N_0 free electrons are originally liberated in the track at distance r_0 from the collector electrode, the number N remaining at position r is, from Eq. (5-11),

$$N = N_0 e^{-huPr/Lw} \tag{5-47}$$

Since the electron current is

$$i = New\frac{E}{V_0} = \frac{New}{d}$$

and the signal at the collector for complete electron collection is

$$V(t) = \frac{1}{C}\int_0^T i\,dt$$

it follows that the signal $V_c(T)$ including the effect of electron capture is

$$V_c(T) = -\frac{N_0 e}{Cd}\frac{Lw}{huP}\left[1 - \exp\left(\frac{huPr_0}{Lw}\right)\right] \tag{5-48}$$

Amplifier Response

The output signal from a fast ionization chamber is so small that it usually requires amplification before analysis. Although it is necessary to employ an amplifier of limited bandwidth to maintain a satisfactory signal-to-noise ratio, and to avoid pileup, the use of such an amplifier results in distortion of the pulse profile from the forms just derived. If the input pulses differ only in amplitude but have the same relative shape, this distortion will not affect the relative pulse height distribution. If the pulse profiles differ, however, as would be the case if short tracks were formed at different distances from the collector or extended tracks had different spatial orientations, understanding of the pulse distortion by the amplifier is important in interpreting the results.

The equivalent circuit of a single stage of the *RC* type of linear amplifier frequently used in such applications is shown in Fig. 5-12. The time constant

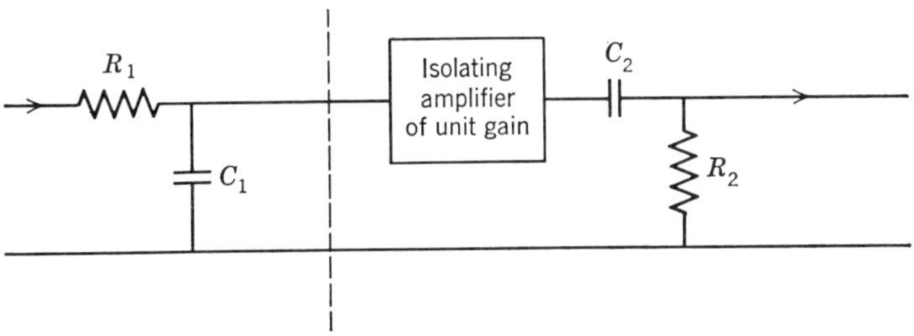

Fig. 5-12. Equivalent circuit of *RC* amplifier.

$T_1 = R_1C_1$ is determined by the plate-load resistor of the amplifier tube and its distributed plate capacitance; the time constant $T_2 = R_2C_2$ is determined by an interstage *RC* coupling network. The response of this circuit to a step function of amplitude V_0 is

$$V(t) = V_0 \frac{T_2}{T_1 - T_2} (e^{-t/T_1} - e^{-t/T_2}) \tag{5-49}$$

If the signal input rise time is much shorter than the amplifier time constants, a maximum signal-to-noise ratio is obtained when $T_1 = T_2$, and Eq. (5-49) reduces to

$$V(t) = V_0 \frac{t}{T_1} e^{-t/T_1} \tag{5-50}$$

If, instead of a step function, the input signal is of the form given by Eq. (5-46) for the parallel-plate chamber, the observed maximum amplitude of the output pulse will depend critically on the collection time T and the amplifier time constant. This has been treated in detail by Gillespie[14] whose results for the percentage drop in pulse amplitude as a function of T/T_1 are shown in Fig. 5-13.

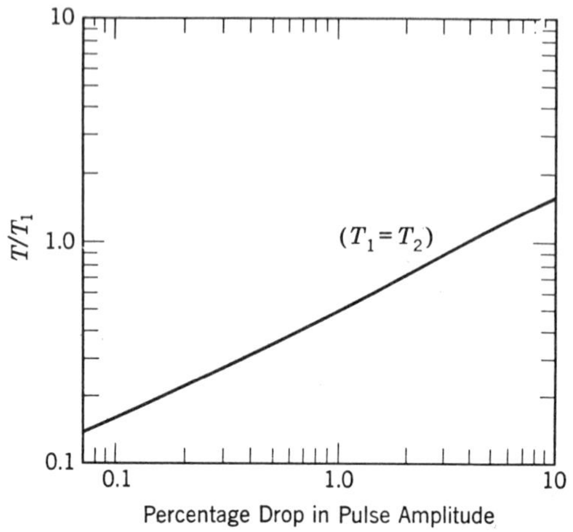

Fig. 5-13. Percentage drop in amplitude of pulse from a parallel-plate ionization chamber as a function of the ratio of electron collection time T to amplifier time constant T_1 of an RC amplifier.[14]

Resolution

A parallel-plate chamber may be used in conjunction with appropriate amplifier and pulse-height-analyzer circuits as a spectrometer to determine the energy distribution of charged particles from a source. Usually, the source is placed on one of the electrodes, in which case it can be shown[28] that the ideal pulse height distribution is a rectangle, with pulse heights ranging in amplitude from a maximum down to $[1 - \frac{3}{5}(R_0/d)]$ times the maximum where R_0 is the range of the particle and d is the separation of the plates. Insertion of a third electrode, in the form of a grid, between the two electrodes will reduce this spread in pulse heights, which is caused by the orientations of the tracks, and will improve the resolution by a considerable factor. With such an arrangement it is possible to reduce the contribution of the chamber itself to the spread in pulse heights to about 2 keV for 5 MeV α-particles.[28]

Noise in the electronic circuits may be reduced to a spread equal perhaps to 10 or 15 keV in the pulse height distribution, and thickness of the source may contribute appreciably to the straggle, although sources of high specific activity may be made sufficiently thin so that this is negligible. Statistical fluctuations in the total ionization produced by the particle also contribute to the width of the pulse height distribution. Fano[11] has shown that the value of the mean square deviation in the number of ion pairs produced is between one-third and one-half the mean square deviation to be expected if the process were governed by a Poisson distribution. This yields a standard deviation of about 9 keV for a 5 MeV α-particle. Since the total standard deviation is the square root of the sum of the squares of the contributions, it is possible to obtain a width of some 15 or 20 keV in such a pulse height distribution.

5-5 PROPORTIONAL COUNTERS

The Amplification Process

Proportional counters are usually made in the form of an outer cylinder of radius b, with a thin coaxial center wire of radius a, and filled with an appropriate gas to a pressure ranging from a few millimeters of mercury to a few atmospheres. In the most common mode of operation the center wire is maintained at a positive potential with respect to the grounded outer cylinder. The ratio b/a might have a value from 50 to 1000. Since the electric field in such a cylindrical geometry is given by

$$E(r) = \frac{V_0}{r \ln (b/a)} \tag{5-51}$$

it is seen that the field in the vicinity of the center wire will become quite large. As a result, electrons moving toward the collector are able to acquire, between successive collisions with the gas molecules, an energy equal to or greater than the ionization potential of the gas. Each primary electron will then, in the course of its motion toward the collector, lead to an avalanche of charge as the secondary electrons, in turn, produce further ionization. Hence, if there are N_0 ionization electrons liberated by the ionizing particle, there will be a total of $2^n N_0$ electrons arriving at the collector, corresponding to n ionizing collisions in the cascade process for each primary electron. The factor 2^n is called the gas amplification.

A true proportional counter will give an output voltage signal proportional to N_0 and there is a somewhat limited range of operating voltages for which any counter is truly proportional. If the gas amplification is less than about 10, the statistical variation in the sizes of the avalanches produced will lead to poor resolution. If the total ionization is sufficiently large, or sufficiently concentrated in space, that the avalanches from individual primary electrons interact appreciably with each other, true proportionality will be lost. Values of gas amplification up to 10^4 can be obtained in a properly designed counter without great difficulty.

Pulse Profile

In contrast to the situation in fast pulse ionization chambers, where the observed voltage pulse at the collector electrode was caused by the motion of the electrons, the signal from a proportional counter is due primarily to the motion away from the collector of the positive ions created in the avalanche. This derives from the fact that most of the electrons and positive ions are liberated in the very near vicinity of the collector electrode. Hence the electrons move through a small portion of the applied potential while the positive ions move through practically the entire applied potential as they drift toward the outer wall.

In the notation of Section 5-4, the electric field in the counter volume can be written $E(r) = kr^{-m}$ with $m = 1$ and $k = V_0/\ln(b/a)$. Equation (5-30) describes the drift velocity of the positive ion as $w = 760\,\mu E/P$. Hence the collection time of ions liberated at $r = a$ and collected at $r = b$ is

$$T = \int_a^b \frac{1}{w}\,dr \approx \frac{P\ln(b/a)}{2 \times 760\,\mu V_0}\,b^2 \tag{5-52}$$

since $b \gg a$ in practical situations. The position of the positive ion of charge q as a function of time is given by

$$r(t) = a\left\{1 + \left[\left(\frac{b}{a}\right)^2 - 1\right]\frac{t}{T}\right\}^{1/2} \tag{5-53}$$

From Eq. (5-44) the current is

$$i(t) = \frac{q}{r\ln(b/a)}\frac{dr}{dt}$$

and the voltage signal is given by

$$\begin{aligned} V(t) &= \frac{1}{C}\int_0^T i(t)\,dt \\ &= \frac{-q}{2C\ln(b/a)}\ln\left\{1 + \left[\left(\frac{b}{a}\right)^2 - 1\right]\frac{t}{T}\right\} \end{aligned} \tag{5-54}$$

where C is the capacitance of the counter. Since $b \gg a$, Eq. (5-54) can be written as

$$V(t) \approx \frac{-q}{2C\ln(b/a)}\ln\left[1 + \frac{1520 V_0 \mu t}{Pa^2\ln(b/a)}\right] \tag{5-55}$$

The signal will change quite rapidly at first, but much more slowly as t approaches the ion collection time T. Since $b/a \gg 1$, it follows from Eq. (5-54) that the signal will reach approximately half its final amplitude $-q/C$ at a time $t = (a/b)T$, which is of the order of 1/100 to 1/1000 of the ion collection time for counters of usual dimensions. Because of this fast initial drop in the negative signal, proportional counters can be used with fast amplifiers which respond only to this fast portion of the signal to form a system capable of high counting speeds.

The pulse profile given by Eq. (5-54) is that of a single ion and, since the avalanche of positive ions is always formed in the very near vicinity of the center wire, the shape of the pulse profile will be independent of the initial position within the counter of the primary electron producing the avalanche. For an extended track of ionization, the primary electrons will drift toward the collector, arriving at times determined by their initial positions. The composite signal observed then will be due to the combination of the pulse profiles resulting from the many separate avalanches and will, in general, decrease more slowly than indicated by Eq. (5-54). An ionization track parallel to the center wire will give the pulse profile described by Eq. (5-54).

Amplifier Response

The general arguments advanced in Section 5-4 for the effect of amplifier response to the pulse profile of a fast ionization chamber are applicable to proportional counters. The range of track orientations from angles of zero to ninety degrees with respect to the center wire results in a range of rise times for the composite pulse profiles, and the problem becomes more difficult to handle analytically. An analysis of this problem has been made by Hurst and Ritchie,[16] who found that the maximum height of the amplifier output pulse is essentially independent of the track orientation, provided the time of collection of the outermost electrons is not longer than the amplifier time constant. For collection times longer than the amplifier time constant, the maximum pulse height will be reduced significantly. As an example, a cylindrical counter with an outer radius $b = 2.1$ cm and a center wire of radius $a = 0.0025$ cm, filled with ethylene at atmospheric pressure, and operated at 3000 volts, has a collection time of about 3 μsec for the outermost electrons. When operated with an amplifier having a time constant of 1.7 μsec, the pulses due to tracks perpendicular to the counter wire will be about 10% smaller than those for tracks parallel to the counter wire.

Resolution

There are statistical fluctuations in the number of ion pairs formed by a charged particle traversing the gas of a proportional counter and the Fano factor[11] for the reduction of this variation is applicable. In addition, there is a statistical fluctuation in the multiplication process which has the effect of doubling the relative standard deviation of the original number of ions. Other factors which contribute to the resolution of proportional counters are amplifier noise, electron capture by electronegative gas contaminants, variation in diameter and positioning of the center wire, drift in gain in both counter and amplifier due to voltage changes, and, as mentioned previously, a range of track orientations. West[27] has presented data for the resolution obtained for particles ranging in energy up to about 200 keV. These data show a relative standard deviation of about 5% for a 10 keV particle and less than 1% for a 100 keV particle (width at half maximum is 2.36 times the relative standard deviation). In the very low-energy region the proportional counter is decidedly superior to the scintillation spectrometer, because the number of ion pairs produced in the counter is large as compared to the number of photoelectrons released in the photomultiplier.

Counter Construction

If proportional counters are to be used as spectrometers with good resolution, it is important to exercise special care in their construction. Since they are usually operated with static gas fillings, it is important to have high-purity gas initially and to clean all surfaces carefully to avoid electronegative gas contaminants from outgassing in continued operation. The center wire should be as smooth as possible to avoid spurious discharges and should be carefully centered for improved resolution. Because of the small output signals, it is important that insulators for the center wire and in the plugs of the high-voltage connections be of high quality to avoid leakage noise in the amplified output.

A special counter developed for application in fast-neutron dosimetry is shown in Fig. 6-4. To make it satisfy the requirements of a Bragg-Gray cavity it is lined with polyethylene and filled with ethylene gas. An internal α-particle source is provided for calibration. The high-voltage lead is a Kovar metal–glass seal and it is fitted with field tubes, after the manner of Cockcroft and Curran,[5] to eliminate end effects and determine the active volume of the instrument. In this arrangement, the potential of the field tube is adjusted to the potential of the same radial position in the gas at the center of the counter length, resulting in radial electric field lines over the region of the counter between the ends of the field tubes. The large diameter of the field tubes prevents any gas multiplication resulting from radiation in the region from the end of the counter shell to the end of the field tube and thus defines the active volume of the counter.

5-6 GEIGER-MÜLLER COUNTERS

If a gas-filled counter is operated at voltages appreciably above those appropriate for the proportional mode of operation described in Section 5-5, the process of gas multiplication will proceed as before, but proportionality between primary ionization and amplitude of signal from the counter will be lost because of interactions between the avalanches from separate primary electrons. For operating voltages sufficient to produce gas amplification above about 10^6, the mechanisms of avalanche formation and discharge termination lead to counter performance characterized as Geiger-Müller operation. In this mode of operation the initial avalanche due to even a single primary electron is followed by a succession of avalanches, propagated probably through photon emission by the excited atoms, molecules, and ions in the avalanche. In counters containing pure gases the photons may produce photoelectrons at the counter wall to initiate further avalanches; in self-quenching counters containing appropriate gas mixtures the photoelectrons come principally from the gas. The result is a discharge which ultimately produces a positive ion sheath along the entire length of the collector.

Multiple discharges caused by the interaction of the positive ions with the counter wall are prevented in two general ways. In counters containing pure gases the discharge is terminated by resistance or electronic quenching circuits which lower the applied potential below that necessary to sustain the discharge. In self-quenching counters containing mixtures of gases and organic vapors such as alcohol or ether, or one of the halogens, the discharge is quenched by the absorption of the excess energy in the dissociation of the molecules of the contaminant. In counters containing organic quenching agents the dissociated molecules do not recombine and the typical counter has a maximum lifetime of about 10^{10} counts corresponding to the depletion of the quenching agent. In halogen-quenched counters there is appreciable recombination of the dissociated molecules resulting in much longer lifetimes.

The Geiger-Müller counter is a degenerate, zero-moment device (Chapter 6) and is used primarily to give a counting rate proportional to the intensity of the radiation field. In this application the counting efficiency will depend on the counter construction and on the nature and energy of the radiation. However, it has several advantageous characteristics that will probably result in its continued use in survey monitoring applications. Since the output pulse of the counter is of the order of one volt, there is need for little, if any, amplification. The typical long plateau of 300 volts or more in which the counting rate is constant does not require great stability in the high-voltage supply, and the high sensitivity, variety of shapes and windows, and simplicity and low cost all contribute to its continued popularity.

By a proper choice of window or wall materials and thicknesses, uniqueness of response to a particular type or energy of radiation may be emphasized. Such a special arrangement is shown in Fig. 6-11, where a small halogen-quenched counter shielded with a lead and tin combination has been used to make an instrument that is energy-independent for photon energy above 150 keV and that is insensitive to neutrons.

REFERENCES

1. Bistline, J. A., *Rev. Sci. Instr.* **19,** 842 (1948).
2. Bortner, T. E., and G. S. Hurst, *Phys. Rev.* **90,** 160 (1953).
3. Bortner, T. E., and G. S. Hurst, *Phys. Rev.* **93,** 1236 (1954).
4. Bortner, T. E., G. S. Hurst, and W. G. Stone, *Rev. Sci. Instr.* **28,** 103 (1957).
5. Cockcroft, A. L., and S. C. Curran, *Rev. Sci. Instr.* **22,** 37 (1951).
6. Crompton, R. W., and D. J. Sutton, *Proc. Roy. Soc.* (*London*) **A215,** 467 (1952).
7. Curran, S. C., J. Angus, and A. L. Cockcroft, *Phil. Mag.* **40,** 36 (1949).
8. Dalgarno, A., and G. W. Griffing, *Proc. Roy. Soc.* (*London*) **A248,** 415 (1958).
9. Druyvesteyn, M. J., *Physica* **10,** 61 (1930).
10. Fairstein, E., Chapter 4 in *Nuclear Instruments and Their Uses* (A. H. Snell, ed.), John Wiley and Sons, New York, 1962.
11. Fano, U., *Phys. Rev.* **72,** 26 (1947).
12. Forester, D. W., and L. W. Cochran, "Diffusion of Slow Electrons in Gases," *Oak Ridge Natl. Lab. Rept.* **ORNL 3091** (1961).
13. Franzen, W., and L. W. Cochran, Chapter 1, in *Nuclear Instruments and Their Uses*, (A. H. Snell, ed.), John Wiley and Sons, New York, 1962.
14. Gillespie, A. B., *Signal, Noise and Resolution in Nuclear Counter Amplifiers*, Pergamon, New York, 1953.
15. Healey, R. H., and J. W. Reed, *The Behavior of Slow Electrons in Gases*, Amalgamated Wireless, Sydney, Australia, 1941.
16. Hurst, G. S., and R. H. Ritchie, *Rev. Sci. Instr.* **24,** 664 (1953).
17. Huxley, L. H. G., and A. A. Zaazou, *Proc. Roy. Soc.* (*London*) **A196,** 402 (1949).
18. Jaffe, G., *Ann. Physik* **42,** 303 (1913).
19. Jesse, W. P., and J. Sadauskis, *Phys. Rev.* **88,** 417 (1952).
20. Jesse, W. P., and J. Sadauskis, *Phys. Rev.* **90,** 1120 (1953).
21. Massey, H. S. W., and E. H. S. Burhop, *Electronic and Ionic Impact Phenomena*, Clarendon, London 1952.
22. Miller, W. F., *Bull. Am. Phys. Soc.* **1,** 202 (1956).
23. *Natl. Acad. Sci. Natl. Res. Council Publ.* **752,** "Penetration of Charged Particles in Matter," (E. A. Uehling, ed.), p. 109, (1960).
24. Northcliffe, L. C., *Ann. Rev. Nucl. Sci.* **13,** 67 (1963).
25. Phipps, J. A., J. W. Boring, and R. A. Lowry, *Phys. Rev.* **135,** A36 (1964).
26. Strickler, T. D., *J. Phys. Chem.* **67,** 825 (1963).
27. West, D., *Prog. Nucl. Phys.* **3,** 18 (1953).
28. Wilkinson, D. H., *Ionization Chambers and Counters*, Cambridge University Press, London, 1950.

PROBLEMS

1. Consider a 1 MeV electron traveling through helium gas until it is stopped. (*a*) How many ion pairs are produced? (*b*) How many helium atoms are left in excited states? (*c*) How much of the energy is left in the subexcitation electrons?

2. It has been found that an ionization chamber filling consisting of 98% argon and 2% nitrogen is rather insensitive to large oxygen contaminations over an extended range of operating voltages. Discuss qualitatively why this should be.

3. Assuming air to be composed of nitrogen and oxygen, compute the W-values for α- and β-particles and compare with the measured values for air.

4. Derive expressions for the pulse profile caused by a single electron in a cylindrical ionization chamber and in a spherical ionization chamber.

5. How many ion pairs will be formed by an α-particle with initial energy of 4 MeV in traveling 2 cm through methane gas at 1 cm Hg pressure? At 20 cm Hg pressure? [The data of Hershfelder and Magee, *Phys. Rev.* **73,** 207 (1948) will be helpful.]

6. Plot the pulse profile for a single electron in a proportional counter filled with nitrogen to atmospheric pressure and operated at 3000 volts. The counter shell has an inner radius of 1 cm and the center wire a radius of 0.0025 cm.

7. A cylindrical proportional counter operated at 3000 volts has an inner radius of 1 cm and thin field tubes of radius 0.2 cm. What potential should be applied to the field tubes?

8. Outline a procedure for calibrating a proportional counter in the 0–100 keV energy region.

9. Design a ^3He proportional counter neutron spectrometer, discuss its efficiency as a function of neutron energy, and estimate the resolution obtainable.

10. (*a*) A step function input signal of amplitude V_0 is applied to an RC amplifier of unit gain having $T_1 = T_2 = 10\ \mu sec$. Plot the output pulse profile. (*b*) If a pulse from a parallel-plate ionization chamber with an electron collection time of 10 μsec is applied to this amplifier, what will be the ratio of output to input pulse amplitudes?

11. A parallel-plate ionization chamber is filled with nitrogen to a pressure of 76 cm Hg. The plates are 10 cm apart and the chamber is operated at 760 volts. (*a*) What is the drift velocity of electrons in the chamber? (*b*) What is the agitation energy of these electrons? (*c*) What is the drift velocity of the positive ions?

12. A parallel-plate ionization chamber has plates 25 cm in diameter separated by 10 cm. The electrometer and associated circuitry has a capacitance of 100 $\mu\mu$f. When filled with nitrogen to atmospheric pressure and operated at sufficient potential to collect all electrons released, a rate-of-drift of 200 millivolts/min is observed with a small source of 5 MeV α-particles on the center portion of the negative electrode. What is the disintegration rate of the source?

CHAPTER 6

Ionization Methods of Mixed Radiation Dosimetry

G. S. Hurst

6-1 INTRODUCTION

This chapter does not attempt to cover all methods of dosimetry based on the ionization method. Instead, a selected set of examples have been chosen, the choices being dictated by two criteria: (1) the methods chosen are most representative of the fundamental approaches to the utilization of the ionization method to measure mixtures of neutrons and γ-radiation and (2) the methods chosen are those which find most general application to radiation shielding, radiation biology, and radiation protection. Methods which are restricted mainly to radiation protection work (for example, survey meters and personnel monitors) are discussed in Chapter 15.

6-2 IONIZATION METHODS USING THE FIRST MOMENT

In Chapters 4 and 5 it was shown that ionization methods are well adapted to the problem of measuring the energy absorbed per gram of a material enclosing a sensitive volume. That is, the response I of a chamber in which the total amount of ionization is collected may be expressed from Eq. (4-18), as

$$I = \sum_{l} \int_0^{\infty} \frac{\epsilon_l}{W(\epsilon_l)}\, n_{l\beta}(\epsilon_l, E)\, d\epsilon_l \tag{6-1}$$

where the symbols in Eq. (6-1) have been defined in Chapter 4 and where it has been assumed that all ion pairs are collected, that is, $B = 0$ in Eq. (4-18). To the extent that W is independent of particle type l and energy loss ϵ_l,

Eq. (6-1) may be written as

$$I = \frac{\epsilon_T}{W} \tag{6-2}$$

where ϵ_T is the total energy absorbed in the gas and, as shown in Chapter 5, W is a slowly varying function of particle type and energy. Therefore the total ionization current provides a fairly accurate measurement of the energy absorbed in the gas. For purposes of distinguishing the method of total ionization from others which also depend on the detection of ionization, it is convenient to recall through comparison of Eqs. (6-1) and (4-9) that total ionization devices respond to the first moment of the energy-loss distribution.

As discussed in Chapter 4, the energy absorbed in the gas may be used to determine the energy absorbed per gram of the material surrounding the gas cavity provided that the conditions stated in connection with the Bragg-Gray principle hold. Such arrangements of gas and solid material usually result in a detector which is too large to serve as a radiation probe.* However, the use of such detectors leads to results which may be interpreted to obtain approximately the absorbed dose in some gases and approximately the first-collision dose in other cases. In either instance, an additional requirement must be imposed on the kind of materials making up the detector if we desire to measure the dose in another medium (for example, tissue) over a range of radiation energies. This requirement may be expressed in the mathematical form

$$\left[\frac{D_{1m}(E)}{D_{1d}(E)}\right]_{E_1}^{E_2} = a(1 \pm \Delta) \tag{6-3}$$

where $D_{1m}(E)$ and $D_{1d}(E)$ are the first-collision doses in the medium of interest and in the detector materials, respectively; a is a constant of proportionality, and Δ is the error which may be tolerated over the energy range $E_1 \leq E \leq E_2$.

Ionization Chambers for the Measurement of Gamma Radiation

An example of an ionization chamber designed to measure the dose arising from γ-radiation is shown in Fig. 6-1. The design follows the Bragg-Gray principle; the sensitive volume (50 cm^3 in this case) is filled with CO_2 gas and is surrounded with graphite walls. The condition expressed by Eq. (6-3) may be examined by calculating the first-collision dose in carbon and in tissue as a function of photon energy. Following the procedures outlined in Chapter 4, the dose curves for carbon and for tissue are obtained. The ratio

* A radiation probe is defined as a device which is small enough to measure the absorbed dose at an arbitrary point in a medium without perturbing the radiation field.

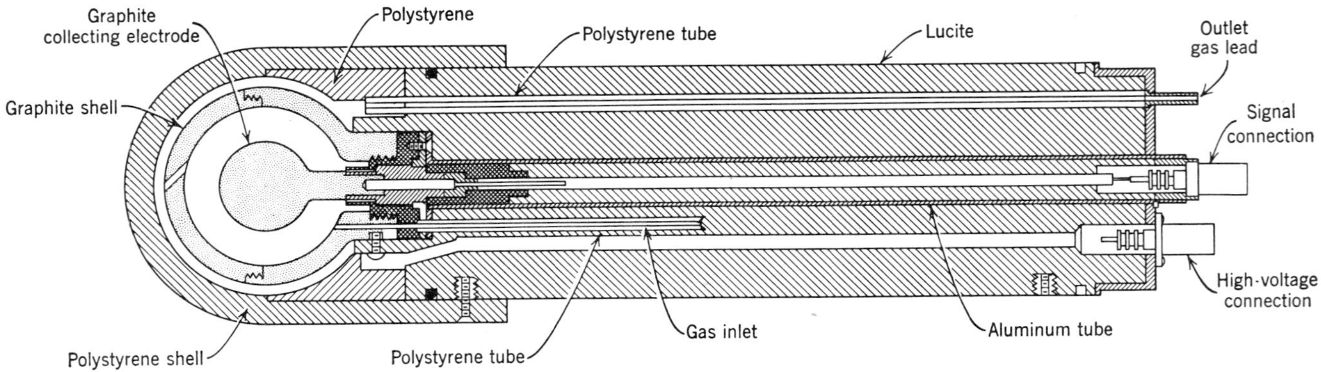

Fig. 6-1. An ionization chamber for γ-ray dose measurements.

of first-collision doses, evaluated from the curves in Fig. 4-1, is

$$\left.\frac{D_{1t}}{D_{1c}}\right]_{E_1=0.2\,\mathrm{MeV}}^{E_2=5\,\mathrm{MeV}} = 1.1(1 \pm 0.05) \tag{6-4}$$

Thus, for photons in the energy range from 0.2 to 5 MeV carbon is "tissue-equivalent" provided that an error of 5% can be tolerated.

The approximate γ-ray sensitivity of the ionization chamber shown in Fig. 6-1 can be calculated from the definition of the roentgen and a few simplifying assumptions. From the definition of the roentgen (R) (Chapter 3), 1 R of X- or γ-radiation will produce in 0.00129 gm of dry air 1 esu or 3.33×10^{-10} coul of charge. We assume (1) that the actual chamber may be approximated by a continuous carbon medium, (2) that charged-particle equilibrium can be established with negligible attenuation or scattering of primary radiation, and (3) that all the ion pairs are collected. Under these conditions we may write

$$I = 3.33 \times 10^{-10}\frac{D_{1c}}{D_{1a}} \times \frac{W_a}{W_g} \times \frac{m_g}{0.00129}\,\dot{r} \tag{6-5}$$

where I is the ionization current in amperes, D_{1a}/D_{1c} is the ratio of the first-collision dose in air and in carbon, W_a and W_g are the W values (eV per ion pair) for electrons in air and in the gas, respectively, m_g is the mass of the gas (in grams) in the sensitive volume, and $\dot{r}$ is the dose rate in roentgens per second. The ratio D_{1a}/D_{1c} may be calculated by the method given in Section 4-2 and the ratio W_a/W_g for various gases may be obtained from data given in Chapter 5.

Evaluation of Eq. (6-5) shows that for a chamber with a volume of 100 cc, containing gas at atmospheric pressure, the ionization current corresponding to a dose rate of 1 R per hr is of the order of 10^{-11} amp. Even simple electrometers,[5] particularly when used as null indicators in a rate-of-drift method, can measure currents on the order of 10^{-15} amp. Thus ionization chambers of reasonable volume can be used to measure dose rates as low as 0.1 mR/hr. Vibrating-reed type electrometers[13] may be used to increase the sensitivity by another factor of 10 to 100.

The most serious limitation of the carbon–CO_2 chamber to the measurement of γ-dose arises out of its response to neutrons. The walls and gases of ionization chambers suitable for dosimeters of X- or γ-radiation must be made of materials having low atomic weights; otherwise the conditions of Eq. (6-4) would not hold over a wide range of energy. Fast neutrons, on colliding with low atomic weight materials, impart appreciable energy. It follows from Eq. (4-4) that the average fraction F_c of energy imparted per elastic collision, assuming isotropic center-of-mass scattering, is $2M/(M + 1)^2$, where M is the atomic mass number of the recoil atom. For purpose of illustration, we approximate the C-CO_2 with a carbon wall and a hypothetical carbon gas and calculate the neutron response of such a chamber. Let $P(E)$ be the neutron response, defined as

$$P(E) = \frac{E_n}{E_\gamma} \tag{6-6}$$

where E_n is the amount of energy absorbed by 1 gm of carbon atoms for 1 tissue rad of incident fast neutrons, and E_γ is the amount of energy absorbed

Table 6-1 Variation of P with Neutron Energy E

E (MeV)	P (%)
0.1	10.9
0.5	14.9
1.0	14.9
2.0	14.5
3.0	15.1
4.0	24.7
5.0	16.8
10.0	34.1
20.0	48.7

by 1 gm of carbon for 1 R of γ-radiation. The quantity E_n may be expressed as

$$E_n = nN_c\sigma_c(E)F_cE \tag{6-7}$$

where n is the number of neutrons per cm^2 in 1 tissue rad of fast neutrons, N_c is the number of carbon atoms in a gram, and $\sigma_c(E)$ is the carbon scattering cross section for neutrons of energy E. Evaluation of P based on tabulated neutron cross sections[1] gives the results shown in Table 6-1.[6]

If the W value for carbon atoms in CO_2 were set equal to the W value for electrons in CO_2, the results in Table 6-1 would be satisfactory estimates of the neutron response of the C-CO_2 chamber. Actually W for carbon recoils is poorly known at present, and the results are perhaps too large by as much as a factor of 2 at some energies. However, it must be concluded that a C-CO_2 chamber may grossly overestimate the γ-dose when fast neutrons are present.

Tissue-Equivalent Ionization Chambers

Ionization chambers[4] may be constructed such that the atomic composition of the materials making up the walls and gas of the chamber is essentially the same as the atomic composition of tissue. Here, one expects that, except for possible differences in the W of the gas for heavy particles as compared to electrons, the ionization current will be proportional to the combined rate of energy absorption caused by neutron and γ-ray interactions with tissue.

For most applications, tissue can be approximated with a plastic[4] having the following composition by weight: hydrogen (10.1%), nitrogen (3.5%), and carbon with traces of oxygen (86.4%). This material differs in atomic composition from tissue mainly in that oxygen is replaced by carbon. In the energy range from 0.5 to 14 MeV, the error in the neutron dose resulting from this substitution is less than 6%.[14] A tissue-equivalent gas mixture which is normally used with this tissue-equivalent plastic is, in terms of percentage partial pressure: methane (64.4), carbon dioxide (32.4), and nitrogen (3.2). If more precise results are required, it is possible to construct chambers lined with tissue-equivalent gels matching a tissue composition ($C_5H_{40}O_{18}N$) exactly.

Figures 6-2 and 6-3 show two designs of tissue-equivalent ionization chambers which have been put to extensive use by the Columbia group.[14]

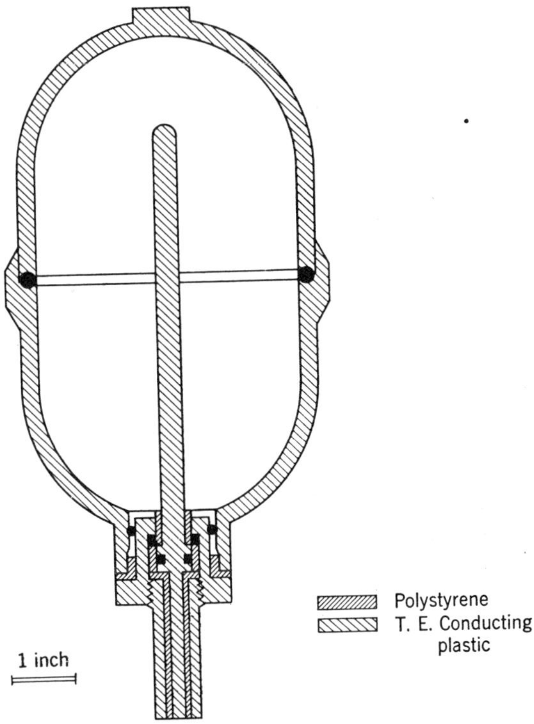

Fig. 6-2. Cross section of a large tissue-equivalent ionization chamber.

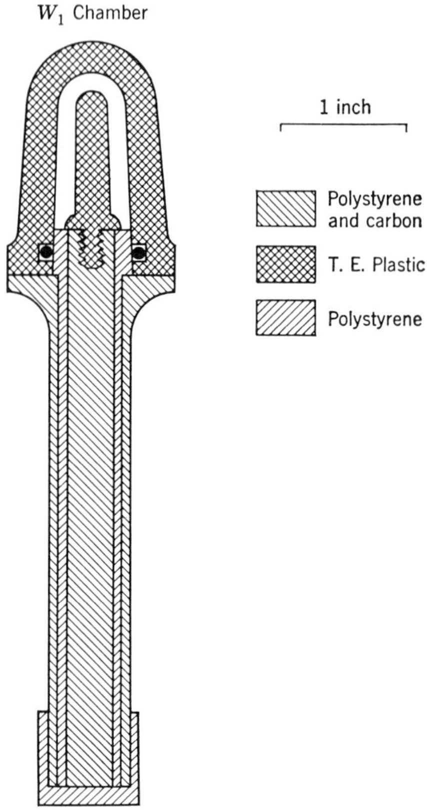

Fig. 6-3. Cross section of a small tissue-equivalent ionization chamber.

Various other designs have been developed,[14] and altogether the tissue-equivalent chamber is a very useful instrument which has found much application in radiobiology and radiation protection. The sensitivity considerations given for the γ-ionization chamber [Eq. (6-5)] apply with minor modification to the tissue-equivalent ionization chambers.

Combined Chambers for the Dosimetry of Mixtures of Neutrons and Gamma Radiation

Consider now the simultaneous use of the tissue-equivalent chamber and a Teflon-CO_2 chamber in a radiation field consisting of neutrons and γ-radiation.* When the tissue chamber is exposed to 1 R of hard X- or γ-radiation,

* This method is discussed in more detail in Ref. 12.

the absorbed dose is approximately 0.97 rad. A neutron dose of 1.03 rad in tissue will produce approximately the same amount of charge as the 1 R of γ-radiation, the larger neutron dose being due to the difference in W. The Teflon-CO_2 chamber absorbs about 0.97 rad for 1 R of γ-radiation, but when exposed to 1 tissue rad of neutrons the reading $\overline{K(E)}$ for a given neutron energy spectrum is much less than the reading produced by 1 R of γ-radiation. Thus, the readings of the tissue-equivalent chamber and the Teflon-CO_2 chamber, R_T and R_c, respectively, will be given by

Table 6-2 Maximum $K(E)$ (R per rad) for a Teflon-CO_2 Chamber

Neutron Energy (MeV)	Observed $K(E)$ (R per rad)	Computed $K(E)$ (R per rad)
0.5	0.08	0.11
1.0	0.08	0.18
2.0	0.09	0.13
3.0	0.12	0.10
4.0	0.15	0.15
6.0	0.20	0.16
8.0	0.24	0.20

$$\begin{aligned} R_T &= \frac{D_n}{1.03} + \frac{D_\gamma}{0.97} = 0.97D_n + 1.03D_\gamma \\ R_c &= \overline{K(E)}D_n + \frac{D_\gamma}{0.97} = \overline{K(E)}D_n + 1.03D_\gamma \end{aligned} \tag{6-8}$$

where D_n and D_γ are the neutron and γ-tissue doses in rad. The quantity $\overline{K(E)}$ may be calculated with equations analogous to Eqs. (6-6) and (6-7). Table 6-2 shows some values of $\overline{K(E)}$ as determined by calculation and by experiment.

Solution of Eqs. (6-8) gives the components D_n and D_γ provided that enough is known about the neutron energy spectrum to allow a reasonable average value, $\overline{K(E)}$, to be selected. If the neutron energy is not known, we may set $\overline{K(E)} = 0.16 \pm 0.08$ over the energy range from 0.5 to 8 MeV. Lack of precise information on $\overline{K(E)}$ results in an error of the order of 10% in D_n, independent of R_T and R_c and hence independent of D_n/D_γ, but may lead to large errors in D_γ depending on D_n/D_γ. However, additional errors are introduced in D_n when $D_n/D_\gamma \ll 1$, because of the problem of subtracting two small numbers which are very nearly equal.

Proportional Counter Methods of Neutron Dosimetry

Even when the response function is chosen to be the integral of the first moment of the energy-loss distribution, considerable advantage ensues from the use of counting methods. In the case of ionization chambers, one can deal only with the total energy loss as expressed in Eqs. (6-1) and (6-2); in particular, the lower limit on the integral sign of Eq. (6-1) must always be set equal to zero. With pulse counting techniques, on the other hand, we may

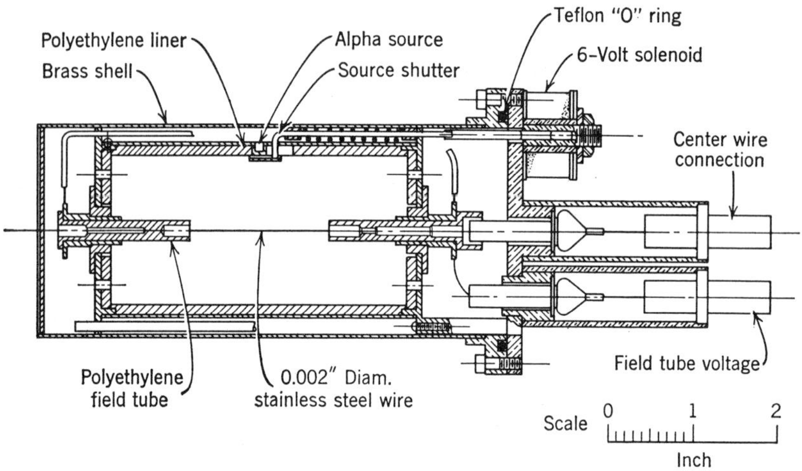

Fig. 6-4. Cross section of a proportional counter for fast neutron dosimetry.

employ a more general response function as follows:

$$R_c(E) = \int_B^\infty \epsilon n(\epsilon, E)\, d\epsilon \tag{6-9}$$

where B is the minimum energy loss which we wish to record. The advantage afforded by the choice of $B > 0$ is that a bias level B may be chosen such that neutrons may be counted even with much larger γ-ray backgrounds. These statements will be illustrated first with the absolute proportional counter.[10]

One design[17] of the proportional counter is shown in Fig. 6-4. The counter is completely lined with polyethylene C_nH_{2n}, and is filled with ethylene, C_2H_4, at a pressure of 75 cm Hg or with cyclopropane, C_3H_6, at a pressure of 50 cm Hg. In either case the walls and gas are matched in atomic composition, which satisfies one of the requirements of the Bragg-Gray principle. Thus, with a response function as specified by Eq. (6-9), we can expect to measure approximately the energy absorbed per gram of ethylene. From this information we may determine the energy absorbed per gram of

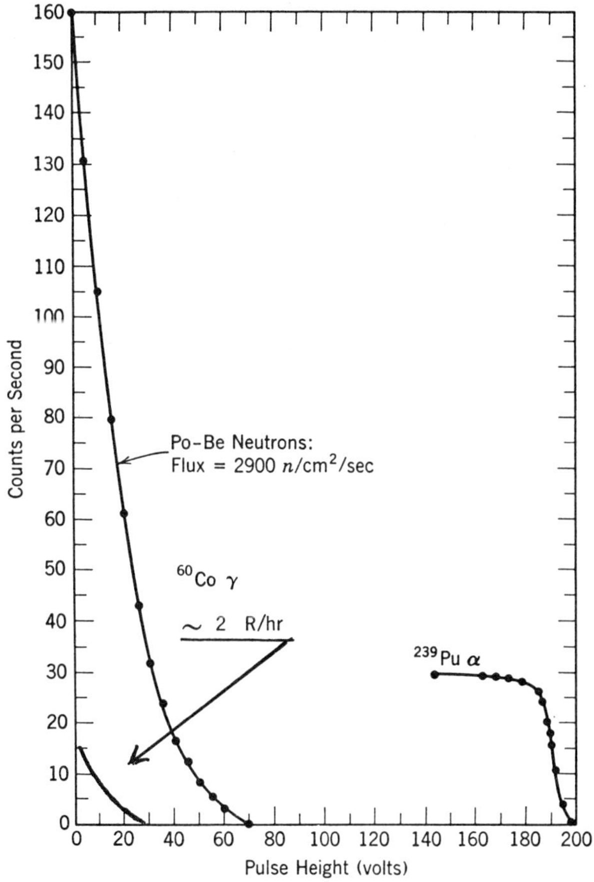

Fig. 6-5. Integral pulse height distribution for Po–Be neutrons, ^{239}Pu α-particles, and ^{60}Co γ-radiation.

tissue; as discussed in Section 4-2, one can show that

$$\left.\frac{D_{1E}}{D_{1T}}\right]_{E_1=0.01\text{ MeV}}^{E_2=20\text{ MeV}} = 1.45(1 \pm 0.10) \tag{6-10}$$

over an energy range from 0.01 to 20 MeV.

From Fig. 6-5 it is seen that γ-radiation, because of relatively low specific ionization by the secondary electrons, produces relatively small pulse heights. Therefore it is easy to reject γ-radiation by pulse height selection. After γ-ray discrimination, the neutron pulses are added linearly to obtain the quantity $\epsilon_T(E)$ which is associated with the Bragg-Gray principle. In practice, γ-ray discrimination causes an amount of energy proportional to the area A_1

(Fig. 6-5) to be lost under the bias; however, the relatively small fraction $f = A_1/(A_1 + A_2)$ can be estimated by a procedure discussed below.

The proportionality implied between pulse height and the number of ion pairs in a single ionizing event depends on two conditions: (1) there must be no electron attachment and (2) the height of the pulse at the output of the linear amplifier must not depend on track orientation. Condition 1 may be fulfilled only by rigorous exclusion from the counter of such gases as water vapor, oxygen, and some of the halogens, which have very large attachment cross sections.[7] Condition 2 may be fulfilled by proper selection of the amplifier rise time and decay time (see Chapter 5). A variation of the angle between the ionizing particle's path and the center wire in a proportional counter causes a variation in the pulse profiles. However,[9] if the rise-time and decay-time constants (assumed to be equal, as they are in many good pulse amplifiers[11]) are greater than the collection time of electrons in the counter, the pulse height at the output of the amplifier depends only slightly on the rise time of the proportional counter pulse.

If it is assumed that the pulse height is proportional to the number of ion pairs, absolute calibration of the neutron dose may be accomplished by means of an α-source, provided that the sensitive volume of the counter is also known. Therefore these two features, α-calibration source and known sensitive volume, have been incorporated into the proportional counter design in Fig. 6-4. The sensitive volume is determined by means of "field tubes" held at the appropriate electrostatic potential (Chapter 5). The neutron dose in ethylene is proportional to $(A_1 + A_2)$; this area may be expressed in MeV per second by means of the ^{239}Pu α-particle pulse height (Fig. 6-5). More specifically, the first-collision dose rate in tissue $\dot{D}_T$ is given by

$$\dot{D}_T\,(\text{MeV sec}^{-1}\,\text{gm}^{-1}) = \frac{1}{1.45}\left(\frac{A_1 + A_2}{A_c}\right) \times \frac{\dot{E}_c}{V\rho} \tag{6-11}$$

where $\dot{E}_c$ is the rate of energy absorption (MeV per second) corresponding to the arbitrarily chosen calibration area A_c, V is the sensitive volume of the counter in cubic centimeters, and ρ is the density of the C_nH_{2n} gas in grams per cubic centimeters. The quantity $\dot{E}_c$ is determined directly from the α-particle energy as follows:

$$\dot{E}_c = \dot{C}_0 \times \frac{V_0}{V_\alpha} \cdot E_\alpha \tag{6-12}$$

where $\dot{C}_0$ is the count rate and V_0 is the pulse height for the arbitrarily chosen area A_c, and V_α is the pulse height produced by α-particles of energy E_α.

When using the absolute proportional counter with conventional electronic apparatus, the value of B may be chosen to suit the particular experimental conditions. Factors governing the choice of B include the following: neutron

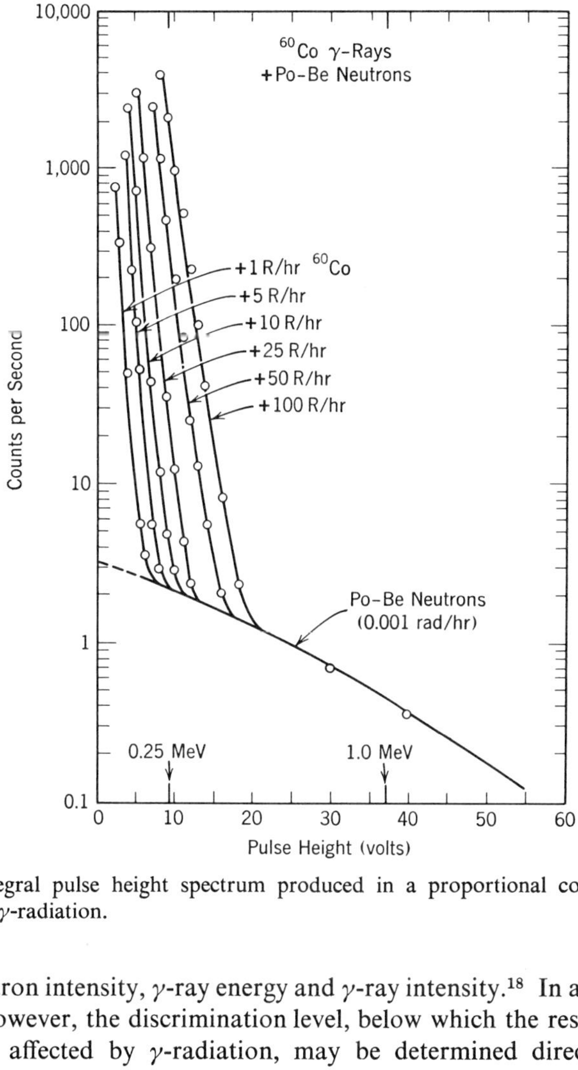

Fig. 6-6. Integral pulse height spectrum produced in a proportional counter by fast neutrons and γ-radiation.

energy, neutron intensity, γ-ray energy and γ-ray intensity.[18] In any practical situation, however, the discrimination level, below which the results may be appreciably affected by γ-radiation, may be determined directly for the particular set of experimental conditions. Figure 6-6 shows how this may be accomplished in a typical example. In this illustration the integral count rate corresponding to just 1 mrad/hr of fast neutrons (Po-Be) plus various γ-intensities ranging from 1 to 100 R/hr of ^{60}Co γ-radiation is plotted on a log scale as a function of pulse height. For example, for a γ-ray dose rate of 1 R/hr the bias level required to discriminate against γ-rays is about 0.20 MeV, and the absorbed neutron energy lost under the bias would be about 10% (see Table 6-3).

Table 6-3 Fraction f of Energy Spent by Recoils Losing Less Than the Bias Energy B in the Counter

Bias		Fast Neutron Energies							
Energy, MeV	PHS, Volts	0.5 MeV	1.0 MeV	2.0 MeV	3.5 MeV	4.8 MeV	14 MeV	Po-B	Po-Be
0.074	2.6	0.089	0.020	0.015	0.013	0.014	0.025	0.006	0.014
0.14	5.3	0.195	0.094	0.041	0.035	0.028	0.085	0.028	0.045
0.21	7.8	0.320	0.129	0.076	0.056	0.059	0.162	0.065	0.101
0.28	10.5	0.526	0.235	0.123	0.091	0.098	0.258	0.099	0.144
0.36	13.2	0.730	0.333	0.188	0.128	0.162	0.369	0.152	0.205

On the other hand, if the ^{60}Co γ-dose rate were increased to 25 R/hr, the bias level would have to be increased to about 0.36 MeV and the neutron energy lost under the bias would be about 20%. This energy lost may be estimated by plotting the data in the form illustrated in Fig. 6-6, extrapolating the count rate versus pulse height curve for values of pulse height $\leq B$, and then integrating the area under the entire curve. In other words, the values for f given in Table 6-3 do not necessarily represent errors in the measured dose, although they do represent the fraction of the total dose which must be approximated by an extrapolation procedure.

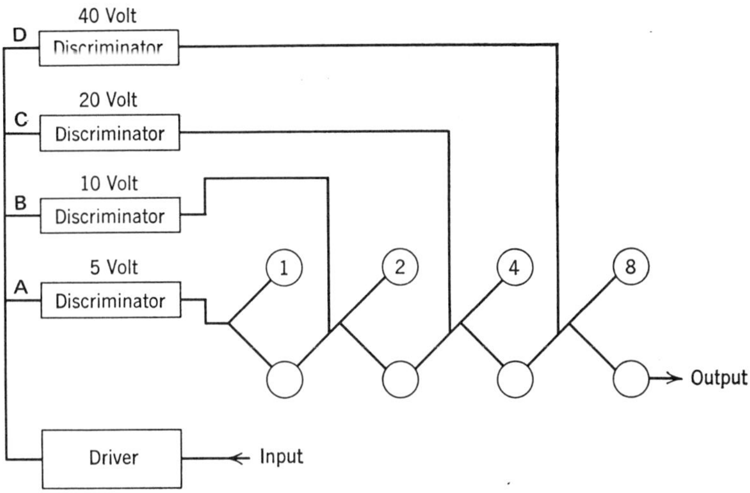

Fig. 6-7. A simple method of pulse integration, using a binary scaler.

In practice, it is convenient to integrate pulses automatically, thus eliminating the requirement for determining the area under the pulse height distributions. Figure 6-7 shows schematically a very simple type of pulse integrator. At A, B, C, and D are discriminator tubes biased so that the pulse heights required to make the tubes conduct are 5, 10, 20, and 40 volts, respectively. The discriminators feed into the 1-, 2-, 4-, and 8-count stages of a usual binary-type scaling unit. Each time Tube A conducts because of the arrival of the 5-volt pulse, one count is added to the scaling unit. A 10-volt pulse causes both A and B to conduct, thus adding three counts to the scaling unit if the No. 1 indicator is not on. If the No. 1 indicator is on, a net of only one count will be added to the system because the pulse which was stored in the No. 1 scaling position is lost in coincidence with the channel which is adding two pulses to the system. Thus the average is two counts for each 10-volt pulse. Similar arguments show that each 20-volt pulse gives an average of four counts while the 40-volt pulse gives an average of eight counts.

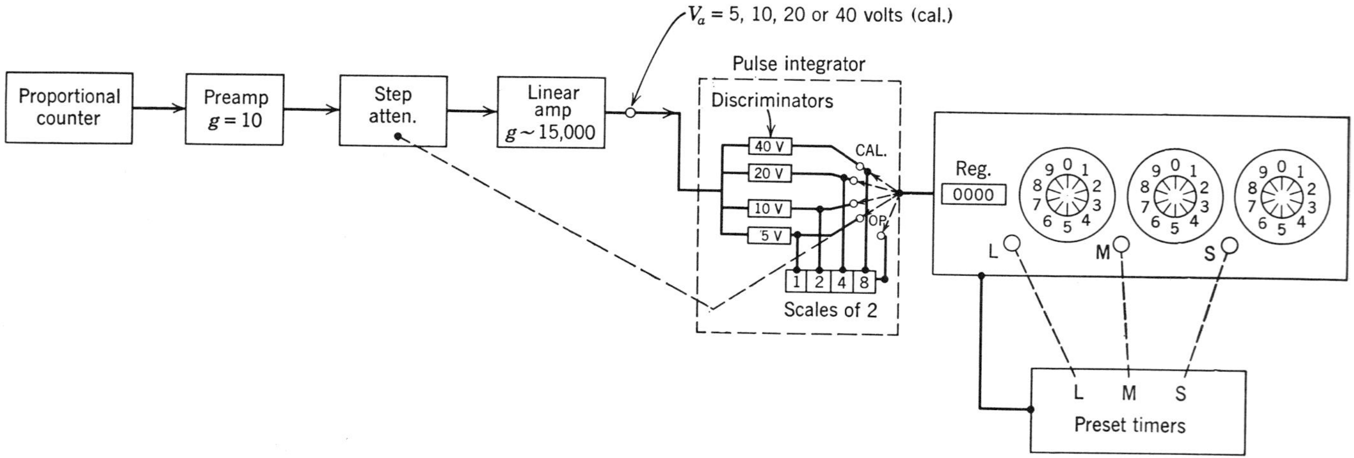

Fig. 6-8. Block diagram of a fast neutron dosimeter (RADSAN).

Therefore it is not necessary to use anticoincidence circuits to prevent the lower level discriminators from adding counts to the system. From studies with monoenergetic neutrons it is seen that the integrator reads to within $\pm 10\%$ of the true area under the pulse height distribution curves when the neutron energy lies between 0.5 and 14 MeV.

A convenient instrument[17] utilizing the counter just described and the four-stage binary integrating circuit has been developed. As shown by the block diagram of Fig. 6-8 the output signals from the pulse height integrator are fed into an indicating system which uses decade scalers, preset timers, and lamps as decimal indicators in such a way that the dose rate is indicated directly in mrad per hour.

6-3 IONIZATION METHODS USING THE ZERO MOMENT

In the general relations stated in Eqs. (4-10) and 4-11), letting $O_i = \epsilon^0$ gives

$$R_i(E) = \int_B^\infty \epsilon^0 n(\epsilon, E)\, d\epsilon = n_T(B, E) \tag{6-13}$$

Here, the response is simply the total number of interactions resulting in an energy loss greater than B. We are describing methods of γ-ray and fast neutron dosimetry that are based on the integral of the zeroth moment of the energy-loss distribution. This response function is the number of counts produced in the test system or detector, which implies considerable technical simplification in its use. We shall see that the energy absorbed per gram of tissue can be measured over a wide range of radiation energy simply by counting the number of pulses produced in the radiation detector. It should be noted that devices of this type, which use the zero moment, are not Bragg-Gray devices, although they do provide a measurement of dose. On the other hand, the pulse devices using the first moment of the energy distribution (discussed in Section 6-2) are based essentially on the Bragg-Gray principle, although they are counting methods.

A Geiger-Müller Tube as a Gamma-Ray Dosimeter

Now we discuss a method of γ-ray dosimetry which uses a commercially available halogen-type Geiger-Müller (G-M) counter as the detector.[19] This illustrates the situation where the response of the test system is $n_T(0, E)$ or simply the total number of interactions with the test system. A single-ion detector has been developed for measurements of γ-dose in the presence of high neutron fields.[2] This counter has a negligible neutron response; however, like other instruments used for this purpose, it generally requires

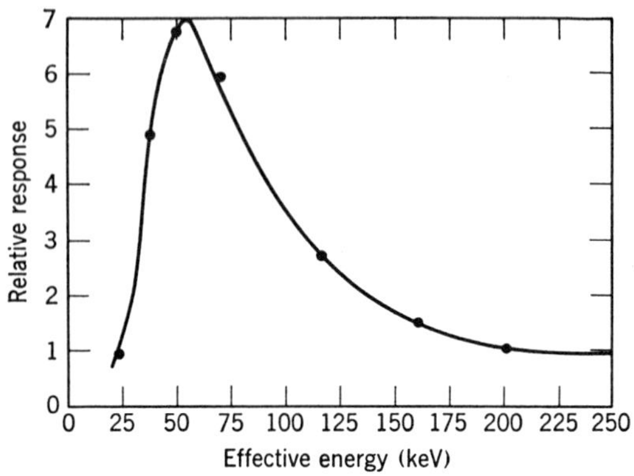

Fig. 6-9. Energy response of a Geiger-Müller counter.

considerable electronic apparatus and operational skill. A small halogen-type G-M counter can be arranged with suitable shields to provide an instrument that is (1) energy independent for X- or γ-radiation above about 150 KeV, (2) insensitive to fast neutrons, and (3) insensitive to thermal neutrons.

The counter used is a Philips No. 18509 micro G-M counter filled with neon, argon, and a halogen quenching agent. The cathode, made of stainless steel, has a wall thickness of 90 mg/cm^2. As shown by Fig. 6-9, the unshielded counter has considerable energy dependence below 200 keV. However, it can be made to furnish readings of exposure dose in roentgens that are essentially independent of γ-ray energies down to 150 keV by shielding the counter with 0.053 in. tin plus 0.010 in. lead on the sides, and 0.022 in. tin plus 0.004 in. lead on the end (Fig. 6-10). A drawing of the probe unit showing the arrangement together with a lithium shield, which will be discussed later, is

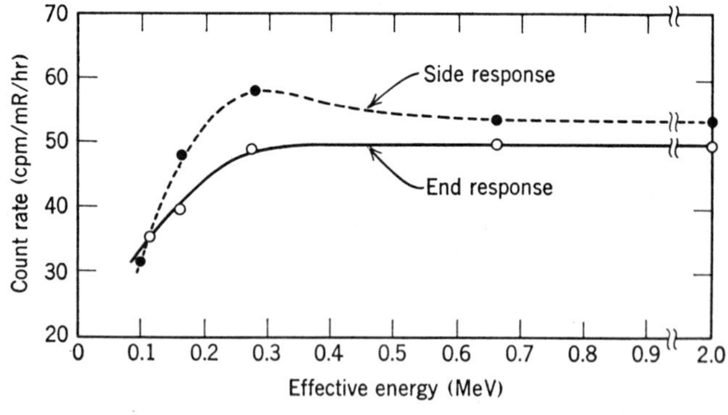

Fig. 6-10. Energy response of a shielded Geiger-Müller counter.

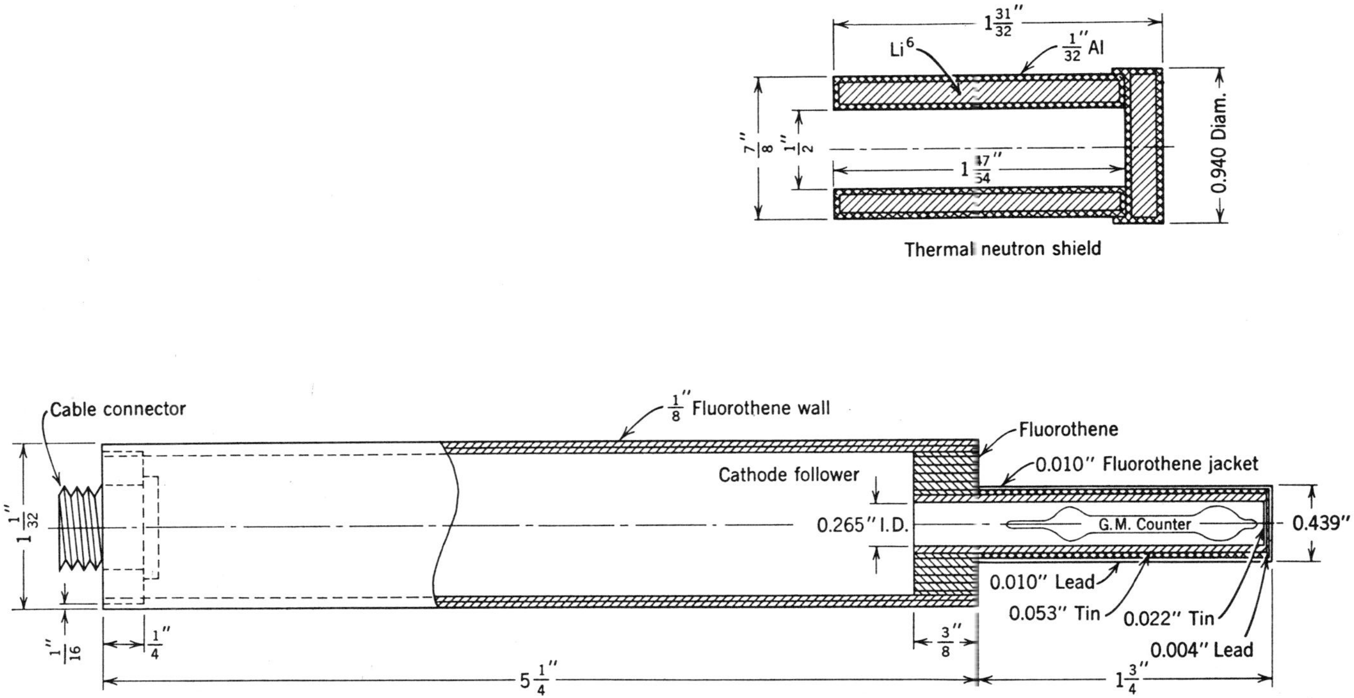

Fig. 6-11. Probe assembly of a Geiger-Müller counter showing provision for the-mal neutron shielding.

shown in Fig. 6-11. The response as a pulse device is essentially linear from 0.1 mR/hr to 5 R/hr, as shown by the curve in Fig. 6-12.

By calculation it can be shown that the main contribution to the fast neutron response derives from inelastic collisions in the γ-energy correction shield and has a maximum at about 5.0 MeV neutron energy. An upper limit to the neutron response at 5 MeV is 0.15%. The assignment of a reasonably

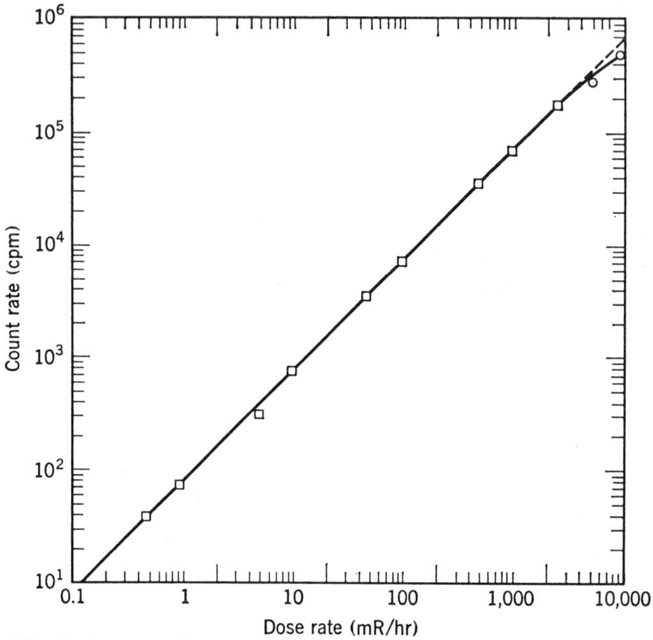

Fig. 6-12. Intensity response curve for the Philips micro G-M counter.

valid neutron response of the G-M counter by experimental methods is not possible at present owing to the unavailability of a facility having a radiation field with a sufficiently large neutron-to-γ-ray ratio. Studies in which the G-M counter was irradiated with monoenergetic neutrons in the energy range from 0.68 to 4.2 MeV indicate that the response in this energy range is less than 0.5%.

The response of the counter to thermal neutrons was determined experimentally, making use of the water thermal column at the Oak Ridge National Laboratory Graphite Reactor. It was concluded that the thermal neutron response of the probe unit is low, requiring about 5×10^9 neutrons/cm^2 to produce the same response as 1 R of γ-radiation. Further, the thermal neutron response may be reduced by using the lithium shield. With the lithium shield (Fig. 6-11), about 1.5×10^{12} neutrons/cm^2 would be required to produce the same response as 1 R of γ-radiation.

Special Counting Methods for Fast Neutron Dosimetry

A proton recoil proportional counter was designed[8] such that the first-collision tissue dose could be obtained simply by measuring the number of counts produced in the counter. The energy response $n_T(B, E)$ of the counter is determined by three sources of recoil proton shown in Fig. 6-13.

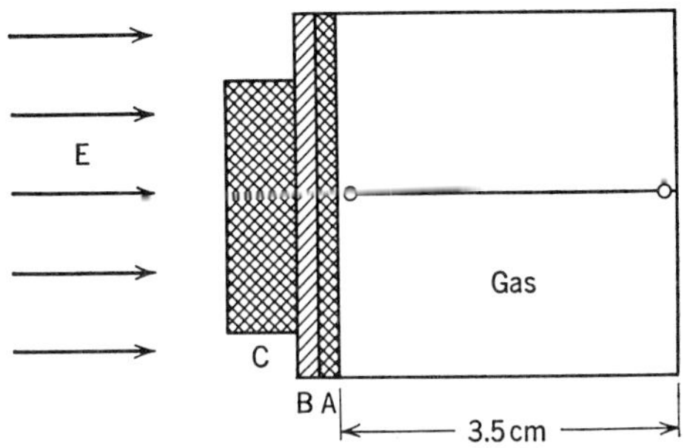

Fig. 6-13. Schematic representation of count-rate dosimeter. Key: A, paraffin (13 mg/cm²); B, Aluminum (29 mg/cm²); C, paraffin (100 mg/cm²). Ratio of paraffin areas 2.9. The gas is methane at 30 cm Hg pressure.

Calculations were made of the probability that neutrons of energy E could cause *a* recoil protons to lose in the counting volume energy greater than a bias energy B, needed to discriminate against γ-radiation. The energy responses (Fig. 6-14) for the three sources of protons are such that when added in the illustrated proportions of hydrogenous materials, the tissue first-collision dose curve is approximated (Fig. 6-15). The response has been checked experimentally with monoenergetic neutrons and agrees well with

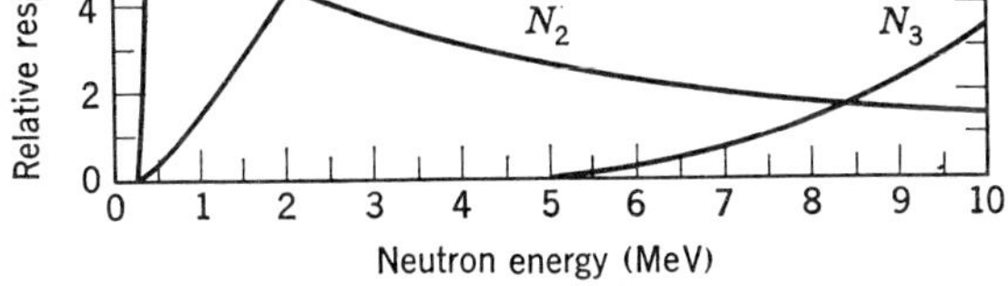

Fig. 6-14. N_1, response of plane slab of paraffin, for which $E_1 = 3$ MeV and $B = 0.2$ MeV; curve N_2, response of hydrogenous gas, where $B = 0.2$ MeV; curve N_3, response of thick paraffin slab, where $E_3 = 5$ MeV and $B = 0.2$ MeV.

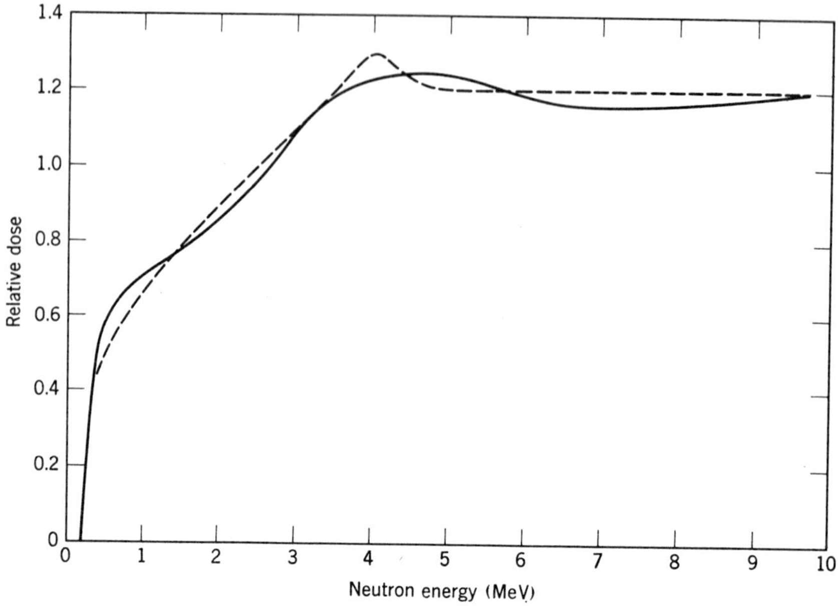

Fig. 6-15. Energy response of count-rate dosimeter. Key: — calculated response; - - dose per unit neutron flux (measured).

theory. The chief disadvantage of this counter is the fact that the response is directional, and the energy response is correct only when a plane beam is normally incident to the end of the counter containing the plane radiators.

Dennis and Loosemore[3] have modified the preceding ideas to develop a proportional counter that is nondirectional in response. In this case the neutron energy response curve follows the recommendations of the ICPR as given in *British Journal of Radiology Supplement* No. 6, 1955. At that time the permissible level was stated in terms of the absorbed energy at a depth 2 cm below the surface of tissue. Skjöldebrand[15] has developed a spherical scintillation detector whose energy response is adjusted to the multiple-collision dose curve of Snyder and Neufeld.[16] Thus this instrument indicates the maximum absorbed dose that a man would receive if located in the radiation field.

REFERENCES

1. *Neutron Cross Sections*, Neutron Cross Section Advisory Group, U.S. Atomic Energy Commission Report AECU-2040, Washington, 1952.
2. Auxier, J. A., G. S. Hurst, and R. E. Zedler, *Health Phys.* **1**, 21 (1958).

3. Dennis, J. A., and W. R. Loosemore, *A Fast Neutron Counter for Dosimetry*, Atomic Energy Research Establishment Report EL/R 2149, Harwell, 1957.
4. Failla, G., and H. H. Rossi, *Am. J. Roentgenal, Radium Therapy Nucl. Med.* **64,** 489 (1950).
5. Glass, F. M., *Nucleonics* **10,** No. 2, 36 (1952).
6. Hurst, G. S., W. A. Mills, F. P. Conte, and A. C. Upton, *Radiation Res.* **4,** 49 (1956).
7. Healey, R. H., and J. W. Reed, *The Behavior of Slow Electrons in Gases*, Amalgamated Wireless Valve Co., Sydney, Australia, 1941.
8. Hurst, G. S., R. H. Ritchie, and H. N. Wilson, *Rev. Sci. Instr.* **22,** 981 (1951).
9. Hurst, G. S., and R. H. Ritchie, *Rev. Sci. Instr.* **24,** 664 (1953).
10. Hurst, G. S., *Brit. J. Radiol.* **27,** 353 (1954).
11. Jordan, W. H., and P. R. Bell, *Rev. Sci. Instr.* **18,** 703 (1947).
12. "Measurement of Absorbed Dose of Neutrons, and of Mixtures of Neutrons and Gamma Rays," *Natl. Bur. Std. (U.S.) Handbook* **75,** (1961).
13. Palevsky, H., R. K. Swank, and R. Grenchik, *Rev. Sci. Instr.* **18,** 298 (1947).
14. Rossi, H. H., and G. Failla, *Nucleonics* **14,** No. 2, 32 (1956).
15. Skjöldebrand, R., *J. Nucl. Energy* **1,** 299 (1955).
16. Snyder, W. S., and J. Neufeld, *Brit. J. Radiol.* **28,** 342 (1955).
17. Wagner, E. B., and G. S. Hurst, *Rev. Sci. Instr.* **29,** 153 (1958).
18. Wagner, E. B., and G. S. Hurst, *Health Phys.* **2,** 57 (1959).
19. Wagner, E. B., and G. S. Hurst, *Health Phys.* **5,** 20 (1961).

PROBLEMS

1. Verify Eq. (6-4) for photon energies 0.5, 1.0, and 2.0 MeV.

2. Calculate the neutron response of a carbon wall–"carbon gas" ionization chamber for neutron energies 1, 5, and 20 MeV.

3. It is observed that in a mixed field of fast neutrons and γ-radiation, 12 times the charge produced by 1 R of γ-radiation is produced in a tissue-equivalent ionization chamber, and 3 times the charge produced by 1 R γ-radiation is produced in a Teflon-CO_2 chamber. Calculate the neutron and γ-ray tissue doses in rad units, assuming the neutron energy to be 1.0 MeV. If the neutron energy is known only to lie between 0.5 and 8.0 MeV, calculate the upper and lower bounds to the neutron and γ-ray tissue doses.

4. Verify Eq. (6-10) at neutron energies 1.0, 2.0, and 10 MeV.

5. Suppose that, in the proportional counter shown in Fig. 6-4, an improper gas-filling procedure leads to the introduction of 2 Torr (1 Torr = 1 mm Hg) of an attaching gas. The electron-attachment coefficient as a function of the ratio of the electric field strength E to the total pressure P is given by $\alpha(E/P) = 3.0\ \text{cm}^{-1}\ \text{Torr}^{-1}$ when E/P lies between 1 and 2 volt $\text{cm}^{-1}\ \text{Torr}^{-1}$ and is zero at all other values of E/P. Calculate the ratio of the observed pulse height caused by a proton moving parallel to the wire at a distance of 1.5 cm to that caused by a proton of the same energy moving parallel to the wire at a distance of 0.25 cm. Assume that the total pressure in the counter is 500 Torr, that a positive potential of 2000 volts is applied to the wire, and that the only effect of the attaching gas is to decrease the number of electrons reaching the wire.

Hint: The attachment is defined by the equation $-dN = \alpha(E/P)Nf_1P\,dx$, where dN is the number of electrons attached when N electrons move a distance dx in the field direction through a gas containing a partial pressure f_1P of an attaching gas. Note that α is a function only of the ratio E/P.

6. Explain the advantage of pulse counting techniques over charge integration techniques for measuring the neutron component in a field of mixed γ-rays and fast neutrons.

7. Show which moment of the energy-loss distribution is measured in the pulse counting technique if the counter is designed according to the Bragg-Gray principle.

8. Analyze the circuit shown in Fig. 6-7 and explain why it measures the first moment of an energy-loss distribution. Derive a discriminator biasing scheme that would measure the second moment of an energy-loss distribution.

9. Explain the electronic simplifications resulting from a detector design such that the radiation dose is determined from the zero moment of the energy-loss distribution in the detector. Is the ability to discriminate neutrons from γ-radiation retained in this case? Explain.

10. Derive Eq. (6-5) including the numerical constant.

CHAPTER 7

Special Methods in Radiation Dosimetry

J. A. Auxier

7-1 INTRODUCTION

Dosimetry is one of the most fundamental requirements in health physics and radiobiology. An understanding of both the interactions of radiations with matter and of the basic concepts of physical instrumentation is necessary to both the accurate measurement and interpretation of radiation exposures. This chapter purports to give a general introduction to dosimetry problems and examples of systems which have been developed to solve them.

The ideal and complete dosimetry in a radiation field would include (at the point of interest in the medium), for each type of radiation present, the following: (1) total energy absorbed, (2) dE/dx distribution, (linear energy transfer, LET_∞, or, better still, LET as a function of track radius), (3) angular distribution, (4) energy absorption rate, and (5) the energy spectrum. Seldom, if ever, is the ideal measurement possible. The quantities measured most often are exposure dose or dose rate for X- or γ-rays and first-collision dose for neutrons. The measurements have often been limited to total fluence or energy fluence measurements. Spectral distributions are sometimes determined for one or all types of radiation present, and occasionally the dE/dx or LET distributions are calculated. One instrument for measuring LET has been reported,[33] and certain emulsions[34] and cloud chambers[48,49] may be used for measuring dE/dx in the constituent media.

In 1962 the ICRU introduced a new unit for "dose," i.e., kerma,* which denotes "the sum of the initial kinetic energies of all the charged particles liberated by indirectly ionizing particles in a volume element of the specified material."[30] Because the term "first-collision dose" had been interpreted in

* See Section 3-2.

more than one way, the term kerma was intended to be more precise and less ambiguous. In general, it is equivalent to one of the widely accepted meanings of first-collision dose, but it does not fill the need for a quantity for "secondary-equilibrium" conditions. The units for kerma are ergs per gram. For γ-energies above 3 MeV and neutron energies above 15 MeV, the ratio of the energy liberated to that absorbed in a small mass of material becomes increasingly greater than unity.

The most widely used methods of "dosimetry" for indirectly ionizing radiation have been based on either (1) the measurement of the energy expended by secondary particles or (2) the computation of dose from fluence measurements. The first approach is typified by an ionization chamber having walls sufficiently thick to establish equilibrium between the indirectly ionizing radiation, for example, γ-rays, and the primary ionizing radiation, for example, Compton electrons. On the basis of the Bragg-Gray relationship, the energy deposited in the cavity E_c, that is, the product of the number of ion pairs per unit mass of gas J and the energy required to produce one ion pair W_c, can be used to compute the energy deposited in the wall medium E_s by the relationship

$$E_s = SE_c = SW_cJ \tag{7-1}$$

where S is the ratio of the mass stopping power of the wall medium to that of the gas for the radiation of interest. The criteria for applying the Bragg-Gray principle are given in Chapter 4 and in *NBS Handbook 75*, but it is essential that the ionizing particles be in equilibrium with the non-ionizing rays and that the chamber be sufficiently small so as not to perturb the incident radiation field.

The other approach to dosimetry is typified by the threshold detector technique.[21,32] Here, the fluence in various neutron energy increments is obtained by analyzing the reaction products in various detectors, for example, the fission product γ-rays from a ^{238}U foil which has an effective threshold for fission, when exposed to neutrons having a fission spectrum, of about 1.5 MeV and an approximately uniform cross section above that energy. From measurements of this type, the energy liberated (or absorbed, under equilibrium conditions) may be computed by the relationship

$$D(E) = 1.602 \times 10^{-8} E_n \sum_i N_i f_i \sigma_i \tag{7-2}$$

where $D(E)$ = neutron dose in rads/neutron/cm^2

E_n = the energy of the neutron (or average energy of the energy increments of neutrons)

N_i = number of atoms per gram of the ith species contained in the medium

f_i = average fraction of the energy of the neutron which is transferred to the interacting nucleus (in general, isotropic scattering in the center of mass system is assumed)

σ_i = elastic scattering cross section in barns (10^{-24} cm^2) for the ith species.

It is not necessary to measure exposure dose with an "air wall" ionization chamber or absorbed dose with a chamber constructed of a tissue-equivalent material. In the strictest sense, all that is required is a detector with associated parameters which permit the response of the detector to be interpreted in terms of the quantities to be measured. Hurst and Ritchie's "generalized principle of radiation dosimetry," discussed in Chapter 4, clarifies this concept.

7-2 PHOTOGRAPHIC FILM AND NUCLEAR EMULSIONS

General Discussion of Response to Radiation

Photographic film was one of the first radiation detectors used after the discovery of X-rays; in fact, natural radioactivity was first noted through this medium. It is still the most widely used detector for personnel monitoring. Photographic emulsions, usually on a plastic film base, are a very versatile medium whose sensitivity and other characteristics are largely controlled by their composition and processing during manufacture. Special techniques have been developed to adapt these emulsions to essentially all types of radiation measurements. The two major categories into which emulsions for radiation measurements are divided, according to intended use, are (1) X-ray film, usually in dental packet size ($1\frac{1}{4} \times 1\frac{3}{4}$ in.), which can be calibrated for X- or γ-ray exposures in terms of blackening; and (2) nuclear emulsions on film or glass plates which are used in studying ionizing particle tracks.

The radiation sensing element in these emulsions is a silver halide, most often silver bromide. Small crystals of the silver bromide, ranging in size from less than 0.5 micron in nuclear emulsions to a few microns in the most sensitive X-ray emulsions, are suspended in a gelatin which is generally deposited on a cellulose acetate film or on a glass plate. The thickness of this layer of emulsion ranges from a few microns for insensitive X-ray film to the order of 2000 microns in the thicker particle detecting emulsions. Electrons traversing the emulsion become trapped in the crystal lattice, causing the reduction of silver ions to atomic silver. These silver atoms, in turn, result in deeper traps, which capture electrons, reducing more silver ions, eventually forming microscopic aggregates of silver atoms which constitute the latent image. Upon development, the latent images are reduced to metallic silver which appears to the eye as the blackening of X- or γ-ray film or microscopic tracks in the nuclear emulsions. Undeveloped silver halide grains are dissolved by the fixer.

Photographic Film As an X- or γ-Ray Dosimeter

Many types of film are commercially available and packaged in the small dental-type packet. Packets are available that contain more than one film type to increase the useful dose range of the packet. Table 7-1 lists some

Table 7-1 Approximate Exposure Ranges of Common Types of Film Using ORNL Processing Techniques

Type	Manufacturer	Description	Approximate Exposure Range (R)
508	DuPont *	Extra fast medical X-ray film	0.03–6
1	Eastman	Single emulsion	0.03–1
555	DuPont *	Single emulsion[a]	0.05–10
2	Eastman	Double emulsion	0.03–1000
834	Eastman	Single emulsion[a]	5–800
1290	DuPont *	Single emulsion	10–3000
548.0	Eastman	Spectroscopic	1000–30,000

[a] Used in 544 film packet.
* No longer available (1973)

common types and their approximate ranges for γ-rays of energy greater than about 300 keV. The exact range of a given type of film depends on exposure, storage, and development conditions.

A typical developing procedure, and that used at ORNL, comprises the following steps: (1) develop in DuPont concentrated X-ray developer (liquid mix) for 3 min at 68 ± 2°F, (2) stop bath for 1 min, (3) fix for 10 min in DuPont X-ray fixer, (4) quick-rinse with agitation for 1 min, and (5) wash for 20 min. The sensitivity of all emulsions can be varied by as much as an order of magnitude by varying the developer and procedures. Typically, films are processed in relatively large batches and a complete set of calibration film is developed simultaneously. In addition to developing time and temperature for a given film and developer type, the dilution and deterioration of the developer by use and exposure to air also affect the blackening.

All photographic emulsions show a marked variation in response per roentgen as a function of photon energy. The major factor in this response function is the relatively high atomic number of silver, that is, the photoelectric cross section for silver for photons of energy less than ~300 keV is large compared with the cross sections of the elements of which air is composed.

A typical set of blackening versus dose curves for DuPont 555 emulsion is given in Fig. 7-1. One curve represents each of four filters; one of these is the plastic holder and is thus essentially unshielded. These and other filters

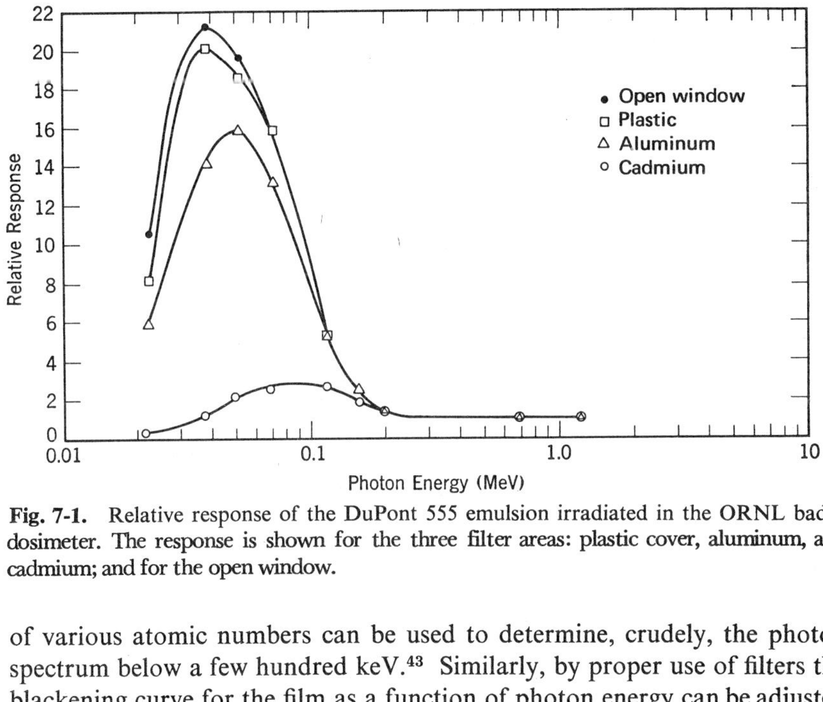

Fig. 7-1. Relative response of the DuPont 555 emulsion irradiated in the ORNL badge dosimeter. The response is shown for the three filter areas: plastic cover, aluminum, and cadmium; and for the open window.

of various atomic numbers can be used to determine, crudely, the photon spectrum below a few hundred keV.[43] Similarly, by proper use of filters the blackening curve for the film as a function of photon energy can be adjusted over a wide range so that it has the same shape as the energy absorption curve for air, tissue, or other medium of low atomic number. Comprehensive studies of such methods have been reported,[13] and filters have been used with film to measure exposure doses of X- and γ-rays over the energy range of about 115 keV to 10 MeV to within $\pm 20\%$ for large dose ranges. The accuracy can be increased for more restricted use.

Beta Radiation Measurements

The response of film to electrons is an insensitive function of the electron energy (comparing blackening to air ionization). This has been demonstrated for the energy region from 0.5 to 1.4 MeV for paper-wrapped film.[22] Typical films exhibit about the same sensitivity to β-rays as to γ-rays, that is, 1 rad of β-rays yields about the same film density as 1 R of γ-rays. A typical film will be surrounded by about 30 mg/cm^2 of paper. When calibrated in the usual manner, that is, by laying the film packet flat on a slab of the β-emitting material that is thicker than the maximum range of the β-particles in it, such film packets exhibit an energy-dependence for β-rays.[40] The shape of this curve is influenced by the type of emulsion and its wrapping. A comprehensive evaluation of some film types in β-ray dosimetry was made by Dudley.[12]

Neutron Measurements

X-ray film has a blackening response to fast neutrons about two orders of magnitude less than that to γ-rays. Therefore, for most applications, a special class of emulsions has been developed. These fine-grain emulsions are coated on the usual acetate film base or on glass plates. Table 7-2 shows some of the commercially available emulsions and their thicknesses and type of bases. In principle, neutron exposures can be determined by the number, length, and angular distribution of recoil proton tracks produced in the emulsion. In practice, neutron exposures can be estimated by simply counting tracks, provided that the emulsion is surrounded by the proper thickness of proton radiators. A track, viewed through a microscope, can be identified only if at least three grains have been developed; this corresponds to a length of about 3 μ and a proton energy of 250 keV. For unidirectional neutron beams, it is possible to use nuclear emulsions to determine fast-neutron

Table 7-2 Nuclear Emulsions for Particle Detection

Emulsion	Available Thickness[b] (Microns)	Type of Base	Approximate Upper Energy Limit (MeV) for Particle Identification: Protons	Particles
Eastman NTA[a]	25–1000	Glass or plastic	3	10
Eastman NTB	50 and 100	Glass or plastic	8	400
Ilford E-1	To 400	Glass	20	250
Ilford G-5	To 2000	Glass	Very high	Very high

[a] Available with aluminum lamination as Kodak personnel neutron monitoring film, type B.
[b] Most emulsions are available in "pellets" without backing.

spectra (in principle, energies greater than 250 keV but, in practice, for energies greater than 500 keV).[9] Special processing techniques must be applied when thick emulsions are used.

By "loading" the nuclear emulsion with boron, lithium, or other isotopes having large (n, α) or (n, p) cross sections, it is possible to use them as sensitive thermal neutron detectors. Because of the nitrogen already in the emulsions, thermal neutrons may also be estimated by covering a part of the emulsion with a thermal neutron shield, commonly 0.040 in. of cadmium, and observing the difference in the number of tracks in the shielded and unshielded areas caused by the $^{14}N(n, p)^{14}C$ reaction in the unshielded areas. The applications of these techniques to personnel monitoring are discussed in Chapter 15.

Advantages and Limitations of Film and Emulsion Techniques

The most outstanding advantages of film include the following: (1) range of sizes down to the order of millimeters, (2) permanent record of measurement, (3) wide range of total exposures, (4) low weight for most applications, (5) excellent spatial resolution, and (6) simultaneous recording of different radiations. Disadvantages of film include (1) photon energy dependence, (2) sensitivity to small changes in processing conditions, (3) the quantity of processing time and equipment, (4) latent image fading, (5) sensitivity to temperature and humidity conditions for most applications, and (6) the tedious effort and skill required for analyzing the nuclear emulsions. Film will be used for many applications in the future, but its role in personnel monitoring is expected to decrease as the use of solid-state devices, for example, photoluminescent systems, increases.

7-3 CHEMICAL DOSIMETRY

General Considerations of the Production of Radicals in Solutions by Radiation

Various chemical systems have been used to detect and measure radiation for many years. Those to be discussed briefly are solution types dependent upon the production of acids or free radicals by radiation. For example, acid generated in an organic solution containing an acidimetric dye produces a color change which can be related to dose qualitatively by inspection.

The detailed processes of irradiation-produced effects in solutions are poorly understood even for water.[17,11,1] Various reaction products are involved in further reactions and back reactions which rapidly become more complex as the number of reagents increases. However, for a given chemical system, the production of radicals to which the sensing element will respond can be empirically established and the system can be calibrated as a radiation detector.

The acid yield is generally specified in terms of "G-values," that is, the number of acid radicals formed per 100 eV of energy absorbed in the medium. The G-values range from near unity to several thousand in various systems; yields higher than about 20 are indicative of chain reactions. In general, long-chain reactions must be stabilized by such reagents as ethyl alcohol, resorcinol, etc., which render the system less sensitive and more stable than the unstabilized solution. A G-value range of 15 to 30 covers most of the commonly used chemical dosimeter systems.

All sensitive chemical systems have one requirement in common: the reagents must be scrupulously pure. The best commercially available reagents must generally be repurified by multiple redistillation or recrystallization. Even slight traces of contaminants result in loss of stability and reproducibility. Containers must be nonreactive; an ampoule of borosilicate glass, such as pyrex or Neutraglas, which has been steam-cleaned and siliconed is best.

Ferrous-Ferric System

The discussions of this dosimetry system are generally applicable to the ceric-cerous system except for obvious differences. A solution of pure ferrous ammonium sulfate in 0.8 *N* sulfuric acid is the most commonly used ferrous-ferric detector. Upon irradiation, ferrous ions are oxidized to ferric ions. The number of ions oxidized is directly related to energy absorbed from X- or γ-radiation, with the exceptions to be discussed. A spectrophotometer can be used to determine the amount of ferrous ion oxidized by measuring the absorption change at the 5100 A peak or the ferric ion absorption change at the 3045 A peak.[18]

Because of the high effective atomic number (relative to air and tissue), the ferrous-ferric system exhibits a marked energy-dependence for photons below several hundred keV. The exact height of the peak of the response curve, yield versus photon energy, is a function of ampoule type and diameter. The most suitable range for the dosimeter is from about 5000 to 50,000 rad. The response as a function of fast-neutron energy is not accurately known; the monoenergetic neutron sources most nearly free of gamma contamination, such as Van de Graaff generators, yield too low doses for most experimenters to conduct studies of response as a function of neutron energy. With proper control and calibration, the effects of room temperature, oxygen concentration, and dose rates on the system are usually negligible.

Chlorinated Hydrocarbon Systems

Two types of chlorinated hydrocarbon dosimeters used frequently in the past are: the "water-equivalent single-phase" dosimeter, consisting of an aqueous dye solution saturated with the chlorinated hydrocarbon, and the "two-phase" dosimeter consisting of a chlorinated hydrocarbon overlaid with a layer of aqueous dye solution.

The single-phase dosimeter is a "short-chain" type which is stable over periods up to a year and, with about 0.02% resorcinol, shows less dose rate and energy-dependence effects than most chemical systems. Due to the hydrogen content, it is sensitive to fast neutrons. The chlorine renders the system sensitive to thermal neutrons. The G-yield of the dosimeter is about 20.

A typical two-phase dosimeter consists of tetrachlorethylene with 0.1% resorcinol overlaid with a layer of water-dye solution (about 25% by volume); a typical dye is chlorphenol red. The G-yield of uninhibited C_2Cl_4 may be as high as 6000, but in the stabilized dosimeter it is about 30.[20] As the pH of the dye solution changes from 6.0 to 5.0, the light transmission at 5800 A increases and the transmission at 4320 A decreases. This corresponds to the usual operating range, and the color changes from red to yellow. The ratio of the transmission at the two wavelengths, 5800 A and 4320 A, is a monotonically increasing function of dose in this pH region as shown in Fig. 7-2. This system exhibits a marked photon energy-dependence of response below 500 keV unless a filter is provided around the detector to "flatten" its response, but it is not rate dependent for dose rates at least as high as 4000 rad/min.[39] It is temperature insensitive from about 5 to 55°C, and the response is linear with total dose to greater than 100,000 rad. Thermal neutrons effect a large response (1 tissue rad of thermal neutrons produces a response equal to about 5 tissue rad of γ-radiation), but 1 tissue rad of fast neutrons yields less than 5% of 1 tissue rad of γ-rays. These dosimeters are stable over periods of years; an exception is the least inhibited, that is, most sensitive ones for less than 50-rad exposures, which are stable for approximately one year.

Advantages and Limitations of Chemical Dosimeters

Perhaps the greatest attribute of chemical dosimeters is the wide range of applications for which one type or another can be made. They are economical when made in large numbers and they, like film, can be made in a great range of sizes. For very high doses ($>10^5$ rad), chemical systems are perhaps the most suitable integrating systems available. For lower doses (<1000 rad), they are extremely sensitive to chemical and ampoule purity and require care and experience for stable reproducible dosimeter readings.

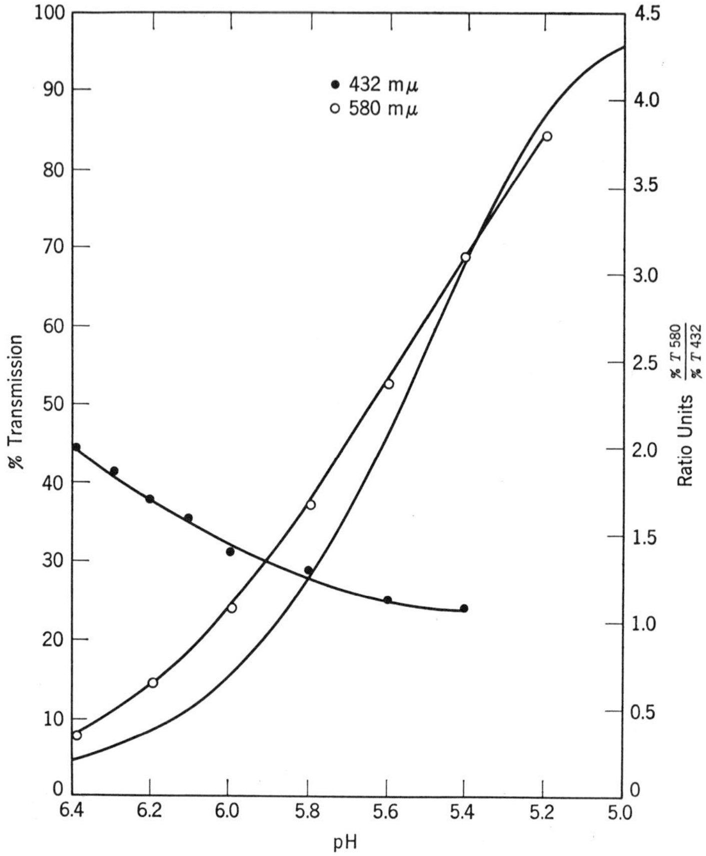

Fig. 7-2. Percent transmission at 580 mμ and 432 mμ (left scale) and the ratio of transmission 580/432 mμ as a function of pH.

7-4 LUMINESCENCE DOSIMETRY

Discussion of the Response to Radiation

Radiation-induced coloration and luminescence of crystals and glasses have been studied for many years (see Weyl[47] and Seitz[38]). Commencing about 1950, a rapidly increasing number of studies of silver-bearing glass for dosimetry purposes was reported; much of the work was done at the U.S. Naval Research Laboratories.[14,35–37,47]

In general, many substances exhibit phosphorescence and coloration when exposed to radiation; in the gaseous state, at sufficiently low pressure, the atoms of such substances exhibit resonance fluorescence, that is, wavelengths characteristic of the atomic-energy levels consisting of a spectrum of discrete lines. These lines represent the permitted transitions from various energy levels of electronic excitation to the ground state.

In ideal crystals there are forbidden energy zones between the permitted energy levels. Because of this periodic field of potential, irradiation in the ultraviolet absorption band has no permanent effect on the crystal. For example, an anion in an alkali halide crystal loses an electron and is changed into a halogen atom or "positive hole." Such an electron immediately returns to its positive hole, and there is no permanent effect on the crystal. The introduction of impurities or imperfections into the lattice structure

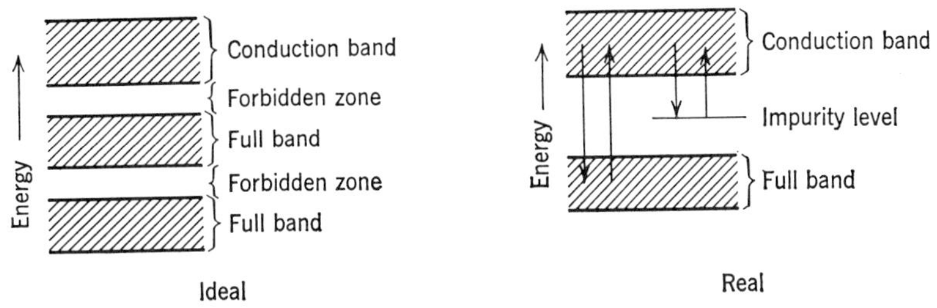

Fig. 7-3. Energy-level diagrams for real and ideal crystals.

disturbs the periodicity of the lattice structure and produces localized energy levels which, in ideal crystals, would lie in the forbidden zone. The energy levels for ideal crystals and for real crystals having impurities are shown in Fig. 7-3. Electrons raised to the conduction band can fall back into these defect levels and become bound to the defect. Hence an impurity-bearing crystal may be changed by radiation and possess a changed absorption spectrum.

When silver-bearing metaphosphate glass is exposed to radiation, two pronounced effects are observed: (1) the absorption peak, which is at 2400 A for the unirradiated glass, is found at 3300 A after irradiation; (2) the emission spectrum peak shifts from 3700 to 6400 A. Either of these effects may be utilized for dosimetric purposes. The change in absorption at one peak or the ratio of absorption at both peaks can be related to X- or γ-radiation dose, with the exceptions noted below. Furthermore, the fluorescence at the 6400 A emission peak may be excited by 3300 A ultraviolet and related to X- or γ-radiation dose. Schulman et al.[36] have described this latter effect as "radiophotoluminescence," a term defined by Przibram[31] as "that phenomenon whereby a material, originally nonluminescent under UV or visible light, is made responsive to such excitation by pretreatment with γ-rays or X-rays," with the further distinction that the newly created stable luminescent centers would not be destroyed by the UV radiation that is used to excite them. Studies of these two effects have centered chiefly on one type of glass, and development of dosimeters has followed similar lines, so the discussion below will be limited to the radiophotoluminescence phenomenon and a metaphosphate glass, generally designated as "high Z," for example, one comprised by weight of 46% $Al(PO_3)_3$, 23% $Ba(PO_3)_2$, 23% KPO_3, and 8% $AgPO_3$.

When such glass is exposed to radiation, loosely bound electrons are released from the negative ions of the glass, and some of them are trapped by interstitial silver ions, forming a special type of F-center, or "luminescent center." Until recently, it was assumed that in the capture of liberated electrons, the Ag^+ ions were in competition with electron traps which arise from irregularities, that is, impurities, in the molecular structure of the glass. The traps would thus provide alternate energy levels which could be occupied by electrons. The distribution of electrons between Ag^+ ions and other traps would control the relationship between electrons liberated in the glass by radiation and the fluorescence. The Ag^0 atoms formed by electron capture (metastable), when exposed to UV light would absorb energy and upon returning to ground state fluoresce with a peak at about 6400 A. Recent work by Yokota and Imagawa,[50] using electron spin resonance (ESR) techniques, have shown that the important luminescence centers are Ag^{++} and Ag^0–Ag^0 pairs. Further theoretical considerations are given by Vogel and Becker.[45]

Silver Metaphosphate Glass as a Gamma Dosimeter

In order to calculate the energy absorbed by a glass detector, assume that the response of the detector (or of the Ag^0 atoms produced) is proportional to the number of electrons freed in the detector. The spectrum of tertiary electrons (the vast majority of the electrons formed by radiation are tertiaries) has been shown to be an insensitive function of photon energy.[16] Then the glass response should be directly proportional to the energy deposited in it by ionizing electrons, which, under electron equilibrium conditions, is equal to the energy deposited by the incident photons. For a uniform radiation field of photons of energy $n(E)$ and a glass detector exposed in it under equilibrium conditions, of such size as not to perturb the field significantly, the energy absorbed per square centimeter of glass can be written as

$$E_a = \sum_i nE_i e^{-[(\mu_i)\tau]\rho} \tag{7-3}$$

where n is the fraction of photons of energy E_i, and μ_i = the average total mass absorption cross section (σ_T, σ_C, and σ_P, the photoelectric, Compton, and pair-production cross sections, respectively) for the elements in the glass for photons of energy E_i, τ is the thickness of the glass, and ρ the density.

For a specific detector, Thornton[42] has written a more detailed equation.

A widely used detector composed of the metaphosphate glass described consists of a small glass rod 1 mm in diameter by 6 mm long. Although several varieties of glass are available commercially, that mentioned above, that is, the high *Z* glass, has been evaluated for the greatest variety of radiations and energies. The characteristics of the response of this detector are given in the paragraphs below.

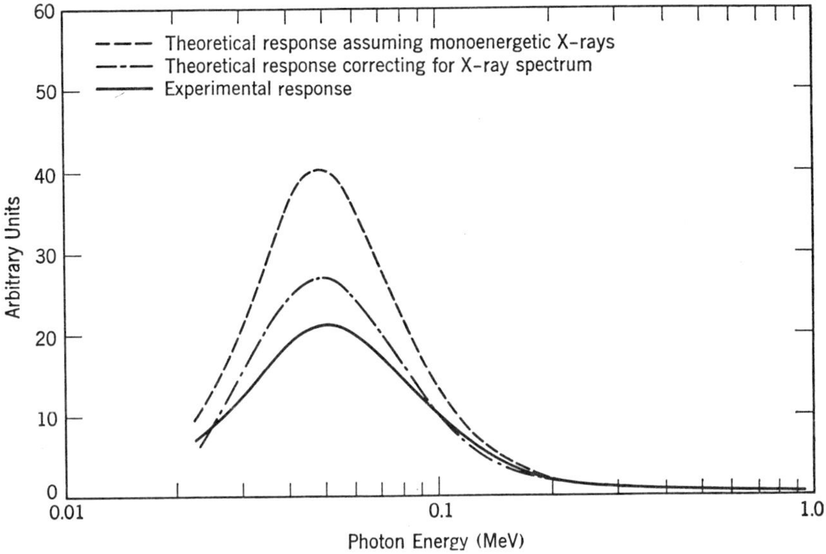

Fig. 7-4. Comparison of the theoretical and experimental photon responses of the unshielded glass.

Figure 7-4 shows the calculated response of the "high *Z*" detector as a function of photon energy. Also, the figure shows a typical measured response curve for various X-ray spectra as a function of "effective" energy. "Effective" energy is defined as the energy of a monochromatic beam which has the same absorption coefficient as the given beam in an incremental thickness of standard filter material. Of course, many of the photons lie well above and below the effective energy causing the peak of the curve to be reduced in height and width. A third curve in this figure represents the response calculated by utilizing Villforth's calculation[44] of Kramer's energy distribution of filtered X-rays; the agreement of these values with the experimental ones is good, if we consider the many approximations used.

Figure 7-5 shows the effectiveness of some typical encapsulation techniques for flattening the peak in the response versus photon-energy curve. For photons greater than 300 keV, the effective dose range of these detectors is

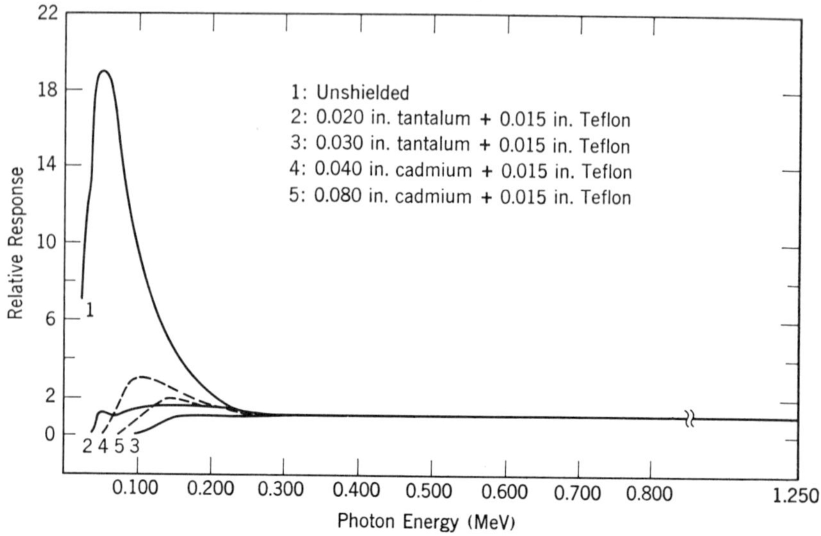

Fig. 7-5. Luminescent response of glass rods as a function of photon energy for various shields.

linear from about 5 to about 10,000 rad. Such detectors have little or no dose rate dependence, but the long-range fading characteristics are not as good as those of the "low Z" glass developed more recently. Cheka[10] has examined the glass characteristics with a view to adapting the detector for personnel monitoring. It appears to be suitable for most applications including those requiring frequent reading, since it may be read at intervals without affecting the integrated response.

Response of the Metaphosphate Glass to Neutrons

The principal thermal neutron absorber in the glass is silver, ^{107}Ag and ^{109}Ag, with cross sections of 30 and 84 barns, respectively. As both ^{108}Ag and ^{110}Ag are β- and γ-emitters with short half-lives (2.4 min and 24 sec, respectively), there is no prolonged buildup after exposure. Because of the small size of the detectors, the β-particles contribute most of the response. The measurement of γ-dose in a field of coexistent thermal neutrons and γ-rays may be accomplished by encasing the rods in a neutron absorber that does not emit β- or γ-rays. Lithium-6 with an (n, α) cross section of 945 barns is commonly used.

Although the energy deposited in the glass by fast neutrons as a function of neutron energy may be calculated, the effectiveness of the ionization produced by the recoil ions cannot be assumed to be the same as it is for fast electrons, that is, "columnar" recombination along the path of such a particle might remove many of the electrons before they could migrate to silver ions. Kondo[26] has indeed found evidence that heavy particles are less effective in producing fluorescent response.

An experimental determination of the response of the glass to fast neutrons has been reported[7] for a limited energy range, and Kondo[25] reported measurement at 14 MeV. Figure 7-6 shows the response for neutrons of 0.5 to 1.5

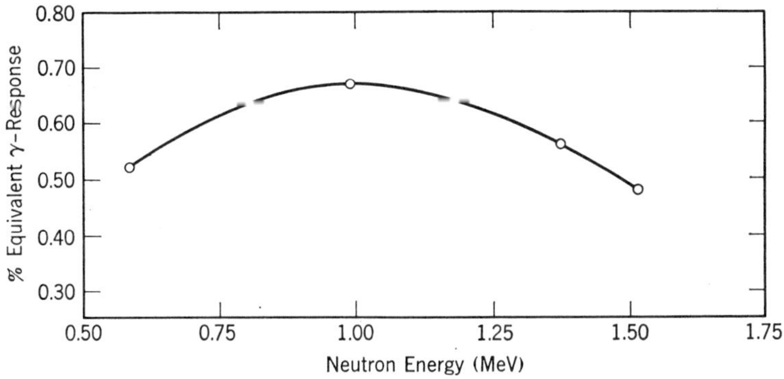

Fig. 7-6. Neutron response of glass relative to ^{60}Co γ-response. (Other more recent data incate a monotonic increase in response with neutron energy).

MeV exposed in a fluorothene capsule and uncorrected for the low γ-radiation background in the Van de Graaff target room. The data are sufficient to indicate the low sensitivity to fast neutrons.

Advantages and Limitations of the Metaphosphate Glass Dosimeter

The commercially available system offers a reliable, economical, and fast neutron-insensitive method for γ-radiation dosimetry. For many applications, such as tissue implantation, the small size is of most importance. As repeated readings can be made without destroying the fluorescence, the detector offers the possibility of a "life-time" integrated reading. The marked dependence of the response on photon energy in the photoelectric region and the sensitivity to thermal neutrons are the greatest limitations of the unencapsulated glass. For the most accurate work, prereadings are required on all dosimeters, and care must be taken to clean the glass thoroughly before evaluation of the fluorescence is attempted.

New types of glass and new readers are being produced more and more frequently. The new "low Z" glasses available in West Germany and Japan make possible measurements of total exposure as low as 10 to 20 mR with a precision of about $\pm 30\%$ or better.[51] They also have exhibited markedly less dependence of response per unit exposure as a function of photon energy, as they are composed of $LiPO_3$ and $Al(PO_3)_3$ with just a few percent of $AgPO_3$.

The presence of large amounts of lithium makes these glasses highly sensitive to thermal neutrons. However, by the use of glass made with depleted lithium (^7Li), along with a specimen of glass made with natural lithium, a differential reading not only enables one to detect the presence of neutrons, but also to evaluate the γ-fraction of a mixed field.

Thermoluminescence Dosimetry

Much of the preceding discussion can be applied to thermoluminescence, the process by which a substance, not normally luminescent when heated to temperatures below its thermally luminescent threshold (optical), can be made to luminesce at these temperatures by pretreatment with ionizing radiation. The chief chemicals used for this application are CaF and LiF, CaF being more sensitive and the LiF exhibiting less "energy dependence." The thermoluminescent systems are suitable for the same applications, in general, as the photoluminescent systems, but there are a few differences which may be noteworthy. The thermoluminescent systems are more sensitive, and exposures of the order of 1 mR have been detected with CaF. This system of dosimetry is less dependent in response on photon energy than the photoluminescent system if LiF is used, but it is also more sensitive to neutrons which hamper its use in coexistent neutron-photon fields. The thermoluminescent system is also, in general, somewhat less convenient to use than the photoluminescent system because of the dry powder form in which the fluoride is used; several researchers are striving to develop a more convenient form of thermoluminescent detector.

A second feature of the thermoluminescent system is the erasure of the response during reading. If repeated use of the detector for unrelated measurements is desired, this feature is an advantage; if long-term accumulation of exposure with intermediate readings is desired, this feature is a disadvantage. This disadvantage can be overcome in many instances by utilizing several detectors simultaneously at the point of interest with sequential removal for evaluation. A comprehensive review of this system is given by Attix.[2]

7-5 SCINTILLATION COUNTER METHODS

General Discussion of Scintillators

The modern scintillation counting system consists of a scintillator, a photomultiplier tube with a magnetic shield, and the associated electronics to amplify, analyze, and display the output. In general, there is radiation shielding for the detector assembly. Two types of associated electronics are common: pulse amplification and pulse height analysis, and current amplification and display.

Scintillation detectors or phosphors of many types are available for many applications. The phosphors include organic solids, organic phosphors in plastic solids, organic solutions, inorganic solids, and inorganic gases.

Much of the discussion of luminescent centers and the excitation of metastable states given in the preceding section parallel the theory of phosphors; a major difference is the decay time of excited states. In the crystalline detectors such as NaI(Tl), the thallium activation or impurity centers are the luminescent centers as well; the impurity center absorbs some of the excitation energy of the atoms of the crystal and emits photons in the wavelength region typical of the NaI(Tl) scintillator. The amount of impurity is small, usually within the limits of 0.01 to 1%. In the various noncrystalline scintillators, some of the excitation energy in the medium, for example, the solvent, is absorbed by the organic molecule which undergoes deexcitation in the wavelengths characteristic of the phosphor.

Table 7-3 lists some typical, commonly used scintillators and some of their characteristics. The values are taken from several of the general references,

Table 7-3 Some of the Characteristics of a Few Typical Scintillators

Name and Composition	Density, gm/cm^3	Peak Emission Wavelength, A	Scintillation Yield Relative to Anthracene[a]	Pulse Decay Time, μsec
Organic				
Anthracene				
$C_{14}H_{10}$	1.25	4450	1.0	0.03
Stilbene				
$C_{14}H_{12}$	1.16	4100	0.7	0.007
p-Terphenyl in Xylene				
C_8H_{10}	0.86	4100	0.2	0.008
Typical commercial				
Plastics	1–1.05	~4500	0.5–0.7	<0.005
Inorganic				
Zinc sulfide				
ZnS	4.1	4500	1 to 2[a]	~10
Sodium iodide				
NaI(Tl)	3.67	4100	~2	0.25
Calcium tungstate				
$CaWO_4$	6.06	4300	~0.75	~1

[a] Zinc sulfide scintillations (yield and decay) are sensitive to the crystals' history, for example, quenching.

most are from References 8, 29, and 41, and some values are averages of reported values.

Once the scintillations are produced, the photons must reach the photocathode of the photomultiplier to produce a response. Typically, solid scintillators are surrounded, except on the phototube side, with thin reflectors of materials such as MgO or Al_2O_3, and the scintillator with a light pipe and phototube are joined either directly or with silicone grease (or a similar viscous liquid) between the faces. In either case, lucite is a common light pipe which transmits a high fraction of the light, and, in addition, is frequently used to avoid having surfaces with large differences in refractive index directly coupled; large differences cause greater reflection losses.

Photons that pass to the photocathode surface eject electrons from it. These electrons are focused by electric fields to the first of several successive multiplying electrodes or dynodes, at which each electron will, typically, cause the ejection of about four others. Since phototubes generally have eight to eleven of these dynodes, the over-all multiplication is of the order of 10^6. A large variety of tubes is available. The range of photocathode sizes is from less than a square inch to hundreds of square inches of area. The peak response wavelengths for the more common tubes range from less than 4000 A to greater than 5000 A. Although most photomultipliers have circular end windows and photocathodes, a few are available with a side window and photocathode. The output of the phototube may be fed into one of a large variety of electronics systems. The electronics systems for application in scintillation work comprise an extensive field, and the reader is referred to many comprehensive works on that subject for example, References 15, 6, and 29.

Scintillation Counters for Beta-Gamma Dosimetry

The scintillation produced by a fast electron in the scintillator is directly proportional to the energy absorbed over a wide energy range. Consequently, in "thick" detectors (thickness greater than the pathlength of the most energetic electrons), the energy distribution may be determined by pulse height analysis. For β-particle and fast-electron counting, anthracene is commonly used. Inorganic phosphors, for example, NaI(Tl), are widely used in γ-ray spectroscopy. The relatively high Z of these phosphors and consequently their high photoelectric cross sections make possible the total absorption of many photons even in relatively small crystals, for example, 2-in. × 2-in. cylinders. An incident photon of 1 MeV, for example, can undergo a Compton interaction in the crystal and impart some fraction of its energy to an electron resulting in a scintillation; the scattered photon can then interact in the photoelectric region which, of course, results in all the remaining energy causing a scintillation which is simultaneous, in the time scale of the apparatus, with that of the Compton electron. The output pulse is then proportional to the total energy of the incident γ-ray.

Scintillation Counters for Neutron Dosimetry

Typically, organic phosphors having a large hydrogen content are used in scintillation counters for neutron measurements. In principle, neutrons collide elastically with the hydrogen nuclei; these recoil protons then expend their energy in the phosphor, producing a pulse which is directly proportional to the energy absorbed provided the crystal is large relative to the proton pathlengths of interest. If the phosphor is "tissue-equivalent," or nearly so, the scintillation output is equal to absorbed tissue dose for a specimen of the size of the detector; for typical detectors, this is equivalent to first-collision tissue dose.

However, there are no pure neutron fields; the γ-radiation in the field will also react with the phosphor, complicating the measurement. Many devices and techniques have been developed to effectively "desensitize" the detector to γ-rays. It is beyond the scope and space of this chapter to list and discuss these techniques of such a rapidly changing field. It is suggested that the reader refer to the large volume of material published annually on neutron spectroscopy by scintillation techniques.

A significant use of scintillation detectors near high-energy accelerators is based on the $^{12}C(n, 2n)^{11}C$ reaction. The carbon-containing crystal can be irradiated and then transported to a counting room, placed on a photomultiplier, and analyzed for ^{11}C.[3,28] The reaction has a threshold at about 20 MeV and an average cross section of about 22 mb in the energy region up to several hundred MeV. Because the short (20.4 min) half-life of the ^{11}C, the technique is best suited for short bursts of radiation.

Thermal neutrons and "slow" neutrons are frequently detected by using phosphors containing atoms having a large cross section for (n_t, P) or (n_t, α) reactions such as the $^{6}_{3}Li(n_t, \alpha)^{3}_{1}H$ reaction.

Advantages and Limitations of the Scintillation System

Perhaps the greatest advantage of the scintillation detector is its high sensitivity. Another feature, also related to the relatively high density, is the applicability to spectroscopy. The wide range of sizes and shapes of the various phosphors is a distinct advantage as is the great resolution of events in time and the range of elemental composition (especially hydrogen). The complexity of associated equipment necessary for many applications is frequently a disadvantage. Sensitivity to electric and magnetic fields is often the limiting factor in using scintillation equipment. In general, scintillation detectors have low sensitivity to low-energy radiation because of the bias against "noise" and poor energy resolution compared with, for example, proportional counters.

7-6 CALORIMETRY

In principle, calorimetric determination of the energy absorbed by the medium of interest from the radiation beam of interest represents the ideal measurement of that one parameter, that is, absorbed dose. This assumes that all of the energy absorbed is converted to heat. If one were interested in the energy that a certain X-ray beam would impart to a cubic centimeter of water, one would simply place the water in the beam in a well-insulated container and measure the temperature rise as a function of time.

The most limiting factor in calorimetric dosimetry has been for many years the inadequacy of thermometry systems. The temperature changes to be measured are generally small, relative to those that can be measured accurately by resistance changes, the thermoelectric effect, or the physical expansion of the absorber. However, since about 1945, a more sensitive temperature detecting element has been available; this is the thermistor.[4]

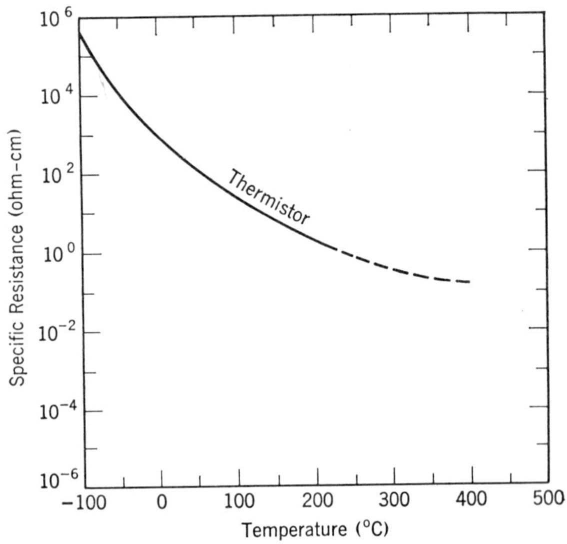

Fig. 7-7. Logarithm of the specific resistance versus temperature of thermistors composed of the oxides of manganese, nickel, and cobalt.[4]

Thermistors are metallic oxide semiconductors which have high negative temperature coefficients. A typical thermistor is the type 12A Western Electric which exhibits a negative temperature coefficient of $-3.9\%/°C$ at 20°C.

The resistance change of the thermistor can be readily measured with a Wheatstone bridge and a galvanometer; most researchers use a d-c amplifier and a recording galvanometer.[27] Figure 7-7 shows a temperature versus resistance plot for a thermistor.

The two most frequently used calorimeter systems are the total-absorption and the low-perturbation types. Laughlin et al.[27] have reported total-absorption calorimeters in which sensibly all of the incident radiation is absorbed. For determining the first-collision energy absorption, a low-perturbation calorimeter is necessary, that is, one that does not significantly alter the radiation field. Such a calorimeter has been reported by Kalil et al.[23]

Total-Absorption Calorimeter

For some applications, the total energy flux in a beam of radiation is desirable, for example, in teletherapy. For measurements of this type, Laughlin et al.[27] designed a total-absorption calorimeter. Basically, a lead cylinder exposed end-on to a collimated beam was used as the total absorber. The lead-filled cylinder was 6.2 cm in diameter by 3.4-cm long and contained a small heating element and a thermistor; it was suspended in an evacuated calorimeter box having thin-walled apertures for the radiation beam entrance, and the box was then surrounded by a controlled-temperature oil bath, except at the radiation entrance ports. The thermistor was incorporated in a Wheatstone bridge circuit, whose output was connected to a d-c amplifier which was followed by a recording potentiometer. Calibration was based on absolute energy input by the heating coil. Measurements of intensity of X-rays (400 kV self-rectifying with 1.75 mm total equivalent filtration) gave excellent agreement with theory and with measurements utilizing ionization chambers.

Low-Perturbation Calorimeters

An example of a low-perturbation calorimeter is that used by Kalil et al.[23] to measure electron stopping power of foils. Here, foils of the order of 100 $\mu g/cm^2$ were the absorbing body and caused little perturbation of a fast-electron beam. For X- or γ-ray measurements, thicker absorbers could be used.

The major components of the calorimeter used by Kalil included the foil, a low-mass mounting ring for the foils and thermocouples, two sets each of six copper-constantan thermocouples, and an aluminum heat sink inside of which the other parts were mounted. These parts were then assembled inside an accelerator tube and the thermocouple leads connected to one of two circuits, those for the first six thermocouples to a "cooling circuit" and the other six to the monitoring circuit. A potential could be applied to the thermocouples in one of the thermopiles so that, by the Peltier effect, the foil could be heated or cooled; these two effects are reversible. The heat generated in or absorbed by the junction is equal to the product of its thermoelectric power, the absolute temperature, and the current flow through it. Consequently, the calorimeter had a built-in absolute calibration device, assuming appropriate circuitry. For low absorption measurements, for example, first-collision dose, such calorimeters hold considerable promise.

Advantages and Limitations of Calorimetric Techniques

Assuredly, the greatest advantage of calorimetry is that it yields energy absorption in absolute units. For facilities where it can be used for routine work, it offers a reliable and reproducible standard. In general, the limiting parameters are relatively low sensitivity and the size and complexity of the apparatus. Calorimetric measurements frequently require more time and analysis than other systems.

7-7 THRESHOLD DETECTORS

General Discussion of the Method

For measuring high neutron dose rates, such as from critical assemblies, the range of proportional counters[46,14] or other rate-type dosimeters is frequently exceeded. A desirable neutron dosimeter feature for these high rates would be that it store a latent image, that is exhibit a reaction effect such as the film, glass, or chemicals do for γ-radiation. An obvious conclusion is that the neutron-reaction products in various elements might be made to suffice. One approach might be to observe neutron reactions that exhibit energy thresholds, for example, the $^{32}S(n, p)^{32}P$ reaction which has a threshold at about 2 MeV and a reaction product which may be analyzed readily by means of the hard β-particle. By utilizing several threshold reactions, the neutron flux as a function of energy may be determined approximately, and the tissue (or other) dose calculated.

The ORNL Threshold Detector System

Hurst et al.[21] and Reinhardt and Davis[32] have utilized four threshold detectors, plus the gold and gold-shielded-by-cadmium detector for thermal neutrons, to measure neutron spectrum and dose. The fast-neutron detectors are ^{239}Pu, ^{237}Np, and ^{238}U fission foils and ^{32}S. The ^{239}Pu is exposed in a ^{10}B shield to effect a threshold at about 4 keV. Figure 7-8 shows the cross-sectional curves of the detectors to an energy of 6 MeV. Table 7-4 shows the effective cross sections and thresholds of the detectors and gives an indication of the sensitivity; the fission foils are commonly evaluated by using 2 NaI(Tl)

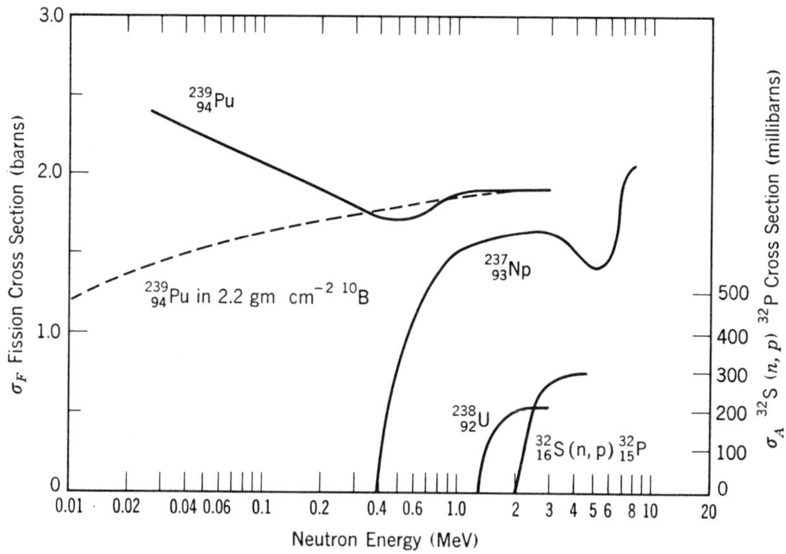

Fig. 7-8. Fission cross section (left scale) and activation cross section (right scale) of threshold detectors.

scintillators, of 4-in. diameter and 2 in. thick, one on each side of the foil. The half-life of the ^{32}P from the $^{32}S(n, p^{32})P$ reaction is 14.3 days and the β-ray energy is 1.71 MeV. However, the delayed fission γ-rays from the fission foils exhibit a rapidly changing half-life; Fig. 7-9 shows the decay curves for these γ-rays. A typical fluence histogram determined by the Hurst system of threshold detectors is shown in Fig. 7-10.

Advantages and Limitations of Threshold Detectors

The major advantages of threshold detectors include the extremely high dose rates and integrated doses at which they can be used and the insensitivity to γ-radiation (negligible response to γ-rays of less than about 20 MeV). It is also advantageous to have an indication of neutron spectrum in addition to dose.

Table 7-4 Effective Thresholds and Cross Sections for the Foils Used in the Hurst Threshold Detector System

Detector	Threshold	Cross Section	Back-ground Counter[a] (cpm)	Foil (cpm/gm)	10-rad Neutron,[b] 1 hr Post-exposure (cpm/gm)
^{239}Pu	~4 keV	2.0×10^{-24} cm^2	280	970	1900
^{237}Np	0.7 MeV	1.6×10^{-24} cm^2	280	0	1440
^{238}U	1.5 MeV	0.55×10^{-24} cm^2	280	80	2520
S	2.5 MeV	0.224×10^{-24} cm^2	22	0	54
Au and Au + Cd	Thermal	Direct calibration	50	0	[c]

[a] Fission foils "counted" by placing between two 4-in.-diameter by 2-in.-thick NaI(Tl) crystals. Sulfur samples "counted" by placing on a $1\frac{1}{2}$-in.-diameter by $\frac{1}{8}$-in.-thick plastic scintillator. Au samples "counted" by placing on a $1\frac{1}{2}$-in. by 1-in. NaI(Tl) crystal.

[b] Exposed to Godiva II at Los Alamos Scientific Laboratory. This neutron spectrum from this reactor is a somewhat softened fission spectrum.

[c] Of the order of several hundred counts per minute (cpm) but sensitive to the exposure environment due to scattering in walls, floors, etc.

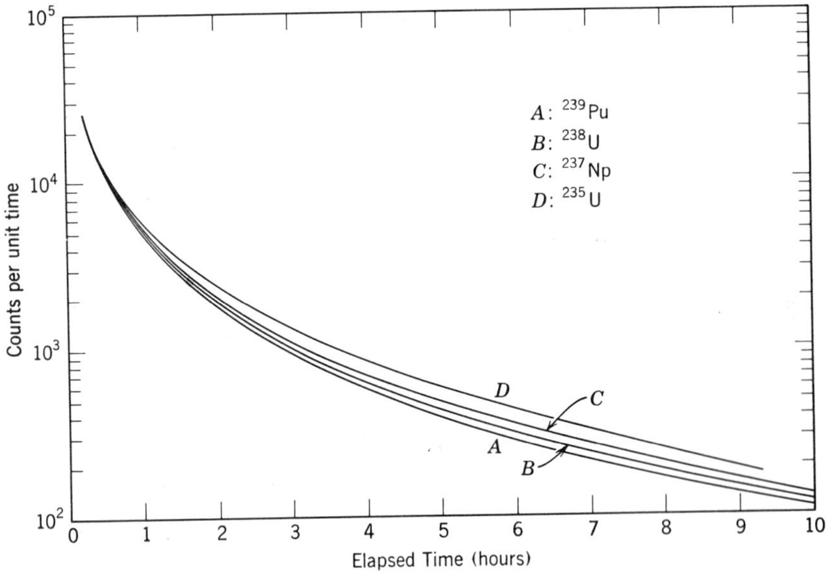

Fig. 7-9. Decay curves showing $C(t)$ as a function of elapsed time for ^{239}Pu, ^{238}U, ^{237}Np, and ^{235}U.

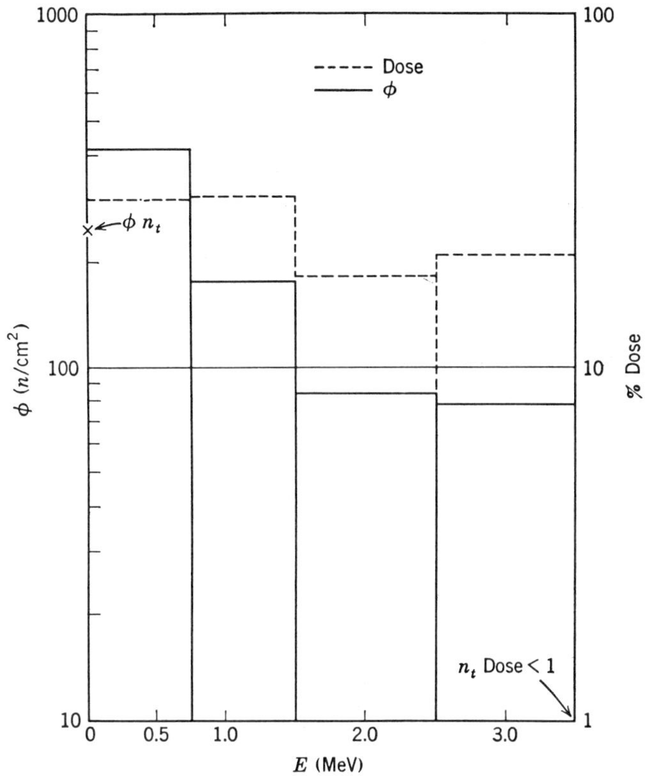

Fig. 7-10. Fluence and dose histogram in air.

The most serious limitations of most threshold detectors result from the rapid decay of the reaction products (fission-product γ-rays) and the relatively complex and expensive equipment for evaluation. Except for installations equipped for automatic or semiautomatc reduction of count rate to flux and dose, these evaluations are somewhat time-consuming. However, a recent innovation has increased the versatility and decreased the limitations of the threshold detector system markedly.[24] By using glass or plastic as a solid-state track detector[52] a sensitive threshold detector has been developed which utilizes thin fissible foils placed in the boron shield with disks of glass or plastic facing the fissile foil. The fission tracks are detected with near-2π geometry without the necessity of quick retrieval. Fission fragment tracks are rendered visible under a simple optical microscope by etching in hydrofluoric acid (glass) or sodium hydroxide (plastic).

Additional applications of the solid detectors for track registration are discussed by Becker[5] with emphasis on applied dosimetry.

REFERENCES

1. Allen, A. O., *Radiation Res.* **1,** 85 (1954).

2. Attix, F. H., "Present Status of Dosimetry by Radiophotoluminescence and Thermoluminescence Methods," *NRL 6145* (September 1964).

3. Baarli, J., K. Goebel, and A. Sullivan, *Health Phys.* **9,** 1057 (1963).

4. Becker, J. A., C. B. Green, and G. L. Pearson, *Elec. Eng.* **65,** 711 (1946).

5. Becker, K., *Health Phys.* **12,** 769 (1966).

6. Bell, P. R., Chapter 5 in *Beta and Gamma-Ray Spectroscopy* (K. Siegbahn, ed.), Interscience Publishers, New York, 1955.

7. Bernard, C. H., W. T. Thornton, and J. A. Auxier, *Health Phys.* **4,** 236 (1961).

8. Birks, J. B. *Scintillation Counters*, McGraw-Hill, New York, 1953.

9. Cheka, J. S., *Nucleonics* **12,** No. 6, 40 (1954).

10. Cheka, J. S., *Health Phys.* **10,** 303 (1964).

11. Dewhurst, H. A., A. H. Samuel, and J. L. Magee, *Radiation Res.* **1,** 62 (1954).

12. Dudley, R. A., *Nucleonics* **12,** No. 5, 24 (1954).

13. Ehrlich, M., and S. H. Fitch, *Nucleonics* **9,** No. 3, 5 (1951).

14. Etzel, H. W., R. D. Kirk, and J. H. Schulman, *Ra-Det* **8,** No. 2, 49 (1955).

15. Gillespie, A. B., *Signal, Noise, and Resolution in Nuclear Counter Amplifier*, Pergamon, London, 1953.

16. Greening, J. R., *Brit. J. Radiol.* **30,** 254 (1957).

17. Hart, E. J., *Radiation Res.* **1,** 53 (1954).

18. Hochanadel, C. J., *J. Phys. Chem.* **56,** 587 (1952).

19. Hurst, G. S., *Brit. J. Radiol.* **27,** 353 (1954).

20. Hurst, G. S., and R. H. Ritchie (eds.), "Radiation Accidents: Dosimetric Aspects of Neutron and Gamma-Ray Exposures," *ORNL-2748, Part A* (November 1959).

21. Hurst, G. S., J. A. Harter, P. N. Hensley, W. A. Mills, M. Slater, and P. W. Reinhardt, *Rev. Sci. Instr.* **27,** No. 3, 153 (1956).

22. Jetter, E. S., and H. Blatz, *Nucleonics* **10,** No. 10, 43 (1952).

23. Kalil, F., W. G. Stone, H. H. Hubbell, Jr., and R. D. Birkhoff, "Stopping Power of Thin Aluminum Foils for 12 to 127 keV Electrons," *ORNL-2731* (September 1959).

24. Kerr, G. D., and T. D. Strickler, *Health Phys.* **12,** 1141 (1966).

25. Kondo, S., *Health Phys.* **4,** 21 (1960).

26. Kondo, S., *Health Phys.* **7,** 25 (1961).

27. Laughlin, J. S., J. W. Beattie, W. J. Henderson, and R. A. Harvey, *Am. J. Roentgenol. Radium Therapy Nucl. Med.* **70,** 294 (1953).

28. McCaslin, J. R., *Health Phys.* **2,** 389 (1960).

29. Murray, R. B., in *Nuclear Instruments and Their Uses* (A. H. Snell, ed.), John Wiley and Sons, New York, 1962, Vol. I.

30. "Radiation Quantities and Units, International Commission on Radiological Units and Measurements," ICRU Report 10a, *Natl. Bur. Std. (U.S.) Handbook* **84** (1962).

31. Przibram, K., *Z. Physik* **20,** 196 (1923).

32. Reinhardt, P. W., and F. J. Davis, *Health Phys.* **1,** 169 (1958).

33. Rossi, H. H., and W. Rosenzweig, *Radiation Res.* **2,** 417 (1955).

34. Rotblat, J., Chapter 3 in *Progress in Nuclear Physics* (O. R. Frisch, ed.), Butterworth, London, 1950, Vol. I.

35. Schulman, J. H., and H. W. Etzel, *Science* **118,** 184 (1953).

36. Schulman, J. H., R. J. Ginther, C. C. Klick, R. S. Alger, and R. A. Levy, *J. Appl. Phys.* **22,** 1479 (1951).

37. Schulman, J. H., W. Schurcliff, R. J. Ginther, and F. H. Attix, *Nucleonics* **11,** No. 10, 52 (1953).
38. Seitz, F., *Rev. Mod. Phys.* **18,** 384 (1946).
39. Sigoloff, S. C., *Nucleonics* **14,** No. 10, 54 (1956).
40. Storm, E., "The Response of Sensitive 552 Dupont Film to Beta Radiation," *LA-1284* (August 1951).
41. Swank, R. K., *Ann. Rev. Nucl. Sci.* **4** (1954).
42. Thornton, W. T., and J. A. Auxier, "Some X-Ray and Fast Neutron Response Characteristics of Silver Metaphosphate Glass Dosimeters," *ORNL-2912* (August 1960).
43. Tochilin, E., R. H. Davis, and J. Clifford, "A Calibrated X-Ray Film Badge Dosimeter, Final Report on Problem Assignment 4P31b NS 083003," *ADP-78, U.S. Naval Radiological Defense Laboratory* (March 1949).
44. Villforth, J. C., R. D. Birkhoff, and H. H. Hubbell, Jr., "Comparison of Theoretical and Experimental Filtered X-Ray Spectra," *ORNL-2529* (July 1958).
45. Vogel, H., and K. Becker, *Nukleonik* **7,** No. 1, 18 (1965).
46. Wagner, E. B., and G. S. Hurst, *Rev. Sci. Instr.* **29,** 153 (1958).
47. Weyl, W. A., J. H. Schulman, R. J. Ginther, and L. W. Evans, *J. Electrochem. Soc.* **95,** 70 (1949).
48. Wilson, C. T. R., *Proc. Roy. Soc.* (*London*) **104A,** 1 (1923).
49. Wilson, C. T. R., *Proc. Roy. Soc.* (*London*) **104A,** 192 (1923).
50. Yokota, R., and H. Imagawa, *J. Phys. Soc. Japan* **20,** 1537 (1965).
51. Yokota, R., and S. Nakajima, *Health Phys.* **11,** 241 (1965).
52. Fleischer, R. L., P. B. Price, and R. M. Walker, *Science* **149,** 383 (1965).

PROBLEMS

1. Scintillation counters are the most sensitive radiation detectors. Why are they generally not used as γ-radiation dosimeters?

2. Compute the kerma (or absorbed dose under equilibrium conditions) for a small sample of tissue exposed to 4 MeV neutrons. Assume a fluence of 10^{10} neutrons/cm^2.

3. Explain why a dosimeter which would measure absorbed dose accurately for all γ-ray energies and for all environmental conditions might yield responses which were not directly related to observed biological response. (Neglect biological differences.)

4. Fast neutrons are always coexistent with some γ-radiation and many neutron detectors are sensitive to γ-rays. Discuss the methods and techniques by which photoluminescent detectors could be rendered either sensitive of insensitive to fast neutrons or to γ-radiation.

CHAPTER 8

Dose from Electrons and β-Rays

R. D. Birkhoff

A. ELECTRON AND BETA SPECTROSCOPY

8-1 RELATIVISTIC EQUATIONS RELATING ENERGY, MOMENTUM, AND VELOCITY

Electrons of more than a few kilovolts energy have velocities comparable to the velocity of light, and relativistic equations are needed to describe their behavior. The kinetic energy T, as a function of momentum p, is given in the first of Eqs. (8-1):

$$T = \sqrt{p^2c^2 + m_0{}^2c^4} - m_0c^2$$

$$T = \frac{m_0c^2}{\sqrt{1 - v^2/c^2}} - m_0c^2 \tag{8-1}$$

$$p = \frac{m_0v}{\sqrt{1 - v^2/c^2}}$$

where m_0 is the rest mass of the electron and c is the velocity of light. The expression for T approaches the classical expression $p^2/2m_0$ if $p \ll m_0c$. Also given in Eqn. (8-1) is the relationship between the kinetic energy T, and the electron velocity v. Again, T equals the familiar $\frac{1}{2}m_0v^2$ only for low velocity. The third relation of Eqs. (8.1) relates the electron momentum to the electron velocity and again $p = m_0v$ only for low velocity. The tedious calculation of electron energy from momentum or velocity may be avoided by use of Reference 23 in which the electron momentum and velocity are given as a function of electron energy from 0.2 eV to 3.3×10^{12} eV.

8-2 PRINCIPLES OF MAGNETIC SPECTROSCOPY

A review of the principles and health physics applications of electron and β-spectroscopy has recently been published by Birkhoff.[6]

An electron with momentum p perpendicular to a uniform magnetic field moves in a circular orbit of radius ρ. The relation connecting these quantities, found by equating the product of the centripetal acceleration and electron mass to the magnetic force on a moving charge, is given in the first of Eqs. (8-2).

$$\begin{aligned} p &= He\rho \\ \frac{\Delta p}{p} &= \text{const} \\ N_1(p) &= \frac{\text{cpm}}{\Delta p} \\ &\approx \frac{\text{cpm}}{p} \approx \frac{\text{cpm}}{H\rho} \end{aligned} \tag{8-2}$$

and this equation is valid both relativistically and nonrelativistically. A differential in p is proportional to a differential in ρ, and this equation implies that, if a group of electrons moving in a constant magnetic field is selected by a slit of width $\Delta\rho$, a band of momentum with width Δp, given above, will pass through this slit. Thus if $\Delta\rho/\rho$ is fixed geometrically, $\Delta p/p$ will be fixed at a constant value also, and the momentum resolution $\Delta p/p$ is a constant as indicated in the second of Eqs. (8-2).

A counter that counts the particles that move through a slit measures the number of counts per minute which lie between p and $p + \Delta p$. The ratio of counts per minute to Δp is the number of electrons per unit momentum, which is defined as the momentum distribution $N_1(p)$. Thus $N_1(p)$ is given by the third of Eqs. (8-2) as cpm/Δp; or a proportional number, cpm/p, or cpm/$H\rho$. We frequently find the label "cpm/$H\rho$" in the literature along the ordinate of a β-ray spectrum. In this instance, the ordinate represents the true momentum distribution of the electrons.

8-3 PRINCIPLES OF ELECTROSTATIC SPECTROSCOPY

If electrostatic rather than magnetic deflection is employed in a spectrometer, the governing relationship equivalent to Eqs. (8-2) is given in

Eqs. (8-3)

$$T = KV$$

$$\frac{\Delta T}{T} = \text{const} \tag{8-3}$$

$$\therefore N_2(T) = \frac{\text{cpm}}{\Delta T} \approx \frac{\text{cpm}}{T} \approx \frac{\text{cpm}}{V}$$

which states that the electron energy T being observed is proportional to the voltage V applied to the deflecting elements. By reasoning similar to that employed in Eqs. (8-2) it may be shown that the energy resolution $\Delta T/T$ is a constant in this case. Thus the energy distribution $N_2(T)$ is given by cpm/ΔT; or (counts per minute)/T; or cpm/V. Electrostatic spectrographs for β-ray spectroscopy are rarely used because the relationships of Eqs. (8-3) are not valid for relativistic electrons. However, they will probably find increasing use in the future, particularly in the low-energy region, because of their simplicity as well as their resolution and transmission characteristics.

8-4 SPECTROSCOPY BY TOTAL ENERGY ABSORPTION

In another class of spectrometers, the energy of an electron is measured by completely absorbing the electron in a scintillating material, a gaseous material, or a semiconductor. The size of the electric pulse in the photomultiplier or other associated circuitry generated by the absorption of the particle is proportional to the energy of the particle, and the distribution of pulse heights gives the energy distribution of the electrons. The resolution of such devices is given by a constant divided by the square root of the energy. This relation results from the fact that the statistical variation is the number of events associated with the absorption of a single electron is governed by a Gaussian distribution of pulses for absorptions of the same amount of energy.

In the scintillation spectrometer the variation occurs primarily in the number of photoelectrons emitted at the cathode of the photomultiplier which views the light from a scintillating crystal. In the gaseous or proportional counter spectrometer, the variation occurs in the number of ion pairs created in the gas of the counter. In the solid-state spectrometer, the variation is in the number of electron-hole pairs which are formed. In all three spectrometers the apparent energy spread may be calculated by first dividing the amount of energy absorbed by the amount required to produce one event, that is, about 1000 eV in a crystal, 30 eV in a gas (the W value of the gas), or 3 eV in a semiconductor. The number of events then has a deviation equal to the square root of this number, and this deviation multiplied by

1000, 30, or 3 gives the apparent energy spread. Although the energy resolution varies as the reciprocal of the square root of the energy, the energy distribution is given directly by the counts per minute observed because the resolution is entirely statistical in nature. These relations are shown in Eqs. (8-4)

$$\begin{aligned} \Delta T &= \text{const}\sqrt{T} \\ \therefore \frac{\Delta T}{T} &= \frac{\text{const}}{\sqrt{T}} \\ \text{but}\quad N_2(T) &\approx \text{cpm} \end{aligned} \tag{8-4}$$

8-5 CONVERSION OF A MOMENTUM DISTRIBUTION TO AN ENERGY DISTRIBUTION

It is often useful to convert a momentum distribution into an energy distribution or vice versa. Eqs. (8-5)

$$\begin{aligned} N_1(p)\,dp &= N_2(T)\,dT \\ \therefore N_2(T) &= N_1(p)\frac{dp}{dT} \\ &= \frac{N_1(p)}{v} \end{aligned} \tag{8-5}$$

state the underlying relationship between the two distributions; the number of particles $N_1(p)\,dp$, having momenta between p and $(p + dp)$ must equal the number of particles $N_2(T)\,dT$, which have energy between T and $(T + dT)$. If, for example, the momentum distribution $N_1(p)$ is known, the energy distribution $N_2(T)$ may be obtained by dividing $N_1(p)$ by the electron velocity, as shown in the third of Eqs. (8-5).

B. NUCLEAR BETA SPECTRA

8-6 ALLOWED BETA SPECTRA

A neutron in a nucleus may decay into a proton, a negative β-ray, and a massless particle called the antineutrino. A proton may decay into a neutron, a positive β-ray (or positron), and a neutrino. In neutron decay, the proton remains behind and the β-ray and antineutrino are emitted. Similarly, the neutron remains in the nucleus after proton decay, and a positron and neutrino are emitted. By the conservation of energy, the sum of the β-ray kinetic energy T and neutrino energy T' is a constant equal to T_0 for any given β-emitter. That is, the β-ray and neutrino share a certain amount of energy T_0 which is the maximum energy that is ever observed in the β-ray spectrum.

Electrons of all energies below T_0 are observed, but with various probabilities determined by statistical considerations, the nuclear Coulomb attraction or repulsion, and the nature and extent of the participation by the nucleus. If only the first two factors are involved, the transitions are said to be "allowed" because they obey certain selection rules for changes in angular momentum and parity.[20] Transitions involving violations of these rules are said to be "forbidden" and the shape of the β-ray spectrum may be slightly or greatly altered. Because a large fraction of all spectra are of the allowed shape and because the essential statistical considerations are easily derived, the following brief description is given.

In statistical mechanics, we consider that a particle of a given position and momentum occupies a cell in "phase space" that is, in a six-dimensional coordinate and momentum space. We may omit from consideration the positional volume of phase space (normalize to one β-ray per cm^3) and consider the momentum space in spherical coordinates. Thus, the momentum p, possessed by the β-ray, may be thought of as the radius of a sphere. Then the number of allowed momentum states, which the electron may occupy after the disintegration, is proportional to the number of available (empty) states, which is the volume $4\pi p^2\,dp$ divided by the volume of each state h^3. If q is the neutrino momentum, then the number of momentum states available to the neutrino is $4\pi q^2\,dq/h^3$. The relationship between momentum q and energy T' for a massless particle like the neutrino, is that $q = T'/c$; that is, the neutrino energy divided by the velocity of light.

We may introduce new units here for momentum: $q = (T_0 - T)/c = (\epsilon_0 - \epsilon)m_0c$ and $\eta = p/m_0c$. That is, ϵ is the β-ray kinetic energy in m_0c^2 units and η is the β-ray momentum in m_0c units.

From the preceding considerations alone, the probability of finding an electron having a momentum between p and $(p + dp)$ and a neutrino having momentum between q and $(q + dq)$ is proportional to

$$p^2\,dp q^2\,dq \tag{8-6}$$

If we consider a particular value of the β-ray momentum p, then the neutrino momentum is fixed, and an integral over the latter (a delta function) yields the neutrino momentum in terms of the chosen β-ray momentum. If we make this substitution, then the β-ray energy distribution (in the new units) is proportional to

$$\eta^2(\epsilon_0 - \epsilon)^2\,d\eta \tag{8-7}$$

Another factor which helps determine the shape of an allowed β-spectrum is the Coulomb factor $F(Z, \eta)$. This Coulomb factor represents the effect of the attraction or repulsion of the nucleus for the emitted β-rays or positrons.

Thus, β-rays emitted from a positively charged nucleus tend to be slowed down in escaping, and the β-spectrum is shifted toward a lower energy level. In a similar fashion, positrons emitted from a nucleus are accelerated away from the nucleus, with the result that a positron spectrum tends to be shifted toward higher energy levels. The allowed momentum spectrum is proportional to the product of the statistical factor above and the Coulomb factor as shown in Eq. (8-8):

$$N_1(\eta)\,d\eta \approx \eta^2(\epsilon_0 - \epsilon)^2 F(Z, \eta)\,d\eta \tag{8-8}$$

where Z is the atomic number of the *daughter* nucleus. It is apparent that if the momentum distribution is divided by $\eta^2 F(Z, \eta)$ and the square root of the quotient is taken, then this quantity will be proportional to $(\epsilon_0 - \epsilon)$, and will yield a straight line having negative slope when plotted against β-ray energy

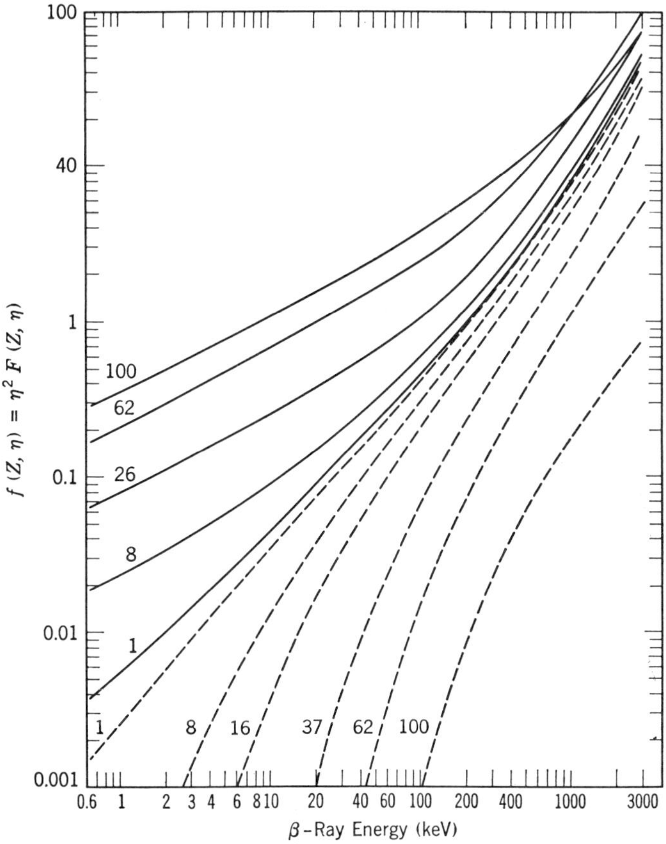

Fig. 8-1. Plot of the function $f(Z, \eta)$ used in calculating β-ray spectra for allowed transitions.

ϵ in m_0c^2 units as shown in Eq. (8-9)

$$\sqrt{\frac{N_1(\eta)}{\eta^2 F(Z,\eta)}} \approx (\epsilon_0 - \epsilon) \tag{8-9}$$

This plot is called a Fermi plot or Kurie plot and its straightness is used as a test of whether or not the β-spectrum is allowed. To simplify making such a plot, the quantity $\eta^2F(Z, \eta)$ has been tabulated.[12] This quantity is denoted by $f(Z, \eta)$, and is shown in Fig. 8-1 for various atomic numbers (left side for β-emitters, bottom for positron emitters).

If we write the quantity η in terms of the electron energy ϵ and express $d\eta$ in terms of $d\epsilon$, we can show that the corresponding *energy* distribution for an allowed β spectrum $N_2(\epsilon)$ is proportional to a function of energy given by

$$N_2(\epsilon) \approx (1 + \epsilon)\sqrt{\epsilon^2 + 2\epsilon}\,(\epsilon_0 - \epsilon)^2 F(Z, \epsilon) \tag{8-10}$$

Alternatively, we may obtain the momentum distribution $N_1(\eta)$ as given in Eq. (8-8) and convert it into an energy distribution by the use of the relation of Eq. (8-5), that is, by dividing it everywhere by v, the β-ray velocity. Representative β-spectra are plotted[21] in Figs. 8-2, 8-3, and 8-4 for atomic numbers Z of 4, 20, and 80 and β-energies 0.2, 0.6, and 2 MeV. A rapid graphical method for obtaining the shape of an allowed β-spectrum has been published by Marshall.[22]

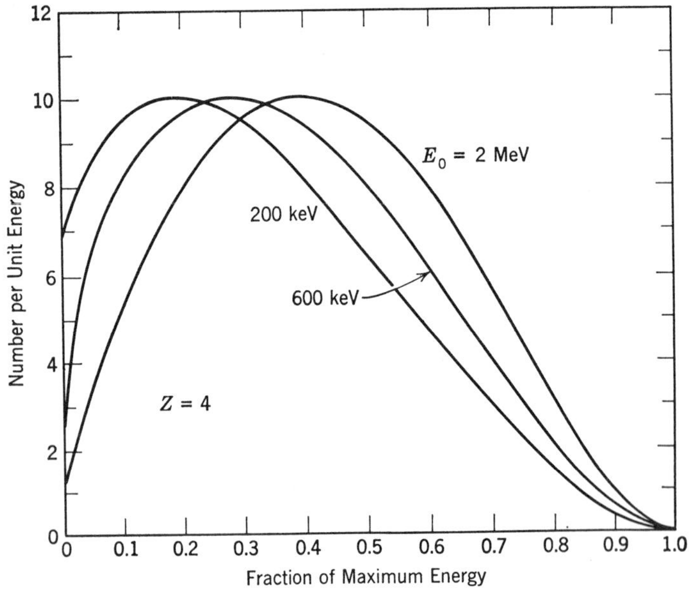

Fig. 8-2. Allowed β-spectra for end point energies of 2 MeV, 600 keV, and 200 keV for $Z = 4$.

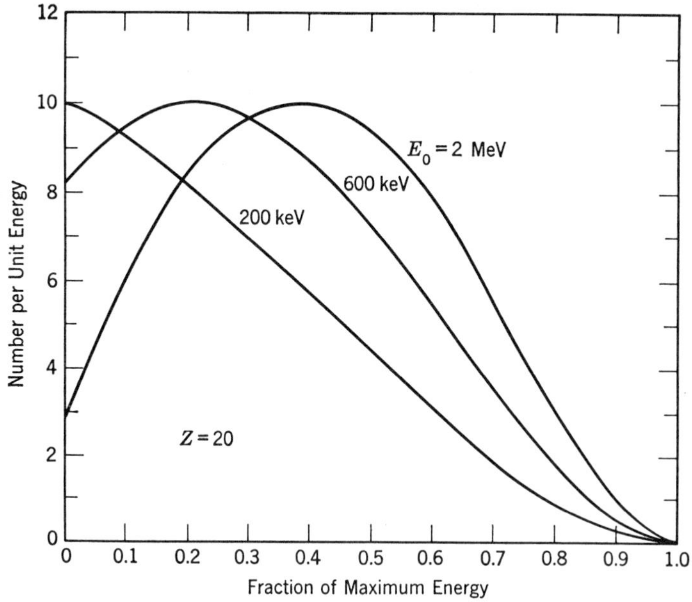

Fig. 8-3. Allowed β-spectra for end point energies of 2 MeV, 600 keV, and 200 keV for $Z = 20$.

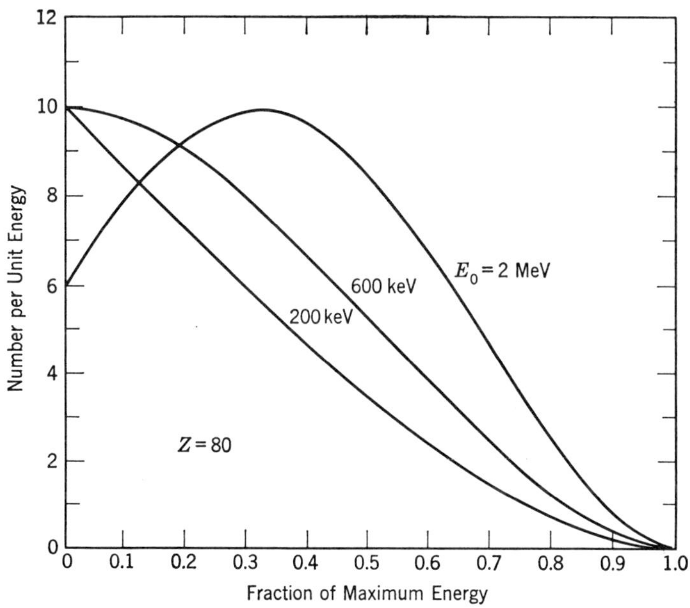

Fig. 8-4. Allowed β-spectra for end point energies of 2 MeV, 600 keV, and 200 keV for $Z = 80$.

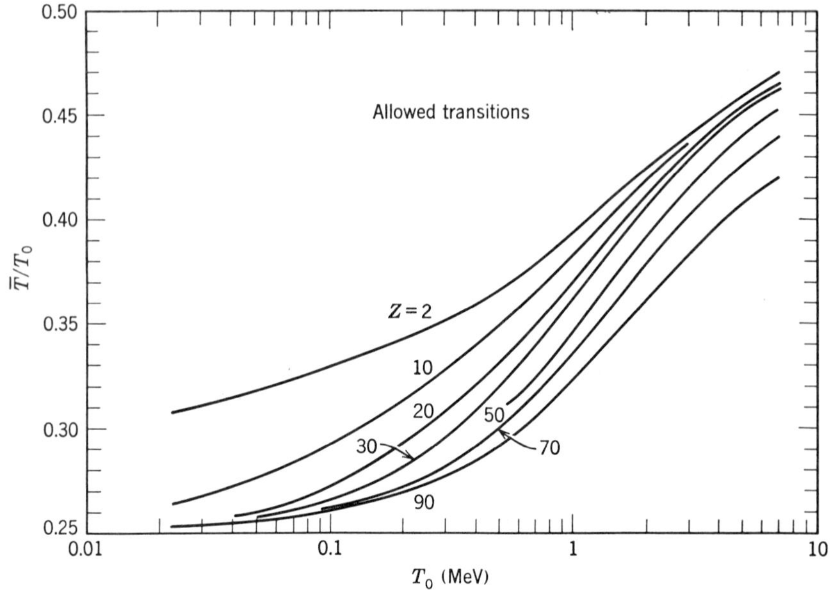

Fig. 8-5. Ratio of average energy to end point energy for allowed transitions and various atomic numbers.

The average energy in an allowed β-spectrum is given by Eq. (8-11)

$$\overline{T} = \frac{\int_0^{T_0} TN_2(T)\,dT}{\int_0^{T_0} N_2(T)\,dT} \tag{8-11}$$

and is plotted in Fig. 8-5. This quantity has been calculated by James, Steel, and Story[18] for β-rays and positrons with energies from 0.0225 to 7.17 MeV for values of Z between 2 and 90.

It can be seen that the ratio of the average β-energy to end point energy varies from a minimum of about 0.26 to a maximum of 0.45. Thus the rule of thumb which states that this ratio is about 1:3 can lead to serious error.

8-7 FORBIDDEN β-SPECTRA

Many β-ray spectra do not have the simple allowed shape which is determined only by the statistics of electron and neutrino emission and the Coulomb factor. In these, the nucleus itself enters into determining the probability of β-and neutrino emission and thus the shape of the β-spectrum. Spectra exhibiting these nuclear effects are called forbidden spectra. The factors that determine these shapes are not considered in this book, since they are not usually needed to obtain a spectrum for practical uses. In general, the shape may be determined by calculating an allowed spectrum for the same atomic number and endpoint energy, and multiplying this allowed shape at each energy by a function of energy called the "shape factor." Shape factors are known theoretically for certain types of β-transitions. Other shape factors have been determined experimentally by measuring deviations between an observed β-spectrum and the allowed shape. Either way, a study of the literature concerning the particular β-emitter in question is necessary. Usually, the actual β-ray spectrum will be plotted or a shape factor given. We should be careful, in case a momentum spectrum is given, to convert it to the energy spectrum which is usually required in health physics calculations. The relation between the momentum spectrum and the energy spectrum was given in Eq. (8-5).

In calculations involving β-spectra, it is important that the correct atomic number be used. The literature is not consistent about using the atomic number of the parent or the daughter nucleus. Care is also necessary in distinguishing between the β-ray kinetic energy and the total energy, the two differing by m_0c^2.

Comprehensive review articles of β-decay theory and experiment which include discussions of the effect of the discovery of nonconservation of parity and charge conjugation in β-decay have been written by Wu[34] and Konopinski and Rose.[20]

C. STOPPING POWER OF MATTER FOR ELECTRONS

8-8 CROSS SECTION FOR ENERGY LOSS TO A FREE ELECTRON

The expression for the momentum transferred from a swiftly moving particle of charge Z_1e to an electron essentially at rest is given in Chapter 2, Eq. (2-9), as

$$p_y = \frac{2Z_1e^2}{bv} \tag{8-12}$$

If the particle is an electron and its deflection may be neglected (a realistic assumption for the vast majority of collisions in which the amount of energy transferred is small), the same equation will hold for this particle as well

($Z_1 = 1$). The energy transferred is then

$$\tau = \frac{p_y^{\,2}}{2m_0} = \frac{2e^4}{b^2 m_0 v^2} \tag{8-13}$$

where T is the kinetic energy of the incident electron. Now any incident electron which passes within a distance between b and $(b + db)$ of the atomic electron will impart an amount of energy between τ and $(\tau + d\tau)$ to it. The area between these radial limits about the atomic electron is said to be the cross section $d\sigma$ of the atomic electron for receiving this energy and is given by

$$d\sigma = 2\pi b\, db = \frac{2\pi e^4}{m_0 v^2}\frac{d\tau}{\tau^2} \tag{8-14}$$

where use has been made of Eq. (8-13).

This cross section was first deduced by J. J. Thomson[33] in 1912 and is basic to an understanding of the exchange of energy between a high-energy electron and matter. In particular, the dependence of the cross section on the reciprocal of the energy of the incident electron indicates that low-energy electrons will be more densely ionizing than high-energy electrons. Because of the dependence on the inverse square of the energy transferred, the energy distribution of the secondary electrons will be strongly peaked at low energy.

8-9 THE CALCULATION OF STOPPING POWER

The total energy loss to all electrons along a segment of track dx may be found by multiplying the energy transferred τ at impact distance b by the number of electrons in the volume $2\pi b\, db\, dx$, and by then integrating over all possible impact distances between b_{min} and b_{max}. The energy transferred to electrons at distances between b and $(b + db)$ is then

$$\frac{e^4}{Tb^2} 2\pi b\, db\, dx \frac{N_A \rho Z}{A} \tag{8-15}$$

where N_A is Avogadro's number and ρ, Z, and A are the density, atomic number, and atomic weight of the absorber, respectively. The integration over all distances b yields

$$-\frac{dT}{dx} = \frac{4\pi N_A e^4 \rho Z}{m_0 v^2 A}\int_{b_{min}}^{b_{max}} \frac{db}{b} \tag{8-16}$$

A suitable value for the maximum distance at which atomic electrons may interact with the incident electron is one in which the impact time is of the order of the period of the electrons in their orbital motion about nuclei. The impact time is of the order of b/v, and the period is of the order of $1/\omega$, where ω is the angular frequency of rotation about the nucleus. Thus the maximum impact parameter is

$$b_{\max} = \frac{v}{\omega} \tag{8-17}$$

The minimum impact parameter will be just the de Broglie wavelength of the incident electron because closer collisions are meaningless in a quantum mechanical sense

$$b_{\min} = \frac{\hbar}{m_0 v} \tag{8-18}$$

and the stopping power becomes

$$-\frac{dT}{dx} = \frac{4\pi N_A e^4 \rho Z}{m_0 v^2 A} \ln\left(\frac{m_0 v^2}{\hbar\omega}\right) \tag{8-19}$$

A more exact treatment has been given by Bethe[2]; this utilizes the Möller instead of the Thomson cross section as well as relativistic quantum mechanics. The binding energy of the atomic electrons represented above by $\hbar\omega$ is replaced with a suitable average excitation potential I. For a relativistic electron of kinetic energy T the result is

$$-\frac{dT}{dx} = \frac{2\pi N_A e^4 \rho Z}{m_0 v^2 A} \left\{ \ln \frac{m_0 v^2 T}{2I^2(1-\beta^2)} - (2\sqrt{1-\beta^2} - 1 + \beta^2) \ln 2 + 1 - \beta^2 + \tfrac{1}{8}[(1 - \sqrt{1-\beta^2}]^2 \right\} \tag{8-20}$$

where β is v/c. The average excitation potential I has been shown by Bloch[8] to be proportional to the atomic number of the absorber with a constant of proportionality equal to the Rydberg energy, 13.5 eV.

$$I = 13.5Z \tag{8-21}$$

Deviations from it have been discussed by Brandt.[9] However, deviations may be neglected in most calculations of interest in health physics because of the position of I inside the logarithm, where such deviations result in only a small change in the stopping power.

For molecular materials the stopping power may be written[7]

$$-\frac{dT}{dx} = \frac{2\pi N_A e^4 \rho}{m_0 c^2 \beta^2} \sum \frac{f_i Z_i}{A_i} [\alpha + f(\beta)] \tag{8-22}$$

where f_i, Z_i, and A_i are the fractional mass abundance, atomic number, and

Table 8-1 Values of Parameters Used in Stopping Power Formula for Various Materials

Parameter	H_2O	Air	Al	Isobutane	Pb	Cu	Sn
$\sum \frac{f_i Z_i}{A_i}$	0.555	0.500	0.482	0.585	0.396	0.456	0.421
α	17.2	16.6	15.4	17.9	12.2	14.1	13.1

atomic weight of the *i*th constituent, and α and $f(\beta)$ are given by

$$\alpha = [\sum(f_i Z_i/A_i) \ln \tfrac{1}{2}(m_0c^2/I_i)^2]/\sum(f_i Z_i/A_i)$$

$$f(\beta) = \ln\left[\frac{\beta^2 - \beta^2\sqrt{1-\beta^2}}{(1-\beta^2)^{3/2}}\right] - [2\sqrt{1-\beta^2} - 1 + \beta^2]\ln 2 + (1-\beta^2) + \tfrac{1}{8}(1-\sqrt{1-\beta^2})^2 \quad (8\text{-}23)$$

Values of $\sum(f_i Z_i/A_i)$ and α for several materials of interest in health physics are listed in Table 8-1. The quantity $f(\beta)$ is given as a function of electron energy and β in Table 8-2. The most recent tabulations of stopping power are contained in *Studies in Penetration of Charged Particles in Matter*, National Academy of Sciences—National Research Council Report 1133 (1964).

Table 8-2 Energy-Dependent Parameter $f(\beta)$ Used in Stopping Power Formula

T (MeV)	β	$f(\beta)$	T (MeV)	β	$f(\beta)$
0.0001	0.0198	−16.1	0.05	0.413	−3.76
0.0002	0.0280	−14.7	0.06	0.446	−3.42
0.0003	0.0343	−13.9	0.08	0.502	−2.87
0.0004	0.0395	−13.3	0.1	0.548	−2.45
0.0005	0.0442	−12.8	0.2	0.695	−1.12
0.0006	0.0484	−12.5	0.3	0.777	−0.298
0.0008	0.0559	−11.9	0.4	0.828	−0.031
0.001	0.0625	−11.5	0.5	0.863	0.815
0.002	0.0882	−10.1	0.6	0.888	1.23
0.003	0.108	−9.31	0.8	0.921	1.93
0.004	0.124	−8.74	1	0.941	2.50
0.005	0.139	−8.27	2	0.979	4.38
0.006	0.152	−7.90	3	0.989	5.53
0.008	0.175	−7.35	4	0.994	6.37
0.01	0.195	−6.89	5	0.996	7.01
0.02	0.272	−5.54	6	0.997	7.56
0.03	0.328	−4.73	8	0.998	8.39
0.04	0.374	−4.19			

8-10 CORRECTION AND TABULATION OF THE STOPPING POWER

The Bethe stopping-power formula holds over a large range of electron energy and atomic number. However, certain corrections must be applied at extremely low and high energies. At low energy it may be energetically impossible for the incident electron to excite or ionize the tightly bound inner electrons. For example, the *K*-shell electrons in aluminum are bound with an energy of 1559 eV, and the Bethe formula overestimates the stopping power below about five times this energy. Similarly, the *K* electrons in lead have a binding energy of 88 KeV, and the Bethe formula requires corrections below about 450 keV. A crude correction is the reduction of the atomic number in the formula by the number of *K* electrons, whenever the energy of the incident electron falls below the *K*-shell energy; for example, the effective atomic number of aluminum changes from 13 to 11. There is no theoretical basis for a better correction to stopping power of electrons at this time.

Another correction originates in the polarization of the stopping medium by the incident electron (the density effect). This leads to a reduction; it was first predicted by Fermi[13,14] and has been evaluated for a large number of materials by Sternheimer.[31,32] A review of the subject has been written by Birkhoff.[5]

In addition to the above effects, which tend to reduce the stopping power, two effects that increase stopping power may be noted. If the velocity of an electron exceeds the velocity of light in a medium, the electron loses energy by means of radiation, in addition to ionization losses. This gives rise to the bluish glow in swimming pool reactors, in which γ-radiation produces high-energy Compton electrons and photoelectrons. The amount of energy going into Cerenkov radiation is small compared to that going into ionization and excitation of the medium. For example, in water the rate of energy loss from Cerenkov radiation is about 2 keV/cm as compared to about 2 MeV/cm lost in atomic collisions.

Another mechanism for energy loss involves the production of electromagnetic radiation. As an electron passes through a medium, it occasionally undergoes a substantial nuclear scattering which results in the emission of radiation called bremsstrahlung or "braking radiation." An estimate of the relative importance of bremsstrahlung loss as compared to loss due to ionization and excitation has been given by Bethe and Heitler[4] who find that

$$\frac{(dT/dx)_{\text{brems}}}{(dT/dx)_{\text{ion}}} = \frac{TZ}{800} \tag{8-24}$$

where T is the electron total energy in MeV and Z is the atomic number of the absorber. Thus, for gold and 1 MeV, the bremsstrahlung losses should be about 10% of the collision losses. A comprehensive summary of the theory of bremsstrahlung production has been published by Bethe and Ashkin[3].

The stopping power of matter for positrons has been calculated by Rohrlich and Carlson[27] who found that the electron and positron stopping powers differ by at most several per cent over the range of applicability of the Bethe-Bloch formula.

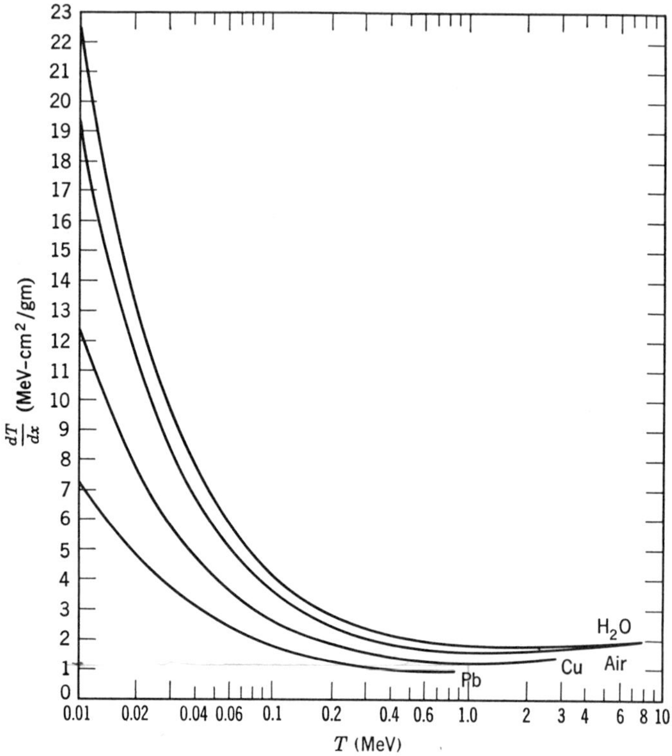

Fig. 8-6. Electron stopping power plotted as a function of energy in water, air, copper, and lead.

An extensive tabulation of electron and positron stopping powers has been given by Berger and Seltzer.[1] Some of these calculations have been plotted in Fig. 8-6 on a semilog graph. It may be seen that the curves all have the same general shape and magnitude, and this allows the introduction of the concept of "relative stopping power." The most commonly used relative stopping power in health physics calculations is that of water to air, which is about 1.14.

The Bethe formula represents the rate of electron slowing down along the electron track when corrected as described above. Agreement between experiment and theory has been reviewed for energies above 150 keV and found to be very good.[5] More recent measurements as low as 5 keV and 12 keV have been reported by Grun[15] and by Kalil et al.,[19] respectively, and are numerically in agreement with the theory.

8-11 ELECTRON SLOWING DOWN SPECTRA IN INFINITE MEDIA

When radiation strikes matter, one of the fundamental effects is the generation of a large number of secondary electrons of various energies. Secondary electrons can diffuse and give rise to electrons of lower energy through successive collisions. For a medium of effectively infinite size, homogeneous, isotropic, and irradiated isotropically, calculations have been made of the energy distribution of the electron flux $y(T)$, defined as the number of electrons per unit energy crossing a sphere having a cross-sectional area of 1 cm^2. A description of the flux distribution is basic to understanding biological and physical effects of radiation, because the probability p_i of an initial atomic or molecular event is proportional to the integral over all electron energies of the product of the cross section for the event $\sigma_i(T)$ and the flux $y(T)$. That is, with proper normalization of $y(T)$, we may write for p_i the equality

$$p_i = \int_0^\infty \sigma_i(T) y(T)\, dT \tag{8-25}$$

A knowledge of the electron flux distribution in irradiated media is as important to the study of radiation damage as the knowledge of the neutron flux is to the operation of a reactor.

To determine a theoretical expression for the electron flux $y(T)$, consider the energy interval between T and $(T + dT)$. If we assume that electrons lose energy continuously as they slow down (the "continuous slowing down" model), every electron having energy above T will at some time pass through the energy level between T and $(T + dT)$, and will thus contribute to the flux at T. The energy delivered to the medium will be equal to the product of the number of such electrons and dT. These electrons will arise from two sources: primary electrons such as β-rays and photoelectrons, and secondary electrons resulting from ionizing collisions by the primary electrons. The energy delivered to the medium per unit volume is then

$$E = dT\left[\int_T^{T_0} N(T')\, dT' + \int_{2T}^{T_0} K(T', T) y(T')\, dT'\right] \tag{8-26}$$

where $N(T')$ is the primary source spectrum per unit volume (for example, a nuclear β-spectrum or a Compton electron spectrum), and $K(T', T)$ is probability per unit path length that an electron of energy T' will undergo a collision in which a secondary electron of energy greater than T is created.

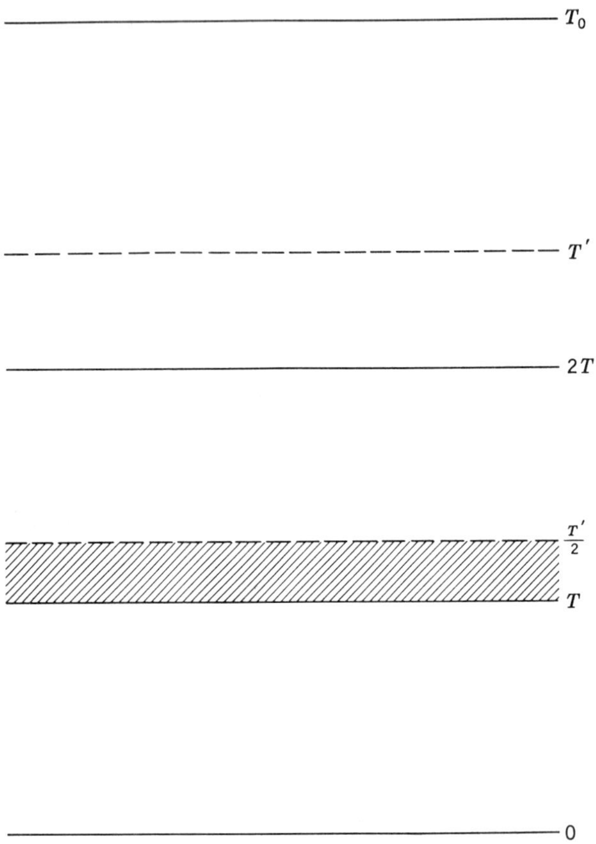

Fig. 8-7. Energy level diagram for calculation of secondary electron production.

The limits on the second integral arise from the definition of a secondary electron as being the electron of lesser energy after a collision ($T' > 2T$). Thus the primary electron must have an energy exceeding $2T$ to generate a secondary above T in energy. The probability $K(T', T)$ is the integral over τ of the probability per unit pathlength $k(T', \tau)$ for a transfer of energy τ from a primary electron of energy T'.

$$K(T', T) = \int_{T}^{T'/2} k(T', \tau)\, d\tau \qquad (8\text{-}27)$$

where for $k(T', \tau)$ the Thomson equation (8-14) or more accurately the Möller expression[26] may be employed. The limits of the integral in Eq. (8-27) reflect the fact that the secondary must have an energy τ between $T'/2$ and T. The energies discussed above may be visualized with the aid of the energy-level diagram given in Fig. 8-7. The cross-hatched region represents the energy interval in which secondary electrons may be produced from a higher energy electron at T'.

The energy absorbed by the medium is the product of the flux and the stopping power at energy T,

$$E = y(T)\frac{dT}{dx}(T)\,dT \qquad (8\text{-}28)$$

If we equate Eqs. (8-26) and (8-28), the flux $y(T)$ is then

$$y(T) = \left[\frac{dT}{dx}(T)\right]^{-1}\left[\int_T^{T_0} N(T')\,dT' + \int_{2T}^{T_0} K(T', T)y(T')\,dT'\right] \qquad (8\text{-}29)$$

Qualitatively, the inverse stopping-power factor indicates that electrons which lose energy slowly (dT/dx small) represent a larger flux than those which slow down rapidly, that is, they traverse longer distances and hence are more readily observed. The desired flux spectrum $y(T)$ appears under the second integral, and numerical methods must be used to obtain a solution. Such solutions have been obtained by McGinnies,[25] using the Spencer-Fano method of calculation,[30] for monoenergetic electron sources. Experimental and theoretical spectra[24] are shown in Fig. 8-8 for ^{64}Cu β-rays slowing down in

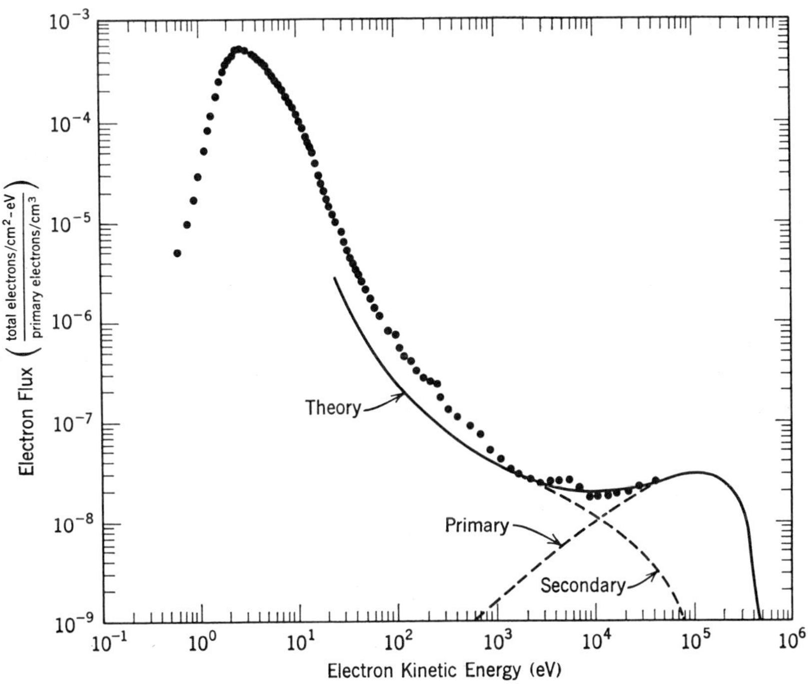

Fig. 8-8. Electron flux spectrum of ^{64}Cu β-rays absorbed in copper; cavity source.

Cu. The flux of secondary electrons increases rapidly at low energies. The action of all ionizing radiation absorbed in a medium therefore will be characterized by the production of large numbers of secondary electrons. The relatively abundant, low-energy electrons are probably of greatest significance from the standpoint of the biological effects of radiation. A bibliography on and calculations of electron flux spectra have been given by Danzker et al.[10]

D. DOSE FROM ELECTRONS AND BETA-RAYS IN UNIFORM INFINITE MEDIA

8-12 THE SPENCER THEORY

The dose derived from electrons and β-rays from sources embedded in a uniform infinite medium has been calculated by Spencer.[28] Important requirements of the theory are that the medium be uniform and of infinite extent, which limits somewhat the applicability of the theory to problems in health physics. Nonetheless the theory has many important applications and represents an important first step in the understanding and the solution of many health physics problems. It is based on detailed knowledge of the fundamental physical processes that occur as electrons slow down. For example, the slowing down is represented by the Bethe stopping-power formula, and the electron scattering is calculated on the basis of the Mott theory of nuclear scattering. The fundamental physical basis for this theory contrasts with that of other theories, which involve poorly defined parameters such as the average electron absorption cross section or the half-value layer or electron mean free path or electron range.

In solving the electron dose problem, Spencer defined electron flux in a medium as a function of position, direction, and energy, in contrast to $y(T)$ defined in Section C, which depends only on energy. Let $I(z, \theta, r)$ represent the number of electrons per unit solid angle at angle θ crossing a small spherical probe which has a cross-sectional area of 1 cm^2 at depth z in the medium. The electrons have an energy characterized by the parameter r, the "residual range" or distance the electrons have to move before they stop. This distance is measured along the electron path and is given in the continuous slowing down model by the integral of the reciprocal stopping power (Bethe formula),

$$r(T) = \int_0^T dT' \left(\frac{dT'}{dr}\right)^{-1} \tag{8-30}$$

We shall express $r(T)$ in units of grams per square centimeter and stopping power in units of MeV per gram-square centimeter (MeV gm^{-1} cm^2).

A brief resume of the Spencer theory for the instance of a plane perpendicular source is next considered. In this geometry, the infinite medium is divided by a plane which emits electrons toward the right, normal to the plane boundary and in the positive z direction. The electron flux per unit solid angle $I(z, \theta, r)$ is then specified for a depth z in centimeters beyond the plane, for an angle θ made with the normal direction, for electrons of energy corresponding to a residual range r. The change in the flux as the particle penetrates to a deeper depth $(z + dz)$ and loses residual range in amount dr while remaining at angle θ is given by Eq. (8-31)

$$\begin{aligned}[I(z + dz, \theta, r - dr) - I(z, \theta, r)]\, d\Omega\, dz \\ = dz\, d\Omega\, dr \int d\Omega' N\sigma(r, \Theta)[I(r, \theta', z) - I(r, \theta, z)] \\ + (2\pi)^{-1}\, \delta(z)\, dz\, \delta(r - r_0)\, dr\, \delta(\cos\theta - 1)\, d\Omega \qquad (8\text{-}31)\end{aligned}$$

The integral on the right side of the equation expresses the change in the number of electrons in the flux at angle θ caused by those electrons which scatter into the angle θ from angle θ' and those which scatter out of that angle θ. Here Θ is the difference between θ' and θ. The quantity $\sigma(r, \Theta)$ is the probability that an electron of residual range r will scatter through an angle Θ and is taken by Doggett and Spencer[11] as the Mott electron scattering cross section. The integral over all θ' is the net number of electrons after traversing a distance dz and losing residual range dr that are still moving in the direction θ with the normal to the plane.

The second term on the right side of Eq. (8-31) is the source term, in which the Dirac delta function is defined by $\delta(x) = 0$ when $x \neq 0$ and $\int_{-\infty}^{\infty} \delta(x)\, dx = 1$. This term implies that one electron is born at $z = 0$ moving normally to the surface ($\theta = 0$) with residual range r_0. The flux which is the solution to Eq. (8-31) may be used to determine the flux at depth z integrated over all solid angles,

$$I_0(z, r) = \int_0^{\pi} I(z, \theta, r) 2\pi \sin\theta\, d\theta \qquad (8\text{-}32)$$

where the flux $I_0(z, r)$ has units of number of electrons per square centimeter per gram per square centimeter of residual range. The absorption associated with this flux may be computed as an integral over all residual ranges (or energies) of the product of the flux and the stopping power,

$$J(z) = \int_0^{r_0} I_0(z, r) \left(\frac{dT}{dr}\right)(r)\, dr \qquad (8\text{-}33)$$

The units of this dose distribution $J(z)$ might be typically MeV per gram if $I_0(z, r)$ has units of electrons per square centimeter per gram per square centimeter of residual range. An extensive tabulation of the dose function

$J(z)$ in a normalized form has been given in terms of a function $J_{pl}(x)$, where $x = z/r_0$.[29] The dose in rad, $D(z)$ is given by

$$D(z) = SJ(z) \times 1.60 \times 10^{-8}$$

$$D(x) = J_{pl}(x)\left(\frac{dT}{dr}\right)(r_0)S \times 1.6 \times 10^{-8} \tag{8-34}$$

and is just S times the dose function of Eq. (8-33), where S is the number of electrons per square centimeter incident on the plane.

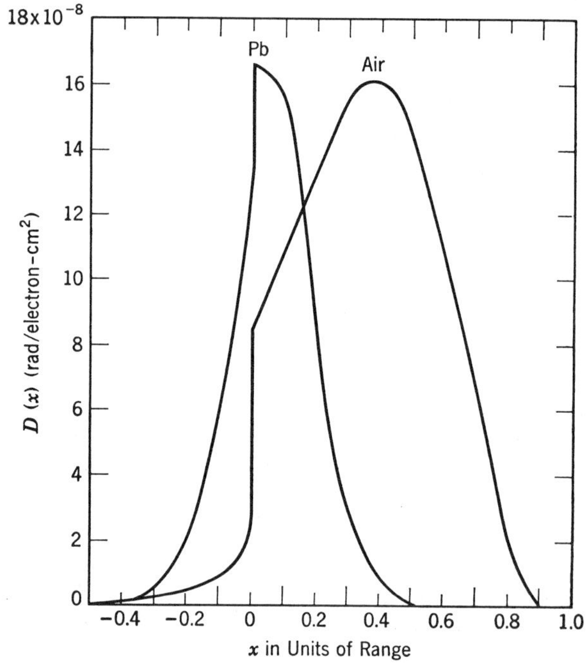

Fig. 8-9. Depth dose distributions for a 0.1 MeV parallel electron beam normally incident on planes in infinite isotropic media of lead and air.

The function $J_{pl}(x)$ is tabulated in Ref. 29 for electrons slowing down in carbon, aluminum, copper, tin, lead, air, and polystyrene, for initial energies ranging from 25 keV to 10 MeV. Dose distributions are shown in Fig. 8-9 for 0.1 MeV electrons in air and lead. The ordinate in each case is the number of rad at depth x per incident electron per square centimeter ($S = 1$) and the abscissa is the depth x. The values for r_0 are 0.0313 and 0.0164 gm/cm² for lead and air, respectively, at 0.1 MeV. Because of scattering, the dose curve in air looks somewhat like the Bragg dose distribution wherein the dose tends to achieve a maximum value at a considerable penetration depth. In lead however, because of large scattering, the maximum dose occurs near the surface. The dose due to backscattered electrons is also shown (negative values of x). It is smaller for air than for lead, as would be expected physically. The discontinuity in the dose at $x = 0$ arises from the energy dissipated by the source electrons introduced with velocities in the positive x direction.

The solution of Eq. (8-31) for a point electron source in an infinite medium is similar, and leads to another function $J_{\text{pt}}(x)$, which is also tabulated in Ref. 29. The dose $D(\rho)$ in rad in terms of $J_{\text{pt}}(x)$ is given by

$$D(\rho) = \frac{J_{\text{pt}}(x)\left(\dfrac{dT}{dr}\right)(r_0)}{4\pi x^2}\left[\frac{S(\rho')^2}{r_0^2}\right] \times 1.6 \times 10^{-8} \tag{8-35}$$

Here x represents the ratio of the radial distance ρ from the point source to the residual range of the electrons at their original energy ($x = \rho/r_0$), and S is the number of electrons born at $\rho = 0$. The quantities $(dT/dr)(r_0)$ and (r_0) have the same units as before and ρ' is the density of the medium. Figure 8-10 gives the dose distribution from point electron sources of energy 0.1 MeV in air and lead.

The numerical tabulations in Ref. 29 apply to monoenergetic electron sources. If the electron source is continuous, it can be divided into monoenergetic groups for which the doses can be calculated individually and then combined.

8-13 THE THEORY OF LOEVINGER, JAPHA, AND BROWNELL

A simpler system for dose determination from β-ray sources has been given by Loevinger, Japha, and Brownell.[21] The procedures for determining the dose in this system are valid for homogeneous media of low atomic number with an arbitrary distribution of sources. All dose distributions are based on a point source distribution obtained from experiment and are given by Eqs. (8-36).

$$J(x) = \frac{k}{(\nu x)^2}\left\{c\left[1 - \frac{\nu x}{c}\, e^{1-(\nu x/c)}\right] + \nu x e^{1-\nu x}\right\} \qquad x \leq \frac{c}{\nu} \tag{8-36}$$

$$J(x) = \frac{k}{\nu x}\, e^{1-\nu x} \qquad x \geq \frac{c}{\nu}$$

Here the product νx is dimensionless; that is, if x, the distance from the point source, is in centimeters, then ν, the apparent absorption coefficient must be in reciprocal centimeters. Or if x is in grams per square centimeter, then ν must be in square centimeters per gram. This function represents an approximation to the dose from a point source from β-emitters varying over a wide range in the spectral shape and energy. Values of ν, the apparent absorption coefficient, are given for air, polystyrene, and tissue.

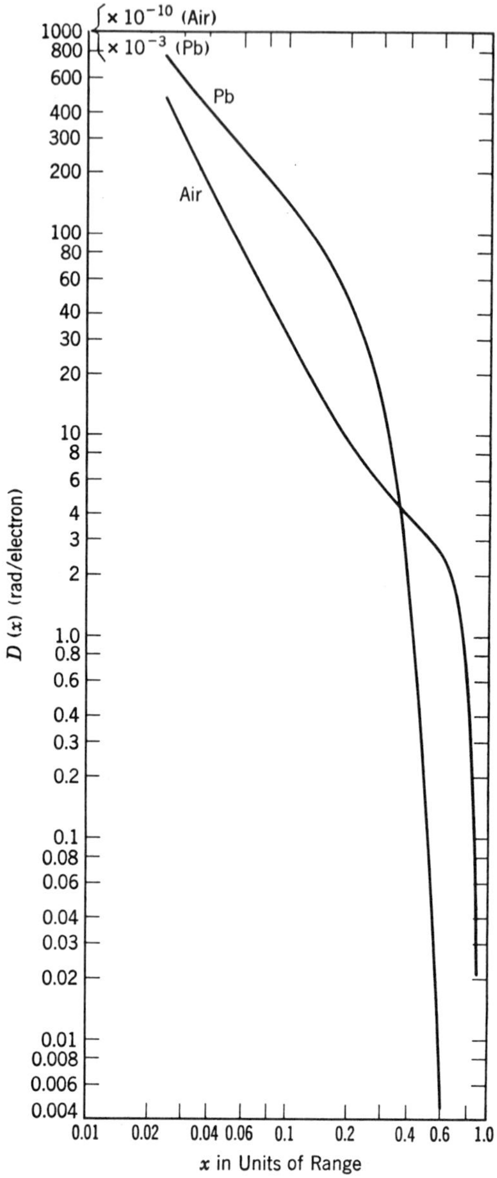

Fig. 8-10. Dose distributions around point sources of 0.1 MeV electrons absorbed in lead and air.

The value for tissue is given in Eq. (8-37)

$$\nu = \frac{18.6}{(E_0 - 0.036)^{1.37}}\left(2 - \frac{\bar{E}_\beta}{\bar{E}_\beta{}^*}\right) \tag{8-37}$$

which contains the parameter E_0, the end point energy of the β-spectrum in MeV, $\bar{E}_\beta$ the average energy under an allowed spectrum from a β-emitter having the same atomic number as the emitter being considered and the same end point energy, and $\bar{E}_\beta{}^*$ which is the actual average energy in the β-spectrum. The ratio $\bar{E}_\beta/\bar{E}_\beta{}^*$ is exactly equal to unity for an allowed β-spectrum and is rarely much different from unity for forbidden spectra. The parameter c is determined by the end point energy of the β-spectrum and is given by

$$c = \begin{cases} 2 & 0.17 < E_0 < 0.5 \text{ MeV} \\ 1.5 & 0.5 \le E_0 < 1.5 \text{ MeV} \\ 1 & 1.5 \le E_0 < 3 \text{ Mev} \end{cases} \tag{8-38}$$

for end point energies in the ranges shown. The quantity k is given by

$$k = 1.28 \times 10^{-9}(\rho')^2\nu^3\,\bar{E}_\beta\alpha \qquad \text{rad per disintegration} \tag{8-39}$$

with

$$\alpha = [3c^2 - (c^2 - 1)e]^{-1} \tag{8-40}$$

and involves the density ρ', the apparent absorption coefficient ν, $\bar{E}_\beta$, the base of the natural logarithm system e, and α, which in turn depends only upon c as shown. This point source solution is used by Loevinger et al. to obtain dose distributions for other geometrical arrangements of β-emitters. In particular, they solve the infinite thin plane source, the infinite plane slab of finite and infinite thickness, the thin plane circular disk source, and the spherical source. Doses are obtained both inside and outside the thick sources. It should be remembered that these calculations apply to an infinite medium, homogeneous except for the distribution of the radioisotope.

Information on the agreement between these dose distributions and experiment may be found in Refs. 5, 16, 17, and 29. Agreement is very good in general but hardly surprising since the basic point source function is empirical.

REFERENCES

1. Berger, M. J., and S. M. Seltzer, Multiple-Scattering Corrections for Proton Range Measurements in *Natl. Acad. Sci. Natl. Res. Council Publ.* **1133** (1964).

2. Bethe, H. A., *Handbuch der Physik* **24,** 273 (1933).

3. Bethe, H. A., and J. A. Ashkin, "Passage of Radiation through Matter" in *Experimental Nuclear Physics* (E. Segrè, ed.), John Wiley and Sons, New York, 1953, Vol. I, p. 166.

4. Bethe, H. A., and W. Heitler, *Proc. Roy. Soc.* **A146,** 83 (1934).

5. Birkhoff, R. D., *Handbuck der Physik* **34,** 53 (1958).

6. Birkhoff, R. D., *Health Phys.* **8,** 973 (1963).
7. Birkhoff, R. D., H. H. Hubbell, Jr., J. S. Cheka, and R. H. Ritchie, *Health Phys.* **1,** 27 (1958).
8. Bloch, F., *Z. Physik* **81,** 363 (1933).
9. Brandt, W., *Health Phys.* **1,** 11 (1958).
10. Danzker, M., N. D. Kessaris, and J. S. Laughlin, *Radiology* **72,** 51 (1959).
11. Doggett, J. A., and L. V. Spencer, *Phys. Rev.* **103,** 1597 (1956).
12. Fano, U., "Tables for the Analysis of Beta Spectra," *Natl. Bur. Std. Appl. Math. Ser.* **13** (1952).
13. Fermi, E., *Phys. Rev.* **56,** 1242 (1939).
14. Fermi, E., *Phys. Rev.* **57,** 485 (1940).
15. Grun, A. E., *Z. Naturforsch,* **12A,** 89 (1957).
16. Hine, G. J., and G. L. Brownell, eds., *Radiation Dosimetry,* Academic Press, New York, 1956.
17. Huffman, F. N., "Spatial Distribution of Electron Depth Dose in Aluminum," *Oak Ridge Natl. Lab. Rept.* **ORNL-2137** (1956).
18. James, M. F., B. G. Steel, and J. S. Story, "Average Electron Energy in Beta Decay," *U.K. Atomic Energy Authority Rept.* **AERE-M 640,** Harwell, Berkshire, England, 1960.
19. Kalil, F., W. G. Stone, H. H. Hubbell, Jr., and R. D. Birkhoff, "Stopping Power of Thin Aluminum Foils for 12 to 127 keV Electrons" *Oak Ridge Natl. Lab. Rept.* **ORNL-2731** (1959).
20. Konopinski, E. J., and M. E. Rose, "The Theory of Nuclear Beta Decay," in *Alpha, Beta, and Gamma-Ray Spectroscopy* (K. Siegbahn, ed.), North-Holland Publishing Co., Amsterdam, 1965, Vol. 2, Chap. XXIII.
21. Loevinger, R., E. M. Japha, and G. L. Brownell, in *Radiation Dosimetry* (G. J. Hine and G. L. Brownell, eds.), Academic Press, New York, 1956, Chap. 16.
22. Marshall, J., *Nucleonics,* **13,** No. 8, 34 (1955).
23. Marton, L., C. Marton, and W. G. Hall, "Electron Physics Tables," *Natl. Bur. Std. (U.S.) Circ.* **571** (1956).
24. McConnell, W. J., H. H. Hubbell, Jr., R. N. Hamm, R. H. Ritchie, and R. D. Birkhoff, *Phys. Rev.* **138,** A1377 (1965).
25. McGinnies, R. J., "Energy Spectrum Resulting from Electron Slowing Down," *Natl. Bur. Std. (U.S.) Circ.* **597** (1959).
26. Möller, C., *Ann. Physik* **14,** 531 (1932).
27. Rohrlich, F., and B. C. Carlson, *Phys. Rev.* **93,** 38 (1954).
28. Spencer, L. V., *Phys. Rev.* **98,** 1959 (1955).
29. Spencer, L. V., "Energy Dissipation by Fast Electrons," *Natl. Bur. Std. (U.S.) Monograph* **1** (1959).
30. Spencer, L. V., and U. Fano, *Phys. Rev.* **93,** 1172 (1954).
31. Sternheimer, R. M., *Phys. Rev.* **88,** 851 (1952).
32. Sternheimer, R. M., *Phys. Rev.* **91,** 256 (1953).
33. Thomson, J. J., *Phil. Mag.* **6,** 23, 449 (1912).
34. Wu, C. S., in *Alpha, Beta, and Gamma Spectroscopy,* Vol. 2 (K. Siegbahn, ed.), North-Holland Publishing Co., Amsterdam, 1965, Chaps. XXII and XXIV.

PROBLEMS

1. Show that the relations for kinetic energy and momentum in Eqs. (8-1) reduce to the classical expressions if $v^2/c^2 \ll 1$.

2. Show that the energy distribution in allowed β-decay, Eq. (8-10), follows from the momentum distribution, Eq. (8-8).

3. Calculate the Thomson cross section, Eq. (8-14), from the expression for the energy transferred in a collision, Eq. (8-13).

4. Consider the stopping power of the conduction electrons in aluminum (3 per atom). Use a maximum impact parameter given by Eq. (8-17) and a minimum equal to the screening length in the conduction electron plasma (about 1 A). The absorption frequency is obtained from the plasma equation $E_p = \hbar\omega_p = \hbar\sqrt{4\pi ne^2/m}$ where n is the density of conduction electrons. Show that about 20% of the total stopping power for 100 keV electrons is due to excitation of plasma oscillations.

5. What is the mean free path for the loss of one quantum of energy to the plasma of Problem 4?

6. A 260 keV electron slows to rest in copper. According to Reference 25, the flux it represents varies from 0.3 g cm^{-2}(m$_0$c^2)$^{-1}$ at 260 keV to 0.1 at 10 keV with an average value of about 0.2. If the K ionization cross section is about 250 barns over this energy interval, calculate the number of K ionizations which take place as the electron slows down past the threshold energy ($\sim$10 keV).

7. Compare the flux at 10 keV cited in problem 6 with the flux at this energy taken from Fig. 8-8.

8. Compute a depth dose curve for 0.1 MeV electrons in air with the assumption that the electrons are emitted and travel to rest along the direction normal to the surface. Compare with the curve of Fig. 8-9.

9. Suppose in Fig. 8-9 that the material on the left of the plane of incidence were removed. Would the dose for positive values of x near zero be greater or smaller? Explain.

10. Show that the total energy delivered to the medium for the two curves of Fig. 8-10 is 0.1 MeV. For air assume that $J_{\text{pt}}(x)$ has an average value of 2.1 for $x < 0.8$ and is zero for $0.8 < x < 1$. For lead assume that $J_{\text{pt}}(x)$ has a value of 4.0 for $x < 0.4$ and is zero for $0.4 < x < 1$.

11. What are the total masses and volumes affected in the two cases of Fig. 8-10?

12. Using the point source dose function of Eq. (8-36), calculate the dose at a distance x from a thin plane source of infinite extent.

13. Show that the definitions of k and α in Eqs. (8-39) and (8-40) are consistent with the requirement that the total energy absorbed around a point β-source must equal $\bar{E}_\beta$, the average energy in the β-spectrum.

CHAPTER 9

Dose from External Sources of Radiation

K. Z. Morgan and L. C. Emerson

9-1 USE OF DISTANCE, MASS, AND TIME FOR RADIATION PROTECTION

The three fundamental units, distance, mass (or shielding), and time of exposure, are involved in radiation protection as in everything else in our existence. We should always strive to keep radiation exposure to a minimum above natural background radiation. When it is necessary to work with radiation sources, we should stay as far away from the source as practical. Shielding can be used with distance to reduce the radiation to acceptable levels, and time of exposure can be reduced by working with sources as rapidly as possible. When not in use, sources should be stored away from personnel. The safety factor of time can often be used by dividing the radiation exposure over an extended period or among a number of persons. This chapter discusses radiation penetration from the standpoint of distance and shielding, this subject being the basis of practical radiation protection, chiefly from external sources of X- and γ-rays and neutrons. Exposure to radiation sources within the body is treated in Chapter 10.

9-2 RADIATION BUILDUP

As pointed out in Section 2-10, simple calculations of the photon flux at some distance away from the source lead to underestimates when the total attenuation coefficient μ is used in the calculation. This is equivalent to assuming that every scattering and absorption event leads to permanent removal of photons from the beam. If the photon beam is sufficiently well collimated, this is not a bad assumption. In this "good geometry" arrangement, which is the way in which attenuation coefficients are measured experimentally, the underestimation of the total photon flux may be quite small. In the more common "poor geometry" situation, broad beam conditions prevail, and scattering events as well as other processes contribute secondary photons to the primary beam at all points.

Buildup Calculations

Calculations of this extra contribution, termed buildup, are at best extremely involved and require the use of digital computers. Basically, the calculation consists of solving the Boltzmann transport equation for photons and keeping track of all absorption and scattering events in which changes in direction and energy occur. The calculations must be carried out in six-dimensional phase space which includes three dimensions for position, two for direction, and one for energy. Three methods for solving the transport equation have been used, each with varying degrees of success: the multiple-scattering method, the method of moments, and stochastic methods. The method of moments, developed by Spencer and Fano (1951) at the National Bureau of Standards (NBS), has been the most widely publicized and used. This seminumerical technique for solving the transport equation has been further developed and applied by Goldstein and Wilkins (1954), who with the cooperation of NBS carried out a comprehensive program of γ-ray attenuation calculations. Their results are available for water and six elements with Z ranging from 13 to 92 for nine photon energies between 0.255 and 10.0 MeV. Since buildup factors are smoothly varying functions of both atomic number and energy, values for other materials and energies can be easily obtained from their results.

The calculations have been carried out for energy buildup, energy-absorption buildup, and dose buildup for both point isotropic sources and plane monodirectional sources. The differences between the results for these different calculations are generally insignificant except in the most critical calculations. The calculated values are valid for infinite media, and the errors involved in using any of the tabulated values for finite geometries may well exceed other errors, for example, those introduced by using the point-source dose buildup factor exclusively. In general, the dose buildup factor is larger than either the energy buildup factor or the energy-absorption buildup factor. The buildup factors for the point source are always larger than the corresponding factor for the plane source.

The calculated values for the point source dose buildup factors for water, aluminum, iron, tin, and lead are listed in Table 9-1.

Table 9-1 Buildup Factors

Material	Energy (MeV)	Number of Mean Free Paths, μr: 1	2	4	7	10	15	23
Water	0.255	3.09	7.14	23.0	72.9	166	456	982
	0.5	2.52	5.14	14.3	38.8	77.6	178	334
	1.0	2.13	3.71	7.68	16.2	27.1	50.4	82.2
	2.0	1.83	2.77	4.88	8.46	12.4	19.5	27.7
	3.0	1.69	2.42	3.91	6.23	8.63	12.8	17.0
	4.0	1.58	2.17	3.34	5.13	6.94	9.97	12.9
	6.0	1.46	1.91	2.76	3.99	5.18	7.09	8.85
	8.0	1.38	1.74	2.40	3.34	4.25	5.66	6.95
	10.0	1.33	1.63	2.19	2.97	3.72	4.90	5.98
Aluminum	0.5	2.37	4.24	9.47	21.5	38.9	80.8	141
	1.0	2.02	3.31	6.57	13.1	21.2	37.9	58.5
	2.0	1.75	2.61	4.62	8.05	11.9	18.7	26.3
	3.0	1.64	2.32	3.78	6.14	8.65	13.0	17.7
	4.0	1.53	2.08	3.22	5.01	6.88	10.1	13.4
	6.0	1.42	1.85	2.70	4.06	5.49	7.97	10.4
	8.0	1.34	1.68	2.37	3.45	4.58	6.56	8.52
	10.0	1.28	1.55	2.12	3.01	3.96	5.63	7.32
Iron	0.5	1.98	3.09	5.98	11.7	19.2	35.4	55.6
	1.0	1.87	2.89	5.39	10.2	16.2	28.3	42.7
	2.0	1.76	2.43	4.13	7.25	10.9	17.6	25.1
	3.0	1.55	2.15	3.51	5.85	8.51	13.5	19.1
	4.0	1.45	1.94	3.03	4.91	7.11	11.2	16.0
	6.0	1.34	1.72	2.58	4.14	6.02	9.89	14.7
	8.0	1.27	1.56	2.23	3.49	5.07	8.50	13.0
	10.0	1.20	1.42	1.95	2.99	4.35	7.54	12.4
Tin	0.5	1.56	2.08	3.09	4.57	6.04	8.64	
	1.0	1.64	2.30	3.74	6.17	8.85	13.7	18.8
	2.0	1.57	2.17	3.53	5.87	8.53	13.6	19.3
	3.0	1.46	1.96	3.13	5.28	7.91	13.3	20.1
	4.0	1.38	1.81	2.82	4.82	7.41	13.2	21.2
	6.0	1.26	1.57	2.37	4.17	6.94	14.8	29.1
	8.0	1.19	1.42	2.05	3.57	6.19	15.1	34.0
	10.0	1.14	1.31	1.79	2.99	5.21	12.5	33.4
Lead	0.5	1.24	1.42	1.69	2.00	2.27	2.65	2.73
	1.0	1.37	1.69	2.26	3.02	3.74	4.81	5.86
	2.0	1.39	1.76	2.51	3.66	4.44	6.87	9.00
	3.0	1.34	1.68	2.43	3.75	5.30	8.44	12.3
	4.0	1.27	1.56	2.25	3.61	5.44	9.80	16.3
	5.1097	1.21	1.46	2.08	3.44	5.55	11.7	23.6
	6.0	1.18	1.40	1.97	3.34	5.69	13.8	32.7
	8.0	1.14	1.30	1.74	2.89	5.07	14.1	44.6
	10.0	1.11	1.23	1.58	2.52	4.34	12.5	39.2

Approximations to Moments Method Results

The point source geometry is the only one for which the results of the moments method calculation can be applied directly with no ambiguity. For all other geometries the calculation involves an integration over the dimensions of the source. Since the buildup factor is dependent upon spatial coordinates, it must also be included in the integrand. Some analytic approximation to the buildup factor must then be used.

In the very early days of health physics it was customary to take account of the scattered radiation by replacing the attenuation coefficient μ by the absorption coefficient $\mu - \sigma_s$ in exposure-rate calculations. This was equivalent to assuming that *all* scattered radiation reached the point of interest with the original photon energy. This procedure, which gives conservative results, is still used occasionally; and its use is warranted in those cases where more detailed information regarding the buildup is unavailable.

One of the most useful approximations to the buildup values listed in Table 9-1 is the linear buildup factor, that is,

$$B(E, \mu r) \simeq 1 + \alpha(E)\mu r \tag{9-1}$$

Here E is the photon energy and μr is the number of mean free paths in the material at which the value of B is needed. For the sake of simplicity, α is sometimes taken to be unity, a practice which is justified only in special instances. In the case of the energy absorption buildup factor, the value of α can be calculated by application of the physical requirement of the conservation of energy. Values of $\alpha(E)$ for water, aluminum, concrete, iron, tin, and lead are listed in Table 9-2 (TR66).

Another analytic approximation that has proved extremely helpful in exposure-rate calculations is that due to Taylor (TA54). In this approximation a sum of exponentials is used

$$B(E, \mu r) \simeq A(E)e^{-\alpha_1(E)\mu r} + [1 - A(E)]e^{-\alpha_2(E)\mu r} \tag{9-2}$$

where A, α_1, and α_2 are the fitting parameters. Values of dose buildup parameters for water, aluminum, concrete, iron, tin, and lead are listed in Table 9-3 (Trubey, 1966). These numbers are accurate to within about 5% for energies > 1 MeV except for lead, where the maximum deviation is only 1.8%.

A four-term polynomial fit has also been useful (Capo, 1959; Trubey, 1966), the buildup factor being given by

$$B(E, \mu r) \simeq \beta_0(E) + \beta_1(E)\mu r + \beta_2(E)\mu^2 r^2 + \beta_3(E)\mu^3 r^3 \tag{9-3}$$

The fitting parameters β are listed for many materials in the references cited.

A two-parameter form of an approximation formula originally due to Berger (1956) has the advantage of fitting the moments-method calculation

Table 9-2 Fitting Parameter for Linear Buildup Factor

Energy (MeV)	Substance					
	Water	Concrete	Aluminum	Iron	Tin	Lead
	7 Mean Free Path Fit					
0.5	4.680	3.744	2.646	1.428	0.5153	0.1549
1.0	1.995	1.906	1.609	1.237	0.7199	0.2990
2.0	1.030	1.023	0.9686	0.8556	0.6731	0.3796
3.0	0.7397	0.7303	0.7197	0.6691	0.5837	0.3810
4.0	0.5884	0.5736	0.5663	0.5403	0.5146	0.3523
6.0	0.4321	0.4329	0.4334	0.4297	0.4153	0.3034
8.0	0.3406	0.3376	0.3476	0.3391	0.3317	0.2419
10.0	0.2877	0.2923	0.2847	0.2681	0.2550	0.1933
	20 Mean Free Path Fit					
0.5	13.093	5.012	5.737	2.377	0.5090	0.1043
1.0	3.479	2.992	2.539	1.864	0.8495	0.2549
2.0	1.255	1.233	1.193	1.119	0.8521	0.3947
3.0	0.7863	0.7857	0.8061	0.8446	0.8509	0.5123
4.0	0.5951	0.5942	0.6075	0.6942	0.8643	0.6378
6.0	0.4030	0.4145	0.4626	0.6134	1.079	1.125
8.0	0.3085	0.3200	0.3697	0.5245	1.191	1.417
10.0	0.2584	0.2737	0.3087	0.4759	1.108	1.237

well over a wide range. It is given by

$$B(E, \mu r) \simeq 1 + C(E)\mu r e^{D(E)\mu r} \tag{9-4}$$

where C and D are the fitting parameters. The values of C and D are listed in Table 9-4 for several materials for fits over 7 mean free paths and 20 mean free paths. The maximum errors for the fit over the shorter range are 3%, except for water at energies below 1 MeV, where the error may reach 10%. The errors are about three times as large for the fit over the larger range. These values, which were calculated by Trubey (1966) using a least-squares procedure, have been reported for other materials and source energies. This reference provides an excellent discussion of the approximations mentioned here as well as a number of others which are important in certain applications.

9-3 CALCULATION OF THE EXPOSURE FROM A POINT SOURCE

Although radiation sources have finite dimensions, they are often sufficiently small compared to the distance from the point of measurement that the source may be treated as a point source. This assumption simplifies

Table 9-3 Parameters for the Taylor Form of the Buildup Factor

Substance	Energy (MeV)	A	$-\alpha_1$	α_2	Substance	Energy (MeV)	A	$-\alpha_1$	α_2
Water	0.5	100.845	0.12687	−0.10925	Iron	0.5	31.379	0.06842	−0.03742
	1.0	19.601	0.09037	−0.02522		1.0	24.957	0.06086	−0.02463
	2.0	12.612	0.05320	0.01932		2.0	17.622	0.04627	−0.00526
	3.0	11.110	0.03550	0.03206		3.0	13.218	0.04431	−0.00087
	4.0	11.163	0.02543	0.03025		4.0	9.624	0.04698	0.00175
	6.0	8.385	0.01820	0.04164		6.0	5.867	0.06150	−0.00186
	8.0	4.635	0.02633	0.07097		8.0	3.243	0.07500	0.02123
	10.0	3.545	0.02991	0.08717		10.0	1.747	0.09900	0.06627
Concrete	0.5	38.225	0.14824	−0.10579	Tin	0.5	11.440	0.01800	0.03187
	1.0	25.507	0.07230	−0.01843		1.0	11.426	0.04266	0.01606
	2.0	18.089	0.04250	0.00849		2.0	8.783	0.05349	0.01505
	3.0	13.640	0.03200	0.02022		3.0	5.400	0.07440	0.02080
	4.0	11.460	0.02600	0.02450		4.0	3.496	0.09517	0.02598
	6.0	10.781	0.01520	0.02925		6.0	2.005	0.13733	−0.01501
	8.0	8.972	0.01300	0.02979		8.0	1.101	0.17288	−0.01787
	10.0	4.015	0.02880	0.06844		10.0	0.708	0.19200	0.01552
Aluminum	0.5	38.911	0.10015	−0.06312	Lead	0.5	1.677	0.03084	0.30941
	1.0	28.782	0.06820	−0.02973		1.0	2.984	0.03503	0.13486
	2.0	16.981	0.04588	0.00271		2.0	5.421	0.03482	0.04379
	3.0	10.583	0.04066	0.02514		3.0	5.580	0.05422	0.00611
	4.0	7.526	0.03973	0.03860		4.0	3.897	0.08468	−0.02383
	6.0	5.713	0.03934	0.04347		6.0	0.926	0.17860	−0.04635
	8.0	4.716	0.03837	0.04431		8.0	0.368	0.23691	−0.05864
	10.0	3.999	0.03900	0.04130		10.0	0.311	0.24024	−0.02783

Table 9-4 Parameters for the Berger Form of the Buildup Factor

Energy (MeV)	Water C	Water D	Concrete C	Concrete D	Aluminum C	Aluminum D	Iron C	Iron D	Tin C	Tin D	Lead C	Lead D
7 Mean Free Path Fit												
0.5	1.325	0.2078	1.449	0.1586	1.244	0.1250	0.9081	0.0752	0.5608	−0.0146	0.2526	−0.0848
1.0	1.062	0.1052	1.045	0.1014	0.9589	0.0864	0.8214	0.0684	0.6219	0.0244	0.3779	−0.0403
2.0	0.8093	0.0408	0.8062	0.0403	0.7267	0.0486	0.7020	0.0319	0.5498	0.0338	0.3862	0.0032
3.0	0.6876	0.0125	0.6267	0.0254	0.6294	0.0227	0.5323	0.0384	0.4379	0.0479	0.3267	0.0253
4.0	0.5800	0.0024	0.6451	−0.0207	0.5253	0.0127	0.4366	0.0358	0.3583	0.0601	0.2530	0.0547
6.0	0.4655	−0.0126	0.5086	−0.0286	0.4177	0.0061	0.3271	0.0457	0.2369	0.0925	0.1622	0.1027
8.0	0.3860	−0.0214	0.4085	−0.0334	0.3371	0.0050	0.2563	0.0464	0.1692	0.1103	0.1220	0.1112
10.0	0.3342	−0.0257	0.3584	−0.0356	0.2752	0.0055	0.1876	0.0592	0.1232	0.1190	0.0939	0.1167
20 Mean Free Path Fit												
0.5	1.804	0.1224	1.518	0.1413	1.441	0.0850	0.9814	0.0548	0.5457	−0.0063	0.2243	−0.0500
1.0	1.228	0.0649	1.221	0.0562	1.083	0.0535	0.8932	0.0460	0.6378	0.0180	0.3530	−0.0211
2.0	0.8594	0.0240	0.8579	0.0231	0.7869	0.0266	0.7173	0.0277	0.5678	0.0254	0.3791	0.0021
3.0	0.7004	0.0074	0.6589	0.0115	0.6504	0.0137	0.5571	0.0261	0.4533	0.0388	0.3244	0.0279
4.0	0.5826	0.0014	0.6056	−0.0018	0.5343	0.0082	0.4518	0.0268	0.3700	0.0518	0.2526	0.0557
6.0	0.4853	−0.0082	0.4769	−0.0093	0.4182	0.0063	0.3381	0.0368	0.2401	0.0891	0.1554	0.1143
8.0	0.3741	−0.0124	0.3789	−0.0113	0.3366	0.0058	0.2603	0.0428	0.1669	0.1145	0.1075	0.1440
10.0	0.3206	−0.0139	0.3318	−0.0128	0.2738	0.0074	0.1902	0.0553	0.1192	0.1278	0.0824	0.1513

analysis. Consistent also with the approximation of a point source is the neglect of photon absorption by the source material itself.

The calculation of the exposure rate from any radioactive source is generally done in two steps: (1) calculation of the photon flux density at the point of interest; (2) the conversion of this quantity to exposure rate. Details of the second step are discussed in Chapter 3. For a point source located a distance r in air from the point of measurement the photon flux density ϕ is given by

$$\phi = BS\frac{e^{-\mu r}}{4\pi r^2} \tag{9-5}$$

where S = the source strength in photons per second

μ = the linear attenuation coefficient of air for the energy of the photons from S

B = the buildup factor.

The term $(4\pi r^2)^{-1}$ represents the inverse-square attenuation and the term $e^{-\mu r}$ is the exponential attenuation discussed in Section 2-9. The quantity $(4\pi r^2)^{-1}e^{-\mu r}$ is frequently referred to as the point kernel. An additional exponential attenuation factor would be used for each layer of absorbing material in the photon path. The exposure rate in roentgens per second is obtained from the flux density by means of the relation

$$\text{ER} = \frac{h\nu}{E_f}\phi \tag{9-6}$$

where $h\nu$ is the photon energy in ergs and E_f is the incident energy fluence per roentgen, as given by Eq. (3-1) and shown in Fig. 3-1.

Expressing $h\nu$ in MeV and putting in the quantities for E_f from Eq. (3-1), we may express the exposure rate as

$$\text{ER} = 1.827 \times 10^{-8} h\nu\frac{\mu_{\text{en}}}{\rho}\phi \tag{9-7}$$

where μ_{en}/ρ is the mass energy absorption coefficient of air for photons of energy $h\nu$ as shown in Figs. 2-18 and 2-19. Equation 9-7 is a general expression for the exposure rate which is independent of the source geometry.

EXAMPLE 1. Calculate the exposure rate at a distance of 1 meter from a 1-Curie point source of ^{60}Co.

Quantities needed:

$$S = 3.7 \times 10^{10} \quad \text{dis/sec}$$

$$r = 100 \text{ cm}$$

Two photons per disintegration:

$$h\nu_1 = 1.17 \quad \text{MeV}$$
$$h\nu_2 = 1.33 \quad \text{MeV}$$

$$\frac{\mu_{en}}{\rho} \quad \text{for} \quad 1.17 \quad \text{MeV} = 0.0265 \quad \text{cm}^2/\text{gm}$$

$$\frac{\mu_{en}}{\rho} \quad \text{for} \quad 1.33 \quad \text{MeV} = 0.0255 \quad \text{cm}^2/\text{gm}$$

From Eq. (9-5)

$$\phi = \frac{3.7 \times 10^{10}}{4\pi(100)^2} = 2.94 \times 10^5 \frac{\text{photons}}{\text{cm}^2 - \text{sec}}$$

for each γ-ray. The product μr is so small that we have assumed $e^{-\mu r} \sim 1$. Adding the contribution to the exposure rate from each γ-ray, we have

$$\begin{aligned} \text{ER} &= 1.827 \times 10^{-8} \times 2.94 \times 10^5(1.17 \times 0.0265 + 1.33 \times 0.0255) \\ &= 3.47 \times 10^{-4} \text{ R/sec} = 1.25 \text{ R/hr} \end{aligned}$$

Here, because of small value for μr, the buildup factor B was taken as unity.

EXAMPLE 2. How much would the exposure rate be reduced by surrounding the source of Example 1 with a lead shield 5 cm thick?

The exponential attenuation factor of Eq. (9-5) now has an additional term $e^{-\mu_{Pb} r_{Pb}}$, and the buildup factor will no longer be unity. We must now calculate ϕ and the exposure rate for each photon energy separately. For the 1.17 and 1.33 MeV γ-rays, we find μr for lead ($\rho = 11.34$ gm/cm^3) to be 3.57 and 3.23, respectively. Using a two-way interpolation in Table 9-1, we find the respective corresponding values for B to be 2.18 and 2.10. Therefore, the flux densities are given by

$$\phi_{1.17} = 2.18 \times 3.7 \times 10^{10} \frac{e^{-3.57}}{4\pi(100)^2} = 1.81 \times 10^4 \text{ photons/cm}^2\text{-sec}$$

and

$$\phi_{1.33} = 2.10 \times 3.7 \times 10^{10} \frac{e^{-3.23}}{4\pi(100)^2} = 2.45 \times 10^4 \text{ photons/cm}^2\text{-sec}$$

The corresponding exposure rates are

$$\text{ER}_{1.17} = 1.827 \times 10^{-8} \times 1.81 \times 10^4 \times 1.17 \times 0.0265 = 1.02 \times 10^{-5} \quad \text{R/sec}$$

and

$$\text{ER}_{1.33} = 1.827 \times 10^{-8} \times 2.45 \times 10^4 \times 1.33 \times 0.0255 = 1.51 \times 10^{-5} \quad \text{R/sec}$$

The total exposure rate, therefore, is 2.53×10^{-5} R/sec = 91 mR/hr.

If the exposure rate were to be corrected for the scattered radiation by using $\mu - \sigma_s$ instead of μ together with a buildup factor, then the answer would have been 160 mR/hr, a value which is 75% too high.

9-4 EXTENDED SOURCES

For radioactive sources of finite dimensions the calculation of the photon flux becomes more complicated. Self-absorption by the source material and the varying paths taken by photons from different parts of the source must be accounted for in the calculation. The primary, or uncollided, flux from extended sources is obtained by integrating the point kernel over the dimensions of the source. The contibution from scattering is accounted for by the use of a buildup factor. Since the buildup factor is different for photons emanating from each region of the source, it must be included in the integration. Only a relatively few forms of the buildup factor meet the dual requirements of being integrable and correctly accounting for the scattering.

Line Sources

We first consider the case of no buildup. In Fig. 9-1 the photon flux at the field point P is given by

$$\phi = \int \frac{S_L e^{-\mu a \sec \theta}}{4\pi r^2}\, dl \tag{9-8}$$

where S_L is source strength per unit length of the line.

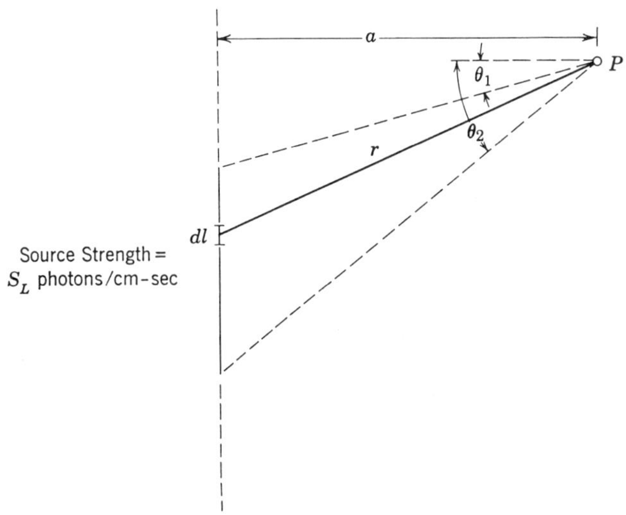

Fig. 9-1. Geometry for flux calculation from a line source; μ = attenuation coefficient of absorber. For more than one absorber replace μa in the point kernel by $\mu_1 a_1 + \mu_2 a_2 + \cdots$.

By setting $l = a \tan \theta$, Eq. (9-8) may be rewritten as

$$\phi = \int_{\theta_1}^{\theta_2} \frac{S_L a \sec^2 \theta}{4\pi(a \sec \theta)^2} e^{-\mu a \sec \theta}\, d\theta$$

$$= \frac{S_L}{4\pi a}\left(\int_0^{\theta_2} e^{-\mu a \sec \theta}\, d\theta - \int_0^{\theta_1} e^{-\mu a \sec \theta}\, d\theta\right) \tag{9-9}$$

The solution may be written in terms of the Sievert integral $F(\theta, \mu a)$ as

$$\phi = \frac{S_L}{4\pi a} [F(\theta_2, \mu a) - F(\theta_1, \mu a)] \tag{9-10}$$

The Sievert integral is discussed in Section 9-6; its values are plotted in Figs. 9-9, 9-10, and 9-11.

When the field point lies on the perpendicular bisector of the line of length L the solution is

$$\phi = \frac{S_L}{2\pi a} F(\theta, \mu a) \tag{9-11}$$

where $\theta = \tan^{-1}(L/2a)$. For an infinitely long source,

$$\phi = \frac{S_L}{2\pi a} F\left(\frac{\pi}{2}, \mu a\right) \tag{9-12}$$

If a linear buildup factor is included, the equation for the flux becomes

$$\phi = \int S_L(1 + \alpha\mu a \sec \theta) \frac{e^{-\mu a \sec \theta}}{4\pi r^2}\, dl \tag{9-13}$$

Part of the integration can be done as before; and this gives

$$\phi = \frac{S_L}{4\pi a}\left[F(\theta_2, \mu a) - F(\theta_1, \mu a) + \alpha\mu a \int_{\theta_1}^{\theta_2} \sec \theta e^{-\mu a \sec \theta}\, d\theta\right] \tag{9-14}$$

In general, the integral must be evaluated numerically. When the source is an infinitely long line, however, $\theta_1 = \theta_2 = \pi/2$; and the integral is equal to

$$2\int_0^{\pi/2} \sec \theta e^{-\mu a \sec \theta}\, d\theta \tag{9-15}$$

By changing variables to $u = \cosh^{-1} (\sec \theta)$, we have $d\theta = (\cosh^2 u - 1)^{-1/2}$ $\tanh u\, du = \cosh u^{-1}\, du$. The integral (9-15) can now be written

$$2\int_0^{\infty} e^{-\mu a \cosh u}\, du \tag{9-16}$$

which is the zero-order modified Bessel function of the second kind, $K_0(\mu a)$ (McLachlan, 1955). The solution for the infinite line source is

$$\phi = \frac{S_L}{2\pi a}\left[F\left(\frac{\pi}{2}, \mu a\right) + \alpha\mu a K_0(\mu a)\right] \tag{9-17}$$

the second term in the brackets representing the contribution due to scattering.

EXAMPLE 1. Cobalt wire is frequently used to measure neutron flux in nuclear reactors. Calculate the expected exposure rate 1 cm from the surface of a 2-cm section of No. 22 B and S gage wire which has been exposed for 1 hour to a thermal neutron flux of 10^{13} n/cm^2-sec. Allow sufficient time for the 10-minute isomer to decay.

The activity (in disintegrations per second) following an irradiation time t is given by (Evans, 1955)

$$\lambda N = P(1 - e^{-\lambda t})$$

where λ = the decay constant
N = the number of atoms present
P = the production rate in atoms/sec

In this example, $P = \sigma N_0 \phi$, where σ is the reaction cross section, N_0 is the number of target atoms, and ϕ is the neutron flux. The wire contains

$$\frac{\pi(0.02535 \text{ in.})^2}{4} \times \left(2.54 \frac{\text{cm}}{\text{in.}}\right)^2 \times 2 \text{ cm} \times 8.9 \frac{\text{gm}}{\text{cc}} \times \frac{6.025 \times 10^{23}}{59} \text{ atoms/gm} = 5.919 \times 10^{20} \text{ atoms}$$

From a table of isotopes we find the thermal neutron cross section of ^{59}Co to be 19 barns = 19×10^{-24} cm^2. The production rate is thus

$$19 \times 10^{-24} \text{ cm}^2 \times 5.919 \times 10^{20} \text{ atoms} \times 10^{13}\ n/\text{cm}^2\text{-sec} = 1.125 \times 10^{11} \text{ atoms/sec}$$

The half-life of ^{60}Co being 5.26 years, its decay constant is

$$\lambda = \frac{0.693}{5.26} \text{ yr}^{-1} \times \left(3.154 \times 10^7 \frac{\text{sec}}{\text{yr}}\right)^{-1} = 4.178 \times 10^{-9} \text{ sec}^{-1}$$

Since $\lambda t \ll 1$ we may expand $e^{-\lambda t}$ and obtain

$$1 - e^{-\lambda t} = 1 - (1 - \lambda t + \cdots) \cong \lambda t$$

Therefore, the source strength is given by

$$\begin{aligned} S_L &= \tfrac{1}{2}\lambda N = \tfrac{1}{2}P\lambda t = \tfrac{1}{2} \times 1.125 \times 10^{11} \times 4.178 \times 10^{-9} \times 3600 \\ &= 8.46 \times 10^5 \text{ dis/sec-cm} \end{aligned}$$

Buildup may be neglected because the wire has a small diameter. Since we are interested in the exposure rate on the perpendicular bisector, $\theta = 45°$. Using Fig. 9-9 we find for the photon flux for each of the two γ-rays

$$\phi = \frac{8.46 \times 10^5}{2\pi \times 1} F(45^\circ, 0)$$

$$= \frac{8.46 \times 10^5}{2\pi} \times 0.787 = 1.06 \times 10^5 \text{ photons/cm}^2\text{-sec}$$

For the exposure rate we obtain, using Eq. (9-7),

$$\text{ER} = 1.827 \times 10^{-8} h\nu \frac{\mu_{\text{en}}}{\rho} \phi$$

$$= 1.827 \times 10^{-8} \times 1.06 \times 10^5 (1.17 \times 0.0265 + 1.33 \times 0.0255)$$

$$= 0.125 \times 10^{-3} \quad \text{R/sec} = 450 \quad \text{mR/hr}$$

Area Sources

The geometry for the area source is shown in Fig. 9-2. The uncollided flux at point P which lies on the center line of the source at a distance a is

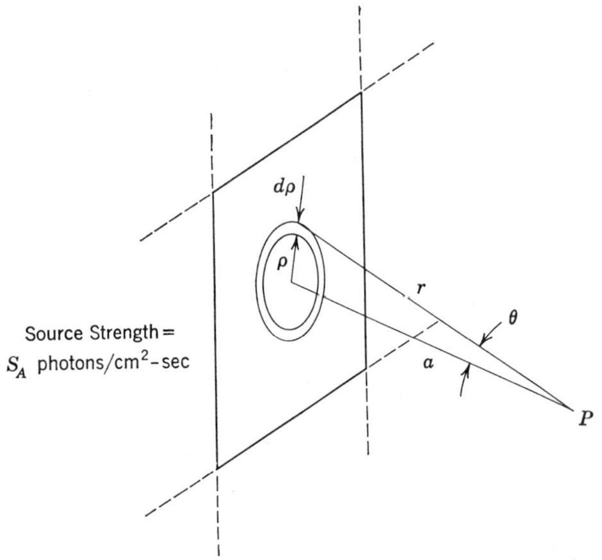

Fig. 9-2. Geometry for flux calculation from an area source; μ = attenuation coefficient of absorber. For more than one absorber replace μa in the point kernel by $\mu_1 a_1 + \mu_2 a_2 + \cdots$.

given by

$$\phi = \int_0^R S_A \frac{2\pi\rho\, d\rho}{4\pi r^2}\, e^{-\mu a \sec\theta} \tag{9-18}$$

where S_A is the source strength in photons per second per unit area. For a circular disk of radius R, Eq. (9-18) can be integrated by making the substitution $y = \mu a \sec\theta$. This leads to

$$\phi = \frac{S_A}{2}\left[\int_{\mu a}^{\infty} \frac{e^{-y}}{y}\, dy - \int_{\mu\sqrt{R+^2a^2}}^{\infty} \frac{e^{-y}}{y}\, dy\right] \tag{9-19}$$

These are the exponential integrals discussed in Section 9-7. Using the notation for these functions, we write

$$\phi = \frac{S_A}{2}\left[E_1(\mu a) - E_1(\mu\sqrt{R^2 + a^2})\right] \tag{9-20}$$

Note that, for a disk of infinite radius, the expression simplifies to

$$\phi = \frac{S_A}{2} E_1(\mu a) \tag{9-21}$$

There are a number of ways in which the correction for scattering can be made. We have already seen, for example, how the linear buildup factor is used. We now use the form suggested by Taylor (1954). Applying Eq. (9-2) to the expression for the uncollided flux, we have

$$\phi = \int S_A[Ae^{-\alpha_1 \mu a \sec\theta} + (1 - A)e^{-\alpha_2 \mu a \sec\theta}] \frac{2\pi\rho\, d\rho}{4\pi r^2}\, e^{-\mu a \sec\theta} \tag{9-22}$$

where A, α_1, and α_2 are the fitting parameters. Using the same technique as before, we combine the exponentials and find that the photon flux can be written

$$\phi = \frac{S_A}{2}\Big\{A\Big[E_1[(1 + \alpha_1)\mu a] - E_1[(1 + \alpha_1)\mu\sqrt{R^2 + a^2}]\Big] + (1 - A) \times \Big[E_1[(1 + \alpha_2)\mu a] - E_1[(1 + \alpha_2)\mu\sqrt{R^2 + a^2}]\Big]\Big\} \tag{9-23}$$

For an infinite plane source this becomes

$$\phi = \frac{S_A}{2}\{AE_1[(1 + \alpha_1)\mu a] + (1 - A)E_1[(1 + \alpha_2)\mu a]\} \tag{9-24}$$

With this form for the buildup factor it is not possible to isolate the terms which account for the scattered radiation as we did with linear buildup. If we were to modify Eq. (9-22) using the linear form, we would find for the finite source,

$$\phi = \frac{S_A}{2}[E_1(\mu a) - E_1(\mu a \sec\theta) + \alpha(e^{-\mu a} - e^{-\mu a \sec\theta})] \tag{9-25}$$

and for the infinite source,

$$\phi = \frac{S_A}{2}[E_1(\mu a) + \alpha e^{-\mu t}] \tag{9-26}$$

Here the scattered radiation terms are clearly the exponential terms.

Volume Sources

With sources having finite thickness it becomes necessary to take into account the self-shielding of the source itself. Consider first the case of a source of infinite planar extent and finite thickness. Referring to Fig. 9-3,

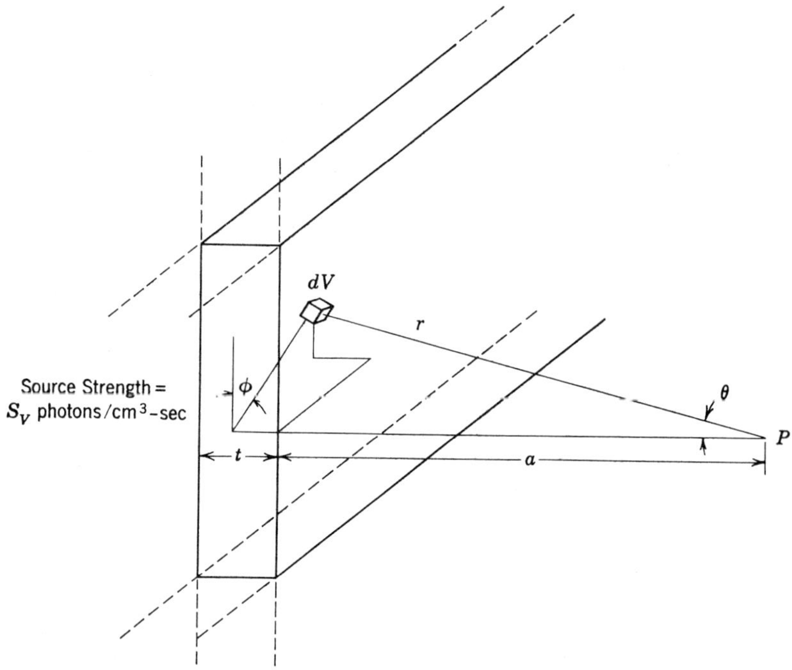

Fig. 9-3. Geometry for flux calculation from a volume source; μ_s = attenuation coefficient of source and μ = attenuation coefficient of absorber. For more than one absorber, replace μa in the point kernel by $\mu_1 a_1 + \mu_2 a_2 + \cdots$.

we can write for the case of unity buildup

$$\phi = \int S_V \frac{e^{-(r-a\sec\theta)\mu_s-\mu a\sec\theta}}{4\pi r^2}\,dV$$
$$= \int_0^{2\pi} d\phi \int_0^{\pi/2} \sin\theta\,d\theta \int_{a\sec\theta}^{(a+t)\sec\theta} \frac{S_V}{4\pi r^2} e^{-(r-a\sec\theta)\mu_s-\mu a\sec\theta} r^2\,dr \tag{9-27}$$

where μ and μ_s are the attenuation coefficients of the shield and source and S_V is the source strength per unit volume.

The integrations over ϕ and r give

$$\phi = \frac{S_V}{2\mu_s}\int_0^{\pi/2} \sin\theta[e^{-\mu a\sec\theta} - e^{-(\mu a+\mu_s t)\sec\theta}]\,d\theta \tag{9-28}$$

Using the substitutions $y = \mu a \sec\theta$ and $y = (\mu a + \mu_s t)\sec\theta$, we write Eq. (9-28) as two second-order exponential integrals. The flux is then given by

$$\phi = \frac{S_V}{2\mu_s}[E_2(\mu a) - E_2(\mu a + \mu_s t)] \tag{9-29}$$

This expression can be rewritten in terms of first-order exponential integrals by using the recurrence relation given by Eq. (9-66). For a source of infinite thickness ($t \to \infty$) the relation becomes simply

$$\phi = \frac{S_V}{2\mu_s} E_2(\mu a) \tag{9-30}$$

This expression can be obtained directly from Eq. (9-27) by setting the upper limit in the integral over r equal to infinity. When no shielding material is present, the flux corresponds to that which exists at the surface; that is, either $\mu = 0$ or $a = 0$, and we have

$$\phi = \frac{S_V}{2\mu_s}[1 - E_2(\mu_s t)] \tag{9-31}$$

If, in this expression, we let $t \to \infty$ we obtain the primary photon flux at the surface of an infinitely thick source of infinite extent, that is,

$$\phi = \frac{S_V}{2\mu_s} \tag{9-32}$$

If a linear buildup factor is used to account for the scattered radiation, Eq. (9-27) is modified to give

$$\phi = \int_0^{2\pi} d\phi \int_0^{\pi/2} \sin\theta\,d\theta \int_{a\sec\theta}^{(a+t)\sec\theta} \frac{S_V}{4\pi r^2}$$
$$\times [1 + \alpha_1\mu_s(r - a\sec\theta) + \alpha_2\mu a\sec\theta] e^{-(r-a\sec\theta)\mu_s-\mu a\sec\theta} r^2\,dr \tag{9-33}$$

Integrating over ϕ and r and collecting terms gives

$$\phi = \frac{S_V}{2\mu_s} \int_0^{\pi/2} [(1 + \alpha_1) \sin \theta (e^{-\mu a \sec \theta} - e^{-(\mu a + \mu_s t) \sec \theta} + \alpha_2 \mu a \tan \theta e^{-\sigma a \sec \theta} - (\alpha_2 \mu a + \alpha_1 \mu_s t) e^{-(\mu a + \mu_s t) \sec \theta}] \, d\theta \quad (9\text{-}34)$$

By making the usual substitution, the integration may be performed, the flux, expressed in terms of first- and second-order exponentials, is given by

$$\phi = \frac{S_V}{2\mu_s} \{(1 + \alpha_1)[E_2(\mu a) - E_2(\mu a + \mu_s t)] + \alpha_2 \mu a E_1(\mu a) - (\alpha_2 \mu a + \alpha_1 \mu_s t) E_1(\mu a + \mu_s t)\} \quad (9\text{-}35)$$

If we let $t \to \infty$, we obtain an expression for the finite volume source. One must exercise caution in dealing with the limits when using exponential integrals. In particular, note that $\lim_{x \to \infty} E_n(x) = 0$ and $\lim_{x \to \infty} x E_n(x) = 0$. In this case, therefore, we have

$$\phi = \frac{S_V}{2\mu_s} [(1 + \alpha_1) E_2(\mu a) + \alpha_2 \mu a E_1(\mu a)]$$

$$= \frac{S_V}{2\mu_s} [(1 + \alpha_1) e^{-\mu a} - (1 + \alpha_1 - \alpha_2) \mu a E_1(\mu a)] \quad (9\text{-}36)$$

This expression could, of course, have been obtained directly from Eq. (9-27) by changing the limits on the integral over r.

On the other hand, if we let $t \to 0$ we should be able to obtain the expression for the infinite area source, that is, Eq. (9-26). Several points must be considered in going to the limit. One must recognize that in the limit of small thickness $S_V t = S_A$ and that $\alpha_1 \equiv 0$, because there can be no buildup in the source. With some manipulation one may show that for small t

$$E_2(\mu a) - E_2(\mu a + \mu_s t) = \mu_s t E_1(\mu a) \quad (9\text{-}37)$$

and

$$\alpha_2 \mu a E_1(\mu a) - \alpha_2 \mu a E_1(\mu a + \mu_s t) = \alpha_2 \mu_s t e^{-\mu a} \quad (9\text{-}38)$$

With the application of these results to Eq. (9-35) the expression for the area source is obtained directly.

Spherical Sources

The spherically symmetric source distribution is often used as an approximation to more complicated geometries. Unfortunately, calculation of the photon flux at an arbitrary distance from a spherical source is somewhat involved and, for this reason, various approximate solutions are generally used (Rockwell, 1956). For certain specific cases, however, explicit solutions for the uncollided flux are possible.

One such example is the calculation of the flux at the center of a sphere of radioactive material. This submersion situation is applicable to finding the exposure rate in a radioactive cloud or in a body of contaminated water. The uncollided flux at the center of a sphere of radius R due to photons emanating from a spherical shell of thickness dr located at distance r is given by

$$d\phi = S_V \cdot 4\pi r^2\, dr \cdot \frac{e^{-\mu_s r}}{4\pi r^2} \tag{9-39}$$

This expression, integrated from $r = 0$ to $r = R$, gives for the flux at the center of the sphere of radius R:

$$\phi = \frac{S_V}{\mu_s}(1 - e^{-\mu_s R}) \tag{9-40}$$

By symmetry, the uncollided flux from a hemisphere is exactly one-half of this. The scattered radiation can be accounted for in several ways. For example, if the Berger form Eq. (9-4) of the buildup factor is used, the photon flux can be shown to be

$$\phi = \frac{S_V}{\mu_s}\left\{1 - e^{-\mu_s R} + \frac{C}{D-1}\left[\frac{1 - e^{(D-1)\mu_s R}}{D-1} + \mu_s R e^{(D-1)\mu_s R}\right]\right\} \tag{9-41}$$

An analytic solution can also be given for the flux at a point P on the surface.

From Fig. 9-4 the uncollided flux at the surface of the sphere is given by

$$\phi = \int_{r=0}^{2R}\int_{\psi=0}^{\theta} S_V(2\pi r \sin\psi) r\, d\psi \frac{e^{-\mu_s r}}{4\pi r^2}\, dr \tag{9-42}$$

The integration over ψ yields

$$\phi = \frac{S_V}{2}\int_0^{2R}(1 - \cos\theta)e^{-\mu_s r}\, dr \tag{9-43}$$

From the cosine law we have $\cos\theta = r/2R$. Putting this into the expression and carrying out the integration over r give the result

$$\phi = \frac{S_V}{2\mu_s}\left(1 - \frac{1 - e^{-2\mu_s R}}{2\mu_s R}\right) \tag{9-44}$$

The exposure rate outside a spherical source is often needed. If the distance from the source is large compared to the source dimensions, then a point source approximation may be used with a modified strength given by

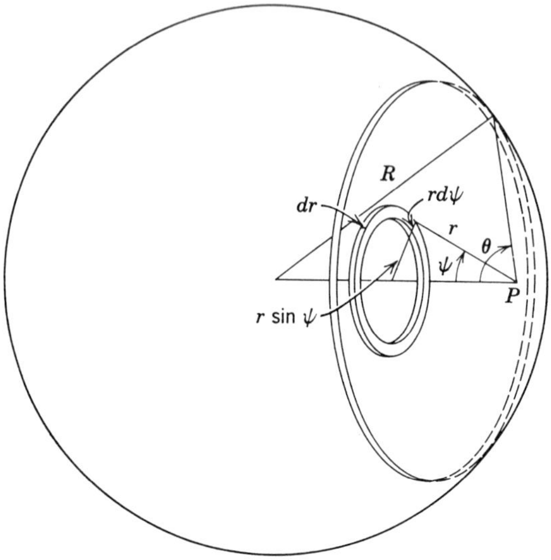

Fig. 9-4. Geometry for flux calculation from a spherical source.

$\frac{4}{3}\pi R^3 S_V L$ where L is the leakage fraction, or the fraction of the source photons that escape self-absorption within the source. In the calculation of L, the quantity needed is not the flux at the surface but its outward component—the outward photon current. This is easily obtained from Eq. (9-42) by the addition of a cosine term:

$$L = \frac{4\pi R^2}{\frac{4}{3}\pi R^3 S_V} \int_{r=0}^{2R} \int_{\psi=0}^{\theta} S_V(2\pi r \sin \psi) r \, d\psi \, \frac{e^{-\mu_s r}}{4\pi r^2} \, dr \cos \theta \qquad (9\text{-}45)$$

Carrying out the integrations indicated leads to the following expression for the leakage fraction:

$$L = \frac{3}{8(\mu_s R)^3} [2(\mu_s R)^2 - 1 + (1 + 2\mu_s R) e^{-2\mu_s R}] \qquad (9\text{-}46)$$

Cylindrical Sources

Cylindrical geometries perhaps are encountered more frequently than any other, and it is unfortunate that analytic solutions are not available for them. Although digital computers have largely alleviated the problem, there is still a need for a simple method whereby the practicing health physicist could obtain approximate answers with manual calculations. As for the spherical source, approximate solutions are available (Rockwell, 1956). One technique uses a truncated right circular cone. An explicit expression for the uncollided flux at the apex of the conical source can be obtained by using the concept of the disk source developed earlier [Eq. (9-18)]. This technique provides an alternative way to calculate the primary photon flux for many of the volume sources previously discussed. The geometry is shown in Fig. 9-5. The primary photon flux at the apex is given by

$$\phi = \int_{t=0}^{h} \int_{\rho=0}^{(a+t)\tan\theta} S_V(2\pi\rho\, d\rho)\, dt\, \frac{e^{-(\mu a+\mu_s t)\sec\psi}}{4\pi[\rho^2 + (a+t)^2]} \tag{9-47}$$

where $S_V\, dt$ is the area source strength S_A, used earlier. By making the usual substitution for the argument of the exponential function, this expression

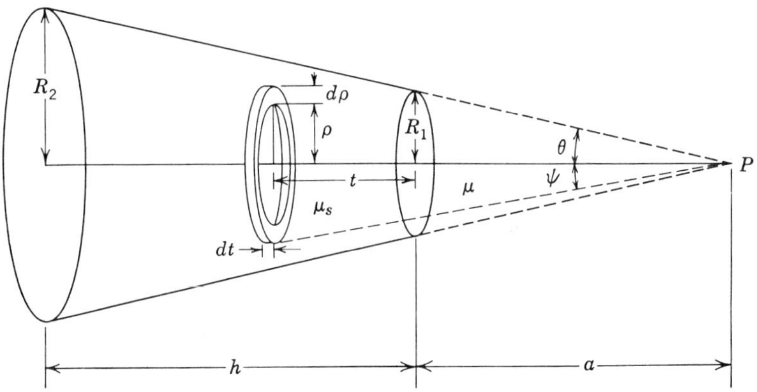

Fig. 9-5. Geometry for flux calculation from a truncated cone source.

may be transformed into

$$\phi = \frac{S_V}{2} \int_0^h \{E_1(\mu a + \mu_s t) - E_1[(\mu a + \mu_s t)\sec\theta]\}\, dt \tag{9-48}$$

With the help of Eq. (9-72) the integration is easily carried out to give

$$\phi = \frac{S_V}{2\mu_s} \Big\{ E_2(\mu a) - E_2(\mu a + \mu_s h) + \cos\theta \Big[E_2[(\mu a + \mu_s h)\sec\theta] - E_2(\mu a \sec\theta) \Big] \Big\} \tag{9-49}$$

The primary photon flux for a point on the axis of a cylindrical source at a distance *a* from one face can thus be approximated from Eq. (9-49). In fact, the correct value will be bracketed by that for a truncated cone source with $R_1 = R$ and one with $R_2 = R$, where R is the cylindrical radius.

The truncated cone source is useful in still another important way. By going to the appropriate limit, may of the previously derived equations for the primary photon flux for area and volume sources can be obtained.

The equations for the slab sources are obtained in the limit $\theta \to \pi/2$, which leads to

$$\phi = \frac{S_V}{2\mu_s}[E_2(\mu a) - E_2(\mu a + \mu_s h)] \tag{9-50}$$

the previously derived expression for the uncollided flux from a slab source of infinite extent and finite thickness. The flux at the surface is obtained by letting $a \to 0$; that for an infinitely thick source is obtained by letting $t \to \infty$.

In the limit $h \to 0$ we obtain the expressions for the primary flux from area sources. Here it is necessary to make use of Eq. (9-37) to obtain

$$\begin{aligned}\phi &= \lim_{h\to 0} \frac{S_V \mu_s h}{2\mu_s}[E_1(\mu a) - E_1(\mu_a \sec\theta)] \\ &= \frac{S_A}{2}[E_1(\mu a) - E_1(\mu a \sec\theta)] \end{aligned} \tag{9-51}$$

where $\lim_{h\to 0} S_V h = S_A$. When we let $\theta \to \pi/2$ we obtain the expression for an area source of infinite extent.

9-5 NONUNIFORM SOURCE DISTRIBUTIONS

The volume source terms S_A and S_V of Section 9-4 were assumed to be independent of the spatial coordinates. In practice, the use of this approximation can lead to large errors in flux calculations.

One example of a nonuniform source is the nuclear reactor where the source distribution is largely determined by the distribution of the nuclear fuel. The fission product source strength is usually greatest in the center, where the fission rate is at its peak, and gradually tapers off towards the sides. A calculation with the uniform source approximation results in an overestimate of the flux.

An opposite situation is encountered with contaminated soil, which can occur as a result of radioactive fallout, waste-disposal operations, or accidents. Because of the ion-exchange capacity of the soil, the source strength is strongest at the surface and gradually decreases with depth. Calculations based on uniform source distributions lead to underestimates of the photon flux.

In some cases it is possible to represent the actual source distribution with an analytic function which can be integrated over the dimensions of the source. If this is not possible, then recourse may be taken to numerical methods. One very useful fitting function is the exponential. The quantity S_V is replaced by $S_{V_0}e^{\pm\eta r}$ where r is the depth within the source, $\eta > 0$ is a fitting parameter, and S_{V_0} is the volume-source strength at $r = 0$.

As an example of the use of the descending exponential source distribution, we consider an infinitely thick volume source of infinite extent. The expression for the preliminary flux with unity buildup is

$$\phi = \int_0^{2\pi} d\phi \int_0^{\pi/2} \sin\theta\, d\theta \int_{a\sec\theta}^{\infty} S_{V_0} e^{-\eta(r\cos\theta - a)} \times \frac{\exp[-(r - a\sec\theta)\mu_s - \mu a \sec\theta]}{4\pi r^2} r^2\, dr \quad (9\text{-}52)$$

The geometry here is the same as that shown in Fig. 9-3 except that t is infinite. Carrying out the integrations exactly as before we find

$$\phi = \frac{S_{V_0}}{2\eta}\left\{E_1(\mu a) - e^{(\mu/\mu_s)\eta a} E_1\left[\mu a\left(1 + \frac{\eta}{\mu_s}\right)\right]\right\} \quad (9\text{-}53)$$

In the limit $\eta \to 0$ we have $S_{V_0} = S_V$ and this expression should reduce to Eq. (9-30). Expanding the exponential function and retaining terms to first order in η, we write

$$\phi = \lim_{\eta\to 0} \frac{S_{V_0}}{2\mu_s}\left\{\frac{E_1(\mu a) - E_1[\mu a(1 + \eta/\mu_s)]}{\eta/\mu_s} - \mu a E_1\left[\mu a\left(1 + \frac{\eta}{\mu_s}\right)\right]\right\} \quad (9\text{-}54)$$

Since

$$\lim_{\epsilon\to 0} \frac{E_1(t) - E_1[t(1 + \epsilon)]}{\epsilon} = e^{-t} \quad (9\text{-}55)$$

it follows that

$$\phi = \frac{S_V}{2\mu_s}[e^{-\mu a} - \mu a E_1(\mu a)]$$

$$= \frac{S_V}{2\mu_s} E_2(\mu a) \quad (9\text{-}56)$$

With the addition of a linear buildup factor the expressions become complicated but still manageable. If we multiply Eq. (9-52) by a buildup factor of the form $[1 + \alpha_1\mu_s(r - a\sec\theta) + \alpha_2\mu a\sec\theta]$ and carry out the integration

$$\phi = \frac{S_{V_0}}{2\eta}\left\{E_1(\mu a) + \frac{\eta}{\eta + \mu_s} e^{-\mu a} - \left[1 - \frac{\mu}{\mu_s}\eta a(\alpha_2 - \alpha_1)\right] e^{(\mu/\mu_s)\eta a} E_1\left[\mu a\left(1 + \frac{\eta}{\mu_s}\right)\right]\right\} \quad (9\text{-}57)$$

For a source with a finite thickness t, the upper limit on the integration over r

is changed to $(a + t)\sec\theta$, and for the uncollided flux

$$\phi = \frac{S_{V_0}}{2\eta}\left\{E_1(\mu a) - e^{-\eta t}E_1(\mu a + \mu_s t) + e^{(\mu/\mu_s)\eta a} \times \left[E_1\left[(\mu a + \mu_s t)\left(1 + \frac{\eta}{\mu_s}\right)\right] - E_1\left[\mu a\left(1 + \frac{\eta}{\mu_s}\right)\right]\right]\right\} \tag{9-58}$$

If a linear buildup factor is applied, for the special case $\alpha_1 = \alpha_2 = 1$, we obtain

$$\phi = \frac{S_{V_0}}{2\eta}\left\{\frac{\eta}{\eta + \mu_s} e^{-\mu a}[1 - e^{-(\eta+\mu_s)t}] + E_1(\mu a) - e^{-\eta t}E_1(\mu a + \mu_s t) + e^{(\mu/\mu_s)\eta a}\left[E_1\left[(\mu a + \mu_s t)\left(1 + \frac{\eta}{\mu_s}\right)\right] - E_1\left[\mu a\left(1 + \frac{\eta}{\mu_s}\right)\right]\right]\right\} \tag{9-59}$$

Other functions may be used to approximate the source distribution, although it may not always be possible to perform the integrations in closed form. A polynomial fit of the form

$$S_V(r) = S_0 + (r + b)S_1 + (r + b)^2S^2 + \cdots \tag{9-60}$$

has been used (Rockwell, 1956).

9-6 NEUTRON SHIELDING

Significant differences exist between the cross sections for photon and neutron interactions. Whereas photon cross sections are usually smoothly varying functions of atomic number and energy, neutron cross sections may vary from one element to the next and they often exhibit resonance structure at particular energies. Photon cross sections are also well known; those for neutrons are not, particularly in energy regions where neutron penetrations are high. Still another difference lies in the nature of the source. Discrete energies are the rule for photon sources, whereas they are the exception for neutron sources.

The concept of buildup is rarely used in computations of neutron penetration. The reason is that calculation of the primary beam penetration, upon which the buildup depends, is itself difficult. Instead, it is more common to calculate tissue kerma directly by using available experimental data.

The physical processes involved in neutron attenuation are elastic and inelastic scattering and capture. The last process is responsible for the final disappearance of neutrons stopped by a shield, an event occurring, as a rule, at thermal energies. It follows that an effective neutron shield should contain appreciable amounts of a material with high capture or reaction cross sections. Cadmium, being such a material, is frequently used as a neutron absorber. It has the disadvantage, however, of emitting high-energy capture γ-rays which themselves must be shielded. Materials which, like ^{10}B and ^{6}Li, undergo an (n, α) reaction are particularly well suited to be thermal neutron shields because the γ-rays are either entirely absent or occur with very small probability.

High-energy neutron shields must contain materials that cause neutrons to lose energy and slow down rapidly to the region where the capture takes place. The energy-loss process usually starts with a single inelastic scattering followed by many elastic scatterings. The cross section for elastic scattering increases as the energy decreases, this process becoming more and more probable as the neutron slows down. At energies below $\sim$1 MeV there is virtually no inelastic scattering: all energy loss takes place by means of elastic scattering.

Because of the importance of elastic scattering, it is advisable to choose a neutron-shielding material with a large cross section and an atomic mass as near that of the neutron mass as possible. This permits a large average energy transfer per scattering event. Hydrogenous matter meets these requirements, and so materials such as water and paraffin are often used in neutron shields. Hydrogen has the additional advantage of having an appreciable capture cross section, that is, that for the (n, γ) reaction, at thermal energies.

Much has been written about the complicated subject of neutron shielding, particularly for the fission-neutron spectrum. Although, as a rule, the health physicist is more occupied with gamma sources, he should be familiar with the principles of neutron shielding. Several excellent references are available (Blizard, 1962; Goldstein, 1957; Rockwell, 1956) for calculational details.

9-7 AUXILIARY MATHEMATICAL FUNCTIONS

In the calculation of the photon flux from extended sources, certain definite integrals appear which are not integrable in closed form. Because of their repeated use in health physics, a brief discussion is included of two or more common types.

Exponential Integrals

The *n*th order exponential integral, also called the *E* function, *F* function, *Ei* function, and the Gold integral, is given by

$$E_n(x) = x^{n-1} \int_x^\infty \frac{e^{-y}}{y^n}\, dy \tag{9-61}$$

By a change of variables the range of integration may be modified to give other forms occasionally encountered. The two most often seen are

$$E_n(x) = \int_1^\infty \frac{e^{-xy}}{y^n}\,dy \tag{9-62}$$

and

$$E_n(x) = \int_0^1 \frac{e^{-x/y}}{y^{2-n}}\,dy \tag{9-63}$$

Direct integration of Eq. (9-61) is possible only in the case of the zeroth order:

$$E_0(x) = \frac{e^{-x}}{x} \tag{9-64}$$

From Eq. (9-63) it can be shown that for $n > 1$

$$E_n(0) = \frac{1}{n-1} \tag{9-65}$$

A recurrence relation for the exponential integral can be obtained by integrating Eq. (9-61) by parts and letting $u = e^{-y}$ and $dv = y^{-n}\,dy$. For $n > 1$ this leads to

$$\begin{aligned} E_n(x) &= x^{n-1}\left[\frac{y^{1-n}e^{-y}}{1-n}\Bigg|_x^\infty + \int_x^\infty \frac{y^{1-n}e^{-y}}{1-n}\,dy\right] \\ &= \frac{1}{n-1}\left[e^{-x} - xE_{n-1}(x)\right] \end{aligned} \tag{9-66}$$

This formula, which permits one to express any exponential integral of order $n > 1$ in terms of exponential integrals of smaller order, is frequently of great help in numerical problems.

Series expansions are useful in providing a convenient means of computing values of the integrals in certain ranges of the argument. A particularly simple form for $E_1(x)$ is given by

$$E_1(x) = -\gamma - \ln x + x - \frac{x^2}{2\cdot 2!} + \frac{x^3}{3\cdot 3!} - \cdots \tag{9-67}$$

where Euler's constant, $\gamma = 0.5772157\ldots$. For small values of the argument, a few terms of this expansion provide an accurate value of $E_1(x)$. For large values of the argument, the following expansion is useful

$$E_n(x) = \frac{e^{-x}}{x}\left[1 - \frac{n}{x} + \frac{n(n+1)}{x^2} - \frac{n(n+1)(n+2)}{x^3} + \cdots\right] \tag{9-68}$$

An inequality which may be used to bracket the value of the exponential integral of any order $n > 0$ is given by

$$\frac{e^{-x}}{x+n} < E_n(x) \leq \frac{e^{-x}}{x+n-1} \tag{9-69}$$

Another useful expression for the exponential integral of any order is the continued fraction

$$E_n(x) = e^{-x}\left[\cfrac{1}{x+n-\cfrac{n}{1+\cfrac{1}{x+n+1-\cfrac{n+1}{1+\cfrac{2}{x+n+2-\cfrac{n+2}{1+3 \cdots}}}}}}\right] \tag{9-70}$$

This relation finds its chief use in digital computer applications in which large numbers of exponential integrals must be calculated.

Occasionally, one has need for the derivative and integral of these functions. By differentiation under the integral sign in Eq. (9-62) one obtains the recurrence relation

$$\frac{d}{dx} E_n(x) = -E_{n-1}(x) \tag{9-71}$$

An extensive table of integrals has been prepared by LeCaine (1945). The integral of the first-order function was used in deriving Eq. (9-49) and is given by

$$\int E_1(ax+b)\,dx = -\frac{1}{a} E_2(ax+b) \tag{9-72}$$

Discussions of these functions appear in Abramowitz and Stegun (1964) and Kourganoff (1952); extensive tables of the exponential integrals are given in Lowan (1940).

The exponential integrals of orders 1 and 2 are shown in Figs. 9-6, 9-7, and 9-8 for values of the argument between 0 and 14. For larger values of the argument, Eq. (9-68) or other published curves Rockwell (1956) can be used.

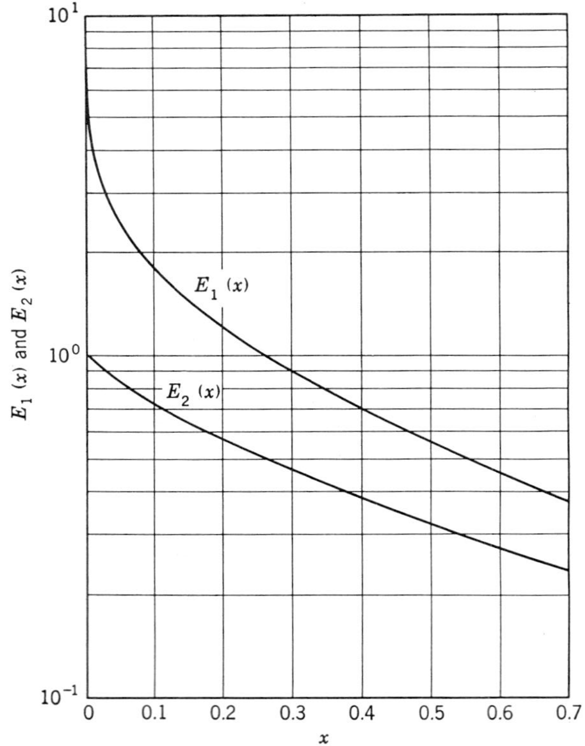

Fig. 9-6. The functions $E_1(x)$ and $E_2(x)$ for $0 < x < 0.7$.

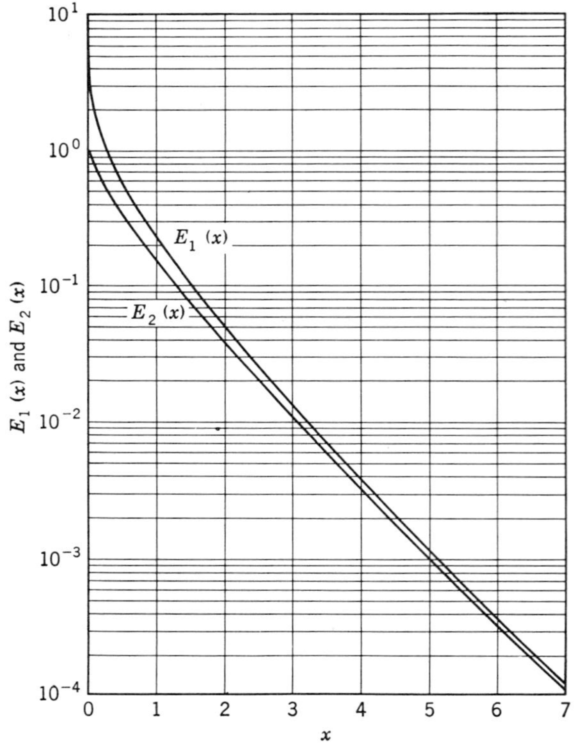

Fig. 9-7. The functions $E_1(x)$ and $E_2(x)$ for $0 < x < 7$.

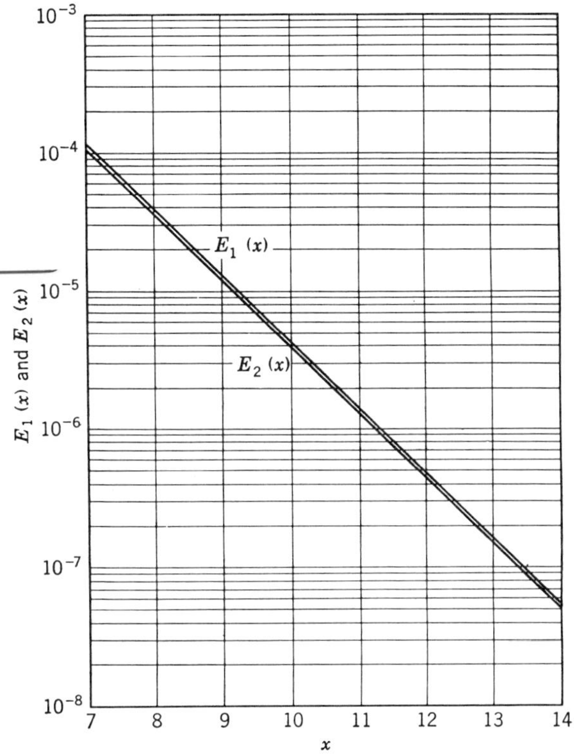

Fig. 9-8. The functions $E_1(x)$ and $E_2(x)$ for $7 < x < 14$.

Sievert Integrals

The Sievert integral, also called the F function and the secant integral, is defined by Sievert (1921) as

$$F(\theta, x) = \int_0^\theta e^{-x \sec \phi} \, d\phi \tag{9-73}$$

Extensive tables of this integral have been prepared.* Figs. 9-9, 9-10, and 9-11 show $F(\theta, x)$ for various values of θ when $0 \leq x \leq 19$. For larger values of θ and x, one can use the approximation (Grotenhuis, 1959)

$$F(\theta, x) \cong e^{-x}\left(\frac{\pi}{2x}\right)^{1/2}\left(1 - \frac{3}{8x}\right) \tag{9-74}$$

For $\theta \leq 5°$, another approximation (Rockwell, 1956) is

$$F(\theta, x) \cong \theta e^{-x} \tag{9-75}$$

* "Tables of the Sievert Integral." *Natl. Bur. Std. (U.S.) Appl. Math. Ser.*, To be published.

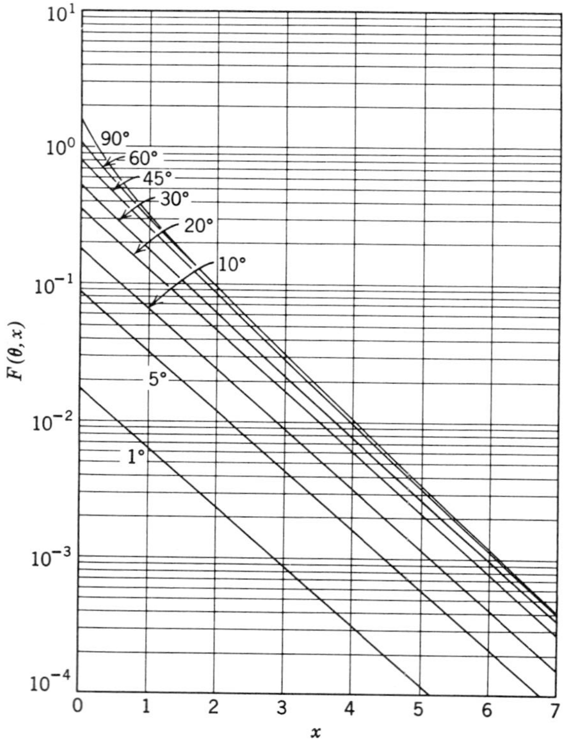

Fig. 9-9. The functions $F(\theta, x)$ for $0 < x < 7$.

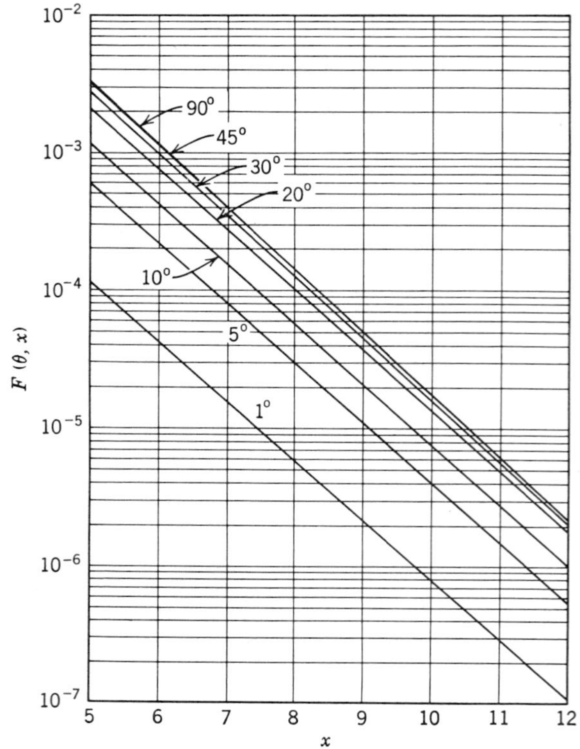

Fig. 9-10. The function $F(\theta, x)$ for $5 < x < 12$.

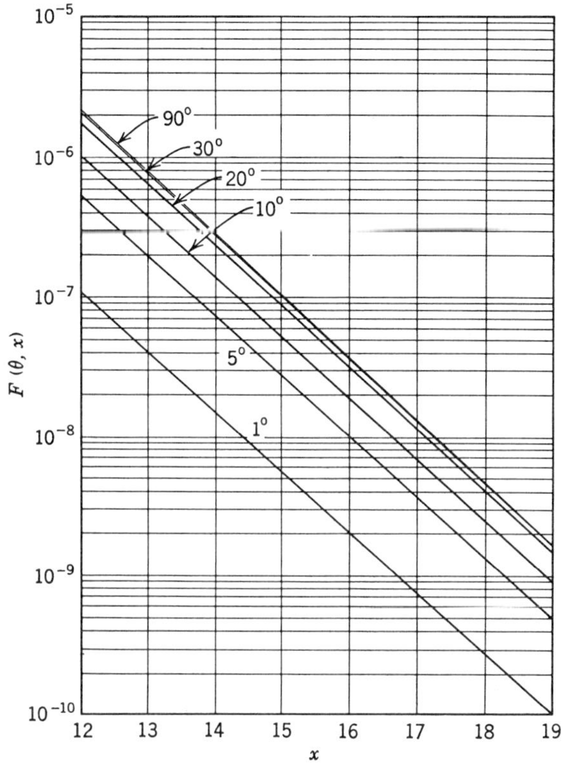

Fig. 9-11. The function $F(\theta, x)$ for $12 < x < 19$.

where θ is expressed in radians. A formula that is useful in interpolation is

$$F(\theta + \delta, x) \simeq F(\theta, x) + \delta e^{-x \sec \theta} \tag{9-76}$$

where δ is expressed in radians (Rockwell, 1956).

REFERENCES

Abramowitz, M., and I. A. Stegun, "Handbook of Mathematical Functions," *Natl. Bur. Std. Appl. Math. Ser.* **55** (1964).

Berger, M. J., *USNRDL Rev. Lectures* **29,** 47 (1956).

Blizard, E. P., "Radiation Shielding" in *The Reactor Handbook* (E. P. Blizard, ed.), Interscience, New York, 1962, Vol. III, Part B.

Capo, M. A., "Polynominal Approximation of Gamma Ray Buildup Factors for a Point Isotropic Source," General Electric Co. Report APEX-510 (1959). Available from the U.S. Govt. Printing Office, Washington.

Evans, R. D., *The Atomic Nucleus,* McGraw-Hill, New York, 1955.

Goldstein, H., *The Attenuation of Gamma Rays and Neutrons in Reactor Shields*, U.S. Govt. Printing Office, Washington, 1957.

Goldstein, H., and J. E. Wilkins, Jr., "Calculations of the Penetration of Gamma Rays," *Nuclear Development Associates, Inc., Rept.*, **NYO-3075** (1954).

Grotenhuis, M., "Lecture Notes on Reactor Shielding," *Argonne Natl. Lab. Rept.* **ANL-6000** (1959).

Kourganoff, V., *Basic Methods in Transfer Problems*, Oxford Univ. Press, London, 1952.

LeCaine, J., "A Table of Integrals Involving the Functions $E_n(x) = \int_1^\infty e^{-xu}u^{-n}\,du$," *Natl. Res. Council Can. Rept.* **MT-131** (1945).

Lowan, A. N., *Tables of Sine, Cosine, and Exponential Integrals*, Natl. Bur. Std., U.S. Govt. Printing Office, Washington, 1940, Vols. I and II.

McLachlan, N. W., *Bessel Functions for Engineers*, Oxford Univ. Press, London, 1955, 2nd ed.

Rockwell, T., III, ed., *Reactor Shielding Design Manual*, Van Nostrand, Princeton, New Jersey, 1956.

Sievert, R. M., *Acta Radiol.* **1,** 89 (1921).

Spencer, L. V., and U. Fano, *Phys. Rev.* **81,** 464 (1951); also see *J. Res. Natl. Bur. Std.* **46,** 446 (1951).

Taylor, J. J., "Application of Gamma Ray Buildup to Shield Design," *Westinghouse Electric Corp. Rept.*, **WAPD-RM-217** (1954).

Trubey, D. K., "A Survey of Empirical Functions Used to Fit Gamma-Ray Buildup Factors," *Oak Ridge Natl. Lab. Rept.* **ORNL-RSIC-10** (1966). Available from the Radiation Shielding Information Center, Oak Ridge National Laboratory, Oak Ridge, Tennessee.

PROBLEMS

1. Find the approximate exposure rate in roentgens per hour at a point 3 ft from a 3 Ci source which emits a 1 MeV gamma ray.

2. Under narrow beam conditions one finds 100 mR/hr at a point, and under wide beam conditions one finds 120 mR/hr at the same point. Find the linear attenuation coefficient (μ) for the shield material which is 3 cm thick.

3. If one out of every 3.7×10^3 α-particles from polonium produces a disintegration in the surrounding beryllium, how many neutrons are emitted per second from a mixture of 10 Ci polonium and excess beryllium? Neutrons are produced by the reaction ${}^9\text{Be}(\alpha, n)^{12}\text{C}$.

4. Given: A 1 Ci point source of an isotope that emits one 3 MeV γ-particle for each disintegration. Find the exact exposure rate in roentgens per hour at 1 meter (Rhm).

(*b*) Given: An infinite line source of the same isotope with a specific activity of 1 Ci/cm. Find the amount of lead that would reduce the exposure rate to 7.5 mR/hr at 1 meter. Neglect buildup and the attenuation in air.

5. (*a*) Find the exposure rate in air at the center of a thin cylindrical shell of activity of radius x cm and half-angle θ. The specific activity is S_{A_N} millicuries per square centimeter. Let R represent the R/hr-mc at a distance of 1 cm from this source. An absorber of thickness t cm and coefficient $\mu(\text{cm}^{-1})$ is between the source and the point of measurement.

(*b*) Repeat, allowing the cylinder to become infinite.

(*c*) Repeat, using only air as the absorber.

6. The exposure rate at a point following an atomic detonation is given by $I = I_0 t^{-1.2}$, where I_0 is the initial exposure rate at 1 hr after the detonation measured in roentgens per hour. Find the total exposure at a point from 1 hr to infinite time (eternity dose) in terms of the initial exposure rate.

7. What thickness of lead is required to reduce the intensity of a beam of 2 MeV γ-rays by a factor of 10^{10}. Repeat using a linear buildup factor.

8. (*a*) Given: An isotope that emits one 1 MeV γ-ray for each disintegrating atom. Find the exposure rate (R/hr) from 1 mCi at 1 cm.

(*b*) Given: A line source of the same isotope with $S_L = 1$ Ci/cm. (1) Find the exposure rate at a distance of 1 meter from an infinite wire, neglecting the attenuation in air. (2) Find the exposure rate at a distance of 1 meter from the midpoint of a 1-meter wire neglecting the attenuation in air. (3) Find the exposure rate at a distance of 1 meter if the wire is of infinite length and 1 cm of lead is used as a shield. Neglect the attenuation in air.

9. One turn of a radioactive ^{60}Co wire is wrapped tightly around a 10-cm OD iron pipe with a wall thickness of 1 cm. The wire has an activity of 1 c/cm. Using a linear buildup factor calculate the exposure rate the at center of the pipe within the plane of the wire. Assume ^{60}Co to have two 1.25 MeV photons/dis.

10. Calculate the absorbed dose rate when a mass of tissue is uniformly exposed to a beam of 2 MeV photons with a flux density of 10^6 photons/cm^2-sec.

11. A thin impervious collimator with a 0.2 cm^2 hole covered by a 1-in. thick Al plate is located 5 ft from a small source of ^{137}Cs with an activity of 10 Ci. Calculate the γ-exposure rate at a point 5 ft behind the collimator along the line joining the source and the hole. Neglect air absorption and scattering.

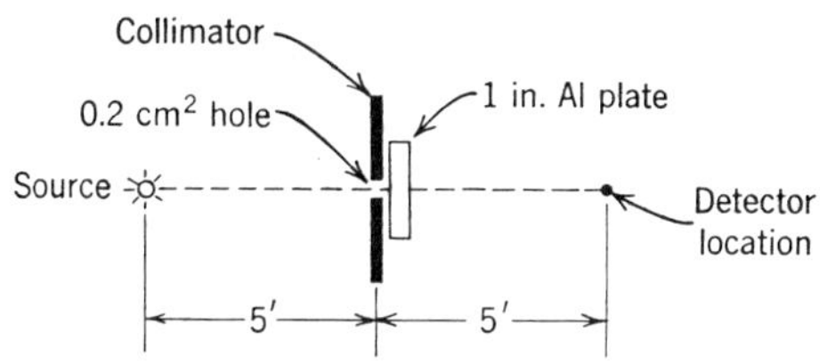

Geometrical Arrangement

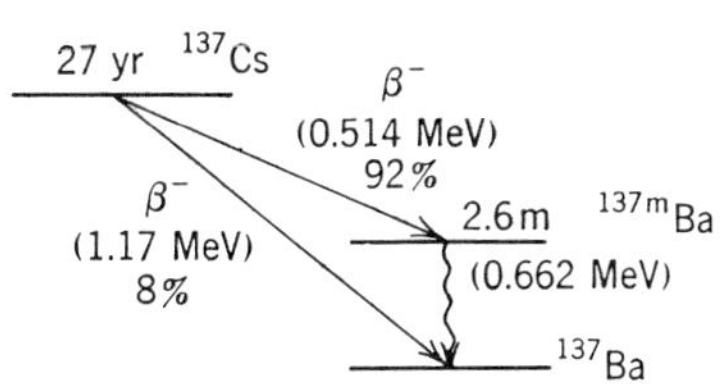

Source Decay Scheme

12. A microcalorimeter is located 20 cm from a point γ-ray source and receives only the 0.5 MeV annihilation photons (two per decay). The calorimeter has an area of 0.3 cm^2 of copper and a thickness of 0.1 cm, and indicates 4 μcal are absorbed per minute. Calculate the activity of the source in Curies.

13. Find the exposure rate at a distance of 1 meter above the center of an annular source with an inner radius of 50 cm and an outer radius of 75 cm. Assume the source to emit 1 photon/dis with an energy of 2 MeV and to have a specific activity of 0.1 Ci/cm^2. Neglect the air attenuation.

14. Show that the photon flux at distance d from the center of a thin spherical shell source of radius r ($r < d$) and strength S_A (photons per unit time per unit area) is equal to the difference in the flux from two infinite plane sources, of strength $(r/d)S_A$, oriented perpendicular to d, one at a distance $(d - r)$, and the other $(d + r)$. Assume all sources are isotropic emitters and that the medium in which the sources are embedded is infinite and homogeneous.

15. Assume that for a particular ecological experiment is it desired to uniformly contaminate with ^{137}Cs the surface of a land mass in the shape of a square with sides 100 meters long.

(*a*) Estimate the activity per unit area which will result in an exposure rate of no more than 100 mR/hr 3 ft above ground at any point on the boundary of the contaminated plot.

(*b*) Estimate the maximum and minimum exposure rates along the boundary, assuming the plot is plowed to uniformly mix the activity with the topmost 6 in. of soil. Take μ_{soil} to be 0.075 cm^2/gm.

CHAPTER 10

Internal Exposure

W. S. Snyder

10-1 GENERAL CONSIDERATIONS

Radioactive materials may be inhaled, ingested, or, occasionally, absorbed either through the skin or via a wound. The problems posed by such exposures are among the most difficult that the health physicist must face, and the methods he must use in coping with them are, for the most part, peculiar to this branch of health physics. With an external source, doses can be limited by control of exposure time or by interposing shielding and distance between the person and the source. But when the source has entered the body, there is most often no practicable way to hasten its elimination significantly. Moreover, once inside the body, some long-lived radionuclides may continue to irradiate the body for the remaining life of the individual. In severe cases treatment may be warranted, but, in general, the health physicist must aim to limit intake to acceptable levels.

10-2 EARLY EXPERIENCE WITH INTERNAL EXPOSURE

Man has always been exposed to external sources as well as to natural radioisotopes. However, he had no knowledge of this very low level of exposure until the twentieth century.

Soon after the discovery of radioactivity, it was found that exposure to what we would now term very large amounts of ^{226}Ra or to high concentrations of ^{222}Rn results in early death. Few people, if any, however, suspected that the ingestion of microgram quantities of radium might be a considerable hazard. During the first 20-odd years of this century, radium was administered to a wide variety of individuals for a wonderfully varied list of reasons. In reviewing the early uses of radium, Looney[1] quotes the following paragraph as indicating this early climate of opinion:

"The value of radium is unquestionably established in chronic and subacute arthritis of all kinds (luetic and tuberculous excepted); acute, subacute and chronic joint and muscular rheumatism (so-called); in gout, sciatica, neuralgia, polyneuritis, lumbago, and the lancinating pains of tabes. 837 or over 80% of the patients were considered to have been improved by the use of radium emanation."

Radium was used instead of radon largely because it provided a relatively constant source of the supposedly beneficial radon within the body. After some 15 years the popularity of radium treatment subsided, and in 1929 the Council on Pharmacy and Chemistry of the American Medical Association announced that radium solutions and radon water were not accepted as therapeutic agents because of failure to demonstrate their value by controlled clinical evidence.

During the years 1917–1924, radium and mesothorium began to be used in manufacturing luminous watch dials, and thus industrial exposure was added to medical exposure or exposure due to scientific work. Even during this period there seems to have been little awareness that microgram quantities of radium might be a serious hazard when deposited in the body. Schubert and Lapp[2] quote a statement appearing in 1916 in the journal *Radium:*

"Radium has absolutely no toxic effects, it being accepted as harmoniously by the human system as is sunlight by the plant."

Continuing to quote from Looney's summary[1]:

"Self luminous paints began to be used in this country in 1913. They were used on instrument panels as well as watch dials. Martland states that an accumulative total of 800 girls were employed by a New Jersey concern between 1917 and 1924. Similar luminous dial plants were in operation in Illinois and Connecticut.

"In 1924 Blum discovered that severe osteomyelitis and necrosis of the mandible occurred with great frequency in the workers of the New Jersey plant. The reports which followed on the clinical course and death of these women from anemias, crippling bone lesions, bone tumors, have now become classics on the destructive effects of radioelements in humans."

Even as late as 1931, after extensive publicity concerning the dial-painter cases, 32 patients in a hospital in Elgin, Illinois, were given doses of from

70–450 μg of radium as treatment for certain mental conditions.[3] Studies of the early elimination of radium from these patients are a valuable source of data concerning its long-term retention in the body.

The National Committee on Radiation Protection (NCRP) was formed in 1929, but at first it was preoccupied by problems concerning the control of exposure to X-rays. In the midthirties the NCRP began to consider the problem of the industrial use of radium, and its first recommendations concerning internal exposure were issued in *Natl. Bur. Std. Handbook* 27 in 1941.[4] The value of 0.1 μg of radium as limiting for occupational exposure to radium was first announced in that *Handbook* in the following words:

> "It is recommended that any worker who shows a deposit of more than 0.1 microgram of radium, as revealed by the expired air test, change his occupation immediately and be treated by decalcification therapy, or any other modality which may have been developed for the purpose."

It is worthy of note that this early recommendation also advised treatment to hasten elimination of the radionuclide. Although the recommended limit on a body burden of ^{226}Ra for occupational exposure is still 0.1 μg, the present versions no longer recommend decalcification therapy.

During World War II the development of the atomic energy program entailed the production of hundreds of new radioactive isotopes in tremendous curie amounts as compared with the quantities of radium previously in use. Under the Manhattan Project, many of the same persons who had been concerned with the radium problem were consulted in devising criteria and techniques for the safe use of the new materials. It quickly became apparent that the health physicists at the various projects could benefit greatly by exchanging information and that one of the pressing needs for communication was establishing a reference individual. Many of the decisions and deliberations of these early health physicists are not readily documented, but after the war many of the more basic results of their work were formally published. In particular, a tripartite conference of health physicists from the United Kingdom, Canada, and the United States was held at Chalk River in 1949. The meeting resulted in agreement on a "Standard Man" as a typical radiation worker, on maximum permissible concentrations (MPC's) in air and water, and on limiting body-burden values for a number of radionuclides. This was followed by the formal publication of recommended values by the NCRP[5] and the International Commission on Radiological Protection (ICRP),[6] the former listing values for 69 radionuclides and the latter for 86.

Both the NCRP and ICRP have established committees for studying the problem of internal exposure. Full reports of these committees,[7,8] appearing in 1959, list recommendations for 236 radionuclides. Since 1959 the Federal Radiation Council (FRC) has issued a number of reports concerned largely with exposure of the population.

10-3 CALCULATION OF DOSE RATES FROM ORGAN BURDENS

The evaluation of an exposure is usually based upon estimates of dose and/or dose rates to tissues of the body. Techniques for estimation of dose have been developed in Chapters 4, 6, and 7, and no new principles are involved when sources are inside the body. Nevertheless, the distribution of a radionuclide in the body is often quite inhomogeneous and sometimes involves geometrical complications not usually present with external sources. In this section, methods and formulas for estimating the dose rates in body tissues due to internal radionuclides are discussed in order to emphasize those aspects of dosimetry of special importance for internal exposure.

The most important case is that in which a source is uniformly distributed in an organ or tissue and the dose rate averaged over that organ or tissue and surrounding tissues is desired. It is important, in part, because most criteria for evaluation of exposure are stated in terms of an average annual or quarterly dose to particular organs or tissues. In a more basic sense, the assumption of a uniform distribution and the estimation of average doses reflect the fact that often the actual distribution of the source is not accurately known and that our lack of knowledge about basic mechanisms of radiation damage usually would not enable us to interpret a very detailed distribution of dose within an organ in any way which would lead to an improved estimation of the biological hazard caused by this dose.

The symbol R_i will be used to denote the retention or organ burden in a tissue i. In general, R_i is a function of time, and when so considered the time dependence is indicated by writing $R_i(t)$. Usually R_i will be expressed in microcuries. The average dose rate to tissue i is then

$$D_i = \frac{51EF_iR_i}{M_i} \quad \text{rad/day} \tag{10-1}$$

in which E = energy (MeV) absorbed per disintegration, F_i = fraction of the energy absorbed in tissue i, M_i = mass (grams) of tissue i, and 51 $\simeq$ (gram-rad per MeV) $\times$ (disintegrations per day per microcurie).

This formulation applies when there is only one type of radiation emitted, for example, a photon or electron of a certain energy. Usually this is not so. When k types of radiation are present, we write in place of Eq. (10-1)

$$D_i = \frac{51\varepsilon_iR_i}{M_i} \quad \text{rad/day} \tag{10-2}$$

in which

$$\varepsilon_i = \sum_{j=1}^{k} f_j E_j F_{ij} \tag{10-3}$$

E_j being the energy (MeV) of the jth type, f_j the probability per disintegration of its production, and F_{ij} the fraction of the energy emitted in the jth mode which is absorbed in tissue i. In Eq. (10-2) the numerator is expressed in gram-rad per day and the denominator in grams. The calculation of formula (10-1) or (10-2) is trivial except for the determination of the fractions f_j and F_{ij}. Almost all the complexities, whether of physics or biology, of estimating the dose rate are inherent in estimating the factors f_j and F_{ij}.

When D_i is specified as the maximum permissible dose equivalent rate to organ i (that is, 5 rem/yr for blood-forming organs, gonads, and lenses of the eye; 30 rem/yr for thyroid, skin, and skeleton; and 15 rem/yr for other organs), Eqs. 10-1 and 10-2 may be solved to determine the so-called maximum permissible organ burden R_i. The value of R_i is obtained by assuming exposure at a constant level, and this organ burden may not be limiting for short-term exposure.

10-4 DECAY SCHEMES

Ideally, the particles or photons emitted per disintegration, together with their probability of occurrence and their energies, may be found by examining a decay scheme. However, frequently the decay scheme is not completely determined and some of the data mentioned are not listed. Usually, these will concern the particles having a low probability of occurrence, and they often do not contribute significantly to the value of ε_i. In such cases we can calculate ε_i under several alternative assumptions and thus determine, plausibly, the uncertainty in the value of ε_i. In most cases the decay energy is carried by α-particles, electrons, positrons, or photons, although neutrons, fission fragments, and recoil nuclei occasionally contribute significantly to the dose.

Electrons arise through β-decay, internal conversion of a photon, or as Auger electrons. Many decay schemes do not give detailed information on the relative numbers of electrons from each of these processes. Wiles[9] discusses the practical use of β-decay theory in assigning proper energy levels to β-transitions, but the procedure is too involved for exposition here. Considerable detailed information can also be found in the review articles cited in Chapter 8. Data on sixteen of the β-transitions of most importance for the health physicist are summarized by Loevinger et al.[10] The average energy may be obtained from graphs and tables given by James et al.[11] for all allowed spectra and for many of the others. Often the maximum and average energies of β-rays are not given in decay schemes, and here the health physicist may have to be content with a rather rough approximation of the average energy or must undertake a literature search and perhaps make detailed calculations.

Positrons are emitted from the nucleus, the usual notation in the decay scheme being a broken line of two segments. The energy is not unique for a given isotope but is distributed over a characteristic spectrum, and the information concerning this spectrum and the average energy is as elusive as that on β-rays. Virtually all positrons produced in tissue become thermalized and then annihilate with an electron to produce two photons, each with 0.51 MeV of energy.

In addition to photons which originate by positron annihilation, photons may arise by a nuclear transition. The more complex decay schemes often do not include precise data on the intensity (probability of emission per disintegration) and the energy of photons that are emitted. Sometimes only relative intensities have been measured and in some instances the very existence of the photon can be inferred plausibly, although it is not shown explicitly on the decay scheme.

Electron capture or internal conversion of a photon produces a vacancy in an electronic shell. This vacancy leads to characteristic X-rays and Auger electrons. A characteristic X-ray may be emitted when an electron from a higher shell fills the vacancy, and is seldom shown on decay schemes since atomic rather than nuclear transitions are involved. The X-ray energy is equal to the difference in electron-binding energies between the two shells involved. An Auger electron, however, may be emitted when an electron from a higher shell fills the vacancy.

In the Auger process a primary vacancy is filled by an electron from some outer shell, but instead of an X-ray being emitted, another outer shell electron is emitted. Thus two vacancies are created from the original primary vacancy when an Auger process occurs, one due to the electron which makes the transition to the primary vacancy and the other due to the Auger electron that is ejected. The energy of the Auger electron is—to a precision quite sufficient for dose calculations—the electron-binding energy in the primary vacancy shell minus the electron-binding energy in the shell associated with the electron that makes the transition into the primary vacancy shell, minus the electron-binding energy in the shell from which the Auger electron is ejected.

The fluorescent yield for a shell, ω_K, for example, is the fraction of K-shell vacancies that give rise to an X-ray. The fractional yield of Auger electron per K-shell vacancy is then $1 - \omega_K$. A more detailed discussion of these and other products of nuclear disintegration is given by Slack and Way,[12] who show, perhaps the most complete decay schemes, such as are needed for internal dose calculations. Unfortunately, Ref. 12 contains only the decay schemes of ^{3}H, ^{14}C, ^{22}Na, ^{24}Na, ^{32}P, ^{35}S, ^{36}Cl, ^{42}K, ^{45}Ca, ^{51}Cr, ^{59}Fe, ^{60}Co, ^{65}Zn, ^{86}Rb, ^{90}Sr, ^{90}Y, ^{131}I, and ^{198}Au so that the scope is quite limited and the decay schemes are among the simpler ones the health physicist needs. However, as far as it goes, this collection is ideally suited for calculating internal exposure.

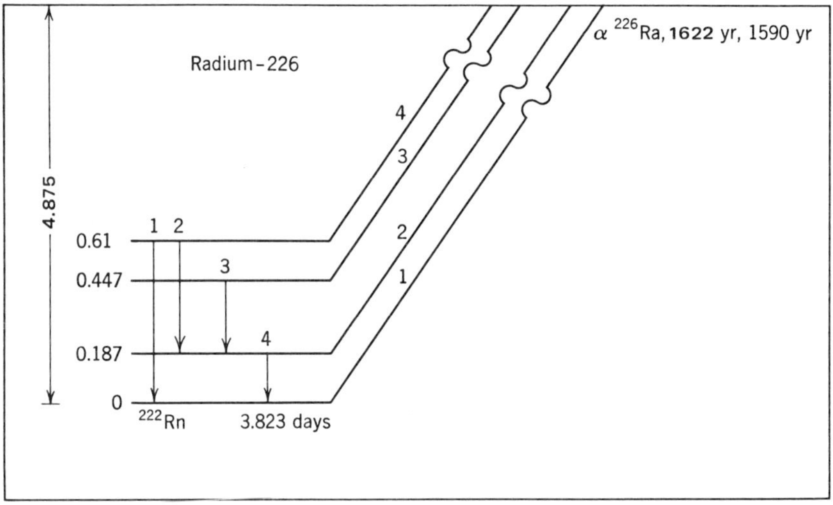

^{226}Ra	α	1	4.777 ± 0.003	94%
		2		5.7%
		3		0.012%
		4		1.7×10^{-3}%
	$\gamma(\alpha)$	1	0.610 ± 0.005	1×10^{-2}%[a]
		2	0.420 ± 0.005	7×10^{-4}%
		3	0.260 ± 0.005	9×10^{-3}%
		4	0.188 – 0.186	3.2%

[a] Intensities of unconverted photons.

Fig. 10-1. Decay scheme for ^{226}Ra adapted from Landolt-Börnstein.[13]

The ratio of the number of internal conversion electrons to the number of escaping photons is called the internal conversion coefficient, usually denoted by the symbol α. When the electrons are ejected from the *K*, *L*, . . . shells, the corresponding ratios are, respectively, denoted by α_K, α_L, . . . , where $\alpha = \alpha_K + \alpha_L + \cdots$.

In Fig. 10-1 the decay scheme for ^{226}Ra is adapted from the compilation by Landolt-Börnstein,[13] with no change in notation. The primary mode of decay is by emission of an α-particle which has an energy of about 4.78 MeV. Three other α-particles are indicated, but their energies are not given. However, they may be closely approximated by taking the difference of the corresponding energy levels. The photons contribute only slightly to the

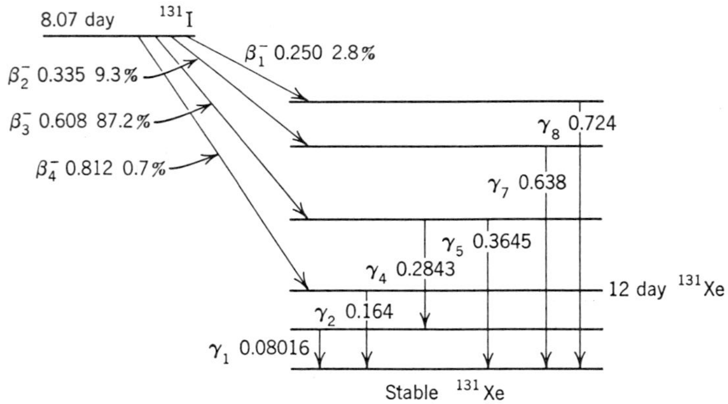

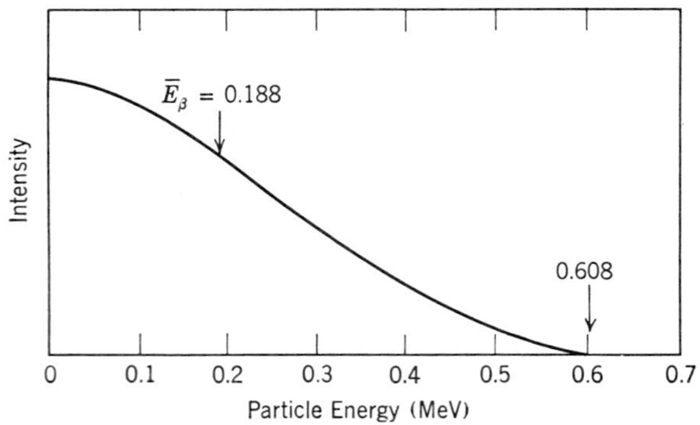

Iodine-131 $_{53}I^{131}$

Half-life: 8.07 ± 0.02 days

Usually obtained by separation from fission products.

Notes: $\bar{E}_\beta$ per disintegration includes contributions from conversion electrons.

Conversion electrons of energies 0.128, 0.249, 0.279, 0.359, 0.603, 0.633, 0.687, and 0.717 all have intensities <0.5%.

Type	Energy (MeV)	$\bar{E}_\beta$ (MeV)	Intensity	Range in Water (cm)
e_{A_L}	~0.003		5.1%	<0.0001
e_{A_K}	~0.025		0.7%	0.001
ce_K	0.045		3.6%	0.003
ce_{LMN}	~0.075		0.7%	0.008
ce_K	0.329		1.4%	0.089
β_1^-	0.250 (max)	0.071	2.8%	0.060
β_2^-	0.335 (max)	0.098	9.3%	0.090
β_3^-	0.608 (max)	0.193	87.2%	0.21
β_4^-	0.812 (max)	0.272	0.7%	0.31
		0.188 per disintegration		

See also notes above.

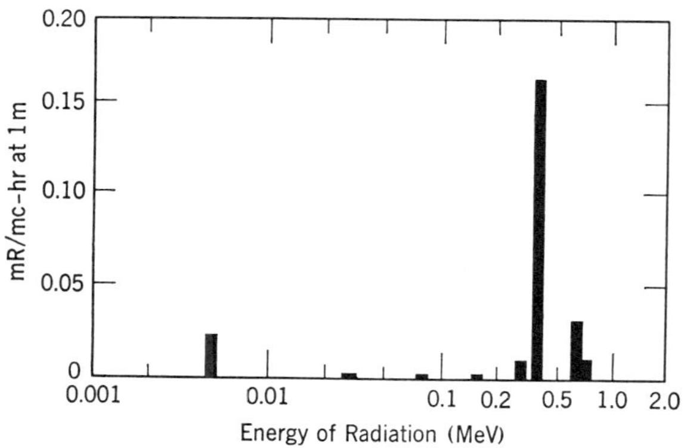

Electromagnetic Radiation

Type	Energy (MeV)	Photon Intensity	Exposure Rate at 1 m (mR/mc-hr)	½ Thickness (cm) Water	½ Thickness (cm) Lead
x_L	~0.005	0.6%	0.021	<0.1	<0.001
x_K	~0.030	5%	0.004	2.1	~0.003
γ_1	0.08016	2%	0.001	4.0	0.044
γ_2	0.164	~0.7%	~0.001	4.8	0.042
γ_3[a]	0.1771				
γ_4	0.2843	6%	0.010	5.8	0.15
γ_5	0.3645	79%	0.166	6.3	0.25
γ_6[a]	0.51				
γ_7	0.638	9%	0.032	7.9	0.58
γ_8	0.724	3%	0.012	8.4	0.67
			0.247		

[a] Not shown on decay scheme

Fig. 10-2. Decay scheme, particle spectrum, and exposure rate for ^{131}I.[12]

local dose, but in tissues farther from the source than the range of the α-particles, the dose will be entirely due to these photons. Anticipating some later material, we might suspect that since ^{226}Ra deposits primarily in bone, its contribution to genetic dose will be due to these photons to a far greater degree than is indicated by their intensity. This illustrates the principle that in judging what particles may be neglected in calculating dose, it is not sufficient to consider only intensities.

It is clear that there is no internal conversion in the γ-rays indexed under 1 and 2 in Fig. 10-1, because the sum of their intensities equals the intensity of the α-particle reaching the level where they originate, that is ($1 \times 10^{-3}\% + 7 \times 10^{-4}\% = 1.7 \times 10^{-3}\%$). However, for the 0.26 MeV γ-ray, internal conversion must occur because its intensity is less than that of the α-particle terminating at the 0.447 MeV level. Thus it is clear that internal conversion does occur and that the intensity of the emitted electron is $3 \times 10^{-3}\%$. Similarly, it is apparent that the intensity of the electrons arising from conversion of the 0.188 MeV γ-ray is 2.5%. In the format adopted in Ref. 13 the α_K, α_L, . . . values are not given, but the percent that are unconverted is shown. If the internal conversion electron originates in the *K* shell, electrons of 0.162 and 0.090 MeV energy are emitted. These energies are obtained by subtracting the binding energy of the electron from the respective γ-ray energies. As the vacancies in the *K* shell created by the emission of these electrons are filled by electrons from outer shells, energy is lost either by X-ray emission or by the ejection of Auger electrons. Since, in this instance, ω_K is greater than 0.95,[12] it can be assumed with very little loss of accuracy that all *K*-shell vacancies result in X-ray emission. These X-rays will be characteristic of the product nucleus and will have the binding energy of the radon *K* shell, that is, 0.098 MeV minus the binding energy of an electron in some outer shell (the shell associated with the electron that made the transition to the vacancy in the *K* shell and thus gave rise to the X-ray). The holes in the outer shells lead to further low-energy X-rays or Auger electrons, but the problem of determining the relative intensities of these various components of radiation is too complex to be discussed here.

By contrast, Fig. 10-2 is from Slack and Way,[12] mentioned previously, and here the tabulation of energies is essentially complete. In this example, the kinds of particles emitted, plus their intensities and energies, are given explicitly.

10-5 ESTIMATION OF THE ABSORBED FRACTION OF ENERGY EMITTED IN AN ORGAN

In general, some fraction of the energy emitted by a radionuclide present in an organ will be absorbed within the organ and the remainder of the energy will escape from the organ. The energy escaping from the organ may produce a dose in other tissues of the body or may escape from the body. These fractions of the energy will depend upon the type of radiation emitted, the size and shape of the organ and body of the individual, and the distribution of the source within the organ or body. The possible situations are so varied and the accuracy of the information available in any usual case so imprecise that only simplified techniques are justified. Generally, we assume that the organ is homogeneous, both in composition and in density, and that the source is distributed uniformly within the organ.

If the radiation has a finite range (α-particles, β-particles, electrons, positrons) as opposed to radiations having an exponential type of attenuation, then any radiation emitted at a site farther from the surface of the organ than this range will be absorbed in the organ. If the organ is large compared to this range and not too irregular in shape, and if the distribution of the radionuclide in the organ is fairly homogeneous, then most of the energy will be absorbed; that is, F_i and F_{ij} are taken as 1 in formulas (10-1) and (10-3).

With an α-emitter, the first two conditions mentioned are usually considered to be applicable, since the range of a typical α-particle (for example, having an energy of 5 MeV) is only about 50 μ. If we remove all material within 50 μ of the surface of most organs, we would expect to have most of the radioactive material remaining. Accordingly, we generally assume that all the energy carried by α-particles is absorbed.

It is often assumed that all the energy of a β-particle is absorbed in the organ where the β-particle originates. If the smallest traverse distances through the organ are large compared to the range of the β-particle this will be acceptable. However, for small organs or rather thin layers of tissue, this assumption may not be very accurate, but will always overestimate the dose from that source. If the distribution of the source is uniform in the organ, then any portion of the tissue which is farther from the surface than the range of the β-particle will get exactly this dose, calculated by assuming total absorption. Perhaps this justifies in some measure the general acceptance of this rough estimate of dose, for even though it may not be the correct average dose in the organ, (1) it does overestimate this average dose and hence is "conservative" and (2) if the distribution of the source is reasonably uniform and if the organ is fairly large compared to the range of the β-particle, a relatively large portion of the tissue of the organ is getting essentially this dose. A more precise calculation of F_i or F_{ij} in formulas (10-1) and (10-3) is not easily realized although computer codes have been developed and tested for special cases of interest.[14]

When a photon is emitted, the assumption of complete absorption is usually unrealistic. The energy absorption coefficient $(\mu - \sigma_s)$ measures the energy absorbed in a material per unit distance in the material (Section

2-8). Thus it does not include the energy scattered by the Compton process, but does include the energy absorbed locally by that process as well as the energy which is taken from the original beam of photons by the photoelectric process or by pair production, and is, to a high degree, absorbed locally. However, it should be noted that if the energy is below 0.01 MeV, about 90 per cent of the energy will be absorbed within 1 cm of the source. This follows because the photoelectric process predominates below this energy and also

$$\exp\left[-(\mu - \sigma_{\text{scatt}})\right] < 0.1 \qquad (10\text{-}4)$$

At higher energies, however, the assumption of complete absorption usually is unrealistic.

Radiologists have devised various rules for estimating the fraction of energy absorbed in regions of various sizes from X-rays and γ-rays as used in medical practice. Until recently there has been little information on the accuracy of these estimates. Recently, the availability of electronic computers has enabled more accurate computation of "absorbed fractions" (A.F.), Ellett et al.[15] have discussed comparisons of absorbed fractions obtained by earlier procedures with those obtained by the newer methods.

Perhaps the procedure most satisfactory for manual calculation is that used in *ICRP Publ.* 2.[8] One makes a simple "first-interaction" calculation of average dose in a volume V by evaluating the integral

$$(\mu - \sigma_s)\,\frac{E}{V}\iiint\limits_{V_R} \frac{e^{-(\mu-\sigma_s)r}}{4\pi r^2}\,dV \qquad (10\text{-}5)$$

where r is the distance to a point source at X_0 (that is, $r = |X - X_0|$, where X and X_0 are position vectors) and the source emits one photon of energy E. The result of this integration provides an estimate of average dose per photon in the sphere V_R of radius R centered at the point source. Analogously, this "straight ahead" approximation attributes a dose $(\mu - \sigma_s)e^{-(\mu-\sigma_s)r}\,E/4\pi r^2$ to a small volume element at distance r from the source. This formula may be regarded as a function of position and may be integrated to obtain the A.F. for an arbitrary distribution of source in V. For simple geometrical figures such as a sphere with a source at the center, this integral is evaluated easily. An "effective radius" of a more general region might be defined as the radius of a sphere having the same A.F. for a point source at the center as does the region for the source distribution in question. Of course, this definition largely evades the issue since we must know the A.F. in order to define r accurately. In earlier practice, r was usually chosen on the basis of experience and judgment with the available estimates of the A.F. in simple cases. Often this effective radius has been mistaken for an equivalent geometrical radius; however, its choice is not dictated by consideration of mass alone but rather by consideration of A.F. Table 10-1 indicates the comparison of the A.F. as given in *ICRP Publ.* 2[8] and the more exact estimates calculated by Ellett et al.[15] The ICRP estimates are about 6 percent lower at the lowest energy and nearly 40 percent greater at the highest energy.

To obtain dose in tissues outside the region of deposition of the source requires still other considerations. Table 10-2 contains values of dose in regions outside a spherical source which is embedded in a homogeneous tissue phantom. The phantom is a right-circular cylinder with the homogeneous

Table 10-1 Estimates of the Absorbed Fraction for Total Body (70 kg)

$E\gamma$ (MeV)	ICRP	Monte Carlo	ICRP/Monte Carlo
0.040	0.812	0.861	0.94
0.080	0.532	0.593	0.90
0.160	0.575	0.513	1.12
0.364	0.632	0.499	1.27
0.662	0.636	0.481	1.32
1.46	0.586	0.434	1.35
2.75	0.504	0.362	1.39

composition of standard man and with a mass of about 42 kg. The gastrointestinal tract is represented by a sphere with a radius of 10 cm. The γ-emitter with an energy of 1 MeV is assumed to be uniformly distributed in the sphere. The sphere and the spherical shells, each 1 cm thick, were divided into zones in order to judge better the importance of backscatter in regions which are near a boundary and in those farther from boundaries (see Fig. 10-3). The dose is highest in the regions of the sphere which contain the source, as would be expected. These are designated as A_0, B_0, C_0, in Table 10-2. The calculation uses the Monte Carlo method which consists essentially of statistical sampling of the population of photon histories. This method is perhaps the most flexible of the available alternatives for allowing us to take account of details of geometry, energy-dependence of cross sections, and averaging of dose over regions of special interest.*

* For a general discussion of techniques that have been or may be used for sampling see *Progress in Nuclear Energy, Series I, Physics and Mathematics*, Vol. 2 (1958), Pergamon Press, "Monte Carlo Methods in Transport Problems," by G. Goertzel and Malvin H. Kalos, or J. M. Hammersley and D. C. Handscomb, *Monte Carlo Methods*, Methuen's Monographs on Applied Probability and Statistics, John Wiley and Sons, New York, 1964.

Table 10-2 Dose from a Spherical Source of Photons Approximating the Gastrointestinal Tract[a]

Region	Mass (gm)	Dose per Photon (rad)	Coeff. of Variance	Region	Mass (gm)	Dose per Photon (rad)	Coeff. of Variance	Region	Mass (gm)	Dose per Photon (rad)	Coeff. of Variance
Photon energy: 1 MeV											
A_0	1.6×10^3	9.7×10^{-13}	0.023	B_0	1.1×10^3	9.4×10^{-13}	0.019	C_0	1.6×10^2	9.1×10^{-13}	0.054
A_1	5.3×10^2	3.5×10^{-13}	0.065	B_1	3.8×10^2	3.7×10^{-13}	0.053	C_1	5.3×10^1	3.7×10^{-13}	0.13
A_2	6.4×10^2	2.4×10^{-13}	0.073	B_2	4.5×10^2	2.7×10^{-13}	0.055	C_2	6.3×10^1	3.1×10^{-13}	0.14
A_3	7.5×10^2	2.1×10^{-13}	0.071	B_3	5.3×10^2	2.2×10^{-13}	0.057	C_3	7.5×10^1	2.5×10^{-13}	0.14
A_4	8.8×10^2	1.6×10^{-13}	0.075	B_4	6.2×10^2	1.9×10^{-13}	0.055	C_4	8.7×10^1	2.2×10^{-13}	0.14
A_5	1.0×10^3	1.4×10^{-13}	0.073	B_5	7.2×10^2	1.5×10^{-13}	0.058	C_5	1.0×10^2	1.3×10^{-13}	0.16
Photon energy: 0.5 MeV											
A_0		5.2×10^{-13}	0.020	B_0		5.0×10^{-13}	0.017	C_0		5.1×10^{-13}	0.047
A_1		1.8×10^{-13}	0.055	B_1		1.9×10^{-13}	0.045	C_1		1.9×10^{-13}	0.12
A_2		1.3×10^{-13}	0.058	B_2		1.7×10^{-13}	0.044	C_2		1.8×10^{-13}	0.11
A_3		1.3×10^{-13}	0.055	B_3		1.1×10^{-13}	0.047	C_3		1.2×10^{-13}	0.13
A_4		9.5×10^{-14}	0.058	B_4		9.9×10^{-13}	0.048	C_4		1.0×10^{-13}	0.12
A_5		7.7×10^{-14}	0.062	B_5		7.4×10^{-14}	0.050	C_5		7.5×10^{-14}	0.13
Photon energy: 0.1 MeV											
A_0		1.2×10^{-13}	0.015	B		1.2×10^{-13}	0.012	C_0		1.2×10^{-13}	0.031
A_1		4.7×10^{-14}	0.033	B_1		5.2×10^{-14}	0.027	C_1		4.9×10^{-14}	0.073
A_2		3.8×10^{-14}	0.033	B_2		3.9×10^{-14}	0.028	C_2		3.8×10^{-14}	0.070
A_3		2.9×10^{-14}	0.035	B_3		3.0×10^{-14}	0.028	C_3		3.0×10^{-14}	0.078
A_4		2.2×10^{-14}	0.037	B_4		2.4×10^{-14}	0.029	C_4		2.7×10^{-14}	0.073
A_5		1.7×10^{-14}	0.039	B_5		2.0×10^{-14}	0.030	C_5		2.0×10^{-14}	0.079
Photon energy: 0.05 MeV											
A_0		9.2×10^{-14}	0.013	B_0		9.2×10^{-14}	0.010	C_0		8.9×10^{-14}	0.029
A_1		3.3×10^{-14}	0.031	B_1		3.5×10^{-14}	0.025	C_1		3.3×10^{-14}	0.069
A_2		2.4×10^{-14}	0.033	B_2		2.5×10^{-14}	0.027	C_2		2.4×10^{-14}	0.076
A_3		1.8×10^{-14}	0.035	B_3		1.9×10^{-14}	0.028	C_3		1.6×10^{-14}	0.077
A_4		1.3×10^{-14}	0.039	B_4		1.4×10^{-14}	0.030	C_4		1.5×10^{-14}	0.074
A_5		9.1×10^{-15}	0.042	B_5		1.0×10^{-14}	0.032	C_5		1.2×10^{-14}	0.082

[a] Regions labeled as in Fig. 10-3.

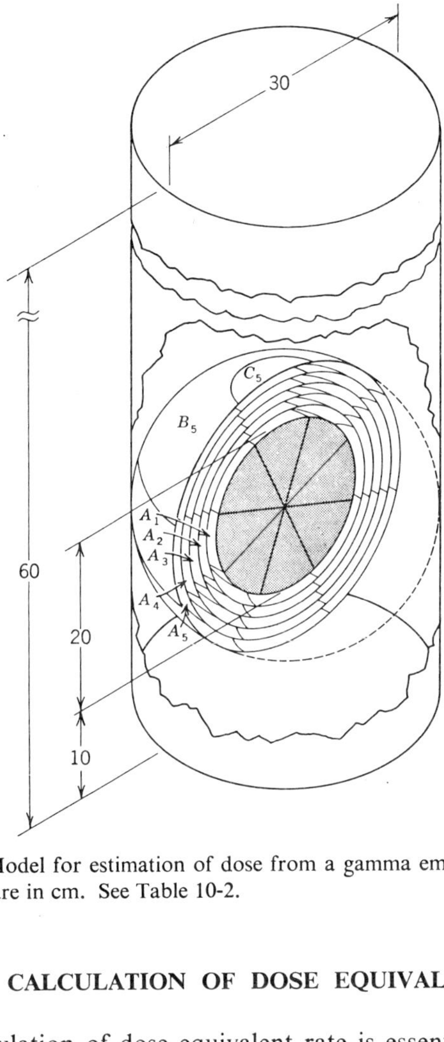

Fig. 10-3. Model for estimation of dose from a gamma emitter present in the GI tract. Dimensions are in cm. See Table 10-2.

10-6 THE CALCULATION OF DOSE EQUIVALENT RATE

The calculation of dose equivalent rate is essentially similar to that for dose rate. We merely insert the proper quality factor or modifying factor into formula (10-3), which now becomes

$$\xi_i = \sum_{j=1}^{k} f_j F_{ij} E_j (QF)_{ij} (MF)_{ij} \tag{10-6}$$

where each of the symbols f_j, F_{ij}, and E_j has the same meaning as in (10-3), and $(QF)_{ij}$ and $(MF)_{ij}$ are appropriate modifying factors. Throughout this chapter the symbol ξ_i will be used with this meaning. With this definition of

ξ_i the formula for dose equivalent rate to organ i is then

$$(DE)_i = \frac{51\xi_i R_i}{M_i} \qquad \text{rem/day} \tag{10-7}$$

The values of $(MF)_{ij}$ may depend on the tissue (here indicated by the index i) as well as on the particle depositing the energy (represented by the index j). The values usually adopted for these quantities have already been discussed to some extent in Chapter 3, and only a few remarks concerning their special use in connection with internal dose are added here.

The symbol $(QF)_{ij}$ is specified, as in Chapter 3, on the basis of purely physical considerations, although evidence of biological effects motivated the definition. Thus the quality factor $(QF)_{ij}$ is almost independent of the tissue considered, for although stopping power does vary somewhat with the composition of the material, the dependence on i as indicated is usually not critical. For most soft tissues the quality factors are much the same, but bone and cartilage have a rather higher average atomic charge per atom (average Z number) and consequently somewhat different values of QF than do most soft tissues.

Animal experiments have indicated, however, that some isotopes, notably, ^{239}Pu and ^{90}Sr, cause a higher degree of biological damage (bone cancers) than does ^{226}Ra when compared on the basis of equal doses to bone. The early work was part of the program of the Manhattan Project and was necessarily based on acute effects, mice and rats being the usual biological material. This work indicated an MF of about 5. Later studies on dogs, extending the exposure period over a span of five to ten years, have been carried out since for ^{239}Pu and some other radionuclides and the studies are still in progress.[16] The specific objective of these studies is to obtain direct biological evidence of the relative effectiveness of ^{226}Ra and other long-lived radionuclides that deposit in the skeleton. Since the accepted limit of 0.1 μc of ^{226}Ra as body burden has until now appeared to be satisfactory for occupational exposure, we might then have a basis for establishing correspondingly satisfactory limits on the skeletal burden of other bone-seekers. The later studies have tended to indicate that the value of 5 for MF is not unrealistic. Figure 10-4 summarizes the data available on the University of Utah studies in 1961.[17] The data indicate a ratio of Ra/Pu = $\sim$5 for the mean skeletal dose in beagles dying with osteosarcomas at about 1300 days, whereas for radiothorium and mesothorium, the ratios are about 15 and 4, respectively. At later times the Ra/Pu ratio is about 7 at 1650 days and about 6 at 2100 days.

It is apparent that no direct and simple conclusion concerning the relative effectiveness of doses from these isotopes can be drawn solely on the basis of the data presented in Fig. 10-4. For example, the $P5$ group (fifth level plutonium group) appears quite out of line with the trend of the data for the other *P* groups of dogs. The published reports of these long-term and important experiments reveal that the dosage levels for groups *P*2, *P*3, *P*4, and *P*5 were approximately in the ratios of 1:3:9:27 and that the cause of death for the *P*5 group was significantly different than that for the other *P* groups. Thus a different mechanism of radiation damage is indicated in this case.

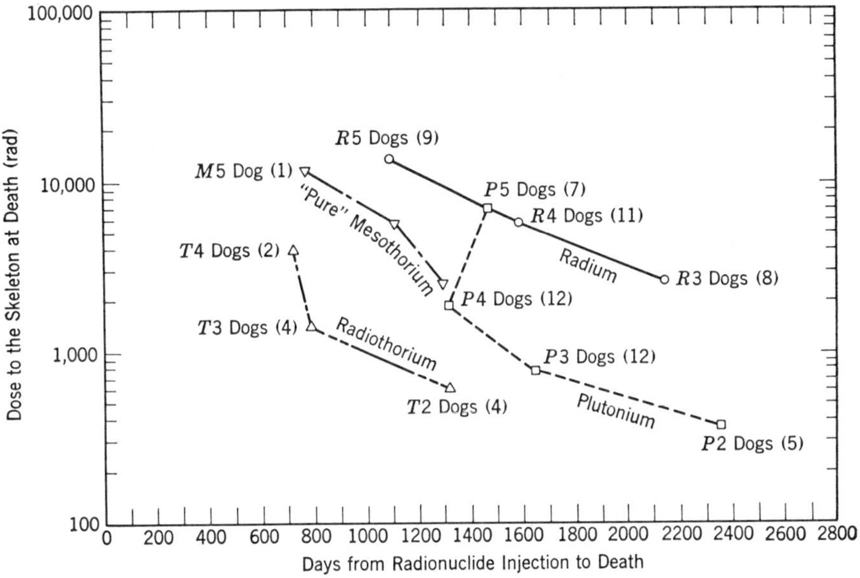

Fig. 10-4. Mean skeletal dose in beagles dying with osteosarcomas.

Similarly, the dose rate to the skeletons of the dogs injected with ^{228}Th (radiothorium) decreased significantly during the period of study because of the 1.9-year half-life for radioactive decay of this radionuclide, whereas for ^{226}Ra and ^{239}Pu there was little change in dose rate after a short initial period. This difference in the time-course of the dose complicates the attempt to draw conclusions concerning the relative effectiveness of these doses. We may speculate that the higher dose rates in the early period of exposure of the *T* groups are more effective in producing osteosarcomas than are the doses delivered at lower dose rates, but the data are not adequate to confirm this conclusively. In spite of all these complicating factors there does seem to be a clear indication that the other radionuclides tested are somewhat more effective, rad for rad, than ^{226}Ra, and in the lower levels of dose where the curves indicate a trend, ^{239}Pu seems to be about five times as effective as ^{226}Ra. The question is too involved to discuss here, but is quite important because the thousands of persons who have accumulated a body burden of ^{226}Ra offer direct evidence for establishing a criterion for protection of man, and if other bone-seekers could be clearly related to ^{226}Ra regarding the hazard of osteosarcoma or other deleterious effects, this would provide a valuable guide for exposure of man to a wide variety of other bone-seeking radionuclides. For ^{90}Sr only comparisons using results of acute studies have been published at this time, though long-term experiments are in progress.

Another modifying factor is used by ICRP and NCRP for β-rays of low energy. The weighing factor here is chosen as 1.7 on the basis of experiments by Furchner, who compared the effectiveness of tritium with ^{60}Co γ-rays in producing lethality in mice.[18] In other studies Furchner et al. had found values of 1.4 and 1.6 respectively, using splenic and thymic atrophy of mice and reduced uptake of ^{59}Fe in red blood cells of rats when comparing tritium with ^{226}Ra and ^{60}Co gamma rays.[18] The use of this factor has been extended to be applicable to all β-particles of maximum energy less than or equal to 0.03 MeV. According to Fig. 8-6 the stopping power of electrons increases sharply when the energy is below 0.1 MeV. Thus this factor might be termed a quality factor, and from this point of view it would be applicable to all energy deposited by electrons with energy below 0.1 MeV. However, the biological data are perhaps the more compelling evidence for using the factor and so the ICRP and NCRP only use the factor when E_{max} of the β-particle is less than or equal to 0.03 MeV. Viewed in this way, it should be termed a modifying factor.

Finally, the question of whether the dose from the internal sources is equally as effective as that from external sources deserved to be noted. All the agencies which have framed regulations have agreed that doses to an organ from external and from internal sources are to be added. In the instances above, that is, for many bone-seekers other than ^{226}Ra and for low-energy β-rays, a modifying factor as well as the quality factor is introduced. Some experiments have been interpreted to indicate that the dose from internal sources is less effective than that from external sources. For example, the FRC cites experimental work indicating that X-ray doses to the thyroid appear to be from five to fifteen times more effective in producing biological changes than doses of ^{131}I.[19] However, the external source irradiated the tissue surrounding the thyroid to a greater degree than did the ^{131}I and the dose-rate was much higher in the exposures from X-rays than from ^{131}I. Whether the differences found are due to these factors or to others not yet identified is not known. The FRC,[20] although noting and citing this experiment, did not base any of its recommendations on this supposed difference, and other recommendatory agencies have likewise continued to assume equal effectiveness of dose from internal and external sources.

10-7 CALCULATION OF TOTAL DOSE AND OF THE TIME COURSE OF DOSE

Formula (10-7) gives the dose-equivalent rate to tissue i when the amount of radioactive material in the tissue is specified. If R_i, which represents the number of microcuries of source present in the organ, is regarded as a function of t, Eq. (10-7) then gives the time course of dose to the tissue. The total dose during any period of time from t_1 days to t_2 days is then given by the integral

$$\int_{t_1}^{t_2} \frac{51\xi_i}{M_i} R_i(t)\, dt \quad \text{rem} \tag{10-8}$$

If the span of time is not too long, ξ_i and M_i may be regarded as constants and these factors may be taken outside the integral sign. However, for long periods of time during which body masses are changing appreciably it may not be accurate to regard ξ_i and M_i as constant. The total dose from a single intake may be had by letting t_1 represent the time of intake and t_2 the extent of life considered to occur after that time. In general, one would take t_1 and t_2 to represent a period of 50 years as the maximum period for occupational exposure, but for population exposure 70 years might be more appropriate.

It may happen that the radionuclide produces a daughter and that the dose delivered by this daughter element must be considered. For example, if ^{90}Sr enters the body, each decay of an atom of ^{90}Sr produces an atom of ^{90}Y and thus a continuous source of ^{90}Y is present in the body. In a matter of a week or ten days this may be expected to be essentially in equilibrium with the ^{90}Sr parent. Not much is known concerning the distribution of these daughter elements produced within the body. It is suggested, for example, that ^{228}Ra produced as a daughter of ^{228}Th does not remain fixed in bone where the Th parent is present. Since the proportions found in bone are not those that would be typical merely from the physical growth of the daughter element it seems more likely that the Ra daughter produced in bone in most instances recirculates to blood before a certain fraction is then redeposited in bone. At any rate this hypothesis seems consistent with the proportions found in experiments.[21] In general, no information on the special character of the distribution of the daughter elements is available, and it is general practice to assume that they remain present in the tissues where they are produced except for elimination by the biological elimination rates typical for the daughter isotope. Thus ^{90}Y produced in bone as a daughter of ^{90}Sr is assumed to remain in bone with an elimination constant typical for that of

^{90}Y administered as parent. For example, if $R_i(t)$ represents the microcuries of parent present in the organ on day t, then the daughter element is being produced at a rate of $3.2 \times 10^9 R_i(t)$ atoms/day which is equivalent to $\lambda^* R_i(t)$ μc/day of the daughter where λ^* is the radioactive decay constant for the daughter element. If this process continues over a period of τ days the number of microcuries of daughter element present in tissue i will be given by the formula

$$\lambda^* \int_0^\tau R_i(t) R_i^*(\tau - t)\, dt \tag{10-9}$$

where R_i^* is the retention function for the daughter when present in tissue i. In some cases the daughter may migrate in significant amounts from other tissues and then R_i would need to be replaced by the retention function for the parent element as present in those tissues and a factor included to represent the fraction transferred to tissue i. However, there are not many cases where sufficient data are available to give a realistic estimate of these transfer retentions. The case of ^{131}Te with ^{131}I as daughter is one such case and it may be assumed that the ^{131}I produced as daughter goes to the thyroid with essentially the same probability as freshly injected ^{131}I. If a second daughter isotope is present, this process may be continued and the formula extended to include production of this second daughter. However, there is little to be gained by writing these very general formulas, provided the general course of the argument is understood. Although the principle involved is clear, there are not many instances where biological data are available to suggest quantitative relationships. When we have the formulas for the retention of daughter elements in the tissue of interest, it is necessary only to compute the dose or dose-equivalent separately for each daughter element and sum these to obtain the total dose or dose rate to the organ.

10-8 BASIS OF MAXIMUM PERMISSIBLE CONCENTRATION VALUES

Maximum permissible concentration (MPC) values are computed on the basis of a constant level of exposure. This is a very artificial premise which is seldom realized in any practical situation. In most practical cases, the exposure occurs sporadically and thus it may be questioned whether the point of view of a constant level of exposure is a realistic one. The interest of calculating on the basis of a constant level of exposure resides in the simplification which this introduces into the calculations as well as the practical importance of having a constant level which is acceptable for routine—and therefore—continuous exposure. If we can establish that in the situation of exposure at a constant level the permissible dose rates to organs are not exceeded, it is clear that any pattern of exposure in which intakes fluctuate but remain below this level will not produce overexposure in the body. Thus the constant level of exposure established by the MPC constitutes a level below which one may allow intakes to occur in any fashion but without producing any overexposure. In addition, the constant level of exposure simplifies the calculation considerably, and enables us to use, in many cases, data on intake of stable isotopes to obtain a realistic estimate of the body burdens and organ burdens that might accumulate under the constant level of exposure of the radionuclide. The formulas expressing the body burden at an arbitrary time following the beginning of exposure at a constant level are readily obtained by modifying the preceding formulas. Thus, if $R_i(t)$ is the retention function for tissue i following an intake of unit amount of the stable isotope, the amount accumulated in tissue i following repeated daily levels is given by the formula

$$I \int_0^t R_i(t - \tau)e^{-\lambda(t-\tau)}\, d\tau = I \int_0^t R_i(\tau)e^{-\lambda\tau}\, d\tau \tag{10-10}$$

In this formula λ represents the radioactive decay constant and I is the daily intake. This organ burden or tissue burden of the radionuclide may then be inserted into any of the formulas given previously and dose rate or total dose calculated. The use of formula (10-10) implies that the retention function does not change appreciably during the period of time under consideration. This will generally not be true if we are dealing with a long period of time and if the individual is growing appreciably during this period. However, for an adult, although the formula still may not rigorously apply, the changes in retention through adult life are generally not so drastic as to appreciably invalidate using this assumption. It is not easy to write a corresponding formula for the case in which the individual is growing and where his retention function is changing rapidly with age. One may, of course, introdude a second variable, the age, into the retention function, but although the mathematical formula is adequate there is little biological information to determine the form of the age-dependent retention function. The simple examples treated here are generally applicable to the adult.

Consider now exposure at a constant level, for example, an intake of 1 μc/day, and suppose that the retention function $R_i(t)$ for the corresponding stable isotope does not change significantly during the period of exposure. The resulting microcurie content of tissue i is then given by the formula (10-10) and the concentration at time t in tissue i by

$$\frac{I}{M_i} \int_0^t R_i(\tau)e^{-\lambda\tau}\, d\tau \tag{10-11}$$

Hence the dose rate at time t to tissue i is given by

$$\frac{51\varepsilon_i I}{M_i}\int_0^t R_i(\tau)e^{-\lambda\tau}\,d\tau \qquad \text{rad/day} \tag{10-12}$$

and the corresponding dose-equivalent rate by

$$\frac{51\xi_i I}{M_i}\int_0^t R_i(\tau)e^{-\lambda\tau}\,d\tau \qquad \text{rem/day} \tag{10-13}$$

where ε_i and ξ_i have the same meaning as in Eqs. (10-3) and (10-6).

As mentioned, it is assumed that the retention function does not change appreciably during this period, thus implying that the individual is not changing very significantly. Hence M_i and ξ_i are regarded as constants and placed outside the sign of integration. It is clear from Eq. (10-13) that the dose rate increases throughout the exposure period and so reaches its maximum at the end of the exposure period, which is here taken as 50 yr. Thus, to compute a constant level so that the limiting dose rate would not be exceeded during occupational life, we need only let t be the number of days in 50 yr, equate this formula to the limiting dose rate, and solve for I. This gives the maximum daily intake which may be maintained constantly throughout the period of exposure. To obtain the MPC in air or water we need only divide this intake by the volume of air or water assumed to be taken in per day to obtain the concentration in this medium such that continued intakes of this material during 50 yr would not cause overexposure. Thus, subject to the assumptions indicated above, the maximum daily intake such that continuous exposure at this level results in a dose level of D_i rem/yr to organ i after t days of continuous exposure is given by

$$I_i = \frac{D_i M_i}{51\xi_i \int_0^t R_i(\tau)e^{-\lambda\tau}\,d\tau} \qquad \mu\text{c/day} \tag{10-14}$$

Consequently, MPC values for continuous occupations exposure (168 hr/wk) are obtained by taking $t = t_0 = 50 \times 365.25$ and setting

$$(\text{MPC})_a{}^i = \frac{I_i}{V_a} \quad \mu\text{c/cm}^3 \quad \text{and} \quad (\text{MPC})_w{}^i = \frac{I_i}{V_w} \quad \mu\text{c/cm}^3 \tag{10-15}$$

In Eq. (10-15) V_a and V_w are respectively the intakes of air and water (cubic centimeters) of the individual. Note, however, that $R_i(t)$ will usually be different when exposure is by inhalation rather than by ingestion. In the above derivation, it has been tacitly assumed that the intake level is constant throughout the full day and week. If exposure is confined to, say, five 8-hr days/wk as will usually be the case for occupational exposure, one need only use $\frac{7}{5}I$ as the intake for a working day and divide this by the intake of air or water during the working day to obtain the corresponding MPC. It is usually assumed that standard man consumes half his daily intake of air and water during the 8 hr of the working day.

It has been mentioned that the integration of $R_i(t)$ to obtain the dose from a continuous exposure involves the assumption that $R_i(t)$ does not change during the period involved and that successive increments of intake may be summed. In particular, this implies a similarity in the chemical form of the intake, at least to the extent that $R_i(t)$ is the same for all intake during the period and that the intake is sufficiently small that neither radiation dose nor chemical effects of these intakes have a significant effect on the retention. This procedure of integration of the single-intake retention function to obtain the retention following continuous intake has been tested a number of times, notably by Cook et al.[22] who used ^{60}Co in mice and Richmond et al.,[23] who used ^{137}Cs in mice. (Compare Ref. 24.) In particular, Richmond et al. found that $R_i(t)$ did change detectably over a period of 107 days, presumably because of the aging of animals. However, when this change in $R_i(t)$ with age was taken into account, the integration of the retention for a single intake provided an accurate estimate of retention for continuous exposure.

Retention may be affected by many other factors besides age; among other influences, the general state of the animal (that is, sick vs. well),[25] temperature and season,[26] deficiency or abundance of other elements in the diet (Sr and Ca or P),[27] and chemical form of the radionuclide[28] have been indicated. A difference in uptake has been found when ^{85}Sr was taken with meals or between meals.[29] In most cases these differences are not extreme, but very special chemical forms are an exception to this and the retention and distribution of the radionuclide in the body may be markedly different when the chemical form differs greatly. (Example: HTO and tritiated thymidine.)

Because of difficulties in experimental procedures most retention functions are determined by injection experiments and by separate studies to measure the fraction of ingested material absorbed to blood from the gut. *ICRP Publ.* 2 denotes this fraction by f_1 and lists values for what are termed "soluble" and "insoluble" compounds. Since the chemical compounds are not explicitly specified, the users must judge whether a given compound of interest may be expected to approximate either of these categories or whether he should interpolate or extrapolate from the tabular values. It goes without saying that the solubility in question here is the solubility of the compound in the relevant body fluids—gastric juice, lung fluid, etc.—that may be appropriate for the route of exposure under consideration and not merely the solubility in water.

10-9 RELATION OF SINGLE INTAKE AND CHRONIC EXPOSURE

According to Section 10-8, the MPC, or equivalently the permissible daily intake I is determined by taking $t = t_0 = 50 \times 365.25$ in Eqs. (10-14) and (10-15). Now, the 50-yr dose D_{ij} resulting from a single intake of isotope i to tissue j can be easily arranged to have exactly the same form as Eq. (10-14). That is, the total dose-equivalent D_{ij} to tissue j during a period of 50 yr following a single intake of I μc of isotope i is given by

$$D_{ij} = \frac{51\xi_{ij} I \int_0^{t_0} R_{ij}(t) e^{-\lambda t}\, dt}{M_j} \tag{10-16}$$

Thus the daily intake at MPC level will deliver a 50-yr dose to the critical organ used as a basis for the MPC which is equal to the daily dose permitted to that organ in establishing the MPC. It is equally easy to see that the rule must hold for any multiple of the daily intake and hence the word "daily" in the rule may be replaced by "weekly," "quarterly," or "annual." This rule is quite obvious if a state of equilibrium is reached, for in such a situation each daily intake must be delivering energy to account for one daily dose. However, this derivation shows the rule to be valid even when a state of equilibrium is not reached, for example, with many of the long-lived bone-seekers. The special significance of the 50-year dose in the rule, as distinct from the dose delivered during the first 5, 10, 15, or 30 years is entirely accidental. Whatever exposure period is selected as representing the period of chronic exposure in estimating the MPC will be the period of dose assessment following single intake which is needed in the rule. If men only began work with radiation at age 25 and if they retired at age 55, occupational MPC values would be based on 30 years of chronic exposure and 30 would replace 50 in the rule.

The rule, while useful, has its limitations. First, it assumes that there is no change, neither in the form of $R_{ij}(t)$ nor in the organ mass, during the period, and this is, at best, only roughly true even for adults. It is entirely unrealistic to apply the rule for exposure of infants or teen-aged individuals. Second, even for adults, the rule predicts only the total dose during the next 50 yr following the single exposure; it offers no clue concerning the distribution of dose within the period of 50 yr. Of course, this can often be inferred from other information, for example, ^{131}I with a radioactive half-life of 8 days will deliver about 99% of this dose within a month, but for ^{90}Sr the dose would be distributed throughout the 50-yr period. However, the rule does not include this information.

10-10 THE LUNG MODEL

Inhalation is a major mode of exposure for many operations with radioactive materials. The control, as well as the assessment of such exposure, is complicated by the fact that the deposition and retention of respirable material are influenced significantly by the particle size and chemical form of the material as well as by the individual's breathing rate and, therefore, by his degree of activity at the time of exposure. In *ICRP Publ.* 2[8] a simple lung

Table 10-3 Particulates in Respiratory Tract of the Standard Man

Retention of particulate matter in the lungs depends on many factors, such as the size, shape and density of the particles, the chemical form and whether or not the person is a mouth breather; however, when specific data are lacking it is assumed the distribution is as shown below.

Distribution	Readily Soluble Compounds (%)	Other Compounds (%)
Exhaled	25	25
Deposited in upper respiratory passages and subsequently swallowed	50	50
Deposited in the lungs (lower respiratory passages)	25 (This is taken up into the body)	25[a]

[a] Of this, half is eliminated from the lungs and swallowed in the first 24 hr, making a total of 62.5% swallowed. The remaining 12.5% is retained in the lungs with a half-life of 120 days, and it is assumed that this portion is taken up into body fluids.

model (as specified in Table 10-3) was used for most of the radionuclides considered.

The 120-day half-life was used in all cases except for isotopes of Pu and Th where 1 yr and 4 yr[30] were used, respectively. Although the report recognized that the deposition and retention might differ widely for various aerosols and individuals and recommended appropriate changes in $(MPC)_a$ in these instances, it did not give specific directions for making such changes.

In 1964 the ICRP appointed a Task Group on Lung Dynamics to review the evidence and revise the lung model as might be warranted by newer evidence. The report of this Task Group[31] is largely concerned with giving a more detailed model for deposition and retention of inhaled aerosols in the respiratory system. In particular, it gives rather detailed recommendations for taking into account the physical and chemical nature of the aerosol in estimating deposition and retention.

The respiratory system is considered to consist of a nasopharyngeal (N-P), a tracheobronchial (T-B), and a pulmonary (P) region. These regions

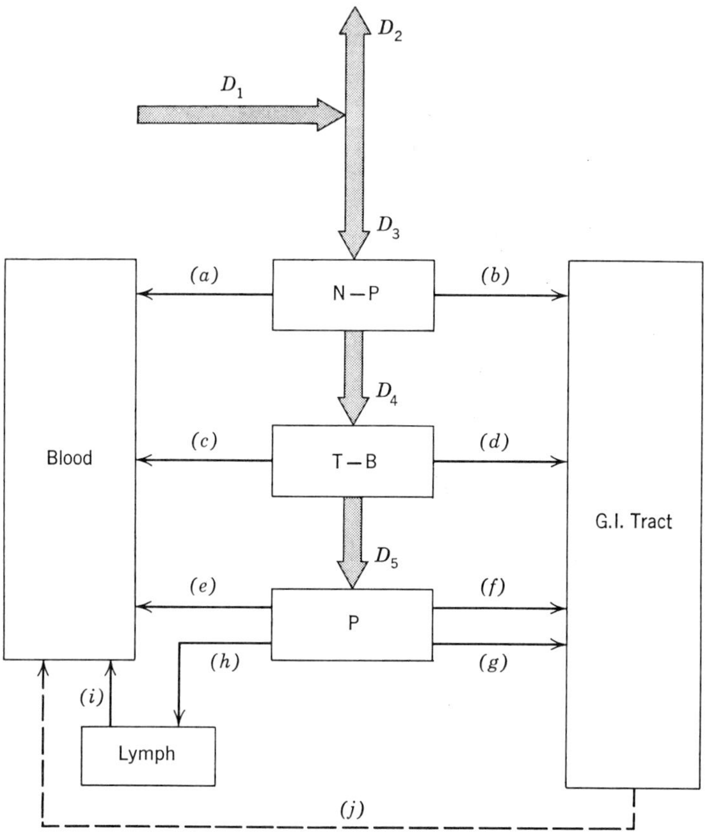

Fig. 10-5. Respiratory tract clearance model.

connect with one another as well as with the blood, the gastrointestinal tract, and the lymphatic system. This model of the respiratory system is shown schematically in Fig. 10-5. The reader may consult the report for the anatomical specifications of the three regions.

Aerosols are grouped into three classes (D, W, and Y) according to the rapidity of clearance of the material from the lung, and fractional amounts of deposited material that are cleared at these various rates are specified in Table 10-4. This constitutes the revised standard lung model for a "standard aerosol." The Task Group also prepared an appendix containing a list of about 60 compounds and assigned a clearance half-time corresponding to pathways *e*, *g*, *h*, and *i* (that is, those pathways with the slowest clearance rate). These values are chosen to correspond to results of experimental work and are to be preferred to those of the standard model for these compounds.

Table 10-4 Constants for Use with Clearance Model

(From Reference 31)

Regions	Pathway	Class D	Class W	Class Y
N-P	*a*	4 min/0.50	4 min/0.10	4 min/0.01
	b	4 min/0.50	4 min/0.90	4 min/0.99
T-B	*c*	10 min/0.50	10 min/0.10	10 min/0.01
	d	10 min/0.50	10 min/0.90	10 min/0.99
P	*e*	30 min/0.80	90 days/0.15	360 days/0.05
	f	n.a.	24 hr/0.40	24 hr/0.40
	g	n.a.	90 days/0.40	360 days/0.40
	h	30 min/0.20	90 days/0.05	360 days/0.15
Lymph	*i*	30 min/1.00	90 days/1.00	360 days/0.10

Note: The first value is the biological half-time; the second is the regional fraction. The lymphatic clearance for Class Y compounds indicates that a 10% regional fraction follows a 360-day biological half-time. The remaining 90% is presumed to be permanently retained in the nodes and is subject only to radioactive decay.

These preferred values replace the clearance half-times only for the four indicated pathways, the other constants being unchanged.

The deposition factors, D_i ($i = 3, 4, 5,$), (see Fig. 10-5) represent the fraction of inhaled material which deposits initially in each of the three regions. These factors depend upon the activity median aerodynamic diameter (AMAD) of the aerosol, and their values are specified as a function of this parameter in Fig. 10-6. The value of D_4 is taken as a constant, namely, 0.08, in this model of the lung. The AMAD is that diameter which separates the distribution of particle activities into two parts of equal activity, and the adjective "aerodynamic" indicates that in assigning a numerical value to a particle diameter the density and shape of the particle are to be taken into account. The aerodynamic diameter of a particle is defined as the diameter of a unit density sphere having the same settling velocity as the particle under consideration.

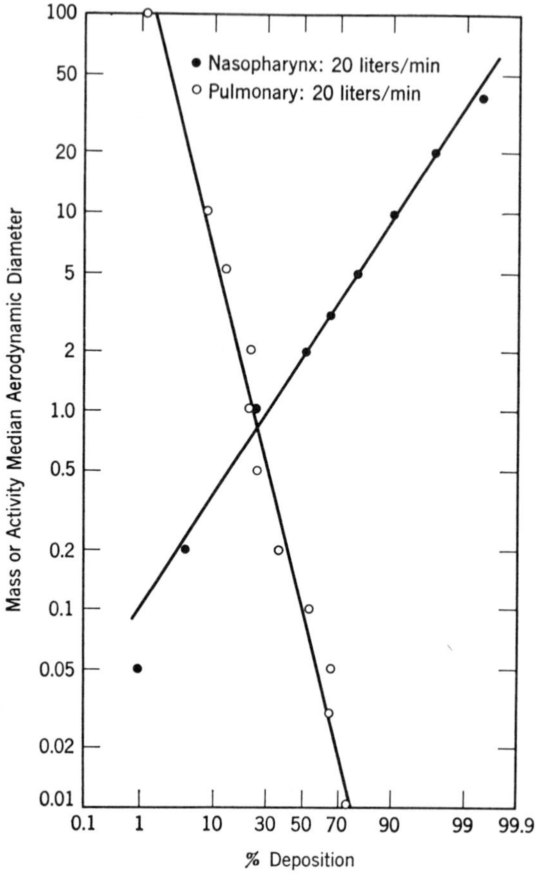

Fig. 10-6. Dependence of deposition on AMAD.

Given an intake I of aerosol in microcuries we easily find the retention in the various regions at time t days as follows:

$$\text{N-P:} \quad R(t) = ID_3(ae^{-\lambda_a t} + be^{-\lambda_b t}) \tag{10-17}$$

$$\text{T-B:} \quad R(t) = ID_4(ce^{-\lambda_c t} + de^{-\lambda_d t}) \tag{10-18}$$

$$\text{P:} \quad R(t) = ID_5(ee^{-\lambda_e t} + fe^{-\lambda_f t} + ge^{-\lambda_g t} + he^{-\lambda_h t}) \tag{10-19}$$

To the T-B region should be added the quantity

$$ID_5T_p(f\lambda_f^b e^{-\lambda_f t} + g\lambda_g^b e^{-\lambda_g t}) \tag{10-20}$$

which represents the amount of material present in the T-B region which is being eliminated from P via routes f and g. T_p represents the fraction of a day required for passing through the T-B region and is of the order of 1 hr, that is, $T_p = \frac{1}{24}$. The above formula for retention due to material eliminated from P neglects radioactive decay during the time of passage, and so if a radionuclide of radioactive half-life of a few hours is in question, the formula requires modification.

The retention function for the lymph is

$$R(t) = ID_5 h\lambda_h^b \left[\frac{0.90}{\lambda_r - \lambda_h}(e^{-\lambda_h t} - e^{-\lambda_r t}) + 0.10te^{-\lambda_h t}\right] \qquad (10\text{-}21)$$

In all these formulas λ_a, λ_b, etc., represent the effective elimination constants for pathways a, b, etc. Thus $\lambda_a = \lambda_a^b + \lambda^r$, where λ_a^b refers only to the biological elimination rate as specified in Table 10-4 or in the Appendix of the report[31] and λ^r is the rate of radioactive decay and hence is independent of a, b, c, etc.

The corresponding retention for an "insoluble" aerosol as estimated by use of the former standard lung model is

$$R = I(0.50e^{-\lambda_u t} + 0.125e^{-\lambda_g t} + 0.125e^{-\lambda_f t}) \qquad (10\text{-}22)$$

In this formula, λ_u is the total elimination rate for the material in the upper respiratory tract, and λ_g and λ_f are the corresponding rates for the fraction of material depositing in the lower respiratory tract which are eliminated via the gut or body fluids, respectively. The biological elimination rates for these three routes are not specified very precisely by the model. In fact, the model as quoted does not assign any value for λ_u^b, but it would be expected to have a half-time of hours or at most a day. The model suggests 8 hr or less as a half-time for λ_g^b, that is, nearly complete elimination within 24 hr, and λ_f^b is taken as 0.693/120 except for Pu and Th as noted previously.

The Task Group's report does not discuss the problem of estimating dose to the T-B or P regions or to the lymph. This is not altogether a straightforward matter. It is not easy to determine the absorbed fraction of energy for such a finely branching system as that consisting of the T-B region even for alpha particles. Moreover, the mass of the T-B region is not easily determined, and the mass of the lymph nodes present in the lung tissue is also uncertain.[32] Figure 10-7 is designed to present the accumulated dose equivalent to the indicated region over any time period (t) following inhalation of 1 μc of the "standard aerosol" that is, AMAD = 1 μ. To make the chart generally applicable, the effective energy equivalent ξ has been taken as 1 MeV, and so the values read from the graph should be multiplied by the appropriate value of ξ. The various curves correspond to the effective elimination half-time for the long-term component of the retention function, that is, pathways, e, g, and h. Only doses from the T-B region and the P region are included, and their dosimetry is rather inprecise both because of uncertainty about the mass of these regions and because it is difficult to estimate ξ accurately.

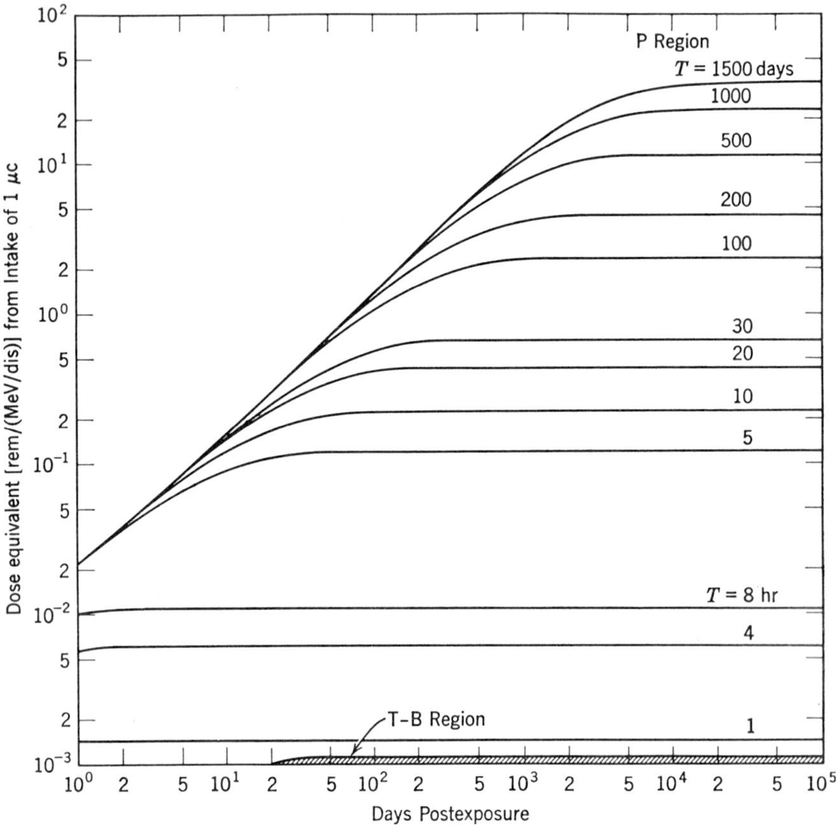

Fig. 10-7. Dose equivalent to the lungs—revised lung model: $\xi = 1$ MeV/disintegration, AMAD $= 1\mu$.

In this example the mass of the P region is taken as 480 gm and that of the T-B region as 400 gm. These estimates are to be regarded as provisional and do not have the endorsement of any recommendatory body, but they are included to provide numerical examples of the type of difficulties encountered in estimating dose to the lung or its subregions. For comparison, a similar chart, Fig. 10-8, is included which provides a comparable estimate of dose to the lung as a whole, based on the former standard lung model. The mass is taken as 10^3 gm and retention is based on Eq. (10-22). In using

Fig. 10-8, the value read from the graph must be multiplied by the appropriate value of ξ; it then represents the dose equivalent accumulated over the indicated period of time following an intake of 1 μc. The various curves are indexed by the long-term effective half-time for elimination from the lung.

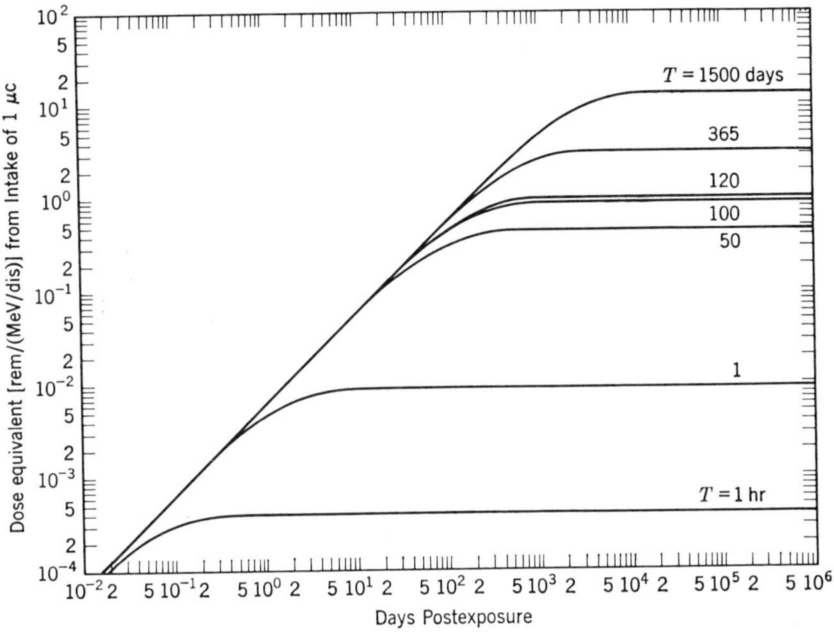

Fig. 10-8. Dose equivalent to lung as a whole from single intake; ICRP *Publ.* **2** model. (ξ = 1 MeV/disintegration.)

10-11 THE MODEL FOR THE GASTROINTESTINAL (GI) TRACT

Because of the very special form of the retention function for the GI tract-or for its various subdivisions, and because of its special functional properties, it is desirable to specialize the general formulas of Section 10-8 in applying them to this organ system. Radioactive material present in the contents of the GI tract is not a part of the systemic body burden; this is also true of particles physically present in the lung passages. Nevertheless, it is necessary to consider such material as part of the body burden both because it may irradiate the body tissues and because it will be included in a measurement by a whole-body counter. Whereas, for the lung, the residence time of some of the material may extend over years, the residence of material in the GI tract is measured in days.

For purposes of dose estimation the GI tract is subdivided into four sub-organs: the stomach (S), the small intestine (SI), the upper large intestine (ULI) and the lower large intestine (LLI). This subdivision is based partly on functional aspects and partly on the mere extent of the tissue and residence time over which the dose is to be averaged and estimated.

The residence time in the various portions of the GI tract may vary greatly for different individuals or even for the same individual from time to time.

Table 10-5 GI-Tract of Standard Man

Portion of GI Tract	Mass of Contents (gm)	Time Food Remains (Days)
Stomach	250	1/24
Small intestine	400	4/24
Upper large intestine	220	13/24
Lower large intestine	135	24/24

For some methods of calculation, the fraction of the 24 hr during which a section of the GI tract is full of food residues is required ("occupancy factor"). Suggested values for this factor are as follows: stomach 6/24, small intestine 14/24, upper large intestine 18/24, and lower large intestine 22/24.

It is suggested that the effective radius of the GI tract as a whole be taken as 10 cm, that the radius of the stomach be taken as 10 cm, and that of the small intestine also as 10 cm; that the upper and lower large intestine each be taken as 5 cm in radius.

I.S.Eve,[33] in a review which was undertaken by request of the ICRP, cites a variety of data to indicate the variability of the occupancy time for the various portions of the GI tract. For example, he cites evidence that 750 cm^3 of water, drunk on an empty stomach, was reduced to half its volume in the stomach in 20 minutes. On the other hand, he notes that fats and proteins particularly delay emptying. Unless exposure is highly correlated with meals (and there might be situations where this is the case!) an average residence time is likely to be more accurate for estimation of the dose from intakes over extended periods of time, that is, if the dose is averaged over a quarter of the year or over a full year. Eve summarizes his recommendations in Table 10-5 which is quoted from his report.[33]

From what has been said it should be clear that these values represent only average or typical values for a grossly normal individual. There is abundant evidence that individuals may vary widely from the above "norm." For example, Hayes et al.[34] studied the range of doses to the lower large intestine in 54 individuals with no apparent abnormalities in GI-tract function.

These doses would be those resulting from the passage of a highly insoluble beta emitter of long radioactive half-life through the tract. They expressed their data in terms of multiples of the dose as computed for "standard man" as defined by *ICRP Publ.* 2[8] and found values ranging from one-fourth the standard-man estimate to seven times that value. The average dose was 70% higher than the standard man value. There is little doubt that if efforts were made to obtain individuals in an extreme state of diarrhea or of constipation, the range of values would be even greater. As with other organs, the dose may differ significantly from one individual to another but in the case of the GI tract these differences are more readily observable and perhaps are more extensive than for many of the other organs.

In general, we are not interested in the precise time interval during which the stomach or ULI, etc. is receiving a dose because of the passage of the material through these sections of the tract but require only an estimate of the total dose. Thus we can use a simple number—the μc-days of residence in each section per unit of intake—for $\int_0^\infty R(t)\,dt$ in formulas (10-12) and (10-13). For a material such that radioactive decay and absorption to blood can both be neglected, $\int_0^\infty R(t)\,dt$ equals the residence time (days) of the material in the section. Thus the μc-days of residence are readily derived from the data in Table 10-5 except for two complications which are discussed below.

First, there may be significant absorption through the walls of the tract and thus some of the material is no longer present in the tract. In the previous GI-tract model used in *ICRP Publ.* 2,[8] all absorption was assumed to occur during the passage of the material through the SI, and Eve endorses this simple assumption as a general practice although recognizing that it is not strictly valid and that there might be situations where it would be inadmissible. In *ICRP Publ.* 2 this absorption was assumed to occur exponentially during the passage through the SI and, in the absence of more precise information, this assumption seems reasonable.

A second complication of the model is due to radioactive decay which must be taken into account if the radioactive half-life is of the order of a day or less. Since these radioactive half-lives are generally well established, this presents no difficulties in selecting the basic data, but it does slightly complicate the formulas.

It is not difficult to show that if an amount A enters a section of the tract and if the time of passage is T days, then the μc-days in the organ is given by

$$\frac{A}{\lambda^r}(1 - e^{-\lambda^r T}) \qquad (10\text{-}23)$$

where λ^r is the decay constant in units of days^{-1}. This formula assumes there is no significant biological elimination. If biological elimination also operates, then we obtain

$$\frac{A}{\lambda^b + \lambda^r}[1 - e^{-(\lambda^b+\lambda^r)T}] \tag{10-24}$$

as the μc-day of residence with λ^b defined by the equation

$$1 - e^{-\lambda^b T} = \alpha_1 \tag{10-25}$$

where α_1 is the fraction absorbed to blood corrected for radioactive decay, that is, the fraction of a stable isotope that would be absorbed. Some experimenters have indicated values of α_1 which are nearly, or even exactly, equal to 1. Since it is unlikely that no material remains in the GI tract, the ICRP has arbitrarily assumed that 5 percent or more will not be absorbed. This represents a negligible inaccuracy if α_1 is, in fact, nearly 1 and does not indicate zero dose for the GI tract.

When the preceding formulas are inserted in formulas (10-12) and (10-13), we also need the effective energy ε and the appropriate mass in order to obtain the dose. For short-range radiation, that is, α- or β-particles we may use the total energy and also use the mass of the contents of the appropriate section of the tract. One-half this dose will be the average dose to the surface layer of this section of the tract. This procedure neglects the cross-irradiation of neighboring sections of the tract, but for short-range radiation this seems sufficiently accurate.

For photons with penetration of more than a few centimeters the assumption of local absorption of energy and neglect of cross-irradiation is unrealistic. It is difficult to be very precise here because the positions of some portions of the tract are not fixed and hence the geometric relationships are changing. For penetrating photons it seems preferable to simply average the dose over the entire tract plus the contents, which may be assumed to form an approximately spherical mass. The example in Table 10-2 provides some data which can be applied here, but it is clear that more refined estimates are desirable, particularly because that calculation assumed a uniform distribution of the source within the spherical mass.

If dose equivalent is required, the appropriate quality factor must be inserted to obtain ξ in place of ε. This is straightforward except that one modifying factor *MF* is used to apply to alpha radiation. Sullivan et al.[35] administered ^{239}Pu orally to rats and estimated nominal surface doses of 10^5 rad to the small intestine but without detectable evidence of biological effect. It appears that the mucosal lining of the tract provides considerable, if not complete, protection for the rapidly dividing basal cells which are known to be rather radiosensitive to beta particles or X-rays. However, the question is not simple, for one cannot completely rule out the possibility of some of the source material entering into the mucosal layer and thus approaching closer to the sensitive cells. The possibility that individuals with some damage to the epithelium might have a high local dose to sensitive cells is also not easy to rule out. On the other hand, the mucosal layer in man should be thicker than in rats and thus provide at least as much protection. These questions are discussed by Eve,[33] who concludes that the use of $MF = 0.01$ for energy from alpha emitters present in the contents of the GI tract seems justifiable in the light of present evidence.

10-12 SPECIAL PROBLEMS AND EXAMPLES

The following sections summarize briefly the more important physical and metabolic data needed to estimate dose for several of the radionuclides frequently of concern to the health physicist. In each case some limiting organ burdens are estimated, a metabolic model is defined, and maximum permissible concentrations in air $(\mathrm{MPC})_a$, and in water $(\mathrm{MPC})_w$, are computed for occupational exposure, that is, exposure during 40 hr per week. These MPC values assume a "standard man" breathes 10^7 ml of air and drinks 1100 ml of H_2O during his 8-hr working day. The student should verify the values in each instance and compute the maximum permissible quarterly intake and the dose to the indicated organ per unit intake.

${}^3_1$H *or* T (*Tritium*)

Physical data: ${}^3_1\mathrm{H} \rightarrow {}^3\mathrm{He}$ (*stable*). A beta particle of maximum energy 0.018 MeV and of average energy 0.0057 MeV is emitted; the radioactive half-life is $T_r = 12.3$ yr. Because the range of the beta is very short and because the maximum energy is less than 0.03 MeV, the effective energy equivalent ξ equals 1.7 times the average energy.[36]

Tritium, in the form of HTO, is assumed to be completely absorbed whether ingested or inhaled, Essentially equal concentrations are found in blood, plasma, saliva, sweat, and urine. Thus, to a first approximation, HTO is assumed to mix freely with all body fluids and to metabolize as does H_2O. *ICPR Publ.* 2 specifies standard man as having 43 liters of H_2O in his body and drinking 2200 ml of H_2O per day. Water of oxidation contributes to raise the total intake of water to 2500 ml/day. Hence the elimination half-time for H_2O in this single-compartment model (assumption of uniform mixing in all body fluids) is

$$T_b = \frac{4.3 \times 10^4}{2500} \times \ln 2 \simeq 12 \text{ days} \qquad \text{and} \qquad R(t) = f_2 e^{(-0.693t/12)}$$

where f_2 is the fraction by weight of total body water in the organ under consideration. The concentration of 0.03 μc/ml of HTO intermixed with body water delivers 0.1 rem/week or about 5 rem/yr to the water.

Many soft tissues have a rather high percentage of H_2O by weight, for example, liver, 75%; brain, 77.5%; spleen, 78.5%; and testes, 84%;[37] thus many soft tissues receive a dose within 30% of that computed above for body water. The total dose equivalent resulting from an intake of I μc is obtained by inserting $R(t)$ in formula (10-16) and the $(MPC)_w$ by dividing the single intake which delivers 5 rem in 50 years by the yearly occupational intake of water. Thus $(MPC)_w = 0.09$ μc/ml for occupational exposure.

This procedure will not be correct in determining $(MPC)_a$ because it has been found that when HTO is present in air there is considerable absorption through the skin. In fact, Pinson and Langham[38] found, by exposing the skin of the forearm, that the intake by absorption through skin was not negligible. Extrapolating to the whole body, they estimated that absorption through the skin produced an intake rate of 1.6×10^4 (μc/min)/(μc/ml), or about 80 percent of that resulting from inhalation at the breathing rate assumed for occupational exposure. This has been confirmed recently by Osborne[39] in a series of experiments involving exposure of the whole body. He found a value of 10^4 (μc/min)/(μc/ml) as an average for 17 subjects when the skin of the whole body was exposed. Osborne also tested the effect of clothing and found intake by absorption through skin was reduced by wearing overalls during 5 to 10-minute exposures but by less than 50 percent. No significant reduction was observed when the exposure was 30 minutes or more. In computing $(MPC)_a$ it was assumed that inhaled HTO is completely absorbed and that intake via skin is 10^4 (μc/min)/(μc/ml). The critical organ was taken as the body water, or essentially equivalent to the testes or all soft tissues. One obtains $(MPC)_a \simeq 7 \times 10^{-6}$ μc/ml.

Problems on Tritium

1. The amount of water in the body is correlated with a number of body parameters such as weight and surface area. For example, Quiring[40] discusses various estimates of surface area and gives preference to the formula first developed by Brody, Comfort and Mathew:[41]

$$A = 0.024M^{0.53} \times H^{0.4} \quad \text{m}^2$$

where the body mass M is in grams and the height H is in centimeters. Osborne[39] used Quiring's formula to derive an estimate of the body water W:

$$W = 0.55M^{0.53} \times H^{0.4} \quad \text{liters}$$

Assume two men *A* and *B* ingest equal amounts of HTO, that *A*'s weight is 50 kg and his height 160 cm, and *B* weighs 75 kg and has a height of 180 cm. If both show the same turnover rate for H_2O, what is the ratio of the doses they receive?

2. A man of height 165 cm and weight 70 kg is found to have 1.5×10^7 d/min/liter of *T* in his urine. Investigation by the health physicist reveals that he was working with HTO two days before submitting the specimen and has had no possible exposure to tritium since that time. Calculate his intake and the resulting total dose, assuming his biological half-time for elimination of H_2O is 12 days. Does the answer depend on the data given for his height and weight? A second sample submitted 36 hr after the first shows 1.35×10^7 d/min/liter. Assuming that these two measurements define T_b for him, correct the previous estimate and compute his dose.

3. An employee is known to have ingested sufficient HTO to be of concern. He agrees to increase his water intake as much as he conveniently can and succeeds in increasing his usual intake by 30%. He continues this regimen for 10 days and then resumes his normal consumption of fluids. Assuming his body-water pool remains constant throughout and that he is a "standard man" what fraction of the total dose did he avoid by increasing his fluid intake? Would the answer be different if he were not like standard man in respect to his water metabolism?

4. Snyder et al.[42] have reported a case of an exposure to HTO where the urine level followed a two-exponential formula:

$$E(t) = E_0[0.9957e^{-(0.693/8.7)t} + 0.0043e^{-(0.693/34)t}]$$

E_0 being the initial level of excretion. What fraction of dose to body water is missed if the second exponential is ignored?

$^{131}_{53}$I (*Radioactive Iodine*)

Physical data: $^{131}_{53}\text{I} \rightarrow {}^{131}_{54}\text{Xe}$ (*stable*). As shown in Fig. 10-2, ^{131}I decays by beta-gamma emission, and $T_r = 8.07$ days. The effective energy equivalent ξ for thyroid as the organ of reference is 0.23 MeV.[8]

Riggs[43] has defined a three-compartment model, illustrated below, for the metabolism of iodine once it has reached the blood in ionic form.

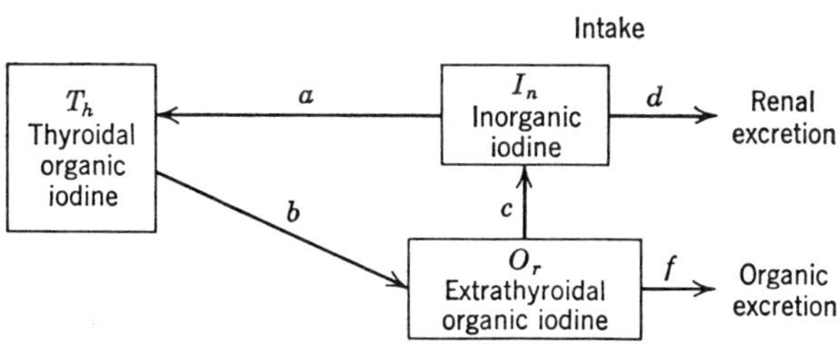

The differential equations representing this sytem are:

$$\frac{d}{dt} I_n(t) = -(a + d + \lambda)I_n(t) + cO_r(t)$$

$$\frac{d}{dt} T_h(t) = aI_n(t) - (b + \lambda)T_h(t)$$

$$\frac{d}{dt} O_r(t) = bT_h(t) - (c + f + \lambda)O_r(t)$$

where T_h, O_r, and I_n represent the amount in the thyroid, and in the rest of the body in organic and inorganic form, respectively. It is somewhat tedious, but straightforward, to verify that the solution of these differential equations for an intake of A μc of inorganic ^{131}I to blood at time $t = 0$ is

$$I_n(t) = \frac{A}{D}[(b - \lambda_1)Ae^{-(\lambda_1+\lambda)t} + (b - \lambda_2)Be^{-(\lambda_2+\lambda)t} + (b - \lambda_3)Ce^{-(\lambda_3+\lambda)t}]$$

$$T_h(t) = \frac{A}{D}[aAe^{-(\lambda_1+\lambda)t} + aBe^{-(\lambda_2+\lambda)t} + aCe^{-(\lambda_3+\lambda)t}]$$

$$O_r(t) = \frac{A}{D}[(b - \lambda_1)(a + d - \lambda_1)Ae^{-(\lambda_1+\lambda)t} + (b - \lambda_2)(a + d - \lambda_2)Be^{-(\lambda_2+\lambda)t} + (b - \lambda_3)(a + d - \lambda_3)Ce^{-(\lambda_3+\lambda)t}]$$

where

$$D = (\lambda_1 - \lambda_2)[\lambda_3(\lambda_1 + \lambda_2 - \lambda_3) - \lambda_1\lambda_2]$$

$$A = (\lambda_2 - \lambda_3)[a + b + d - (\lambda_2 + \lambda_3)]$$

$$B = (\lambda_3 - \lambda_1)[a + b + d - (\lambda_1 + \lambda_3)]$$

$$C = (\lambda_1 - \lambda_2)[a + b + d - (\lambda_1 + \lambda_2)]$$

and λ_1, λ_2, and λ_3 are the three solutions of

$$-\lambda^3 + (a + b + c + d)\lambda^2 - (ab + ac + af + bc + bd + bf + cd + df)\lambda + (abf + bad + bdf) = 0$$

Riggs gives values of $a = 0.0389$, $b = 0.000364$, $c = 0.00222$, $d = 0.0800$, and $f = 0.000208$ as typical for euthyroid subjects. These solutions, that is, $I_n(t)$, $T_h(t)$, and $O_r(t)$ for the above values of the parameters and normalized for a unit intake are shown graphically in Fig. 10-9.

The forms of ^{131}I occurring in most exposure situations are quickly taken up in the blood or other body fluids whether the material is inhaled or ingested. For example, ^{131}I in milk, whether organically bound or not, as well as methyl iodide, silver iodide, and sodium iodide are quickly and essentially completely absorbed in blood when they are ingested or inhaled.[44]

Because the single exponential term $0.312\, e^{-0.00384t}$ adequately represents the retention function for stable iodine in the thyroid except for a few hours immediately following intake the other two terms are often omitted.

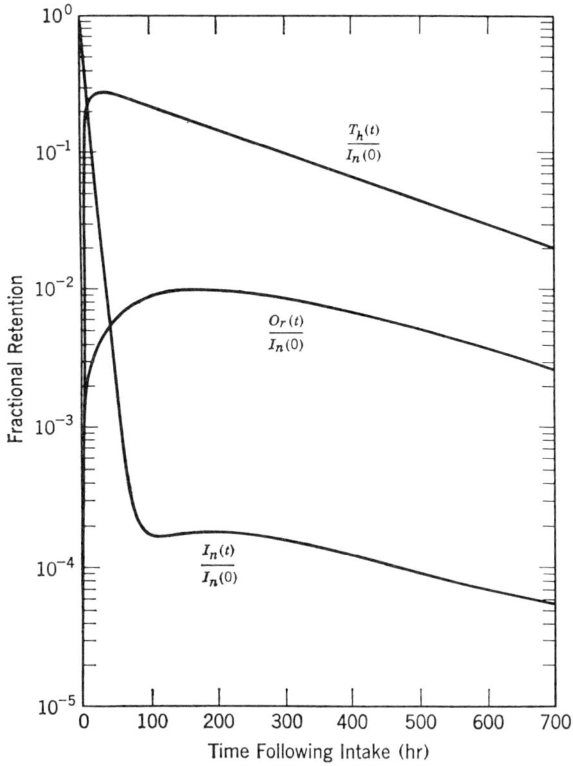

Fig. 10-9. Fractional retention of thyroidal, inorganic, and extrathyroidal organic ^{131}I following single intake.

The permissible thyroid burden is found to be 0.14 μc. Assuming the adult thyroid has a mass of 20 gm the $(MPC)_w$ and $(MPC)_a$ for occupational exposure are 5.4×10^{-5} and 5.9×10^{-8} μc/ml, respectively.

Problems on Iodine

1. A vial of ^{131}I contains 1 mc. How careful must one be that this material is not dispersed to air or water where someone might be exposed to a significant fraction of the activity? How careful need one be of the contents 6 months later?

2. What percent of error is made in the total dose to the thyroid by omitting the two exponential terms of shortest half-time from the retention function?

3. Using the single exponential model estimate the intake to the thyroid which would deliver a dose of 7.5 rem to the thyroid in 13 weeks and compare this initial burden and the resulting dose rate to the thyroid with the so-called maximum permissible thyroid burden and the maximum permissible dose rate to the thyroid.

4. An infant on the average has a daily intake of milk which is almost that of a typical adult. If the infant's thyroid has a mass of 2 gm and if he has the same retention function for his thyroid as the adult, what will be the ratio of total doses to the adult and infant thyroids if all their milk comes from the same contaminated supply? There is some evidence[45] to indicate that the infant's thyroid has a shorter T_b, perhaps as low as 10–15 days, and that the initial uptake to thyroid may be about the same as for the adult. Recompute the ratio of total doses to the infant and adult thyroids using these latter assumptions for the child. Also compute this ratio for ^{129}I which has $T_r = 6.3 \times 10^9$ days and $\xi = 0.068$ MeV.

5. There is evidence to suggest that the infant thyroid is no more resistant to radiation damage than are most other tissues of the body,[20] and the FRC has therefore taken 1.5 rem/yr as the RPG for the child's thyroid instead of the usual $\frac{1}{10}$ of the occupational limit. Compute the limiting concentrations in milk which, for continuous consumption of 1 liter per day, would deliver 30 rem/yr to the adult's thyroid; also for 1.5 rem/yr to the child's thyroid.

6. Assume that following a reactor incident, ^{131}I is deposited on pasturage to the extent of C μc/kg of grass and that, due to weathering as well as radioactive decay, the concentration will be $Ce^{-\lambda t}$ μc/kg on day t thereafter. Assume further that cows eat the grass and that the concentration in milk is $fe^{-\mu t}$ μc/l on day t following a unit intake. Assume an infant consumes 1 liter of this milk per day with a time lag of one day from the production of the milk to its consumption. Using the adult retention function for the infant, compute the total dose received by the child's thyroid if he continues to drink the milk indefinitely. Assume now that on the nth day following the incident, the cows are taken off the pasture and are fed uncontaminated feeds so that there is no further intake to the cow. What fraction of the total dose to the infant's thyroid does this action avert? In FRC Report No. 5[20] a more realistic model is described for such a situation and Protective Action Guides (PAG) for such a situation are given. The FRC assumes the elimination half-time due to weathering and radioactive decay is about 5 days.

7. A radiation worker is found to have a thyroid burden of 0.1 μc. Upon investigation, the health physicist concludes the exposure probably did not occur more than 3 days ago. Assuming the exposure did occur at sometime during the last 3 days, estimate the range of doses to the thyroid corresponding to the time of occurrence within this range.

Strontium Isotopes: $^{85}_{38}$Sr, $^{89}_{38}$Sr, *and* $^{90}_{38}$Sr

Decay schemes:[46]

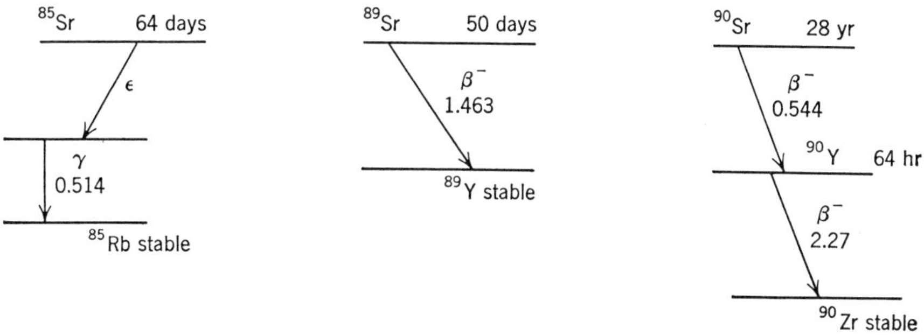

Strontium-85 decays by electron capture ε and the emission of 0.514-MeV gamma rays to produce stable ^{85}Rb. Its radioactive half-life is 64 days. Values of effective energy equivalent ξ computed for various body organs are: 0.091 MeV for bone, upper large intestine, and lower large intestine; and 0.16 MeV for lung, stomach, and small intestines.

Strontium-89 with a half-life of 50 days decays to stable ^{89}Y by emission of beta particles of maximum energy 1.463 MeV. This is a first forbidden unique transition and the average energy as read from the graphs of James, Steel, and Story[47] is 0.55 MeV. Thus, applying the factor of MF = 5 for bone, the effective energy equivalent in bone is 2.8 MeV, and in soft tissue where MF = 1 the effective energy equivalent is 0.55 MeV.

Strontium-90 with a half-life of 28 yr decays by emission of 0.544-MeV beta particles to produce ^{90}Y, which in turn decays with a half-life of 64 hr to stable ^{90}Zr by emission of betas of maximum energy equal to 2.27 MeV. These are both unique forbidden transitions of the first order and the average energies are 0.21 MeV and 0.89 MeV for ^{90}Sr and ^{90}Y, respectively. When considering the MF factor for bone, the effective energy equivalents in bone are 1.1 MeV for ^{90}Sr and 4.4 MeV for ^{90}Y. The corresponding organ burdens are given in Table 10-6.

The skeleton has been estimated to contain 99 percent or more of the stable strontium normally present in the body.[48] However, at short times after an intake of ionic strontium, there will be a considerable fraction of the intake present in soft-tissues, but for chronic or long-term exposure one may assume all the systemic body burden is in bone. For example, following IV injections of $^{85}SrCl_2$ the early retention has been described by

$$R(t)^{(4)} = 0.73e^{-(0.693/3)t} + 0.10e^{-(0.693/44)t} + 0.17 \qquad (10\text{-}26)$$

or alternatively by

$$R(t)^{(5)} = 0.28e^{-(0.693/3)t} + 0.51t^{-0.215} \qquad t < 300 \text{ days} \qquad (10\text{-}27)$$

For oral intake of the same compound, a factor designated as f_1 in *ICRP Publ.* 2[8] must be applied. There are indications that this factor, f_1, rather loosely described as "the fraction absorbed from gut to blood," may vary widely among individuals and there are indications that it may be influenced

Table 10-6 Maximum Permissible Organ Burdens, μc for ^{85}Sr, ^{89}Sr, and ^{90}Sr

Radionuclide	Organ of Reference	
	Bone	Lung
^{85}Sr	62	5.0
^{89}Sr	4.0	1.5
$^{90}Sr^{a}$	2.1	0.73

[a] Values computed assuming equilibrium of parent and daughter.

by the residence time of the material in the GI tract, by the state of positive or negative balance with respect to Ca, by other characteristics of the diet, and by the age of the individual. Values ranging from 0.17 to 0.5 have been suggested for grossly normal adults.[49] For patients with osteoporosis and other clinical conditions or for individuals on special diets other values might be appropriate. *ICRP Publ.* 2 uses $f_1 = 0.3$ as the value for standard man. This factor should be used with the retention functions given above to obtain the retention function for intakes by ingestion of a compound of strontium which ionizes readily in blood or other body fluids.

The formulas for retention as given suggest the existence of more than one "pool" of the strontium in the body: that present in blood, in an "exchangeable pool" with rather rapid turnover, and "one or more pools" with turnover rates which are long compared to 100 days. If we assume a single exponential term, $0.17e^{-0.693t/T}$, where T is the effective half life, for the 17% indicated above for this pool of slow turnover rate and use the formula to calculate the number of daily intakes present at equilibrium we find

$$\frac{0.3\text{I}}{0.693} \times (0.73 \times 3 + 0.10 \times 44 + 0.17T) = \text{I}(2.9 + 0.074T) \qquad (10\text{-}28)$$

According to Snyder et al.[49] a typical United States diet might contain 1.8 mg Sr per day and an average body burden of Sr might be 330 mg. This gives an estimate for T of 2500 days. British sources[50] generally report a value of $f_1 = 0.2$ for Sr in the diet. If this value is used in place of $f_1 = 0.3$, we obtain an estimate of T which is of the order of 4000 days. In this way data on a stable element may sometimes be used to supplement data on the metabolism of a radionuclide although it is clear that the above use of "averages" can only provide a rough estimate of the turnover rate. Nevertheless, this method of assigning an average turnover rate for the skeleton as a whole is, perhaps, as accurate a means as is now available for completing the formula for $R(t)$ (see preceding paragraph). This method has the advantage of being based on diet and bone levels observed in man and resulting from strontium present in the environment. As so completed the formula becomes

$$R(t) = 0.73e^{-(0.693/3)t} + 0.10e^{-(0.693/44)t} + 0.17e^{-(0.693/4000)t} \qquad (10\text{-}29)$$

For a very poorly soluble compound of Sr, such as $SrTiO_3$ ($f_1 \approx 1\%$),[51] the GI tract and lung likely will be the organs receiving the highest doses. Table 10-7 includes the 50-year doses to bone, GI tract, and, in case of inhalation, to lung of standard man from intake of 1 μc of strontium-85-89-90 in a "soluble," and again in an insoluble form. The doses to the pulmonary region of the lung are computed using Fig. 10-7.

The chemical similarity of Sr and Ca leads to another metabolic model. Comar and others[52] have found that the ratio of Sr/Ca in diet and the same ratio in newly formed bone are related. Under grossly normal conditions of diet and status of the animal there is a discrimination against utilization of Sr and favoring utilization of Ca. This leads to the concept of the observed ratio (O.R.) which is defined as the double ratio

$$\text{O.R.} = \frac{(\text{Sr}/\text{Ca})_{\text{bone}}}{(\text{Sr}/\text{Ca})_{\text{diet}}}$$

and its reciprocal is often termed the discrimination factor. Some of the discrimination may be effected by the GI tract, but there is also evidence of discrimination by the kidney. The O.R. metabolic model has been found to be roughly consistent with data on ^{90}Sr present in the food chain as a result of fallout from weapons testing and with the observed levels in human bone.[53,54,55] However, the O.R. reported in British studies[56] is usually about 0.25, whereas Rivera[57] in studies of three United States cities finds values of about 0.16. There is also considerable evidence that during the first year or so of life the O.R. is higher, perhaps almost 1, and there is some reason to believe it might be high also during the teen-age period of rapid bone growth.

Unfortunately, there is not much known concerning the turnover rate of various bones and thus, although the O.R. concept provides a simple model for

Table 10-7 Fifty-Year Dose Equivalent from Intake of Compounds of ^{85}Sr, ^{89}Sr, and ^{90}Sr

Compound	Inhalation			Ingestion	
	DE to Bone (rem/μc)	DE to GI Tract (rem/μc)	DE to Pulmonary Region (rem/μc)	DE to Bone (rem/μc)	DE to GI Tract (rem/μc)
$^{85}SrCl_2$	6.7×10^{-3}	S: 1.3×10^{-4} SI: 2.7×10^{-4} ULI: 7.6×10^{-4} LLI: 2.3×10^{-3}	1.3×10^{-4}	4.4×10^{-3}	S: 6.8×10^{-4} SI: 1.4×10^{-3} ULI: 4.0×10^{-3} LLI: 1.2×10^{-2}
$^{89}SrCl_2$	0.17	S: 4.4×10^{-4} SI: 9.3×10^{-4} ULI: 4.6×10^{-3} LLI: 1.3×10^{-2}	4.4×10^{-4}	0.11	S: 2.3×10^{-3} SI: 4.9×10^{-3} ULI: 2.4×10^{-2} LLI: 7.1×10^{-2}
$^{90}SrCl_2$	14	S: 1.7×10^{-4} SI: 4.2×10^{-4} ULI: 2.7×10^{-3} LLI: 1.2×10^{-2}	8.8×10^{-4}	8.4	S: 8.9×10^{-4} SI: 2.2×10^{-3} ULI: 1.4×10^{-2} LLI: 6.1×10^{-2}
$^{85}SrTiO_3$	1.5×10^{-4}	S: 3.3×10^{-4} SI: 8.3×10^{-4} ULI: 2.8×10^{-3} LLI: 8.3×10^{-3}	0.22	1.5×10^{-4}	S: 6.8×10^{-4} SI: 1.7×10^{-3} ULI: 5.7×10^{-3} LLI: 1.7×10^{-2}
$^{89}SrTiO_3$	3.8×10^{-3}	S: 1.1×10^{-3} SI: 2.8×10^{-3} ULI: 1.7×10^{-2} LLI: 4.9×10^{-2}	0.55	3.8×10^{-3}	S: 2.3×10^{-3} SI: 5.8×10^{-3} ULI: 3.4×10^{-2} LLI: 0.10
$^{90}SrTiO_3$	0.70	S: 5.1×10^{-4} SI: 1.3×10^{-3} ULI: 1.1×10^{-2} LLI: 4.8×10^{-2}	12	0.28	S: 8.9×10^{-4} SI: 2.3×10^{-3} ULI: 1.9×10^{-2} LLI: 8.5×10^{-2}

a near equilibrium situation, it is not easy to use it to estimate dose from a single intake of radioactive strontium. Rivera[54,55] has developed a detailed model which uses an average rate of turnover for the skeleton as a whole. The model neglects recycling of Sr leaving bone and entering blood but does provide for variation of the metabolic parameters with age. His equation may be written:

$$\frac{dSr}{dt} = -(\lambda_r + f)Sr + \left(fCa + \frac{dCa}{dt}\right)(\text{O.R.})\left(\frac{\text{Sr}}{\text{Ca}}\right)_{\text{diet}} \tag{10-30}$$

In this equation *Sr* and *Ca* are the body burdens of ^{90}Sr and stable Ca, respectively; f is the fraction of the bone mass remodeled per unit time and the O.R. and the diet ratio are as indicated above. In principle, all these quantities except the radioactive decay constant λ_r, are considered as functions of time. In using the equation to estimate the body burden of ^{90}Sr we need information on the other quantities involved and, in general, this information is very scanty. Actually, Rivera[54,55] has used the data on ^{90}Sr in diet and in bone together with data of Mitchell[58] on the body content of Ca to estimate f and O.R. as functions of time, or equivalently, of age. Although Rivera's results suggest that the model may adequately explain the diet and bone relationships observed in fallout studies he regards his present estimates of f and of the O.R. as preliminary at the time of writing. The model is, however, the only detailed treatment available for individuals of all ages since the retention functions given in Eqs. (10-26) and (10-27) are based entirely on data for adults.

The O.R. model has the advantage that it provides a means of examining the dose to small regions of bone whereas the approach by a retention function for the whole body or skeleton provides no information on this point. The basis of the microscopic dosimetry is the fact that newly formed bone shows the same ratio of Sr/Ca as was present in blood when the bone was formed. On this basis we can estimate an approximate dose rate to this bone. This estimate, as generally made, is "conservative" because, in the absence of any information on the spacing of these regions where new bone has been formed, we do not know what allowance to make for the irradiation of one region by the radiation emitted from a neighboring region. In a near-equilibrium situation involving ^{90}Sr most of the bone would be "labeled" to about the same degree, and the assumption of complete absorption of energy is more nearly correct. However, as remarked above, the total dose to a site cannot be estimated without knowing when remodeling will occur. There have been a number of attempts to make detailed models which would be amenable to computation of dose and yet include to some degree the very complex structure of trabecular bone and the marrow spacings.[59,60,61] The general result is that the dose to the red marrow is usually taken as one-third the dose to bone, this being computed on the basis of complete absorption of energy.

Problems on Strontium

1. Data of Bishop et al.[50] indicate that all but about 5% of ^{85}Sr injected as the chloride is cleared from the blood within about 10 hr, the rest having been excreted or, for the most part, deposited in bone. Assuming all the ^{90}Sr present in the body is in the skeleton, that is, neglecting the initial phase when a considerable fraction of the ^{90}Sr is in blood or in soft tissues, calculate the dose to the skeleton represented by each of the three terms of formula (10-29) during (*a*) the first 13 weeks (*b*) the first year, (*c*) the first 50 years following intake. Is it justifiable to omit the two exponential terms of short half-time as in the case of ^{131}I? Examine the same matter for ^{85}Sr and ^{89}Sr.

2. Show that the expected excretion from an oral intake of 1 μc of ^{90}Sr by use of Eq. (10-29) is 5.4×10^{-3} μc/day at time $t = 10$ days and 8.9×10^{-4} μc/day at time $t = 20$ da. Fujita[62] has estimated that the ratio of urinary excretion of Sr to fecal excretion, that is, the so-called urine-fecal ratio, is about 2 after exogeneous excretion is no longer significant. Suppose a radiation worker has ingested some ^{90}Sr which was soluble in body fluids and suppose his urinary excretion on day 3 following intake was 0.1 μc. Estimate his body burden at the end of day 3, assuming exogenous excretion is negligible and that retention is given by Eq. (10-26) or (10-27).

3. In general, a good bioassay laboratory will be able to measure 5×10^{-6} μc of ^{90}Sr in a 24-hr urine sample but may have difficulty in going much below this level unless very special methods are used. If employees working with ^{90}Sr are required to submit a sample every 30 days and if the above detection limit is valid, what is the intake to blood which the model indicates would just be detectable by such a laboratory in the "worst" situation, that is, assume the employee submits the urine specimen 30 days post-exposure?

4. Two workers are exposed to an aerosol which is essentially $SrTiO_3$, one being considerably closer to the source than the other. If the closer of the two inhales 1 μc with an AMAD of 1.5 μ, and the other inhales 0.5 μc with an AMAD of 0.5 μ, compute the ratio of the lung doses according to the revised lung model. What would the older "standard lung" model predict?

5. A radiation worker is involved in an incident where he was exposed to $^{90}SrTiO_3$ by inhalation. The health physicist requests samples of urine and feces for the next few days. The urines are negative but feces show only back ground on the first day and 2×10^4, 1.8×10^4, 5×10^3, 2×10^3, and 1×10^3 dis/min for the samples passed on successive days. What might you plausibly conclude concerning this exposure? Discuss the various uncertainties involved.

6. Estimate the (Sr/Ca) diet ratio that would produce 30 rem/yr to bone allowing an O.R. of 0.25 and assuming all bone is labeled at this level. Assume the daily intake of Ca is 1 gm and obtain the corresponding $(MPC)_w$. How does it compare with that implied by the estimates in Table 10-7? Discuss the assumptions involved, particularly as to whether they tend to increase or decrease the estimated dose. What would be the dose rate to the red marrow?

7. Integrate Eq. (10-30) for an adult on a steady diet, that is, assume $dCa/dt = 0$ (no growth), the body burden of Ca is constant at 10^3 gm, a constant daily intake of 1 gm Ca per day, O.R. $= 0.25$, $\lambda_r = 6.93 \times 10^{-5}$, and $f =$ constant. Average rates of turnover have been estimated at 13 percent per year. Find the equilibrium level and compute the daily intake of ^{90}Sr that, at equilibrium, would just produce a dose rate of 30 rem/yr averaged over the skeleton. Compare this with the corresponding daily intake obtained by use of the data in Table 10-7.

Radium: $^{226}_{88}$Ra

The complete decay scheme for Ra and its daughter elements is quite complicated and is not given here. The effective energy ε in either bone or soft tissue is 11 MeV, and the effective energy equivalent ξ in either bone or soft tissue is 110 MeV.

At early periods after radium enters the blood in ionic form, a considerable fraction of the intake is present in blood and in soft tissues. However, one-half or more of the intake to blood is excreted during the first day and at later times an increasing fraction of the body burden is present in bone.[63] Norris et al.[64] suggested a power-function model for retention based on a study of retention in hospital patients who had received IV injections of the chloride. However, only the data during the first year or so of the treatment and data or retention after some 20 years are available from this source. Several recent studies have confirmed the validity of the power-function representation during the early period.[65] There is evidence that retention at much later times does not continue to follow the power law, but that an exponential type of retention may be valid beyond, say, approximately 20 years.[66] Norris et al.[64] suggested the specific function

$$R(t) = 0.54t^{-0.52} \qquad t \text{ in days} \qquad (10\text{-}31)$$

and this seems consistent with the known data calculated to perhaps 20 years.

The fact that the power function (10–31) becomes infinite at $t = 0$ is a minor complication. Actually, unless one wishes to study retention during the first day or so the misrepresentation of the retention for a period of a few days is of little importance since the function is integrable and the first few days do not make a significant contribution to the dose estimate over periods of a year or more.

The body burden of ^{226}Ra can be estimated by whole-body counting, by measurement of ^{222}Rn in breath, or by analysis of excreta. The first of these methods is the most accurate and the third the least accurate. Measurement of radon in breath is found to vary somewhat with the breathing rate and, more importantly, with the residence time of the radium in the body. Mays et al.[67] have estimated that the fraction of ^{222}Rn escaping from the body varies from about 0.90 for very early deposits in the body to values of about 0.50 for late deposits, that is, of the order of 25 years. A value of 0.70 is used in *ICRP Publ. 2.* Unlike Ca and Sr there is considerable endogenous fecal excretion of Ra, the generally accepted value of the urine-fecal ratio being about 0.05.[63]

The value of f_1 is taken as 0.3 in *ICRP Publ. 2* for soluble compounds such as $RaCl_2$, but as with Sr, may vary considerably among individuals or for other compounds. A value of 0.2 has been reported for a form of "mock dial paint."[63] There are a number of communities where the concentration of ^{226}Ra in the water supply is exceptionally high. Lucas[68] has reported the body burdens accumulated by these populations in contrast to the body burdens found in communities where ^{226}Ra is present in water in very low concentrations. In these communities the intake of ^{226}Ra is largely through foodstuffs, whereas in the communities where the water supply is high the intake from water is the major source. Because of the high endogenous excretion the daily intake can be reliably estimated by analysis of feces. The body burden of adults who are long-term residents of these communities are of the order of 50 daily intakes when water is the main source (assuming 1 liter of water per day) of the intake, but is about 25 daily intakes when the intake is largely via the food supply. This suggests a difference in the retention function for these two types of exposure. No large age differences of ^{226}Ra per gm of Ca (or of wet bone) have been reported[69,70] and thus the dose rate to bone should be almost independent of age. These population studies provide a striking instance of how an MPC may be estimated without direct knowledge of the retention function.

The permissible body burden of ^{226}Ra has been set at 0.1 μc by both the NCRP[7] and the ICRP.[8] At times long after exposure nearly all the body burden will be in bone, but for continuous exposure the soft tissues may contain a significant fraction of the body burden. Table 10-8 contains the values of the 50-year dose equivalent to bone and the pulmonary region of the lungs for intake of 1 μc of ^{226}Ra sulfate or ^{226}Ra chloride. These were computed using Norris's power-function model for systemic retention and the

Table 10-8 Fifty-Year Dose Equivalent from Intake of Compounds of Radium

Compound	Inhalation		Ingestion, DE to Bone (rem/μc)
	DE to Bone (rem/μc)	DE to Pulmonary Region (rem/μc)	
^{226}Ra Sulfate	2.6	430	1.0
^{226}Ra Chloride	50	0.044	30

revised lung model. ^{226}Ra sulfate was assumed to be of Class Y and ^{226}Ra chloride of class D. The value of f_1 was taken as 0.01 for the sulfate and as 0.3 for the chloride.

Problems on Radium

1. Calculate the accumulated dose to bone over t days assuming

$$R(t) = at^{-n}$$

if (1) exposure is by IV injection of A μc and (2) if exposure is continuous at the rate of I μc/day to blood. Adjust A and I so that the same body burden results on day t. What is the ratio of total doses? What is the ratio of maximum dose rates averaged over the first month? The first year?

The limiting body burden of ^{226}Ra was selected on the basis of data on medical exposures and exposure of dial painters, and the latter usually did not work for more than a few years at this trade. The worker who would accumulate his body burden gradually would have a smaller accumulated dose than the medical patient or dial painter if both had the same terminal body burden after, say, 10 or 20 years. The above type of calculation provides an indication of the difference in total dose or in dose rate that might be expected to occur in such cases.

2. Use the relationship of daily intake and body burden in the communities mentioned above to estimate the total dose received by the skeleton of an adult from 1 μc of ^{226}Ra ingested in food or alternatively in water. Can this method be used to estimate the dose a child would receive from a single intake?

3. Estimate the dose to bone from ingestion of 1 μc of ^{228}Ra. Make the same estimate for ^{224}Ra. *Hint:* Use the effective energy equivalent $\xi = 190$

MeV and $\xi = 280$ MeV in the two cases. The integral may be evaluated by use of tables of the incomplete gamma function[71,72] or by use of the series

$$\int_0^T t^{-n} e^{-\lambda t}\, dt = \sum_{j=0}^{\infty} \frac{(-\lambda)^j T^{j-n+1}}{(j-n+1)(j!)}$$

which is valid for $0 < n < 1$ and will converge rapidly if λT is small.

Uranium: $^{234}_{92}U$, $^{235}_{92}U$, *and* $^{238}_{92}U$

The uranium decay schemes including daughter radionuclides are quite complicated and are not given here. The effective energies ε in either bone or soft tissue are 4.8 MeV, 4.6 MeV, and 4.3 MeV for ^{234}U, ^{235}U, and ^{238}U, respectively. The effective-energy equivalents ξ are:

for ^{234}U	49 MeV in soft tissue 240 MeV in bone
for ^{235}U	46 MeV in soft tissue 230 MeV in bone
for ^{238}U	43 Mev in soft tissue 220 MeV in bone

Uranium is poorly absorbed from the GI tract, f_1 being taken as 10^{-2} in *ICRP Publ. 6*.[73] Data on retention show a rather wide dispersion, the best-controlled study being one with terminal brain-tumor patients whose excretion was followed for periods of up to a year and on whom some autopsy data were also secured.[74] The retention function, obtained from this excretion data over the first 600 days has been estimated to be $R(t) = 0.6t^{-0.5}$ with t in days. The ICRP assumes that 11 percent of the uranium reaching blood deposits in bone and 11 percent deposits in kidney. Since kidney is assigned a relatively short elimination half-time of 15 days[8] and bone an elimination half-time of 300 days, the dose equivalent to bone is greater than the dose to kidney.

Exposure to uranium isotopes of low specific activity are often limited by the chemical toxicity rather than by the radiation dose delivered to tissue Eve[75] has reviewed the problem and recommended a limit of 2.5 mg of soluble uranium compounds in any one day and a limit of 150 mg by ingestion in any two days. It is readily found that for ^{238}U, the principal constituent of natural uranium by weight, the dose to bone per μc ingested is only 0.015 rad and thus the toxicity is limiting and not the dose to bone. In the case of natural uranium, which consists of 99.2830% ^{238}U + 0.0054% ^{234}U + 0.7114% ^{235}U, toxicity is also limiting. However, with various degrees of enrichment a continuous scale of dose and toxic potential is possible and these mixtures must be considered from both points of view. The problem is discussed by Ford.[76]

Values of dose equivalent over the first 50 years following a single intake of uranium dioxide are listed in Table 10-9. These have been computed using the revised lung model (see Appendix of Ref. 31 for values appropriate for a single exposure). The value of f_1 has been taken to be 0.01 although it is possible this might be too high, particularly for the ingested material. The deposition in bone and the elimination half-time were taken from *ICRP Publ. 2*[8] as mentioned above.

Table 10-9 Fifty-Year Dose Equivalent from Intakes of Uranium Dioxide

Compound	Inhalation: DE to Bone (rem/μc)	Inhalation: DE to Pulmonary Region (rem/μc)	Ingestion DE to Bone (rem/μc)
$^{234}UO_2$	2.2	157	0.83
$^{235}UO_2$	2.1	147	0.80
$^{238}UO_2$	2.0	138	0.76

Problems on Uranium

1. Suppose an intake of 1 μc of radionuclide i results in a 50-year dose of D_i rem to a certain organ and that a mixture of these radionuclides is such that P_i is the fraction of the total activity due to radionuclide i. Show that an intake of the mixture of total activity A μc delivers a 50-year dose of $A \sum_i P_i D_i$ to the organ. Hence if the same organ is limiting in defining the MPC of all these radionuclides, show that the MPC of the mixture is defined by

$$\frac{1}{\text{MPC}} = \sum_i \frac{P_i}{(\text{MPC})_i}$$

Discuss the significance of this relation if the various radionuclides do not have the same organ as limiting in all cases. Is it correct to apply the above principle to compute the MPC for a soluble compound of natural uranium?

2. In working with natural uranium, chemical toxicity is more limiting than is radiation dose and hence the limits are often specified in micrograms per liter of air rather than in units involving microcuries. Show that for a single radionuclide the conversion from microcuries per liter of air to micrograms per liter of air is given by $C \cdot N$ where $N = \mu\text{c/liter}$ of air and $C = 2.79 \times 10^{-6} \times AT$ is the micrograms per microcurie of radionuclide, A is the atomic mass and T is the radiological half-life in years. Will this same formula apply for natural uranium? Show that if 1 μc of natural uranium is defined as consisting of

$$3.7 \times 10^4 \; d/s \;\text{ of }\; {}^{238}\text{U} + 3.7 \times 10^4 \; d/s \;\text{ of }\; {}^{234}\text{U} + 1.7 \times 10^3 \; d/s \;\text{ of }\; {}^{235}\text{U}$$

then the conversion from microcuries per liter to micrograms per liter is given by $C \cdot N$ where $C =$ micrograms per microcurie of the mixture which is the same formula as applies for a single radionuclide. Motivated by this it has been customary to use this "artificial" curie for natural uranium as is done, for example, in *ICRP Publ. 2*.[8]

3. Compute the $(\text{MPC})_w$ for a mixture containing ^{90}Sr and ^{234}U with relative activities in the ratio of 2:3.

4. Suppose the level in a certain source of drinking water is above the MPC of Problem 3, and that remedial action might be taken to reduce either or both of the levels of the two radionuclides, but that the costs of reducing the levels by $X\%$ are $3/(100 - X)$ and $2/(100 - X)$, thousands of dollars per day of operations, respectively. What action will reduce the level to that of an acceptable MPC and at minimal cost?

Plutonium: $^{239}_{94}$Pu

Decay Data[13]

$$^{239}\text{Pu} \xrightarrow{24413 \text{ yr } \alpha,\gamma} {}^{235}\text{U} \; 6.8 \times 10^8 \text{ yr.}$$

Ten different alpha particles of ^{239}Pu have been identified, but only three of these contribute significantly to dose. These have energies and intensities of 5.147 MeV, 72.5%; 5.134 MeV, 16.8%; and 5.096 MeV, 10.7%, respectively. There are a variety of photons, internal conversion electrons, X-rays, and Auger electrons emitted, some of which are of importance in attempting to measure ^{239}Pu in the body, for instance, for measurements with a wound counter. However, their contribution to the effective energy is very small, particularly because the alpha energy is weighted with a QF of 10, and in bone it is also weighted by the MF of 5. Thus the effective energy ε for ^{239}Pu in either bone or soft tissue is 5.2 MeV and the effective energy equivalent ξ in bone is 270 MeV and in soft tissue is 53 MeV.

The chemistry and metabolism of Pu are quite complex. One principal source of variability is the fact that Pu can exist and react rather readily in the plus IV and VI valence states but can also exist in the III and V valence states. It also has a strong tendency to become polymeric under conditions of interest, for example when injected into blood. The absorption to blood and the uptake by bone and liver vary greatly with these chemical conditions.

The inorganic compounds of Pu typically show a value f_1 of 10^{-4} or less.[8]

However, when Pu is complexed with citrate or DTPA or other sequestering agents, values of f_1 of 10^{-2} or more have been achieved.[77] In more usual exposure situations the value of 3×10^{-5} for f_1 seems appropriate.

Once Pu has entered the blood the deposition is largely influenced by the degree to which it is or becomes colloidal. The colloidal forms tend to deposit largely in liver and the complexed forms in bone. In actuality, there will be some of each form present and it is a matter of degree and not of absolutes. Typically, highly colloidal Pu deposits as much as 80 % or more in liver and with complexed forms as much as 70–80 % may deposit in bone. It is expected that when exposure is at low levels and when the Pu enters the blood in trace quantities, as by absorption from the GI tract or from lungs or by solubilization from a wound site, the Pu will complex with the blood protein and that deposition will favor bone rather than liver. *ICRP Publ. 2*[8] lists a value of 3×10^{-5} for absorption from the GI tract to blood with 80 % of this material depositing in bone and 15 % in liver.

Once the Pu reaches blood the retention has been described in terms of a power function[8] $R(t) = 0.99t^{-0.01}$. This is based on studies of 12 hospitalized patients who were given IV ^{239}Pu complexed with citrate.[78] Actually, the retention is almost complete, the equation predicting only 10 percent elimination in 50 years so that for estimation of dose it is consonant with the accuracy of the model to neglect elimination once the material enters the bloodstream.

However, the elimination is important because it is the only basis generally useful for estimating the body burden. Since ^{239}Pu cannot be measured by *in vivo* counting at levels below the accepted organ burdens the estimation of dose is almost entirely on the basis of interpretation of excretion data. Hence there have been a number of methods proposed for estimation of the body burden. Langham[78] fitted the excretion data of the patients mentioned above with power functions and these remain the basis of most methods of estimation of the body burden. The urinary excretion (E_u), fecal excretion (E_f), and total excretion (E_τ) on day t are, by Langham's equations[78]

$$E_u(t) = 0.0023t^{-0.77}$$

$$E_f(t) = 0.0063t^{-1.09}$$

$$E_\tau(t) = 0.0079t^{-0.94}$$

where t is in days and the excretion is expressed as a fraction of that entering blood at time $t = 0$. It is clear that the fraction of the body burden cleared per day decreases with time and thus the estimate of the body burden corresponding to a given level can only be derived on the basis of a time at which the exposure occurred.

Generally, plutonium workers submit urine samples regularly and the record of the concentrations found is one basis for estimation of the worker's body burden. Various methods of such estimation are discussed by Langham,[79] Lawrence,[80] Snyder,[81,82] and Beach and Dolphin.[83] None of these methods can claim any great accuracy and it is difficult to estimate the uncertainty involved in any one of them.

In 1958 a plutonium worker was killed in a criticality accident which did not involve internal exposure to Pu. The autopsy findings revealed that after some 6 years of work with Pu the distribution in the body was rather more typical of what might be expected following inhalation of a very insoluble aerosol. The concentration in the lungs was much greater than that in either

Table 10-10 Fifty-Year Dose Equivalent from Inhalation of Plutonium Compounds

Compound	DE to Bone (rem/μc)	DE to Pulmonary Region (rem/μc)
^{239}Pu dioxide	570	610
^{239}Pu nitrate	2.5×10^3	47

liver or bone and the pulmonary lymph nodes showed the highest concentration of all. In their report of this case, Foreman et al.[84] note that the ratio of ^{238}Pu to ^{239}Pu had changed during the period of 11 years when this exposure was received. The ratio of ^{238}Pu/^{239}Pu found in bone and in lymph nodes corresponded with that in the materials worked with in the early part of the period, while the ratio found in lungs corresponded with the ratio present in the materials used in the latter part of the period The ratio of ^{238}Pu/^{239}Pu in liver was between these extremes. This suggests that bone and lymph nodes have a slower elimination rate than liver which has, in, turn, a slower elimination rate than the lungs. Beach et al.[85] have discussed the practical aspects of a personnel-monitoring program for plutonium workers.

Table10-10 lists the limiting organ burdens and the 50-year dose equivalent from an intake by inhalation of 1 μc of each of two typical plutonium compounds. Figure 10-7 has been used to estimate the dose to the pulmonary region in each case. ^{239}Pu dioxide was assumed to be of Class Y with a long-term elimination half-time of 500 days from the pulmonary region and ^{239}Pu nitrate was taken to be of Class W with the longest elimination half-time as 38 days. These data are those recommended in the Report of the Task Group on Lung Dynamics.[31] It was assumed that 80% of the plutonium reaching the blood deposits in bone and that elimination is negligible. The limiting organ burdens for bone and for lungs are found to be 0.042 μc and 0.01 μc, respectively.

Acknowledgements. The author, who is entirely responsible for the material of this chapter and for its presentation, gratefully acknowledges the very substantial assistance of the editors and many helpful suggestions made by his colleagues. More particularly, he wishes to recognize the inspiration of K. Z. Morgan, whose knowledge and wisdom guided his first efforts in this subject area, and to acknowledge the extensive technical assistance of M. R. Ford, who might well be considered a joint author of the chapter.

Problems on Plutonium

1. Assume that urinary excretion following unit intake to blood is given by Langham's formula

$$E_u(t) = 0.0023t^{-0.77}$$

If a worker averages a certain daily urinary excretion E and it is believed the exposure occurred 3 to 5 years ago, what is the uncertainty in the estimate of body burden due to the uncertainity concerning the time of exposure? Would it be more, or less, if the time of the exposure was believed to be 10 to 15 years ago? How much more or less?

2. A Pu worker suffers a puncture wound which contains 0.01 μc of Pu which solubilizes slowly, 1% entering blood each day until the wound site is cleared. What would be the pattern of urinary excretion predicted by the Langham equation? Consider the same problem if 0.01 μc of Pu is inhaled and deposits and clears according to the revised lung model.

3. Assume Pu workers submit a urine sample once a month routinely. What would be the amount in the urine according to the model if an intake to blood corresponding to 13 weeks of exposure at MPC level occurred?

PROBLEMS

Problem on Section 10-2

1. Assume p is the probability that a person with a chronic body burden of 0.5 to 1.5 μc of radium will develop a bone sarcoma within 20 years. Show that

$${}_nC_k\, p^k(1 - p)^{n-k}$$

is the probability that exactly k among n such individuals will develop bone sarcomas within 20 years. Here

$${}_nC_j = \frac{n!}{k!\,(n-k)!}$$

is the binomial coefficient. Hence show that the probability of k or more of the n persons developing a bone sarcoma within 20 years is

$$\sum_{j=k}^{n} {}_nC_j p^j (1 - p)^{n-j}$$

2. Although precise numbers of cases at risk are not determined easily, it seems likely that the number of persons on record with a body burden in the range given in problem 1 is, perhaps, a few hundred; the number of cases of bone sarcoma recorded in this group during the first 20 years postexposure is probably in the range of 10 to 20. (*a*) If the probability of a bone sarcoma within 20 years postexposure is $p = 0.01$, is it cause for surprise that 10 or more cases are found among 100 persons at risk? (*b*) Would your answer be the same if an incidence of 5 or more cases were found among 50 persons at risk? (*c*) two cases among 20 persons at risk? (*d*) Assuming the above estimates of incidence and persons at risk to be correct, would you be surprised if $p = 10^{-4}$, 10^{-3}, or $p = 0.03$?

Note. It may be convenient to use the algebraic identity

$$\sum_{j=k}^{n} ({}_nC_j) p^j (1 - p)^{n-j} = k({}_nC_k) \sum_{j=0}^{n-k} (-1)^j \frac{({}_{n-k}C_j) p^{k+j}}{k + j}$$

for very small values of p. The value of the sum for two-digit values of p and $1 < n < 150$ is tabulated in Reference 86. The ratio of the $(j + 1)$th term to the ith term has absolute value

$$\frac{(n - k - j)(k + j)p}{(j + 1)(k + j + 1)}$$

and this will be less than 1 if

$$p < \frac{(k + j + 1)(j + 1)}{(k + j)(n - k - j)}$$

This last expression is an increasing function of j, and once the inequality is satisfied for the jth term, the error made in neglecting further terms is less than the first term neglected.

3. Discuss the reasons why a strict application of probability to such a situation as that mentioned in problem 2 is questionable, particularly with regard to (*a*) nonrandomness of the sampling, (*b*) nonconstancy of p within the range of body burdens, (*c*) a "background" or natural incidence of bone sarcomas, and (*d*) incomplete recording of data. Would certain of these biases tend to increase—or decrease—the estimate of p from such a study?

4. The natural incidence of bone sarcomas is about 1 per 10^5 persons per year.[87] A typical body burden of radium may be about 10^{-7} μc ^{226}Ra/gm of Ca in the skeleton of the general population,[69,88] the concentration, according to Hallden et al.[70] being nearly independent of age. If all bone sarcomas are attributed to the presence of radium, show that the probability a person having this constant level of ^{226}Ra present in his skeleton throughout life will develop a bone cancer sometime during his life is $L \times 10^{-5}$ where L is the average life span in years of this population.

Hint: Consider a stable population with a death rate $D(x)$; that is, the probability of death of an individual of age x in the small time interval dx is $D(x)\,dx$. Let the birth rate in such a population be N persons per year, N being constant. Then the number of persons with age between y and $y + dy$ is $N\,dy \int_y^\infty D(x)\,dx$ and hence the population size is $N \int_0^\infty dy \int_y^\infty D(x)\,dx$. Integrating by parts one obtains $N \int_0^\infty y\,D(y)\,dy = NL$ since $L = \int_0^\infty y\,D(y)\,dy$ is the definition of the mean life-span.

5. It is sometimes assumed that at low dose levels the probability of an effect is in proportion to the dose, or—what is much the same in the case of radium—to the chronic concentration of the radionuclide in the tissue of interest. If the incidence of bone sarcoma mentioned in problem 4 is attributed to the natural background of ^{226}Ra present in the skeleton and if the linear hypothesis is assumed to hold, what would be the probability that a person constantly exposed at a level equivalent to 0.1 μg of ^{226}Ra in the adult skeleton would develop bone cancer? Discuss this estimate in the light of the fact that there are some hundreds of persons who have a chronic body burden of ^{226}Ra between the limits of 0.1 to 0.5 μc and at most 1 or 2 cases of bone cancer have been found in this group.

6. In some studies of radiation effects the hypothesis has been made that the probability of a given effect per year is constant after exposure, that is, following an exposure of R rad, the probability that a malignancy due to this exposure will occur in any subsequent year is pR, p being a constant. Of course, if subsequent exposure occurs the corresponding risk is added. Assuming this theory is correct, and that the population is that described in problem 4 show the following: (*a*) the probability that an individual will develop a malignancy during his Kth year due to his body burden of ^{226}Ra is about $3pK \times 10^{-3}$. Recall a body burden of 0.1 μg of ^{226}Ra produces a dose rate of about 3 rad/yr. (*b*) Show that the expected number of malignancies in the population per year due to the specified level of ^{226}Ra is given by

$$3pN \times 10^{-3} \int_0^\infty dx \int_x^\infty D(y)\,dy = \frac{3}{2} pN \times 10^{-3} \int_0^\infty x^2\,D(x)\,dx = \frac{3pN}{2} 10^{-3}(L^2 + \sigma^2)$$

where $L = \int_0^\infty x\,D(x)\,dx$ is the mean life-span of the population, N = total number of births per year, and $N\sigma^2 = \int_0^\infty (x - L)^2\,D(x)\,dx$ is the variance of the distribution of deaths. (*c*) Assuming the incidence level of bone cancer as in problem 4 and taking $L = 70$ years and $\sigma = 20$ years show that $p \cong 10^{-4}$ malignancies per rad per year according to this theory. (*d*) Does the above derived value of p seem consistent with the data on the radium patients and dial painters, that is, about 200 with a body burden of 0.5–1.5 μc of ^{226}Ra and with an incidence of 10–20 bone cancers in 20 years or more?

Hint: If $Q(x)$ is the number that have not developed a malignancy x years after exposure, and if we neglect other causes of death, we have $[dQ(x)/dx] = -15pxQ(x)$ if all had a body burden of 0.5 μc. Solve this differential equation for $Q(0) = 200$ and evaluate for $p = 10^{-4}$ and $x = 20$ years.

Problems on Section 10-3

7. Compute the average dose in a tissue cylinder of unit density from an α-particle whose track is the axis of the cylinder with radius equal to the maximum range of a δ-ray from the α-particle.

8. A spherical particle of $^{239}PuO_2$ of diameter 1 μ is embedded in tissue of unit density. Neglecting absorption of energy in the particle, (*a*) calculate the average dose in a surrounding tissue sphere of radius equal to the range of the α-particle. (*b*) How does the answer vary with the diameter of the particle? (*c*) Does the answer depend on the density of the tissue? (*d*) How serious is the neglect of self-absorption in the source?

Hint: The range of an α-particle of E MeV in tissue (H_2O) is given approximately by $R = 6E^{3/2}$ microns.

Problems on Section 10-4

9. Calculate the intensities of the converted photons from the data in Fig. 10-1.

10. Use the data shown on the decay scheme below for ^{64}Cu to compute the energy emitted per disintegration.

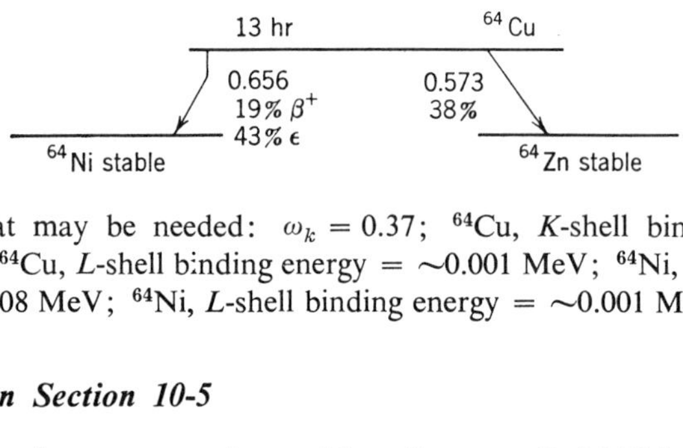

Other data that may be needed: $\omega_k = 0.37$; ^{64}Cu, K-shell binding energy = ~0.009 MeV; ^{64}Cu, L-shell binding energy = ~0.001 MeV; ^{64}Ni, K-shell binding energy = ~0.008 MeV; ^{64}Ni, L-shell binding energy = ~0.001 MeV.

Problems on Section 10-5

11. A source of monoenergetic particles of energy E_0 MeV is uniformly distributed in an infinite slab of thickness T. Assume that the track of a particle is a straight-line segment of length r. Show that if I particles are emitted per unit volume the expected number of particles crossing a unit area of slab surface is given by $rI/4$ if $r < T$ and by $TI(2r - T)/4r$ if $T < r$. The particles emitted from a column of material of volume T (that is, those for a column having a base of unit area) have a total potential track length of TIr. Show that the fraction of this total track length within the slab is given by $1 - (r/4T)$ if $r < T$ and by $[2T \ln (r/T) + 3T]/4r$ if $T < r$. The corresponding calculations for a sphere are much more tedious but are essentially elementary. The results are the following: Assuming an intensity of I particles emitted per unit volume in a sphere of radius a, the number of particles crossing a unit surface area is $rI(12a^2 - r^2)/48a^2$ if $r < 2a$ but is, obviously, $aI/3$ if $2a < r$. The fraction of the total potential path length (that is, assigning a length r for each particle emitted) lying within the sphere is given by $(64a^3 - 24a^2r + r^3)/64a^3$ if $r < 2a$ but is given by $3a/4r$ if $2a < r$.

12. Assume that each particle crossing the boundary in problem 11 carries away the energy E_0. By this assumption, estimate the average dose in the slab for the following cases: (*a*) α-particles of energy 5 MeV and (*b*) electrons of energy 2 MeV, 1 MeV, 0.5 MeV, 0.11 MeV, 0.01 MeV?

13. The estimates in problem 12 are lower bounds for the dose. Obtain upper bounds by assuming that all of the energy of the particles is absorbed in the slab.

14. (*a*) Prove that if the tracks of particles are straight and if the thickness of a slab is greater than the range of the particles, then regardless of the distribution of the source in the slab, the energy absorbed in the slab, E_{ab}, and the energy emitted, E_{em}, satisfy the relationship

$$\frac{1}{2} \leq \frac{E_{ab}}{E_{em}} \leq 1$$

Hint: Show that not more than half of the particles emerge from the slab. (*b*) Can both bounds be realized? (*c*) Generalize the result in (*a*) for a large body with a surface which is not too irregular, for instance, a sphere or cube.

15. Use the formulas of problem 11 to obtain numerical estimates for the cases of an α-emitter (energy of 5 MeV) and an electron emitter (energies 1, 0.5, 0.11, 0.05 MeV) in a spherical organ of the size of the kidney (mass of one kidney is about 150 gm) and for a clavicle (the clavicle is assumed to be a slab of bone 1 cm thick and of density 1.5 gm/cm^3). Edge effects are to be neglected. Assume the range of the electron is the same number of gm/cm^2 as in soft tissue and water.

16. Compute the energy escaping from the slab in problem 12 under the assumption that rate of energy loss along the track is constant.

Problems on Sections 10-7 and 10-8

17. It has been suggested[49] that the retention of ^{90}Sr in adult bone can be represented by

$$R(t) = I \times [0.22e^{(-0.693/3)t} + 0.03e^{(-0.693/44)t} + 0.05e^{(-0.693/4000)t}] \times e^{(-0.693/10^4)t}$$

with I as the number of μc ingested and $e^{-(0.693/10^4)t}$ representing radioactive decay. On the basis of this hypothesis, compute the dose during (*a*) the first quarter-year, (*b*) the first year, and (*c*) the first 50 years following intake. What percentage of these doses is produced by each of the three exponential terms? Which term has most influence on the quarterly dose? The 50-year dose? Make the same estimation for ^{89}Sr.

18. It has been suggested[64] that the retention of Ra in bone is describea by the formula $R(t) = I \times 0.15t^{-0.5}$ where I is the amount ingested. The infinite value of $R(0)$ is, of course, spurious and may be eliminated by modifying the formula in various ways.[89] Without modification, compute the dose to bone from ^{226}Ra during: (*a*) the first quarter, (*b*) the first year, and (*c*) the first 50 years following intake of 1 μc. What fraction of these doses is delivered during the first day? Make the corresponding estimates for ^{224}Ra.

19. Retention, corrected for radioactive decay, is often described as being of the form $at^{-\alpha}$. Thus

$$R(t) = at^{-\alpha}e^{-\lambda t}$$

With λ as the radioactive decay constant insert this function in formula (10-12) and show that the integral can be reduced to the difference of two incomplete gamma functions. Tables may be used to evaluate these.[71,72]

20. Assume a parent radionuclide has a succession of daughter radionuclides. Show that the microcurie amounts present in the tissue are given by the solution of the differential equations

$$\frac{dA_0}{dt} = -(\lambda_0 + \mu_0)A_0$$

$$\frac{dA_1}{dt} = \lambda_1 A_0 - (\lambda_1 + \mu_1)A_1$$

$$\frac{dA_2}{dt} = \lambda_2 A_1 - (\lambda_2 + \mu_2)A_2$$

Solve these, assuming $A_0(0) = 1$, $A_1(0) = A_2(0) = 0$, and compare with the formulas in *ICRP Publ. 2.*[8] In these formulas λ_i is the radioactive decay constant for the ith daughter and μ_i is the constant for biological elimination.

21. Prove that in the case of a radionuclide of long half-life in the body, for example, ^{90}Sr, ^{226}Ra, ^{239}Pu, the dose during the first year will be at least one-fiftieth of the fifty-year dose.

Hint: One may assume $R(t)$ to decrease monotonically with time.

Problem on Section 10-9

22. Estimate the 50-year dose to the thyroid (bone) from intake of 1 μc of ^{131}I (^{90}Sr, ^{226}Ra, ^{239}Pu) assuming the intake is via inhalation and that the $(\text{MPC})_a$ for continuous exposure (168 hrs/wk) based on this organ is 3×10^{-9} μc/cc (4×10^{-10}, 10^{-11}, and 6×10^{-13}) respectively.

Problems on Section 10-10

23. Compute the total microcurie days of residence in the *N-P*, *T-B*, and *P* regions and in the lymph for the first t days following an intake of I μc. Also, compute the number of microcuries present in the three regions of the lung at time t days if exposure has been at a constant level of I μc/day during the time 0 to t days. Compare the above two sets of formulas.

24. The area of the *T-B* region is estimated to be about 10–20 m^2. If a layer of tissue 45-μ thick and of unit density be based on this surface, the total mass might be approximately 450–900 gm. This is much too large for the mass of the *T-B* region. What do you conclude concerning the fraction of energy absorbed in the *T-B* region from an alpha emitter (range 45 μ) uniformly distributed over the surface of the region?

25. The minute volume of air breathed varies with the activity of the individual. Typical rates are 11 liters/min for a resting state up to 32 liters/min corresponding to a state of vigorous activity. What other characteristics of individuals significantly affect the dose they receive when they are exposed to supposedly identical situations? What can or should the health physicist do to take account of the effect of individual differences on the doses radiation workers receive?

26. Experimental data suggest that deposition on a surface is related to the velocity of impaction, and this, in turn, is a function of the shape of the particle. Bagnold[90] has indicated that for desert sand the relative terminal velocity is about 50% of that of a spherical particle of the same mass and density. Assuming this relation, calculate the aerodynamic diameter of a grain of sand of mass 1 mg. Assume the density of sand (SiO_2) is 2.6 gm/cm^3.

27. Given an intake of I μc to the lungs show that the amount transported from the lymph nodes to the blood according to the revised lung model is

$$aD_3\frac{T^a}{T_b^{\,a}} + cD_4\frac{T^c}{T_b^{\,c}} + eD_5\frac{T^e}{T_b^{\,e}} + ihD_5\left(\frac{T^h}{T_b^{\,h}}\right)^2 \mu\text{c}$$

Problems on Section 10-11

23. Assume the GI tract is the critical organ for ^{90}Sr titanate and that the (MPC)$_a$ is 4×10^{-6} μc/cc. Is the rule developed in Section 10-11 valid with the GI tract as critical organ? If so estimate the dose from an intake of 1 μc of ^{90}Sr titanate.

29. Many experimenters have demonstrated that when a radionuclide is injected some fraction of the injected dose is excreted in feces and this is termed the endogenous excretion. Assume the ratio of urinary excretion to endogenous fecal excretion is found to be r and is reasonably constant in time. Further assume that a state of equilibrium is reached for ingestion of a certain nuclide, e.g., one may obtain data on a stable element with which an adult individual of regular habits may be presumed to be in equilibrium. If f_1 is defined as the fraction of ingested nuclide which is absorbed from gut to blood and if the individual excretes amounts E_f and E_u per day in feces and urine, respectively, show that $f_1 = (I - E_f + rE_u)/I$ where I is the daily intake. Note this argument assumes ingested material reaching blood metabolizes as does the material injected intravenously.

30. Assume an intake of I μc is ingested. The amount leaving the stomach according to the model is then $Ie^{-\lambda_r/24}$ μc where λr is the radioactive decay constant. This quantity is then the input to the *SI*, that is, the quantity A of formulas (10-23) and (10-24). Also the dose to the stomach wall is given by

$$\frac{51I \times (1 - e^{-\lambda_r/24})\varepsilon}{2M\lambda_r} \text{ rad}$$

The factor 2 in the denominator is used because ε is taken to estimate dose to the contents of stomach and the stomach wall would only get half this dose. Complete this calculation to derive formulas for average dose to each section of the GI tract.

31. Use the above formulas to estimate the dose to each section of the GI tract from an ingestion of 1 μc of (*a*) $^{89}SrCl_2$ ($f_1 = 0.3$), (*b*) $^{89}SrTiO_3$ ($f_1 = 0.01$), (*c*) $^{239}PuO_2$ ($f_1 = 10^{-4}$).

32. Assume I μc of ^{90}Sr are ingested. Derive formulas for the dose delivered to the various sections of the GI tract due to the daughter ^{90}Y ($f_1 = 10^{-4}$).

33. Assume an insoluble (i.e. $f_1 \approx 0$) β-emitter of fairly short radioactive half-life is ingested. How short must this half-life be so that the average dose to the *SI* equals the average dose to the *LLI*? Assuming the material moves at a uniform rate along the *SI*, how much greater is the dose at the entrance than at the exit portions of the *SI*? For what radioactive half-life will the dose to the stomach just equal the average dose to the *SI*?

34. Assume radionuclide (2) is a daughter of radionuclide (1) and these subscripts are used to indicate the constants to each of them. If amounts A_1 and A_2 (μc) enter the *SI*, show that the microcurie-days of residence in the section is given by $A_1[(1 - e^{-\lambda_1 T})/\lambda_1]$ for the parent and

$$\frac{A_1\lambda_2^r}{\lambda_2 - \lambda_1}\left[\frac{1 - e^{-\lambda_1 T}}{\lambda_1} - \frac{1 - e^{-\lambda_2 T}}{\lambda_2}\right] + \frac{A_2}{\lambda_2}(1 - e^{-\lambda_2 T})$$

for the daughter. These quantities may be used in formulas (10-12) and (10-13) in place of $I\int_0^t R(\lambda)\,d\lambda$, In these formulas λ_1 and λ_2 are the total elimination rates in the *SI*, that is, $\lambda_1 = \lambda_1^r + \lambda_1^b$ etc. The amounts entering the next section of the tract equal $A_1e^{-\lambda_1 T}$ for the parent and $[A_1\lambda_2^r(e^{-\lambda_1 T} - e^{-\lambda_2 T})/(\lambda_2 - \lambda_1)] + A_2e^{-\lambda_2 T}$ for the daughter. For other sections of the tract absorption is considered negligible and so $\lambda_1{}^b = \lambda_2{}^b = 0$. Generalize these formulas for the case of two or more daughters. A general formulation is given in *ICRP Publ. 2*, p. 19, but the procedure there is formulated to give the dose at a specified site in the section and not the average dose over the section.

REFERENCES

1. Looney, W. B., *Am. J. Roentgenol. Radium Therapy Nucl. Med.* **72,** 838 (1954).

2. Schubert, J., and R. E. Lapp, *Radiation: What It Is and How It Affects You*, p. 112, Viking Press, New York, (1957).

3. Norris, W. P., T. W. Speckman, and P. F. Gustafson, *Am. J. Roentgenol. Radium Therapy Nucl. Med.* **73,** 802 (1955). See p. 785.

4. "Safe Handling of Radioactive Luminous Compounds," *Natl. Bur. Std. Handbook* **27,** U.S. Dept. of Comm., Washington (1941).

5. "Maximum Permissible Amounts of Radioisotopes in the Human Body and Maximum Permissible Concentrations in Air and Water," *Natl. Bur. Std. Handbook* **52,** U.S. Dept. Comm. Washington (1953).

6. "Recommendations of the International Commission on Radiological Protection," *Brit. J. Radiol. Suppl.* **6** (1955).

7. "Maximum Permissible Body Burdens and Maximum Permissible Concentrations in Air and in Water for Occuptional Exposure," *Natl. Bur. Std. Handbook* **69,** U.S. Dept. Comm. Washington (1959).

8. Report of Committee II on Permissible Dose for Internal Radiation, *ICRP Publ. 2*, Pergamon, London (1959).

9. Wiles, Donald, R., *Nucleonics* **11** (11), 32 (1953).

10. Loevinger, R., E. M. Japha, and G. L. Brownell, "Discrete Radioisotope Sources," pp. 701–703 in *Radiation Dosimetry*, G. J. Hine and G. L. Brownell, eds., Academic, New York (1956).

11. James, M. F., B. G. Steel, and J. S. Story, "*Average Electron Energy in Beta Decay*", *U.K.A.E. Authority Rept.* AERE-M 640, Harwell, Berkshire (1960).

12. Slack, L., and K. Way, *Radiations from Radioactive Atoms in Frequent Use*, USAEC, Washington, (1959).

13. Landolt-Börnstein, *Numerical Data and Functional Relationships in Science and Technology*, New Series, Vol. 1, Energy Levels of Nuclei: A = 5 to A = 257, A. M. Hellwege and K. H. Hellwege, eds. Springer-Verlag, Berlin (1961).

14. Boyett, R. H., and B. R. Fish, Abstr. P/22, *Health Phys.* **11,** 803 (1965).

15. Ellett, W. H., A. B. Callahan, and G. L. Brownell, *Brit. J. Radiol.* **37,** 45 (1964).

16. Ann. Progr. Rept., COO-119-232, Radiobiology Div., University of Utah, Salt Lake City, (1965).

17. Dougherty, T. F., W. S. S. Jee, C. W. Mays, and B. J. Stover, eds. *Some Aspects of Internal Irradiation*, Pergamon, London (1962), p. 50.

18. Pinson, E. A., and W. H. Langham, *J. Appl. Physiol.* **10,** 108 (1957) see p. 125.

19. Doniach, I., *Health Phys.* **9,** 1357 (1963).

20. "Background Material for the Development of Radiation Protection Standards," *Federal Radiation Council Rept.* No. 5, U.S. Govt. Printing Office, Washington, July, 1964.

21. Dougherty, T. F., W. S. S. Jee, C. W. Mays, and B. J. Stover, eds., *Some Aspects of Internal Irradiation*, Pergamon, London (1962), pp. 499–516.

22. Cook, M. J., K. Z. Morgan, and A. G. Barkow, *Am. J. Roentgenol. Radiotherapy Nucl. Med.* **75,** 1177 (1956).

23. Richmond, C. R., J. E. Furchner, and G. A. Trafton, *Health Phys.* **7,** 219 (1962).

24. Snyder, W. S., M. J. Cook, and M. R. Ford, *Health Phys.* **10,** 171 (1964).

25. Dow, E. C., and J. B. Stanbury, *J. Clin. Invest.* **39,** 885 (1960).

26. Butler, H. L., and J. H. Leroy, *Health Phys.* **11,** 283 (1965).

27. Samachson, Joseph, *Clin. Sci.* **25,** 17 (1963).

28. Bair, W. J., D. H. Willard, J. P. Herring, and L. A. George, II, *Health Phys.* **8,** 639 (1962).

29. Barnes, D. W. H., Margaret Bishop, G. E. Harrison, and Alice Sutton, *Intern. J. Radiations Biol.* **3,** 637 (1961).

30. Rept. of Comm. II Permissible Dose Internal Radiation, *ICRP Publ. 2*, p. 34, Pergamon, London (1959).

31. Task Group on Lung Dynamics, "Deposition and Retention Models for Internal Dosimetry of the Human Respiratory Tract," *Health Phys.* **12,** 173 (1966).

32. Pochin, E. E., *Health Phys.* **12,** 563 (1966).

33. Eve, I. S., *Health Phys.* **12,** 131 (1966).

34. Hayes, R. L., J. E. Carlton, and W. R. Butler, Jr., *Health Phys.* **9,** 915 (1963).

35. Sullivan, M. F., P. L. Hackett, L. A. George, and R. C. Thompson, *Radiation Res.* **13,** 343 (1960).

36. Rept. Comm. II Permissible Dose Internal Radiation, *ICRP Publ. 2*, p. 29, Pergamon London (1959).

37. Spector, W. S., *Handbook of Biological Data*, WADC Tech. Rept. 56-273; ASTIA Document No. AD 110501, pp. 70–75, 1956.

38. Pinson, E. A., and W. H. Langham, *J. Appl. Physiol.* **10,** 108 (1957).

39. Osborne, R. V., *Health Phys.* **12,** 1527 (1966).

40. Quiring, Daniel P., "Surface-Area Determination," in *Medical Physics*, Vol. 1, Otto Glasser, ed., pp. 1490–1494, Year Book Publ., Chicago, Illinois.

41. Brody, S., J. E. Comfort, and J. S. Mathews, *Res. Bull.* **115,** Univ. of Missouri, Columbia, Coll. Agr., Agr. Expt. Station (1928).

42. Snyder, W. S., B. R. Fish, S. R. Bernard, Mary R. Ford, and J. R. Muir, in Oak Ridge National Laboratory Rept., ORNL-3697 (1964).

43. Riggs, D. S., *Pharmacol. Rev.* **4,** 284 (1952).

44. Morgan, A., D. J. Morgan, and G. Arkell, Proc. Second Intern. Conf. Inhaled Particles Vapors, Pergamon, New York (1965).

45. Cook, Mary J., and W. S. Snyder, in Oak Ridge National Laboratory Rept., ORNL-3849 (1965).

46. Nuclear Data Sheets of the Natl. Acad. Sci.–Natl. Res. Council: NRC 60-3-35, NRC 60-3-78, and NRC 60-4-26, Washington (1960).

47. James, M. F., B. G. Steel, and J. S. Story, "Average Electron Energy in Beta Decay," *U.K. At. E. Authority Res. Group Rept. No. AERE-M-640* Harwell, Berkshire (1960) pp. 9–10.

48. Snyder, W. S., and M. J. Cook, in Oak Ridge National Laboratory Rept., ORNL-3697 (1964).

49. Snyder, W. S., M. J. Cook, and M. R. Ford, *Health Phys.* **10,** 171 (1964). See p. 176.

50. Bishop, Margaret, G. E. Harrison, W. H. A. Raymond, and Alice Sutton, *Intern. J. Radiation Biol.* **2,** 125 (1960).

51. McClellan, R. O., and L. K. Bustad, Hanford Laboratory Rpt. HW-80500, p. 100 (1964).

52. Comar, C. L., R. H. Wasserman, and M. M. Nold, *Proc. Soc. Exptl. Biol. Med.* **92,** 859 (1956).

53. Kulp, J. L., *Proc. Conf. Germantown, Maryland, Nov. 15–17, 1961*, TID-7632 (1962), Book 2, p. 457.

54. Rivera, Joseph, *Proc. Second AEC Conf. Radioactive Fallout Nucl. Weapons Tests, Germantown, Maryland, Nov. 3–6, 1964*, CONF-765 (1965), Paper No. 7.

55. Rivera, Joseph, and John H. Harley, Health and Safety Laboratory Rept. HASL-163, U.S. Atomic Energy Commission, New York Operations Office, New York (1965).

56. Bryant, F. J., A. C. Chamberlain, and G. S. Spicer, *Brit. Med. J.* p. 1371 (1958).

57. Rivera, Joseph, Health and Safety Laboratory Rept. HASL-138, U.S. Atomic Energy Commission, New York Operations Office, New York (1963).

58. Mitchell, H. H., T. S. Hamilton, F. R. Steggerda, and H. W. Brown, *J. Biol. Chem.* **158,** 625 (1945).

59. Engstrom, A., R. Bjornerstedt, C. J. Clemedson, and A. Nelson, *Bone and Radiostrontium*, John Wiley and Sons, New York, 1957.

60. Spiers, F. W., *Brit. J. Radiol.* **39,** 216 (1966).

61. Spiers, F. W., *Radiation Res.* **28,** 624 (1966).

62. Fujita, M., A. Yabe, K. Ueno, M. Oshine and N. Okuyama, *Health Phys.* **9,** 407 (1963).

63. "Evaluation of Radiation Doses to Body Tissues from Internal Contamination due to

Occupational Exposure." Rept. prepared by a Task Group of Comm. 4 of the Intern. Comm. Radiol. Protect. (1966), in press.

64. Norris, W. P., T. W. Speckman, and P. F. Gustafson, *Am. J. Roentgenol. Radium Therapy Nucl. Med.* **73,** 785 (1955). See p. 797.

65. Mays, C. W., oral communication.

66. Marshall, J. H., Abstract P9, *Health Phys.* **12,** 1161 (1966).

67. Mays, C. W., M. A. Van Dilla, R. L. Floyd, and J. S. Arnold, *Radiation Res.* **8,** 480 (1958).

68. Lucas, H. F., Jr., in Argonne National Laboratory Rept. ANL-6297 (1960).

69. Rajewsky, R., V. Belloch-Zimmerman, E. Löhr, and W. Stahlhofen, *Health Phys.* **11,** 161 (1965).

70. Hallden, N. A., I. M. Fissene, and J. H. Harley, *Science* **140,** 1327 (1963).

71. Pearson, K., *Tables of the Incomplete Γ-Function,* Cambridge Univ. Press, London (1946).

72. Rowland, R. E., and C. J. Leuer, Argonne National Laboratory Rept. ANL-6398, p. 24 (1961).

73. "Recommendations of the International Commission on Radiological Protection," ICRP Publication 6, Pergamon, London (1964) (as amended 1959 and Revised 1962).

74. Bernard, S. R., *Health Phys.* **1,** 288 (1958).

75. Eve, I. S., *Health Phys.* **10,** 773 (1964).

76. Ford, M. R., Oak Ridge Natl. Lab. Rept. ORNL-3697, pp. 202–207 (1964).

77. Weeks, M. H. et al., *Radiation Res.* **4,** 339 (1956).

78. Langham, W. H., S. H. Bassett, P. S. Harris, and R. E. Carter, Los Alamos Sci. Lab. Rept. LA-1151 (1950).

79. Langham, W. H., *Assessment of Radioactivity in Man,* Vol. 2, IAEA, pp. 565–581 (1964).

80. Lawrence, J. N. P., *Health Phys.* **8,** 61 (1962).

81. Snyder, W. S., Argonne National Laboratory Rept. ANL-6637, p. 13 (1962).

82. Snyder, W. S., *Health Phys.* **8,** 767 (1962).

83. Beach, S. A., and G. W. Dolphin, *Assessment of Radioactivity in Man,* Vol. 2, IAEA, pp. 603–615 (1964).

84. Foreman, N., W. Moss, and W. Langham, *Health Phys.* **2,** 326 (1960).

85. Beach, S. A., G. W. Dolphin, K. P. Ducan, and H. J. Dunster, *Health Phys.* **12,** 1671 (1966).

86. "Tables of the Cumulative Binomial Probabilities," Ordance Corps Pamphlet ORDP 20-1 (1952).

87. "Rept. U.N. Sci. Comm. Effects Atomic Radiation," *Official Records, Nineteenth Session,* Suppl. No. 14 (A/5814) (1964).

88. Hursh, J. B., and A. A. Gates, *Nucleonics* **7**(1), 47 (1950).

89. Marshall, J. H., *J. Theoret. Biol.* **6,** 392 (1964).

90. Bagnold, R. A., *The Physics of Blown Sands and Desert Dunes,* Methuen, London (1954).

CHAPTER 11

Radiation Biophysics

*P. R. J. Burch**

11-1 INTRODUCTION

In radiation protection we are concerned with the biological effects of ionizing radiation in both the individual and the species. We must prevent the harmful effects to the individual from radiation rising above certain acceptable limits, and we must also safeguard the genetic heritage of the species by ensuring that our children, and their successors, will not be burdened with a heavy and avoidable load of genetic defects. In both areas of responsibility we are committed to long-term predictions. Thus a moderate dose of radiation to the individual may have carcinogenic or leukemogenic consequences that are delayed by a period of years or decades. The time lag in the expression of genetic damage is generally much greater. At the earliest the damage will appear in the next generation, but where recessive traits are involved, manifestations may be delayed for many generations.

To make wise decisions now—and to know that they are soundly based—we require accurate and detailed knowledge of every stage of those intricate processes that start with the absorption of ionizing radiation in the tissues of an individual and which may culminate either in a leukemia or shortened life-span in the irradiated person, or, possibly, in a mentally defective baby several generations hence.

The following stages in the development of radiation-induced biological damage may be delineated:

1. The initial interaction of photons, charged particles, and uncharged particles with body tissues.

* I am indebted to my colleague, Miss Susan Chesters, for helpful comments on my manuscript and for providing a formal analysis (not illustrated here) of the problem of multiple dependent-type hits.

2. The spatial and temporal distribution in the organism—from the macroscopic down to submicroscopic levels—of the energy transferred from the radiation.
3. The production in tissues of chemically active species such as electrons, ions, radicals, and molecules.
4. The interaction of such physically and chemically active agents with biologically important molecules or macromolecules such as DNA.
5. The effect of damage to biologically important molecules on the economy of the cell.
6. The influence of endogenous protective, repair, and recovery processes.
7. The influence of damaged cells, their daughters, and their products on the economy of the body.
8. The effect of damaged genes on their inheritors.
9. The impact of injured individuals on society.

Unfortunately the only stage about which our knowledge and understanding is reasonably satisfactory and complete is stage 1. Although we may in fact succeed in making fairly accurate predictions about the biological outcome of an irradiation without fully understanding all the details of the intermediate stages, we shall generally not *know* how reliable our calculations are until our ignorance of detail has been eliminated.

Radiation biophysics is concerned with the physical basis of the biological changes produced by ionizing radiation. The determination of the radiation dose to tissues, and to substructures within tissues, may be regarded as the first objective in radiation biophysics; this aspect of the subject has been covered in previous chapters.

However the radiation dose to a tissue—that is, the radiation energy dissipated per unit mass of the tissue—is not the only physical factor influencing the extent of biological change. Another important factor is the *quality* of the radiation.

11-2 RADIATION QUALITY

Quality is a term that refers to the way in which radiation energy is distributed at the track level. The energy of natural α-particles is densely distributed along a column surrounding the path of the primary particle, and along the paths of the many δ-tracks branching off from the main column. Much of the energy of a high-energy electron is, by contrast, sparsely distributed along its path, although, occasionally, much more densely ionizing secondary or δ-tracks are produced as the result of more energetic collisions. If a biological (or other) change should require a high (or sparse) localization of energy in a particular region or regions, then the dose of radiation required to accomplish the change will depend upon its quality. The parameter *relative biological effectiveness* (RBE) is used to describe the effectiveness of one radiation relative to that of another of a different quality. If the dose of radiation A required to produce a specific biological end point is D_A rad, and if the dose of radiation B required to produce the same end point, in an identical system, is D_B rad, then the RBE of radiation B in comparison with radiation A is defined by $RBE_{B/A} = D_A/D_B$.

X-rays generated by 200 to 300 kvp X-ray machines have often been used as a reference radiation (radiation A) but because of the lack of precision in this specification, it has been proposed[70] that ^{60}Co γ-radiation—with its greater reproducibility and, apart from scattered radiation, its defined spectrum—would provide a more appropriate reference standard.

In certain test systems, the amount of change produced by radiation of a given quality is directly proportional to the amount of radiation energy delivered to the test system. This linear dose–response relationship sometimes holds over a very wide range of dose and of dose rate. Thus, when the Fricke ferrous sulfate chemical dosimeter is irradiated by ^{60}Co γ-rays, about 15.5 ferrous ions are oxidized to ferric ions for every 100 eV of energy dissipated, over a wide range of dose and dose rate. If the same kind of chemical solution is irradiated by ^{210}Po α-particles, a linear relationship between dose and ferrous ion oxidation is still obtained, but the yield is drastically reduced to about 6 ferrous ions oxidized for every 100 eV of α-particle energy dissipated. In this dosimeter, therefore, the effect of radiation measured in terms of the yield of ferric ions is quality dependent. That is, when the energy is largely confined to a dense column along the track of the α-particle, it is less effective at oxidizing ferrous ions than when it is more sparsely spread along the tracks of electrons.

If, on the other hand, we measure the yield of molecular hydrogen in this chemical dosimeter, we find that ^{210}Po α-particles are more efficient in generating this product (about 1.5 molecules H_2 per 100 eV) than ^{60}Co γ-rays (about 0.4 molecules H_2 per 100 eV.) This example from radiation chemistry illustrates a further complication: there is no simple, universal relationship between the efficiency of a radiation and its quality. Densely ionizing radiations can be more efficient than lightly ionizing ones in producing certain effects, but they are less efficient at producing others. This inconvenience is not confined to chemical systems; radiobiologists are equally well acquainted with it. Thus, if we consider the cytotoxic (cell-killing) effect of radiation, we find that α-particles are in general conspicuously more effective (per rad) than high-energy β-particles or γ-rays.[4,5] However, in radiation cytotoxicity the RBE generally changes with the level of survival that is chosen as the biological end point.

Whereas the survival of mammalian cells irradiated by natural source α-particles can be represented by a negative exponential function over a wide dose range, the survival curve obtained when the same types of cells are irradiated by X- and γ-rays, generally shows a shoulder at low dose levels, although at high doses, a linear log (survival) versus dose relationship is often attained. Hence at low levels of cytotoxicity the RBE of α-particles relative to γ-rays tends to a maximum as $D \rightarrow 0$, but at high levels of cell killing, the RBE often tends to a constant (minimum) value. However when certain types of gene mutation are induced by radiation, the efficiency can be in the reverse order to the cytotoxic effect, and radiations can be listed in the following sequence of descending RBE: high-energy β-, γ-, and X-rays; low-energy X-rays; fast neutrons; natural α-particles.[54]

It is interesting to refer to the recommendations of the ICRP in this context.[51] The Commission assumed that natural α-particles are always more hazardous than fast neutrons and that fast neutrons are always markedly more hazardous than high-energy X-rays, γ-rays, or β-particles. This assumption may not be valid in all circumstances. Thus, in several important biological situations (mutagenesis, leukemogenesis, carcinogenesis, and aging phenomena) we may be concerned with specific gene mutations that have been induced in those cells that survive the cytotoxic action of radiation.[13,16] It could well happen that at certain ranges of dose and/or dose rate, radiations of low average linear energy transfer (LET) are more efficient than those of high LET at inducing pathological lesions. Indeed, Upton et al.[75] have found that fast neutrons are generally *less* effective than ^{60}Co γ-rays at inducing ovarian tumors in RF mice over fairly wide ranges of dose and dose rate. Theoretical expectations are therefore given direct experimental support.

Other experiments by Upton et al., in which the induction of myeloid leukemia in the RF strain of mouse was observed, show that doses of neutrons of energy from 1 to 14 MeV generally have a rather similar effect to comparable *acute* doses (0–500 rad) of X- or γ-rays[74] although the detailed picture is too complicated for easy generalization. These various observations[74,75] also confirm that *dose rate* can be of great importance. Where lightly ionizing radiations are concerned, and for doses of up to about 500 rad, X- and γ-rays are much more effective under acute conditions ($\sim$100 rad/min) than at low dose rates ($\sim$0.1 rad/min or less). Fast neutrons, on the other hand, often maintain their effectiveness down to low dose rates; indeed at high doses ($>$500 rad) they can be more effective at low dose rates than at high.[74]

It is doubtful whether the situation is any less complex in man and it must be concluded that, in addition to dose, the quality and the dose rate of an irradiation are both exceedingly important factors in the context of radiation protection.

Because the formal definition of RBE assumes a particular end point, it is illogical to use this parameter in the context of radiation protection when we wish to compare and/or summate the hazards of radiations of different qualities and various dose levels. The parameter QF, meaning *quality factor*, has been recommended by the ICRP to meet this need. An attempt has also been made to relate the QF value of a radiation to its average LET in water.[51] The *dose equivalent* of an irradiation is defined by the product of the dose in rads, and the *quality factor*, QF. The many complications described above are perforce ignored in the practical context of radiation protection, but in most instances, the simplifications and the neglect of dose-rate effects probably lead to very generous safety margins.

Specification of Radiation Quality

Although any general attempt to specify a concept such as radiation quality in terms of a *single* parameter must inevitably suffer from severe scientific limitations, there are certain situations—and notably that of radiation protection—where some relaxation of scientific rigor becomes necessary in the interests of practical convenience.

Gray[45] introduced the concept of *mean linear ion density* in 1947. For X-, γ-, and β-radiations, this parameter was defined as follows:

$$\text{Mean linear ion density} = \frac{\bar{T}}{R_T \cdot W} \tag{11-1}$$

where $\bar{T}$ = average energy of electrons liberated
R_T = range of electrons of energy $\bar{T}$
W = average energy required to form an ion pair.

Because the excitation—in addition to the ionization—of molecules by radiation may often result in physical, chemical, or biological change, the concept of *linear ion density* has, in recent years, been replaced by that of *linear energy transfer* (LET).

Gray's definition of mean linear ion density [Eq. (11-1)] can be readily converted into a definition of mean LET by simply omitting W from the denominator:

$$\text{Mean LET} = \frac{\bar{T}}{R_{\bar{T}}} \tag{11-2}$$

In 1952 Cormack and Johns[35] offered a revised definition of mean linear ion density: "The mean value is obtained by dividing the total number of ion pairs produced in 1 cm^3 by the total length of the electron tracks in 1 cm^3." No cognizance was taken of δ-tracks in applying this definition, but the change in averaging procedure leads to mean values that are somewhat different from Gray's.[22]

Lea (1946) stressed the importance of δ-tracks in connection with certain biological actions of radiation, and he presented tables of δ-track production in tissue by electrons, protons, and α-particles.[54] Any definition of mean LET is arbitrary, but there is some justification for adopting a definition that takes into account δ-tracks and the variation of local LET along primary and other tracks. Accordingly, a mean LET weighted by energy dissipation has been defined[22] as follows:

$$\text{Mean LET} = \frac{\sum_{0}^{T\max} Q_T \, \Delta T \, dT/dx}{\sum_{0}^{T\max} Q_T \, \Delta T} \tag{11-3}$$

where $Q_T \, \Delta T$ = total amount of energy dissipated locally in the system by particles in the kinetic energy range T to $(T + \Delta T)$; dT/dx = average energy dissipated locally, per unit length of track, by a particle of kinetic energy T to $(T + \Delta T)$.

The definition of *local energy dissipation* is also arbitrary, and it can be adjusted to suit the model of interest; in certain calculations, energy transfers of up to 100 eV have been regarded as *local energy transfers* [Refs. 10, 22–24, 37, 48, 54, 60].

It should be stressed that a "mean LET" has restricted value as a quantitative tool because it would be possible to devise two different types of radiations, with the same mean LET, but with different efficiencies in chemical or biological systems. As a rough guide to the quality of a radiation and its *approximate* effectiveness in comparison with other radiations, a mean LET has a certain practical utility, but in view of the great divergence of values calculated by different procedures it is clearly important to specify the averaging procedure that has been adopted whenever values are quoted.

In more rigorous applications, the distribution of energy dissipation as a function of $(\text{LET})_{\text{local}}$, or of particle kinetic energy, is of far greater importance and Tables 11-1 and 11-2 show these relations for several radiations of practical interest. In these tables, local collisions refer to energy transfers up to 100 eV.

If particles in the kinetic energy range T to $(T + \Delta T)$ produce a yield g_T of some specified radiation product (per 100 eV) through local energy transfers, then the over-all yield G of the product will be given by

$$G = \frac{\sum_{T=0}^{T\max} Q_T \, \Delta T \cdot g_T}{\sum_{T=0}^{T\max} Q_T \, \Delta T} \tag{11-4}$$

Table 11-1 dT/dx and $Q_T\,\Delta T$ Values for β-, γ-, X- and Monochromatic Electron Radiations in Water[22]

Electron Kinetic Energy Range (keV)	dT/dx Local Collisions keV/μH_2O	$Q_T\,\Delta T$ Values for 1 Rad Energy Dissipation in Water (Units: 10^{10} eV)							
		Betatron 25 MVp X-rays	2 MeV Electrons	1 MeV Electrons	^{60}Co γ-rays	220 kVp Primary X-rays	200 kVp Total X-rays	100 kVp Primary X-rays	Tritium β-Particles
819.2–T_{max}	0.103	2482	2048	637	262				
614.4–819.2	0.109	209	363	721	509				
409.6–614.4	0.118	258	379	727	726				
307.2–409.6	0.131	158	202	371	443				
204.8–307.2	0.151	187	220	385	505				
153.6–204.8	0.178	112	123	205	282				
102.4–153.6	0.217	133	143	223	312	6	3		
76.8–102.4	0.268	81	87	125	173	37	34	1	
51.2–76.8	0.345	105	110	147	198	207	178	82	
38.4–51.2	0.447	68	72	89	115	295	253	220	
25.6–38.4	0.595	93	97	113	140	576	444	615	
19.2–25.6	0.782	64	66	75	88	450	322	400	
12.8–19.2	1.05	91	94	102	115	608	466	466	42
9.6–12.8	1.39	65	67	71	78	379	345	299	222
6.4–9.6	1.86	94	96	101	108	451	487	432	634
4.8–6.4	2.46	69	70	73	77	264	320	301	548
3.2–4.8	3.30	102	104	108	112	311	400	394	751
2.4- 3.2	4.34	76	77	79	82	186	243	246	460
1.6–2.4	7.00	113	116	120	122	211	299	307	543
1.2–1.6	11.0	84	86	89	90	148	185	191	311
0.8–1.2	16.6	129	131	135	138	201	241	248	372
0.6–0.8	25.3	98	100	103	105	141	163	166	228
0.4–0.6	36.6	153	156	160	164	210	232	236	301
0.3–0.4	50.4	120	122	125	128	158	170	172	207
0.2–0.3	66.6	187	190	197	200	242	256	258	296
0.15–0.2	81.0	140	142	148	149	179	186	188	209
0.1–0.15	90.2	216	219	228	230	275	285	287	315
0–0.1	93.0	556	563	586	592	708	731	734	804
	$\Sigma Q_T\,\Delta T$	6243	6243	6243	6243	6243	6243	6243	6243

Table 11-2 dT/dx and $Q_T\,\Delta T$ Values for ^{210}Po and 1 MeV α-Radiations in Water[22]

Kinetic Energy Range (keV) α-Particles	dT/dx Local Collisions keV/μH_2O	$Q_T\,\Delta T$ Values in Water for 1 rad Energy Dissipation (Units: 10^{10} eV) ^{210}Po α^a	1 MeV α
5000–6000	47.4	195	—
4000–5000	51.5	651	—
3000–4000	63.4	656	—
2000–3000	83.7	663	—
1000–2000	111.1	679	—
600–1000	141	283	1503
300–600	154	229	1215
100–300	153	183	971
0–100	133	118	626
Electrons			
2.4–3.2	4.34	1	—
1.6–2.4	7.00	16	—
1.2–1.6	11.0	32	—
0.8–1.2	16.6	86	—
0.6–0.8	25.3	94	—
0.4–0.6	36.6	186	2
0.3–0.4	50.4	170	38
0.2–0.3	66.6	297	135
0.15–0.2	81.0	244	177
0.1–0.15	90.2	398	487
0–0.1	93.0	1062	1089
	$\sum Q_T\,\Delta T$	6243	6243

^{a} ^{210}Po α-particle, energy assumed to be 5.3 MeV.

The equation will apply when the yields g_T, are simply additive; this requires that once the product is formed at one section of a track, no subsequent decomposition or change of the product shall be produced by other products in the irradiated system. The quantity $Q_T\,\Delta T$ can be calculated fairly reliably because, apart from computational approximations, it depends only on the accuracy of the stopping-power equations. Except for low-energy transfers, and for low-energy primaries, these equations are fairly reliably based and reasonably well tested against experimental data. This distribution can also be derived from Spencer and Fano's calculations[71] of the slowing-down spectra of electrons and their analysis and mine yield very similar numerical results.[10,37] However, in most systems, the evaluation of g_T (or of the equivalent *quality factor* in radiation protection) presents formidable difficulties that have not yet been fully overcome. An approximate analysis of the dependence of chemical yield on radiation quality in the ferrous and ceric sulfate dosimeters has been described elsewhere.[21] Calculated and experimental yield values are, in view of the approximations, in surprisingly good agreement.

Another approach to the problem of radiation quality in radiobiology has been worked out by Rossi.[61–63] He directs his attention to the determination of the amount of energy dissipated within a small sphere located in the irradiated medium. The symbol Z is used to describe the energy dissipated per unit mass, or the *local energy density*. If $P^1(Z)$ represents the probability that a particular Z will be recorded in the test sphere, then a graph of $P^1(Z)$ versus Z can be constructed for different irradiations. The form of the distribution curve depends on the average dose to the irradiated medium and the diameter of the test sphere; examples of experimentally determined distributions have been published.[61,64] The *incremental local energy density*, ΔZ, from individual events in the test sphere may also be considered; $\Delta Z = KY/d^2$ where K is a normalization factor which depends upon the units in which Y and Z are measured, d is the diameter of the test sphere, and Y is termed the individual *event size*, defined as the energy expended in the event divided by the diameter of the test sphere.

One radiobiological model in which a knowledge of the Z distribution would be of value is the following[64]: "Inactivation of a cell will occur if Z is higher than some minimum value, Z_0 in any one of an unknown number, n, of spherical volumes of unknown diameter, d, within each cell."

It should perhaps be noted that the experimental determinations of the distribution of Y or Z have been based upon measurements of ionization. Generally, energy deposition will be directly proportional to the amount of ionization produced, although excitation can occur without ionization. It is distinctly possible that excitation is comparable in importance with ionization in producing chemical, and hence biological, change.[9,21,58] A further interpretational difficulty is presented when the diffusion of chemically active species to the "target zone" produces biological change.

11-3 DIRECT AND INDIRECT ACTION

Two broad modes of radiation action have been considered—*direct* and *indirect*.[54] In the simplest radiobiological model of *direct action*, it is assumed that the solute, or target molecule, undergoes biologically significant chemical change if it is hit directly by an ionizing particle. Under these conditions, the amount of solute (or the number of target molecules or structures) surviving a given dose of radiation without being hit, diminishes exponentially with dose. The sensitivity of the target to change can be specified by the dose (D_0) that reduces the *surviving fraction* to e^{-1} (approximately 37%) of the starting amount. D_0 is sometimes described as the 37% dose. In the simplest model of *indirect action* it is postulated that solute molecules are not excited or ionized directly by ionizing particles. Instead they are changed chemically through reaction with excited or ionized *products* of the irradiated *solvent* molecules. In the absence of complications, the number of chemically changed solute molecules is directly (linearly) proportional to dose and independent of solute concentration.

Under favorable conditions, the differences in the kinetics of the two classes of action enable the relative contribution of each to be assessed.[54]

A more complicated (but perhaps more realistic) model that shows the *kinetics* of direct action but involves, in part, the *mechanism* of indirect action, can be postulated: A target in an aqueous medium might be changed or inactivated, either through a direct hit, or through chemical reaction with a diffusible radiation product that is generated within a defined zone surrounding the target.[78]

The yield in systems where indirect action predominates can be greatly modified by introducing agents that compete with the solute for the chemically active species generated in the solvent. When a competing solute is present in large enough excess, the kinetics of inactivation of the solute of interest can simulate those of direct action; in other words, the concentration of the solute is then found to diminish exponentially with dose, instead of linearly, as with uncomplicated indirect action.[54]

11-4 POSSIBLE MECHANISMS OF BIOLOGICAL CHANGE

The striking effectiveness of radiation in imparing the integrity of biological systems can be judged from the following example. If mammalian cells are irradiated, then an appreciable proportion of them—say from about 10 to over 50%—may be made to lose their reproductive integrity by a dose as small as 100 rad. In chemical systems, a yield of 10 ions or molecules per 100 eV is above average; 100 rad corresponds to slightly more than 6×10^{15} eV/gm., and hence a change of 6×10^{14} atoms would represent an above average yield. However, the number of atoms in 1 gm of cells is $\sim 10^{23}$. Hence it would appear that a chemical change in as few as 1 in $\sim 10^{8}$ of the atoms of a cell can have profound biological consequences! In principle, there are, of course, many ways whereby a very small percentage chemical change might give rise to serious biological effects.

A chain reaction might be initiated so that the yield of some active species is very much greater than 10 per 100 eV. However, biological systems abound in molecules with chemically reactive groups that terminate chain reactions and this interpretation is therefore improbable. Alternatively, a toxin might be generated that is capable of poisoning a cell even at very low concentrations. The toxin hypothesis is by its very nature difficult to test; it has not been disproved. But the hypothesis that commands the greatest interest and attention at the present time is the one that postulates that damage to one or more critical macromolecules (such as RNA or DNA) is chiefly responsible for such effects as the inactivation of viruses, gene mutation, and the killing of cells. Lack of redundancy, especially of genetic material, is likely to be a crucial factor in the death of a cell. In the normal diploid mammalian cell, autosomal genes are represented by only two units on homologous chromosomes. However, there are many cells in men and women where only one *active* representative of a gene, on the *X* or *Y* chromosomes, is present. The destruction of any gene that is essential to the economy of the cell will, by definition, lead to the loss of cell function.

Target Theory

Target theory has been applied to the kind of action, where, for example, a single hit or ionization in a critical structure is sufficient to produce detectable biological change.[54] Under these circumstances, a graph of the logarithm of the surviving unchanged fraction (of cells or viruses), against dose, on a linear scale produces a straight line of negative slope (see Fig. 11-1). This assumes that the targets are homogeneous. If ionizations are produced singly and at random (to take an extreme and unrealistic model) then at the 37% dose, D_0 there would be an average of one ionization per sensitive target. By determining D_0, the size of the sensitive target can be calculated. At the opposite extreme, it can be postulated that a target will be inactivated only when it is *traversed* by an ionizing particle, such as an electron, proton, or α-particle. Such a model estimates the projected area of a target.

To formulate a comprehensive target theory it will be necessary to define the chemical and steric structure of the target, the mechanism (direct and/or indirect) of chemical change that results in the biological effect, the influence of protective and repair mechanisms, together with the physical details (spatial and temporal) of energy distribution at the submicroscopic level.

One of the outstanding barriers to a rigorous analysis of target problems is the uncertainty surrounding the history—range and scattering—of low-energy electrons.

The cytotoxic (cell-killing) effect of radiation on mammalian cells has been explored in considerable detail since the pioneering experiments of Puck et al.

in 1956.[59] Nevertheless the extensive experimental evidence is still waiting to be interpreted in terms of a comprehensive theory of radiation cytotoxicity. Mammalian cells generally exhibit a simple exponential form of survival only when they are irradiated by particles of high average LET—for example,

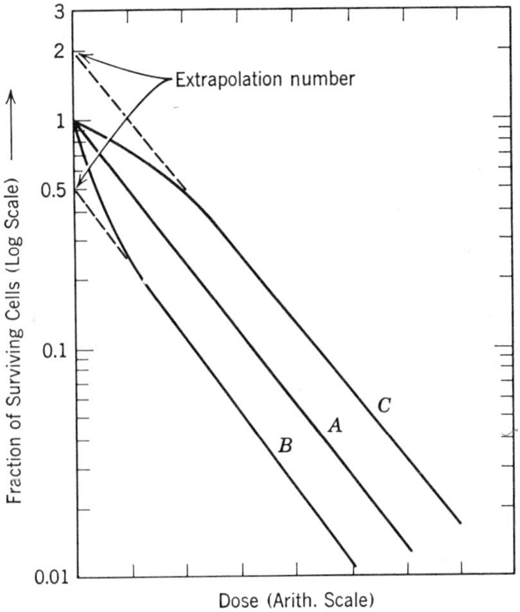

Fig. 11-1. Types of mammalian cell survival curve. Logarithm of fraction of cells S surviving the (acute) irradiation against dose D. Curve A: $\ln S = -\lambda D$. Observed when homogeneous cells are irradiated by particles (such as low-energy natural α-particles) of high LET. Curve B gives an extrapolation number of less than unity (0.5 in example); observed when heterogeneous mixtures of cells, of low and high radiosensitivity, are irradiated. Curve C gives an extrapolation number of more than unity (2 in example); observed when most types of mammalian cells are irradiated by, for example, ^{60}Co γ-rays—that is, by radiations of low average LET.

in the region of 100 keV/μ, or above. When the average LET is low—as, for example, with 200 KVp X-rays and ^{60}Co γ-rays—the curve of log survival against dose (linear scale) usually shows a shoulder at low dose (curve C, Fig. 11-1) although it generally becomes a straight line (corresponding to a negative exponential curve) at high dose. If the exponential portion of the survival curve is extrapolated back to cut the ordinate (zero dose), the intercept generally has a value equivalent to a surviving fraction of more than unity. When 100% survival is designated as unity on the ordinate scale, the value of this intercept is described as the *extrapolation number* (Fig. 11-1).

Various interpretations of this type of survival curve have been proposed.[44,47,57,78] If the killing of a cell requires n successive random identical hits on a single target of size λ, then the surviving fraction of cells S_D, for a dose D, will be given by the Poisson function:

$$S_D = e^{-\lambda D}\left[1 + \lambda D + \frac{(\lambda D)^2}{2!} + \cdots \frac{(\lambda D)^{n-1}}{(n-1)!}\right] \tag{11-5}$$

From biological and molecular considerations, Atwood and Norman concluded that this equation is untenable.[1]

In another model, it is also considered that there are n sensitive targets in a cell, each of size λ but that cellular inactivation requires that each sensitive target should receive at least one independent random hit in any sequence. For these postulates, survival is given by:

$$S_D = 1 - (1 - e^{-\lambda D})^n \tag{11-6}$$

The extrapolation number for this type of curve is equal to n. It is doubtful, however, whether this model is generally relevant. It fails to account for the characteristics of some survival curves[44,57] and for the degree of complexity of the phenomenon of cytotoxicity. The radiosensitivity of a cell, from the point of view of colony-forming ability, depends upon many factors such as (1) LET; (2) dose fractionation; (3) dose rate; (4) oxygen tension and other metabolic factors; (5) the substitution of base analogs, such as 5-bromodeoxyuridine for thymidine, in the nuclear DNA; and (6) the position of the cell in the mitotic cycle.

Barendsen[3] and Deering and Rice[38] have studied cytotoxicity for mammalian cells as a function of LET and their results (replotted in Figs. 11-2 and 11-3) are of the utmost importance to the understanding of the biophysical basis of cytotoxicity. From this general kind of evidence, Howard-Flanders and Barendsen previously attempted to calculate the number and linear density of ionizing events involved in the inactivation of viruses, bacteria, yeast cells,[50] and human cells.[3]

Cytotoxic Events: Biological Character

Figures 11-2 and 11-3 suggest a somewhat more specific interpretation than has been formerly proposed. At high LET (where, conveniently, the spectrum of $Q_T\,\Delta T$ versus local LET is fairly narrow—see Table 11-2), the relative cytotoxic effectiveness per particle (P) conforms to the relationships $P \propto 1 - e^{-k_3(L_{200})^3} \cdots$ (Fig. 11-2) and $P \propto 1 - e^{-k_3'(L_\infty)^3} \cdots$ (Fig. 11-3). This suggests that at high LET the effective *cytotoxic event* generally involves a dependent-type triple interaction between the primary ionizing particle, or

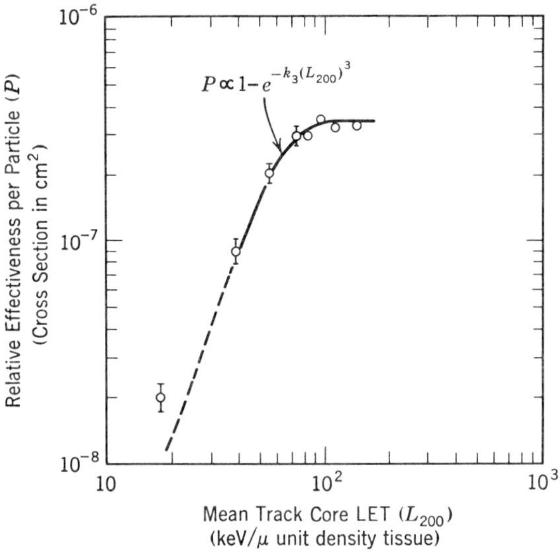

Fig. 11-2. Radiation cytotoxicity for human kidney cells as a function of track core LET. Cytotoxic effectiveness of ionizing particles (P) in relation to the mean track core LET. The "track core LET" (L_{200}) refers to the linear density of energy transfers from the primary particle up to and including 200 eV. Results of Barendsen[3] for human kidney cells. For $L_{200} > 40$ keV/μ unit density tissue, the experimental results fit the theoretical relationship $P \propto 1 - e^{-k_3(L_{200})^3}$.

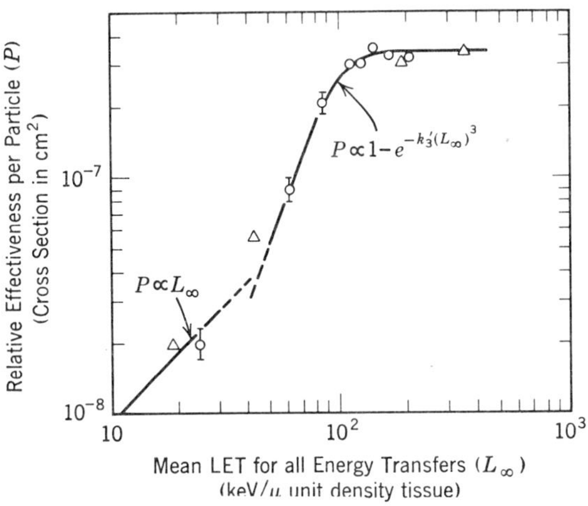

Fig. 11-3. Radiation cytotoxicity as a function of $\overline{\text{LET}}$. Relationship between cytotoxic effectiveness of ionizing particles (P) and mean LET for all energy transfers from primary particles (L_∞). Key: ○, results of Barendsen[3] for human kidney cells; △, Deering and Rice[38] for HeLa cells; both normalized to Barendsen's results at high $\overline{\text{LET}}_\infty$. For $L_\infty >$ 60 keV/μ unit density tissue, the experimental results fit the theoretical curve $P \propto 1 - e^{-k_3'(L_\infty)^3}$. It is probable that, at low L_∞, $P \propto L_\infty$ and that there is an intermediate range of L_∞ where both single- and triple-interaction cytotoxic events are effective.

its products, and a critical complex target structure. The value of $k_3\,(L_{200})^3$ is unity at $L_{200} = 70$ KeV/μ, and the value of $k_3'(L_\infty)^3$ is unity at $L_\infty \simeq 100$ KeV/μ.

Assuming, for simplicity, that the average energy expended per single cytotoxic interaction is in the region of 70 to 100 eV, respectively (corresponding to the approximate energy per primary ionization), for a unit density medium the average separation between the successive cytotoxic interactions should be about $10^{-3}\,\mu$ or 10 A. For a medium of effective density ρ, the separation will be about $10/\rho$ A. Although the concept of physical density within, for example, a chromosome becomes somewhat nebulous, if we assume (see below) that double-stranded DNA constitutes a part of the target material, then ρ is likely to have a value in the region of 1.2.

The original concept of hit is somewhat misleading in this context. The expression hit suggests a single, uncomplicated event, and it arose from the interpretation of exponential survival curves. The current interpretation indicates that a single effective cytotoxic event at high LET corresponds to a transit of the primary particle through, or very near to, a complex, triple-element target structure. Because the average spacing between successive cytotoxic interactions is in the region of $10/\rho$ A and, because a vast amount of independent evidence implicates DNA as the target structure, it is possible that two of the three interactions produce breaks in, or damage to, the twin strands of the Watson-Crick DNA double helix. That a third interaction should be required is most intriguing. It might produce a break in a polypeptide chain lying in the minor groove of the double helix (Fig. 11-4) or it might affect such a chain running along the axis of a modified DNA duplex. A fuller understanding of euchromatin and heterochromatin will be needed to elucidate the biophysical nature of the triple-interaction cytotoxic event.

At lower $\overline{\text{LET}}$ the interpretation of the graphical representation of the experimental data in Figs. 11-2 and 11-3 becomes somewhat more uncertain. In this region the spectrum of $Q_T\,\Delta T$ versus local LET becomes broader—in other words, the LET of interacting particles is much more heterogeneous than at high $\overline{\text{LET}}$. Superficially, the change in slope of the curves of Figs. 11-2 and 11-3 at low $\overline{\text{LET}}$ suggests that single-interaction cytotoxic events may become important in that region. This approximate form of analysis does not enable us to decide whether double-interaction events also contribute to cytotoxicity. But despite the approximations, this interpretation of the LET data,[3,38] involving at least two types of cytotoxic events is in good accord with independent evidence and other considerations.

From dose-fractionation experiments using X-rays and mammalian cells, Elkind and his colleagues have shown that there are at least two categories of cytotoxic injury.[39–42] One form is irreversible, whereas the other may be repaired under appropriate conditions. The extent of cellular recovery after a first acute irradiation depends upon many factors including the size of the first dose and the duration of the interval between the first and second radiation fractions. With radiations of high LET no significant recovery between successive fractions is observed; this suggests that the triple-interaction cytotoxic event may actually break the thread of deoxyribonucleoprotein, producing ends that do not usually rejoin.

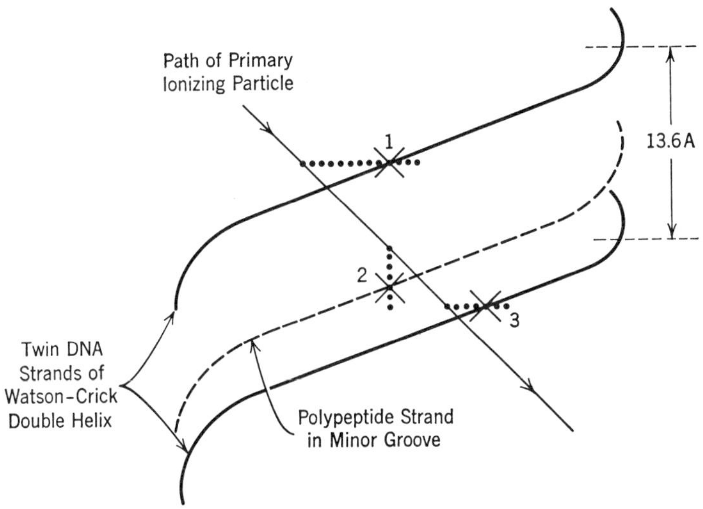

Fig. 11-4. A possible model of triple-interaction cytotoxic event, effective at high LET. Dependent-type triple interactions 1, 2, and 3 occur close to one another with a typical separation of about $10/\rho$ A, where ρ is the effective mean density of the medium between successive interactions. Interactions 1 and 3 may break the twin strands of DNA, whereas interaction 2 may break a polypeptide strand in the minor groove or, say, along the axis of the duplex. For the induction of certain point-type mutations in yeast[56]—see Section 11-5—a double-interaction dependent-type event is probably implicated at high LET. The typical spacing between these interactions (producing chemical change rather than strand breaks?) is in the region of $14/\rho$ A, and this suggests that this type of complex event affects the two strands of DNA.

If single-interaction events break only one strand of DNA, the intact polypeptide chain, combined with the unbroken strand of DNA, probably confer sufficient stability on the macromolecule to allow repair mechanisms to operate. However, the proportion of these lesions that is repaired decreases with increasing (acute) dose. Moreover, in unsynchronized cell cultures, the delay in cell division after an acute dose is proportional to dose.[42] Dose-fractionation experiments indicate that cell division is delayed until the repair effort is completed. The total number of single-interaction lesions per cell will be proportional to dose, and because division delay is also proportional to dose, it is probable that repair of single-interaction lesions is a *sequential* process. It appears that the time spent in the attempt to repair a lesion is long compared with the interval between the repair of one lesion and the next. It seems that monitoring agents, such as enzymes, enzyme complexes, or ribosome-like particles, traverse chromosome threads continuously, and are capable: (a) of recognizing damaged DNA; (b) of excising damaged regions; and (c) of sometimes restoring the chromosome structure using the intact DNA strand as a template, and Watson-Crick base pairing, to preserve the genetic information content during DNA repair synthesis.

Haynes[47] states that all survival curves (presumably for homogeneous cells) can be represented by the following equation:

$$S_D = 1 - e^{-[kD+\alpha(1-e^{-\beta D})]} \tag{11-7}$$

If we associate k with the irreparable triple-interaction cytotoxic events and $\alpha(1 - e^{-\beta D})$—or a term of broadly similar form—with the single-interaction, potentially reparable lesions, then the preceding biophysical model of cytotoxicity is consistent with the form of cell-survival curves, as well as with the LET evidence. It is necessary to postulate that β is a function of dose rate, increasing with it. (The possibility must be borne in mind that another class of reparable lesions—involving double-interaction events—may also contribute to the phenomenon of cytotoxicity.)

It follows from this model that at low dose rates the $\alpha(1 - e^{-\beta D})$ term—or its equivalent—becomes unimportant, and, neglecting the complication of mitosis during irradiation, exponential survival should be observed. (At sufficiently low dose rates, almost all single-interaction lesions will be repaired, and the only effective cytotoxic events will be those of the triple-interaction kind.) In fact, Bedford and Hall[6,46] find that the survival of HeLa S3 cells becomes approximately exponential at dose rates (for ^{60}Co γ-rays) of the order of 1 rad/min or less.

Oxygen generally enhances the cytotoxic action of radiation. The oxygen enhancement ratio (OER) is close to unity for low-energy, high LET α-particles, but it increases to a factor of about 2 to 3 with decreasing LET.[2] Elkind has shown that oxygen has relatively little effect on repair processes.[41] These facts indicate that single-interaction lesions are not produced exclusively by uncomplicated direct hits. Oxygen probably reacts with one or more of the radical products of the radiation (for example, hydrogen radicals, carbon radicals) to produce an even more reactive product.[21,50] The absence of an oxygen enhancement effect in the cytotoxic action of 2.5 MeV α-particles[2] suggests, however, that the triple-interaction irreversible event may involve an uncomplicated direct action. Both strands of the DNA duplex, and a polypeptide chain(?) associated with it, are perhaps broken mainly by energy transfers that cause direct ionization (and/or excitation) in the macromolecule.

Despite its great theoretical interest, the phenomenon of radiation cytotoxicity should not be of major importance to health physicists—provided the exposure of individuals to radiation does not exceed maximum permissible levels. At the maximum permissible levels of radiation exposure—with small doses and low dose rates—only a very small proportion of cells will lose reproductive capacity. Of those that do, the great majority in most tissues will be replaced through mitosis of viable cells. However, in some situations—when, for example, internal body burdens of α-emitting radioisotopes are present—the high cytotoxic effectiveness of α-particles might create a continuous state of replacement hyperplasia and this condition could have neoplastic consequences.

Consideration of radiation cytotoxicity is essential when interpreting the leukemogenic and carcinogenic effects of rather large, acute doses of radiation,[13,16] but cell death is likely to be of the greatest practical consequence when people are accidentally exposed to lethal, or near-lethal doses of radiation. The acute radiation syndrome and acute radiation death probably arise from the cytotoxic action of radiation on particularly vulnerable cell systems.[73]

11-5 MUTAGENESIS

The induction by radiation of gene mutation (including gross chromosomal changes) is a phenomenon that is likely to be of the utmost importance to health physicists. When mutation occurs in germ cells it may produce abnormalities in the offspring, but when it occurs in somatic cells, it may contribute to the induction of neoplasms, and, perhaps, to morbidity and aging in the exposed individual. Although mutation in somatic cells is generally difficult to investigate directly—except where microscopically visible aberrations of chromosomes are concerned—mutation in germ cells has been investigated in many species.

Of the species that have been investigated in detail, the one that is phylogenetically closest to man is the mouse. It was once thought that the yield of radiation-induced mutations is simply proportional to dose, over a very wide dose range, and that it is independent of dose rate and dose fractionation. It is now known that these generalizations do not apply to the induction of mutations in premeiotic mouse germ cells.[66–68]

In spermatogonia, it appears that for exposure rates of less than $\sim$0.01 R/min (and possibly $<$0.8 R/min) the relation between induced mutations and dose is a linear one, but that the average radiomutability at seven specific loci is about one-third of that found for acute (90 R/min) exposures of 300 to 600 R.[66] It appears that the mutation yield at 9 R/min, (and at 600 R total exposure) is probably intermediate between the high and low dose-rate yields for the same total dose.[66]

In oocytes, the dose-rate effect appears to be even more pronounced.[67] There are powerful reasons for concluding that the dose-rate effect—both in spermatogonia and in oocytes—is an intracellular phenomenon that involves either the repair, or the prevention of genetic damage.[66] Russell suggests that repair of *premutational* damage probably occurs because a completed mutation appears to be quite stable.

Even more complicated effects are observed when large acute exposures ($\simeq$1000 R) are delivered singly, or in fractions,[66] but these are unlikely to be of much concern to health physicists. However, they show that radiation-induced mutation is a complicated and imperfectly understood phenomenon. It is reassuring that under the dose and dose-rate conditions encountered in supervised occupational exposure, to low LET radiations, the radiomutability of genes in spermatogonia and oocytes of mice is at its minimum.

The average sensitivity of spermatogonial genes at seven specific loci to acute exposures of 250 kvp X-rays (300 to 600 R) is about 2.5×10^{-7} mutations per gene per R.[66] If Lea's target theory[54] is applied to this result to give some idea of target size, it indicates that the average target has a molecular weight $\sim 3 \times 10^5$. This is of the same order as the molecular weight of the DNA of a typical gene (cistron) and it suggests that the whole, or the greater part of the DNA of a gene may be the principal target for radiation-induced mutation. (It would be of considerable interest to know the RBE of fast neutrons and helium ions in connection with the radiomutability of these genes.) Because the molecular weight of the typical gene (cistron) in man is almost certain to be closely similar to that of the typical mouse gene, the radiomutability of comparable genes should also be similar in the two species. It is difficult, however, to predict the extent of the repair of premutational damage at low dose rates, because this evidently varies in the mouse with the type and stage of the cells.

An attempt has been made to measure the radiomutability of genes in mouse somatic cells during fetal life, at four of the seven loci studied in germ cells.[65] The mean value was found to be 7×10^{-7} R, with 95% confidence limits of (2.5 and 11.3) $\times 10^{-7}$ R. It seems likely therefore that the absolute radiomutability of a particular gene does not vary greatly from cell type to cell type, although repair mechanisms can introduce some variation. The extreme limits for a specific gene would appear, in general, to lie within about an order of magnitude of one another. These findings may have an important bearing on mechanisms of radiation carcinogenesis and leukemogenesis.

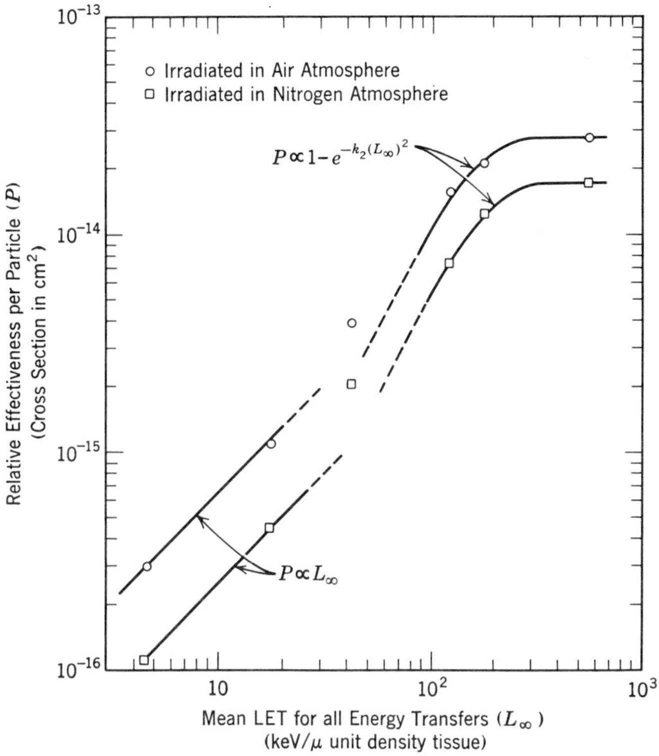

Fig. 11-5. Mutagenic efficiency of ionizing particles as a function of LET_∞. Mutagenic effectiveness of ionizing particles (*P*), in relation to their mean LET for all energy transfers (L_∞). Results of Mortimer et al.[56] for the induction of Tr^+_{NF} mutations in the yeast *Saccharomyces cerevisiae*. At high L_∞, the results for irradiation both in air and in nitrogen fit the theoretical relationship $P \propto 1 - e^{-k_2(L_\infty)^2}$. (The value of k_2 may differ slightly for the two conditions—see text.) At low L_∞, the results for air and nitrogen fit the relationship $P \propto L_\infty$. There is an intermediate range of L_∞ where both single- and double-interaction events contribute to mutagenesis.

Recent observations[56] on the radiomutability of certain genes in yeast have interesting biophysical implications. Mortimer et al. studied the effectiveness of heavy ions, covering the $\overline{\text{LET}}_\infty$ range of 4.5 to 600 keV per μ water, for inducing specific mutations in diploid yeast. They also examined the effect of oxygen tension on radiomutability. The results for the induction of gene mutation at three loci were very similar, except that the absolute cross sections at a given LET and oxygen tension, differed (by a factor of up to two) from locus to locus. Their results[56] for one specific mutation ($\text{Tr}_{\text{NF}}{}^+$) are reproduced in Fig. 11-5. There are important similarities and differences between this figure and Fig. 11-3, which relates to cytotoxicity.

At high average $\overline{\text{LET}}_\infty$, the mutational data for irradiation in air and in nitrogen (anoxic) fit the equation

$$P \propto 1 - e^{-k_2(L_\infty)^2} \tag{11-8}$$

This relationship indicates that the induction of gene reversion at high LET is mainly contingent upon a specific double-interaction event. The value of $k(L_\infty)^2$ is unity at $L_\infty = 145$ keV/μ water for irradiation in air, and $L_\infty \simeq 135$ keV/μ water under anoxic conditions. For a medium of effective average density ρ, the typical spacing between the two successive dependent-type mutagenic interactions is therefore in the region of $14/\rho$ A. This is greater than the spacing ($\simeq 10/\rho$ A) between successive triple-interaction cytotoxic events (see Section 11-4 and Fig. 11-4). Conceivably, both strands of the DNA duplex are affected in this type of mutational event at high LET. At relatively low $\overline{\text{LET}}$ (below a $\overline{\text{LET}}_\infty$ of about 10 keV/μ water) it appears from the changing slope of the curves in Fig. 11-5, that a single-interaction mutational event predominates over the double-interaction class. (The slope of the curve in Fig. 11-5 tends towards unity at low $\overline{\text{LET}}$).

There is one other important difference between this mutational situation and cytotoxicity. The oxygen enhancement ratio for mutagenesis is about 1.6 at high $\overline{\text{LET}}_\infty$, whereas it is unity for cytotoxicity.[2] This indicates that some degree of indirect action is involved in the double-interaction mutagenic event. At low $\overline{\text{LET}}_\infty$, the oxygen enhancement ratio is nearly 3—which is comparable to the equivalent value for cytotoxicity.

Another important feature of the findings of Mortimer et al. is the absolute target cross section at high $\overline{\text{LET}}_\infty$. This is of the order of 10^{-14} cm^2 and "On a direct action consideration, this would correspond to a target with linear dimensions in the order of 10 A (approximately half the diameter of a DNA molecule)."[56] The discrepancy between the target size for this type of mutational event in yeast, and the target size for "visible mutations" in mice (corresponding approximately to the whole physical gene) is therefore enormous. The mutations studied in yeast are believed to correspond either to base change, or to base deletion.[56] It has been proposed that a certain class of mutation in multitissue organisms (including mammals) may involve an entirely different mechanism from the point-type mutation familiar to students of bacterial genetics.[25]

At many loci involved in mitotic control and spontaneous disturbed-tolerance autoimmunity in man, only two major alleles can be discerned.[14,27,28] It has been suggested[25] that one major allele corresponds to messenger RNA (mRNA) transcription from one strand of the DNA duplex, while the second major allele at the same locus corresponds to mRNA transcription from the complementary, antiparallel strand. DNA strand determination may depend upon the polypeptide chain product of the active allele (DNA strand); this chain may associate with the DNA helix and block mRNA transcription from the antiparallel strand.[25] On this model "mutation" (apart from point-mutation) involves a spontaneous or induced switch of mRNA transcription from the regular strand of DNA over to the complementary strand.[25]

This switchover probably requires the dissociation of the polypeptide chain—DNA complex. One of the many advantages of this model is that it makes intelligible the finding that the target cross section for certain radiation-induced mutations in mice corresponds, approximately, to the entire gene. (This is a paradox which has perplexed radiobiologists for many years.) A mutational event anywhere along the gene may block mRNA transcription from the regular strand of DNA and dislodge the polypeptide chain from the Watson-Crick DNA double helix. It must be emphasized that this model awaits experimental confirmation. Nevertheless, it shows that we must be cautious in applying the findings for point-type mutation (base transition, transversion, or deletion) in bacteria and yeasts to certain types of mutation in man.

11-6 CARCINOGENESIS AND LEUKEMOGENESIS

That moderate and large doses of internal or external radiation can cause malignant growth in man, as well as in the lower mammals, is not in dispute. However, the carcinogenic effects of small doses of radiation are inevitably much more difficult to investigate; consequently opinion is divided over this issue. The complex mechanisms of carcinogenesis—both "natural" and radiation—have been widely investigated and debated but no general consensus of opinion has yet emerged. It is not even clear whether radiation induces malignant change through somatic gene mutation, through the activation of latent oncogenic viruses, or by some other process.[53] However, the general consistency observed in the relation between radiation dose and leukemia induction in widely diverse human populations is difficult to accommodate to the "virus-activation" hypothesis.

It seems highly improbable that the average number of latent leukemogenic viruses carried per person, in different populations, is approximately constant. This does not deny that viruses may play a part in the pathogenesis of malignant disease in man. However, it is unlikely that in man they are directly "activated" by radiation. An attempt has been made to interpret a wide range of evidence relating to natural and radiation-induced malignancies in man in terms of a single, though somewhat complicated hypothesis [Refs. 11–13, 15, 16, 18, 20, 21, 25]. In the simplest situation, and in the absence of oncogenic viruses, it has been proposed that malignant growth develops from one or more cells with a complement of four specific mutant genes or alleles. Two of the mutations may well affect homologous autosomal *structural* genes, and the remaining two may affect homologous autosomal *regulator* genes.[17]

When the pathogenesis of "natural" cancers and leukemias in analysed, it is concluded that these are mainly, though not exclusively, prezygotic and are largely confined to genetically predisposed individuals carrying a single dominant predisposing gene. (Rare leukemias and lymphomas in childhood can apparently involve recessive inheritance of major homologous autosomal alleles—especially where there is parental consanguinity—possibly at a regulator locus.[17])

In prezygotic malignancies in adults where a major dominant predisposing autosomal allele is inherited, it is assumed that the remaining three carcinogenic mutations are accumulated in one or more somatic cell lines through the process of random somatic mutation. With most prezygotic childhood leukemias, it appears that only two somatic mutations are accumulated in the cell line that becomes malignant and that an oncogenic or leukemogenic virus may often be required to promote malignant growth. Postzygotic malignancies involve no major predisposing inheritance, and in the absence of extrinsic carcinogens, their incidence should be low; however in the presence of powerful extrinsic carcinogens the ratio of postzygotic to prezygotic malignancies should increase.[13,16]

When this hypothesis of natural carcinogenesis is used to interpret the phenomena of radiation leukemogenesis and carcinogenesis, it is considered that radiation is capable of inducing mutations that are entirely comparable in their effects to those that occur spontaneously. (Where gene mutations in germ cells are concerned, there is much evidence to support an analogous assumption.) Four broad categories of response must generally be considered: the "short-term" and the "long-term" responses in predisposed and nonpredisposed individuals. The short-term response is defined as that which results from the formation of a "carcinogenic nucleotype"—a cell with the full complement of mutant carcinogenic nuclear genes—at the time of the irradiation. There is still a lag or latent period before the transformation to malignant growth actually occurs, but in the short-term response it is postulated that all the nuclear changes necessary to subsequent malignant growth are present at the termination of the irradiation.

The long-term response is defined as that which results when the last initiating somatic mutation occurs spontaneously, subsequent to the irradiation. Frequently, the number of precarcinogenic cells containing three specific mutations (L_3-type cells) will be greatly increased by an irradiation *without* an L_4-type "carcinogenic nucleotype," with a full complement of carcinogenic mutations, being formed. A subsequent spontaneous carcinogenic mutation in an L_3-type cell, giving an $L_3 \rightarrow L_4$ transition, could yield the carcinogenic nucleotype at any interval after the irradiation. Because of the long latent period that intervenes between the completion of the initiation phase of carcinogenesis and the actual clinical detection of a tumor, there will generally be an appreciable time overlap between the short-term and the long-term responses.

With this approach, the function describing the carcinogenic response to radiation should involve polynomials in dose, and where the short-term response is concerned, a cytotoxicity function will also be implicated. For, if an L_4-type cell is generated by mutation at the time of irradiation, but is also killed—by the same, or a different event—then clearly it cannot subsequently develop into a tumor; only viable L_4-type cells are capable of that. If a genetically-predisposed individual is irradiated, then L_4-type cells could be generated along any of the following pathways:

$$L_1 \rightsquigarrow L_2 \rightsquigarrow L_3 \rightsquigarrow L_4 \qquad \text{(3-hit)}$$

$$L_2 \rightsquigarrow L_3 \rightsquigarrow L_4 \qquad \text{(2-hit)}$$

$$L_3 \rightsquigarrow L_4 \qquad \text{(1-hit)}$$

The average carcinogenic short-term response to an acute dose of D rem will be described by a function of the following kind:

$$L_{4,D} = f_c(D) \cdot (\alpha^3 D^3 L_1 + \alpha^2 D^2 L_2 + \alpha D L_3) \qquad (11\text{-}9)$$

where $L_{4,D}$ is the average number of L_4-type cells in a carrier that survives the irradiation in a viable condition; $f_c(D)$ in a cytotoxicity function describing the fraction of L_4-type cells that survives a dose of D rem; α is the average radiation mutation coefficient, defined in terms of the average number of carcinogenic mutations, per gene at risk, per cell at risk, per rem; L_1 is the number of cells at risk at the commencement of the irradiation with one specific carcinogenic mutation; L_2 is the corresponding number of cells containing two carcinogenic mutations, . . . , etc. These cell numbers are, of course, a function of age and they are obtained from the theory of natural carcinogenesis and mortality and biological statistics. When the numerical value of $L_{4,D}$ approaches or exceeds unity, the carcinogenic response is described by the probability function:

$$1 - \exp(-L_{4,D}) \qquad (11\text{-}10)$$

When considering the response in noncarriers (individuals without a specific genetic predisposition) the extra pathway

$$L_0 \rightsquigarrow L_1 \rightsquigarrow L_2 \rightsquigarrow L_3 \rightsquigarrow L_4 \qquad \text{(4-hit)}$$

must be taken into account, and this involves the further generating term: $\alpha^4 D^4 L_0$.

To predict the long-term response, it is necessary first to calculate the over-all spectrum of mutant cells produced by prior spontaneous mutation and the irradiation, and then to compute the changing spectrum with time resulting from subsequent spontaneous somatic mutation. In these calculations, no allowance for the cytotoxic effect of radiation will generally be required, because after the irradiation, cell recovery and tissue regeneration will usually go to completion, and to a first approximation the spectrum of mutant cells after recovery should be closely similar to that of cells surviving the irradiation. If, in the short-term response, an L_4-type cell in an individual survives the irradiation and after recovery increases in number to, for example, ten cells, only one malignancy will be recorded.

This general scheme gives a consistent interpretation of all the more reliable quantitative data relating to radiation carcinogenesis and leukemogenesis both in childhood and maturity.[13,16,19,20] Certain complications arising from ambiguities in spontaneous somatic mutation rates make it difficult to estimate accurately the absolute values of α, the average radiation mutation coefficient for "carcinogenic genes," although typical values for their upper limits can be derived. Except for the possible induction of childhood malignancies through fetal irradiation, the deduced α values for the induction of chronic granulocytic and acute leukemias are somewhat higher than they are for typical nonleukemic malignancies.

The highest α value found is for the induction of granulocytic leukemias in the most heavily irradiated survivors at Hiroshima (<1000 meters from hypocenter). This value is $\simeq 2 \times 10^{-6}$ rem (upper limit). For prenatal X-rays and the induction of all types of childhood malignancies, the corresponding upper limit for α is found to be $\simeq 5 \times 10^{-7}$ rem. There is a general suggestion that α may vary with absolute dose and dose rate in a manner similar to that observed or deduced by Russell in the premeiotic germ cells of mice.[66–68] Radiation is generally less effective in inducing nonleukemic malignancies in adults and the corresponding α value (for acute irradiation at Hiroshima) is about 2×10^{-7} rem (average, upper limit). It is interesting that these estimates of the upper limits for the absolute values of α lie in the same kind of range as that found for germ and somatic cells of mice (mean, absolute values). Theory also accounts satisfactorily for the age dependence of radiation-induced malignancies and for various biological and immunological aspects of cancer.[11–13,16,17,25]

The implication of Eq. (11-9) and the corresponding equations for other short- and long-term responses is that at low enough dose the terms in D^2 and higher powers will become relatively unimportant. Furthermore, $f_c(D) \rightarrow 1$ as $D \rightarrow 0$, and the response to radiation should then be directly proportional to dose. (As $D \to 0$, even at an acute dose rate, α should tend to a constant value equal to the "chronic irradiation" value.) If this general hypothesis should prove to be essentially correct, this predicted response behavior at low dose and low dose rate will greatly facilitate radiation protection calculations.

It should be remarked that the formalism outlined above will not be directly applicable to the induction of tumors by α-emitting radioisotopes. A particular α-particle that induces a carcinogenic gene mutation is also quite likely to kill the cell through a cytotoxic interaction. The value of the coefficient α, determined in mutational studies with α-particles, would need to be reduced to allow for cytotoxic competition. A further complication could arise from the constant turnover of cells arising from the cytotoxic action of α-particles; this hyperplastic situation could of itself accelerate tumor formation. Unfortunately our knowledge of the influence of LET on the process of radiation carcinogenesis is very scanty. It is not yet possible to derive valid RBE or QF values for the induction of bone tumors in mice[43] or in other animals, because the distribution of different types of radioisotopes relative to the cells at carcinogenic risk is unknown.

There is no reason to suppose that one RBE or QF value can be attached to a specific quality of radiation that is valid for all forms of radiation carcinogenesis; in fact, the LET dependence for the induction of ovarian tumors in RF mice differs from that found when myeloid leukemia is the end point.[74,75] It is doubtful whether the situation in man is simpler than that in mice. At the theoretical level, much depends upon whether radiation-induced carcinogenic "mutations" involve "point" mutations, small deletions, chromosome breaks, DNA strand switching,—or any combination of these and other lesions. The importance and imponderability of repair mechanisms has already been stressed.

11-7 OTHER FORMS OF MORBIDITY AND LIFE-SHORTENING

Although the irradiation of mice and dogs early in the life-span indubitably increases subsequent age-specific mortality rates from nonmalignant diseases,[32,36,55,76] it has not been clear until recently whether the same aging phenomenon is important in man. However, the most recent studies of mortality in American radiologists[69] and in the Hiroshima and Nagasaki survivors[7,52] indicate that radiation exposure not only increases the mortality from leukemia and cancer in man, but also that from certain nonmalignant diseases as well. The most important group of disorders in this latter category comprises the degenerative diseases of the cardiovascular and renal systems. Here the mortality ratio (US. Radiologists)/(Unirradiated

Medical Specialists) is 1.2 for the age range 35 to 79 and 1.4 for the age range 65 to 79 years. For all nonleukemic cancers, the comparable ratios are 1.6 and 1.5. Because the absolute mortality rate for the cardiovascular-renal group of diseases is much higher than that for all cancers, the actual surplus of deaths among the United States radiologists is higher in the cardiovascular-renal than in the cancer group.[69]

Many authors have recently suggested that various idiopathic nonmalignant morbid and mortal diseases in man may be fundamentally autoimmune in character [Refs. 8, 14, 18, 25–30, 33, 34, 77]. Detailed analysis suggests that these diseases have a spontaneous disturbed-tolerance autoimmune etiology and that they are initiated by the formation of "forbidden-clones" of lymphoid cells.

"Forbidden-clone" is the expression coined by Burnet[31] to describe descendant cells from a mutant lymphoid progenitor stem cell. He postulates that these mutant descendant cells synthesize autoantibodies (cell-bound or humoral) that attack "self" tissues as though they were foreign or "not-self." The condition of self-tolerance that normally exists between the lymphoid system and self-tissues is abrogated with the formation of forbidden-clones. It is widely suspected that Burnet's[31] forbidden-clone concept may prove to be one of the most important ideas in the history of medicine. The statistics of the sex-specific and age-specific onset rates for diseases of known or suspected autoimmune etiology are in complete accord with the postulate that the growth of forbidden-clones is initiated by the occurrence of one or more gene changes (some form of mutation) in lymphoid stem cells.[14,26–29,49] When such diseases are confined to one or more subpopulations within the general population (and they often are), the predisposed subpopulations are characterized by specific genotypes.

A remarkable finding is that the average rate of the random initiating events (a form of somatic gene mutation) is effectively constant, in ordinary environments, from birth to death. The average rate of a specific event in predisposed females is either the same as or double that in similarly predisposed males. No significant departures from this simple rule have been observed so far and it seems probable that when the rate in females is double that in males an X-linked gene is implicated. A latent period inevitably elapses between the occurrence of the last somatic mutation and the symptomatic or clinical onset of the autoimmune condition. In one class of autoimmunity (where the target tissue is normally freely infiltrated by small lymphocytes) the average latent period, for a given environment, is twice as long in females as in males. In another class of autoimmunity (where the target tissue lies behind a blood-tissue barrier) the average latent period for a given environment is of the same duration in the two sexes.[25–28]

When a correction is made for latent period, it is found that the sex-specific and age-specific prevalence of chronic autoimmune conditions is generally described by one of the following equations:

$$N_{n,t} = F_0(1 - e^{-kt})^n \tag{11-11}$$

$$N_{r,t} = F_0(1 - e^{-kt^r}) \tag{11-12}$$

$$N_{n,r,t} = F_0(1 - e^{-kt^r})^n \tag{11-13}$$

In these equations, N_t is the sex-specific and age-specific prevalence of the disease (the proportion of a population by age and sex, with the disease initiated, at any given instant of time); t is the age at initiation in years (the age at which the last random initiating event occurs) and F_0 is the fraction of the general population (by sex) that is predisposed to the disease at birth. In Eq. (11-11), k is the average rate of somatic mutation for the induction of each of the n (dissimilar) independent forbidden-clones. If L is the number of cells at risk with respect to the initiation of each type of independent clone, and if m is the average mutation rate per gene at risk, per cell at risk, per year, then $k = Lm$. Equation (11-12), describes the model where the disease is initiated by r dependent-type random events, such as r gene mutations, in a single lymphoid stem cell, out of many similar cells at risk. If there are L such cells, r genes at risk in each, and an average mutation rate m, per gene at risk, per cell at risk, per year, then if the sequence of mutation is irrelevant, $k = Lm^r$. For this equation to hold, it is essential that $mt \ll 1$ for all t of interest. Equation (11-13) refers to a disease initiated by n independent clones, where each distinctive clone is initiated by r dependent-type events. In all examples, the fit of data to the equations depends upon there being no appreciable secular change in F_0, and no selective mortality in the predisposed subpopulation, as compared with the general population.

Sex-specific and age-specific initiation rates (onset rates corrected for latent period) are given by the differential coefficient of Eq. (11-11), (11-12), and (11-13) with respect to age:

$$\frac{dN_n}{dt} = nkF_0e^{-kt}(1 - e^{-kt})^{(n-1)} \tag{11-14}$$

$$\frac{dN_r}{dt} = rkF_0t^{(r-1)}e^{-kt^r} \tag{11-15}$$

$$\frac{dN_{n,r}}{dt} = nrkF_0t^{(r-1)}e^{-kt^r}(1 - e^{-kt^r})^{(n-1)} \tag{11-16}$$

These equations are relevant even when the disease is a fatal one, provided the remaining restrictions apply. Because they contain more structure than the integral (prevalence) forms they are the more suitable equations for determining n, r, and k.

In the special case of a mortal disease that is initiated by q independent-and/or dependent-type random events, and where the whole population is at risk, dN/dt will remain proportional to $t^{(q-1)}$ at all t provided dN/dt and mt both remain appreciably less than unity.[18]

If radiation is capable of inducing the same type of mutation as that which occurs spontaneously, then the problem of assessing the effects of radiation on the initiation of autoimmune diseases is entirely analogous to the problem of radiation carcinogenesis described in Section 11-6. The crucial ratio to be assessed is α/m (where α is the radiation mutation coefficient) with its dependence (if any) on factors such as absolute dose and dose rate. Seltser and Sartwell's observations[69] suggest that the value of α/m for the cardiovascular-renal group of diseases is comparable to that for nonleukemic malignancies, but clearly, many more data are needed to confirm this indication. Nevertheless, it seems quite likely that rather similar spontaneous and radiation mutational mechanisms (? spontaneous and induced DNA strand switching) are implicated in the pathogenesis of both autoimmune and malignant conditions.[25]

The latent period in autoimmunity—the interval during which the forbidden-clone proliferates and attacks the target tissue—is affected by various environmental factors which interfere with an endogenous defense mechanism.[14,27,28,30] Certain infections, certain drugs, and mental stress all impair defense (which is probably mediated through immunoglobulins) and hence it is important to determine whether radiation can also impair this system. It is a fair assumption that large acute doses will almost certainly be detrimental—both to the clone and to the defense—but there is as yet little evidence which bears directly on this problem.

11-8 CONCLUSIONS

From the viewpoint of health physics the most serious gap in our current biophysical knowledge is that concerning the effectiveness, and the mode of action of small doses and low dose rates of ionizing radiation in producing malignant and life-shortening (? mainly autoimmune) diseases. The closing of this gap is not a simple or easy exercise. Results for experimental animals can only be extrapolated to man when the pathogenesis of disease in both the species investigated and in man is thoroughly well understood. Investigations of the effects of small doses of diagnostic X-rays in man[72] are much more direct, but they are often difficult to interpret unambiguously, because generally, the irradiated individuals represent a selected group; it then becomes possible that the group receiving diagnostic X-rays is more prone to leukemia, for example, than are the control groups. Extrapolation of the effects at moderate and high doses to those at low is only possible when the details of the dose–response relationship are known independently. The possible variation of α with dose—in the low dose region, at low or high dose rate, and at low LET—are particularly difficult to predict. Even the estimation of upper limits requires some assumptions regarding mechanisms.

Nevertheless, some of these uncertainties should be reduced with the help of fundamental radiobiological, biophysical, and genetic investigations. Such studies should enable us to test models of radiation action—such as those outlined in this chapter—and to refine or replace them. The elucidation of the mechanisms of repair of premutational and sublethal damage would be especially valuable because it would then enable us to predict the pathological consequences of radiation at low doses and low dose rates with far greater confidence than we can at present.

Obviously, the exploration of the fundamental basis of radiation action at the genetic and cellular levels will of itself be insufficient for our purposes; in addition we have to unravel the complex biology of human cancer, autoimmunity, and aging. It is encouraging that intensive efforts are being directed to these ends and we can confidently anticipate that the next edition of this book will be more definitive in this respect than the current one.

The task of the health physicist, however, will not be ended by the acquisition of fundamental knowledge regarding the biological hazards of ionizing radiation. The equally difficult problem of applying such basic knowledge remains. Simple and effective monitoring of workers exposed to radiation represents a constant challenge and this must be done in such a way that the health hazards can be readily computed. Although the definition of an "acceptable" hazard is a social and political responsibility, the health physicist will be required to furnish some of the expert advice on which such decisions have to be made.

REFERENCES

1. Atwood, K. C., and A. Norman, *Proc. Natl. Acad. Sci. U.S.*, **35**, 696 (1949).
2. Barendsen, G. W., C. J. Koot, G. R. van Kersen, D. K. Bewley, S. B. Field, and C. J. Parnell, *Intern. J. Radiation Biol.* **10**, 317 (1966).
3. Barendsen, G. W., *Intern. J. Radiation Biol.* **8**, 453 (1964).
4. Barendsen, G. W., *Nature*, **193**, 1153 (1962).
5. Barendsen, G. W., T. L. J. Beusker, A. J. Vergroesen, and L. Budke, *Radiations Res.*, **13**, 841 (1960).
6. Bedford, J. S., and E. J. Hall, *Intern J. Radiation Biol.*, **7**, 377 (1963).
7. Beebe, G. W., M. Ishida, and S. Jablon, *Radiation Res.*, **16**, 253 (1962).
8. Blumenthal, H. T., and A. W. Berns, *Advan. Gerontol. Res.*, **1**, 289, (1964).

9. Boag, J. W., *Am. J. Roentgenol. Radium Therapy Nucl. Med.* **90,** 896 (1963).
10. Bruce, W. R., L. M. Pearson, and H. S. Freedhoff, *Radiation Res.* **19,** 606 (1963).
11. Burch, P. R. J., *Proc. Roy. Soc.* (*London*), **B162,** 223 (1965a).
12. Burch, P. R. J., *Proc. Roy. Soc.* (*London*), **B162,** 240 (1965b).
13. Burch, P. R. J., *Proc. Roy. Soc.* (*London*), **B162,** 263 (1965c).
14. Burch, P. R. J., *Brit. J. Psychiat.*, **110,** 808, 818, and 835 (1964a).
15. Burch, P. R. J., *Nature*, **202,** 711 (1964b).
16. Burch, P. R. J., *Ann. N.Y. Acad. Sci.*, **114,** 213 (1964c).
17. Burch, P. R. J., *Nature*, **197,** 1145 (1963a).
18. Burch, P. R. J., *Lancet*, **2,** 299 (1963b).
19. Burch, P. R. J., *Discovery*, **24,** 2, 16 (1963c).
20. Burch, P. R. J., *Nature*, **195,** 241 (1962).
21. Burch, P. R. J., *Radiation Res.*, **11,** 481 (1959).
22. Burch, P. R. J., *Brit. J. Radiol.*, **30,** 524 (1957).
23. Burch, P. R. J., *Radiation Res.*, **6,** 289 (1957).
24. Burch, P. R. J., and P. M. Bird, *Progress in Radiobiology*, p. 161. The Kynoch Press, Birmingham, England (1956).
25. Burch, P. R. J., and R. G. Burwell, *Quart. Rev. Biol.*, **40,** 252 (1965).
26. Burch, P. R. J., and D. Jackson, *J. Brit. Dent. Assoc.*, **120,** 127 (1966).
27. Burch, P. R. J., and N. R. Rowell, *Am. J. Med.*, **38,** 793 (1965a).
28. Burch, P. R. J., and N. R. Rowell, *Acta Dermato.-Venereol.*, **45,** 366 (1965b).
29. Burch, P. R. J., and N. R. Rowell, *Lancet*, **2,** 507 (1963).
30. Burnet, F. M., *Brit. Med. J.* **1,** 338 (1965).
31. Burnet, F. M., *The Clonal Selection Theory of Acquired Immunity*, Cambridge University Press, Vanderbilt University Press (1959).
32. Casarett, G. W., *Radiation Res. Suppl.*, **5,** 246 (1964).
33. Comfort, A., *Ageing, the Biology of Senescence*, Routledge and Kegan Paul, London (1964).
34. Comfort, A., *Lancet*, **2,** 138 (1963).
35. Cormack, D. V., and H. E. Johns, *Brit. J. Radiol.*, **25,** 369 (1952).
36. Cosgrove, G. E., A. C. Upton, C. C. Congdon, D. G. Doherty, K. W. Christenberry, and D. G. Gosslee, *Radiation Res.*, **21,** 550 (1964).
37. Danzker, M., N. D. Kessaris, and J. S. Laughlin, *Radiology*, **72,** 51 (1959).
38. Deering, R. A., and R. Rice, *Radiation Res.*, **17,** 774 (1962).
39. Elkind, M. M., *J. Cellular Comp. Physiol.*, **58,** 113 (1961).
40. Elkind, M. M., and H. Sutton, *Radiation Res.*, **13,** 556 (1960).
41. Elkind, M. M., H. Sutton-Gilbert, W. B. Moses, T. Alescio, and R. W. Swain, *Radiation Res.*, **25,** 359 (1965).
42. Elkind, M. M., A. Han, and K. W. Volz, *J. Natl. Cancer Inst.*, **30,** 705 (1963).
43. Finkel, M. P., *Radiology*, **67,** 665 (1956).
44. Fowler, J. F., *Phys. Med. Biol.*, **9,** 167 (1964).
45. Gray, L. H., *Brit. J. Radiol. Suppl.*, **1,** 7 (1947).
46. Hall, E. J., and J. S. Bedford, *Radiation Res.*, **22,** 305 (1964).
47. Haynes, R. H., *Physical Processes in Radiation Biology*, p. 51. Academic Press, New York (1964).
48. Haynes, R. H., and G. W. Dolphin, *Phys. Med. Biol.*, **4,** 148 (1959).
49. Holmes, M. C., and F. M. Burnet, *Heredity*, **19,** 419 (1964).
50. Howard-Flanders, P., *Advan. Biol. Med. Phys.*, **6,** 533 (1958).
51. ICRP, *Recommendations of the International Commission on Radiological Protection*, Pergamon Press, London (1964).

52. Jablon, S., M. Ishida, and M. Yamasaki, *Radiation, Res.* **25,** 25 (1965).
53. Kaplan, H. S., *Am. J. Roentgenol. Radium Therapy Nucl. Med.*, **90,** 907 (1963).
54. Lea, D. E., *Actions of Radiations on Living Cells*, 1st Ed., Cambridge University Press (London) (1946) and 2nd Ed. (London), (1955).
55. Lindop, P. J., and J. Rotblat, *Proc. Roy. Soc. (London)*, **B154,** 332 (1961).
56. Mortimer, R., T. Brustad, and D. V. Cormack, *Radiation Res.*, **26,** 465 (1965).
57. Oliver, R., and B. J. Shepstone, *Phys. Med. Biol.*, **9,** 167 (1964).
58. Platzman, R. L., *Radiation Res.*, **17,** 419 (1962).
59. Puck, T. T., P. I. Marcus, and S. J. Cieciura, *J. Exptl. Med.*, **103,** 653 (1956).
60. Randolph, M. L., *Ann. N.Y. Acad. Sci.*, **114,** 85 (1964).
61. Rossi, H. H., *Ann. N.Y. Acad. Sci.*, **114,** 4 (1964).
62. Rossi, H. H., *Radiation Res. Suppl.*, **2,** 290 (1960).
63. Rossi, H. H., *Radiation Res.*, **10,** 522 (1959).
64. Rossi, H. H., M. H. Biavati, and W. Gross, *Radiation Res.*, **15,** 431 (1961).
65. Russell, L. B., and M. H. Major, *Genetics*, **42,** 161 (1957).
66. Russell, W. L., *Repair from Genetic Radiation*, p. 205. Pergamon Press, London (1963).
67. Russell, W. L., L. B. Russell, and E. M. Kelly, *Intern. J. Radiation Biol. Suppl.*, p. 311 (1960).
68. Russell, W. L., L. B. Russell, and E. M. Kelly, *Science* **128,** 1546 (1958).
69. Seltser, R., and P. E. Sartwell, *Am. J. Epidemiol.* **81,** 2 (1965).
70. Sinclair, W. K., *Radiation Res.*, **16,** 394 (1962).
71. Spencer, L. V., and U. Fano, *Phys. Rev.*, **93,** 1172 (1954).
72. Stewart, A., W. Pennybacker, and R. Barber, *Brit. Med. J.* **2,** 882 (1962).
73. Till, J. E., and E. A. McCulloch, *Ann. N.Y. Acad. Sci.*, **114,** 115 (1964).
74. Upton, A. C., V. K. Jenkins, and J. W. Conklin, *Ann. N.Y. Acad. Sci.*, **114,** 189 (1964).
75. Upton, A. C., M. L. Randolph, E. B. Darden, Jr., and J. W. Conklin, *Biological Effects of Neutron and Proton Irradiations*, Vol. 2, p. 337, I.A.E.A., Vienna (1964).
76. Upton, A. C., M. A. Kastenbaum, and J. W. Conklin, *Cellular Basis and Aetiology of Late Somatic Effects of Ionizing Radiation.* p. 285, Academic Press, New York (1962).
77. Walford, R. L., *J. Gerontol.*, **17,** 281 (1962).
78. Zimmer, K. G., *Studies on Quantitative Radiation Biology*, Oliver and Boyd, Edinburgh (1961).

QUESTIONS

1. What methods have been used to specify radiation quality? What are their shortcomings?

2. Why should the biological effect of an ionizing radiation depend on its quality? Is the relationship between radiation quality and biological effectiveness the same for all biological end points?

3. What do you understand by the terms "direct action" and "indirect action"?

4. In what circumstances is "target theory" applicable? What information would you need in order to calculate "target size"?

5. Why should the effect of a radiation sometimes depend upon dose rate?

6. What effects of radiation are of particular importance in radiation protection?

CHAPTER 12

Radiation Biology

A. C. Upton and R. F. Kimball

12-1 INTRODUCTION

The first biological effect of ionizing radiation is the random absorption of radiation by atoms and molecules. From the resulting molecular alterations come biochemical disturbances that lead to functional and morphological changes in the cell. Cellular injuries, in turn, cause the pathological and physiological changes associated with "radiation sickness" and other large-scale manifestations of radiation damage to the entire animal. Finally, some of the initial molecular lesions can eventually affect whole populations of individuals by way of the self-reproducing mechanisms of heredity. Thus, the central feature of all radiation biology is the biological "magnification" at various levels of complexity of what would be very minor molecular damage when viewed from the standpoint of the radiation chemist or physicist.

In this chapter we consider effects at all these levels and, whenever possible, relate one to the other. More extensive accounts of radiation biology are contained in the following references: Hollaender (1954–1956, 1961), Lea (1955), Bacq and Alexander (1961), Claus (1958), and Errera and Forssberg (1960). General summaries are also contained in the reports of the National Academy of Sciences—National Research Council, of the British Medical Research Council, and of the United Nations Scientific Committee on the Effects of Atomic Radiations. Fairly general presentations, as well as more detailed research papers, are to be found in the published proceedings of the two International Conferences on Peaceful Uses of Atomic Energy held under United Nations auspices in Geneva. Review articles on special topics can be found in *Advances in Radiation Biology* (Academic Press, New York) and *Current Topics in Radiation Research* (North-Holland Publishing Co., Amsterdam).

12-2 EFFECTS ON CELLS

Radiation Chemistry of Cellular Components

We emphasize in this chapter only a few aspects of radiation chemistry that are of particular interest to the radiation biologist. The cell is composed of water, inorganic salts, a wide variety of small organic molecules, and a number of macromolecules, that is, proteins, nucleic acids, lipids, and carbohydrates. The smaller molecular species are sufficiently numerous so that doses of a few hundred rad could not appreciably alter their concentration directly. Some of the macromolecular species, however, are present in much smaller numbers, so that even very small doses of radiation could destroy or seriously alter all or a large fraction of some of them. The consequences of such a loss depend on the immediate importance of the affected molecules to the cell and the possibility of sufficiently rapid replacement or repair. The cellular effects of low doses (no more than a few hundred rad) of radiation, therefore, probably arise primarily from damage to macromolecules individually important to the cell and not readily replaceable or repairable.

Although the radiation chemistry of macromolecules in complex aqueous mixtures is, therefore, the most important system for radiation biology, it is not very suitable for studies of radiation chemistry. Consequently, it is necessary to rely heavily on simpler systems, in which the following findings are noteworthy:

1. Radicals produced through the radiolysis of water play a major role in causing radiation effects on solutes in dilute aqueous systems.
2. Molecular oxygen enhances many of these radical reactions.
3. Molecular oxygen also interacts with irradiated dry materials.
4. Radiation produces long-lived electron spin resonance centers in dry materials.
5. Energy absorbed within macromolecules is transferred to and concentrated in particular bonds.

The first of these findings has led radiation chemists and biologists to emphasize the distinction between *direct* and *indirect* actions of radiation. Most cells contain some 80 to 90% water by weight; therefore a very large fraction of all the radiation energy deposited in the cell is absorbed by water molecules. The discovery that the removal of molecular oxygen protects cells against radiation damage and the demonstration that this is a widespread phenomenon (Gray, 1957–58) led radiation biologists, on the basis of finding 2, to suggest that a large part of the biological effect of radiation was indirect. This conclusion was weakened, however, by the demonstration that oxygen also enhances radiation damage in dry materials and by studies that suggested possible roles of oxygen other than the radiolysis of water. Studies on radiation inactivation of specific macromolecules *in vivo* (Hutchinson and Arena, 1960) suggest that only radiation absorbed within some 10 to 30 A of a macromolecule has a good chance of altering it. Thus transfer of energy from regions outside macromolecules can occur and may account for a considerable fraction of the energy reaching these molecules. Since, however, the distances *in vivo* are much smaller than in dilute aqueous solutions, the distinction between direct and indirect effects probably does not imply a great difference in spatial distibution of the initial absorption of the radiation. Moreover, the distinction is hard to make in practice because most tests cannot distinguish between the two in biological systems.

Perhaps it is more useful to distinguish the three major stages in the evolution of cellular damage from the atomic and molecular levels: (1) initial absorption of radiation; (2) radiation-chemical reactions (usually rapid), leading to a more or less stable alteration of some molecules; and (3) biochemical and physiological reactions (sometimes requiring hours or even days), by which molecular alterations from stage (2) are converted into biochemically, morphologically, or physiologically detectable changes in the cell. The various phenomena often attributed to an indirect effect (enhancement by oxygen for example), concern reactions in the second stage.

Studies on electron spin resonance and on transfer of energy within macromolecules suggest that ionizing radiation may have greater chemical specificity in such systems than had earlier been assumed (that is, whereas randomness of initial absorption had led earlier investigators to assume a rather nonspecific damage to all molecules, studies with macromolecules have suggested that the energy tends to concentrate preferentially in certain bonds). This partial chemical specificity, combined with the biological "magnification" of particular types of damage, helps to account for the considerable specificity of ionizing radiation in inducing certain kinds of effects and not others.

Effects on Chromosomes

One of the most thoroughly investigated effects of radiation is the production of chromosome damage. Detailed discussions of this effect are to be found in all the books and reports listed at the beginning of this chapter, and these should be consulted for more information than can be given here.

The chromosomes are elongated, fiberlike structures in the cell nucleus and are composed largely of two kinds of nucleic acid, deoxyribonucleic acid (DNA) and ribonucleic acid (RNA), and two or more kinds of protein. Genetic investigations have established that chromosomes are a linear array of heredity-determining units, or genes, each gene carrying the information needed by the cell to make a specific protein or polypeptide. This information is encoded in the order of purine and pyrimidine base pairs in the DNA molecule of the gene. The encoded information is used to determine the order of bases in RNA; the RNA, in turn, determines the specific order of amino acids in proteins. In this way, the information in the genes is used to determine the properties of the enzymes, structural proteins, and other constituents of the cell.

The cells of many tissues are continuously multiplying by dividing in two. At each division each daughter cell receives a complete or nearly complete set of genes. This is achieved by two successive processes, chromosome duplication and mitosis.

Chromosome duplication, which occurs during interphase (the stage extending from one mitosis to the next), involves, among other things, duplication of the sequence of bases in DNA. The double-helix model of DNA and studies on its *in vitro* synthesis have gone far toward elucidating the biochemistry of DNA duplication. This duplication is amazingly precise. A gene with a slight internal change, even a change in a single base pair, or a chromosome with an altered arrangement of a whole group of genes is duplicated just as precisely as the original gene or gene order. Such self-reproducing changes are often referred to broadly as *mutations*. These mutations are of distinct types: *chromosomal aberrations*, which are changes in the order (or losses) of whole blocks of genes; and *gene* or *point mutations*, which are changes within a single gene. The larger chromosomal aberrations can be directly observed under the microscope, but it is usually very difficult to distinguish small chromosomal aberrations from gene mutations; consequently, the word mutation is often used to cover both kinds of events.

Mitosis, the second process, is described in every elementary textbook on biology. At its start the chromosomes have already duplicated and consist of two parallel strands or *chromatids*. The strands line up along the equatorial plane of the *spindle;* the two chromatids of each chromosome then separate, going to each pole of the spindle. The two chromatid (now chromosome) groups reconstitute new nuclei, one for each daughter cell. Thus, each daughter cell gets one, and only one, replica of each of the chromosomes in the parent cell. The proper alignment and movement of each chromosome and chromatid depend on a body known as the *centromere*. Some of the serious disturbances caused by chromosome aberrations come from misdistribution of chromosomal material at mitosis occasioned by rearranged chromosomes with more than one centromere or fragments of chromosomes with no centromeres. The processes of precise self-duplication of chromosomes and mitosis make the cell the example par excellence of a system that

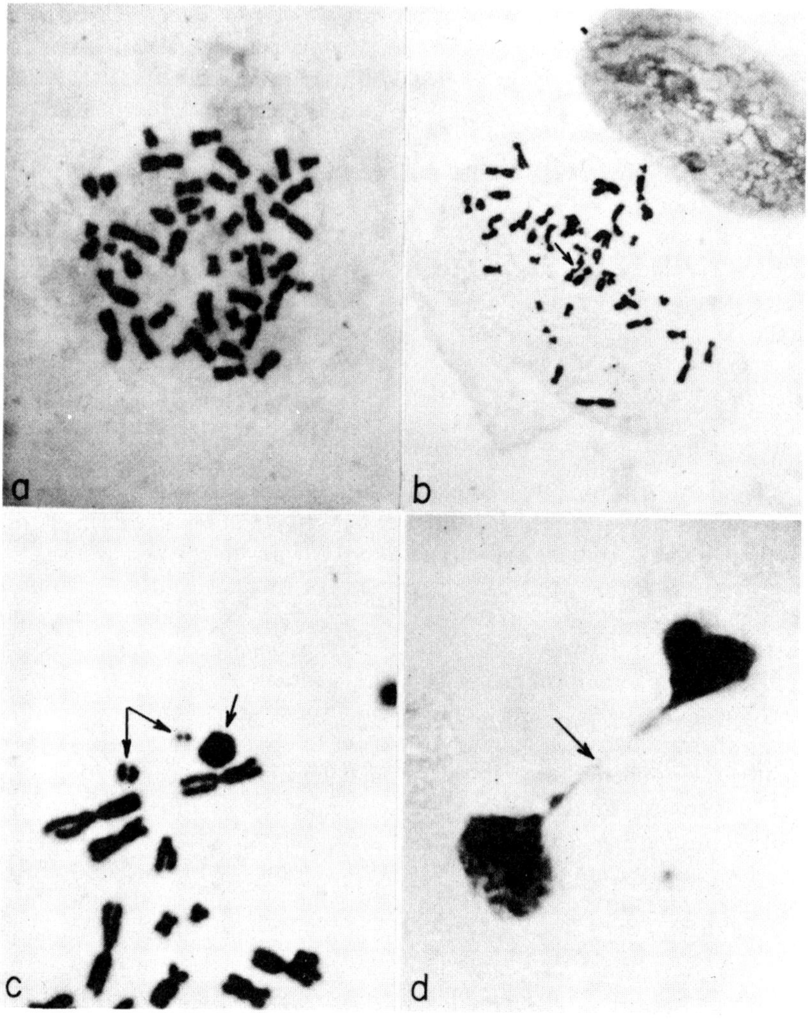

Fig. 12-1. X-ray-induced chromosome aberrations in human cells. a, chromosomes of male, unirradiated; b, chromosomes of female, showing deletion (arrow); c, part of figure from irradiated male, showing ring chromosome (single arrow), and fragments (double arrow); d, very late telophase of mitosis, showing bridge (arrow) from irradiated male. (Illustrations kindly supplied by Dr. M. A. Bender.)

can magnify a minor molecular alteration so greatly that a single initial radiation event eventually may be spread by replication through many individuals in a population.

A large number of studies on the induction of chromosome aberrations has led to a more detailed theory for this form of radiation damage than for any other (see especially Lea, 1955; Giles, 1954; Kaufmann, 1954; and S. Wolff, 1960, 1961). The initial damage is believed to be a break in the chromosome thread caused by radiation absorbed in or near the region of the break. The broken ends may then reunite with each other (*restitution*), with ends from other breaks (*two-break aberrations*), or may fail to reunite (*one-break aberrations*).

Restitutions cannot be directly observed, but indirect evidence suggests that the overwhelming majority of all breaks reunite in this manner. One- and two-break aberrations produce a great variety of abnormal chromosomes (Fig. 12-1): fragments with no centromere, dicentrics (two centromeres), rings, and chromosomes with one centromere and various genic rearrangements. Some of the first three types are easily seen at the metaphase stage of mitosis and are thus scored most often in quantitative work. Because of their abnormal structure, they are not distributed properly at mitosis. The result is that the daughter cells have abnormal numbers of chromosomes and genes and therefore may not live.

In the much-studied plants *Tradescantia* and *Vicia*, two-break aberrations increase nearly as the square of the dose when radiation of low linear energy transfer (LET) is administered at high intensity, but increase at a lower rate and almost linearly with the dose when the plants are irradiated at low intensity (Fig. 12-2). For radiation of high LET, two-break aberrations increase almost linearly with dose and independently of intensity. One-break aberrations, on the other hand, are intensity independent and increase linearly with the dose for radiations of either low or high LET. When radiation of low LET is given at high intensity but in several fractions separated by rest periods, the number of two-break aberrations is nearly proportional to the square of the total dose if the rest periods are short. It is proportional to the sum of the squares of the fractional doses, however, if the rest periods are long. In the sperm of the fruit fly *Drosophila*, two-break aberrations increase roughly as the two-thirds power of the dose for low LET radiation and linearly with the dose for high LET radiation; there is no intensity effect for either radiation.

These relations are consistent with the breakage and reunion hypothesis, with the provision that the two breaks required for rearrangement must be produced close together in time and space to exclude restitution. Because individual ionization tracks of radiations of high LET have an appreciable chance of producing two breaks in neighboring chromosome threads by a single radiation event (hit), the resulting one-hit, two-break effects would be expected to (1) increase linearly with dose, (2) be intensity independent, and (3) comprise most of the rearrangements produced by such radiations at the intensities usually available. The individual ionization tracks of radiation of low LET, on the other hand, have a much smaller chance of producing two neighboring breaks. Consequently, at high intensity most of the two-break aberrations caused by such radiations come from two independent radiation

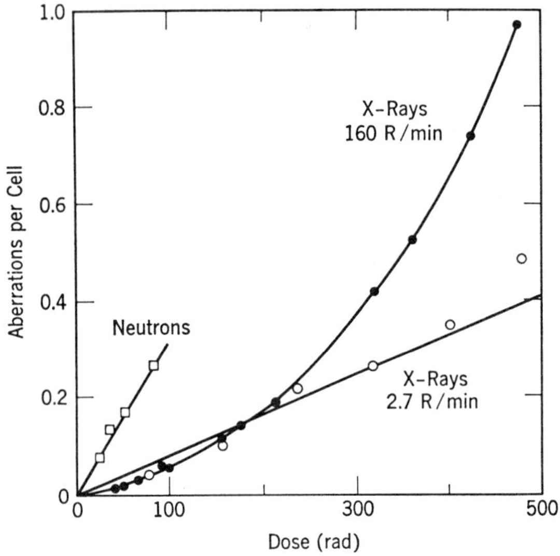

Fig. 12-2. Dose curves for two-break chromosomal aberrations. The X-ray data are replotted from Sax (1941); the neutron data from Giles (1943). The neutron doses were given in *n* units and have been converted to rad by multiplication with the factor 2.5 (this conversion is very approximate and should not be relied upon).

events (two-hits). Consequently these aberrations increase approximately as the square of the dose. At low intensity a large proportion of breaks restitute before others can be formed nearby. Therefore rearrangements come mainly from the small number of one-hit, two-break events that are produced by such radiations. Nearly complete interaction occurs between breaks produced by two successive exposures if the rest period is short compared to the reunion time; there is little or no interaction if the rest period is long.

There is no reason to assume from theory or from experimental findings that there is a true threshold dose. Any exposure, no matter how low the dose or the intensity, has a finite, if small, probability of producing some aberrations The mechanism for repairing chromosome damage by restitutional reunion is efficient but under no known conditions is it perfect.

Thus, the relative biological effectiveness (RBE, discussed later in this chapter) of radiations of different LET will vary, depending on the conditions under which the low LET radiation is given. There is indirect evidence, moreover, that radiations of high LET are more efficient in producing the initial breaks than those of low LET, evidence suggesting that several ionizations close together are usually required to produce a break. The actual RBE's found will differ greatly, depending upon the radiations used and the conditions of the experiment. Most reported RBE values for comparisons between radiations of high and low LET, however, are in the range of 3 to 6.

Various auxiliary treatments have been used to modify the production of chromosome aberrations (reviewed by Sheldon Wolff, 1960). Such studies were stimulated by the discovery that molecular oxygen, when present during irradiation, increases the number of aberrations. Thus the number of breaks produced by radiation of low LET is decreased by a factor of 2 to 3 if the material is irradiated in the absence of molecular oxygen; however, the effect is much smaller with radiations of high LET. Various other substances have shown protection similar to anoxia. Some substances probably remove molecular oxygen; others may take part in the rapid radiation-chemical reactions occurring at the time of irradiation.

Workers more recently have concentrated on the reunion process. The time for restitutional reunion varies greatly in different organisms. Estimates of a few minutes have been made for reunion in the microspores of the plant *Tradescantia*, and a few hours for cells of the soaked seed of the bean *Vicia*. In *Drosophila* sperm, there seems to be no reunion of some, perhaps all, of the breaks until after fertilization—sometimes many days after irradiation. On the other hand, fairly rapid reunion occurs in immature germ cells of the same species. Thus, the time available for modification of the reunion process varies considerably with the type of cell and its condition at irradiation.

Reunion, in many instances at least, seems to involve cellular metabolism. It can be delayed by blocking either the reactions that provide energy-rich compounds, such as adenosine triphosphate (ATP), or the reactions leading to protein synthesis. In some cases radiation itself can inhibit the reunion mechanism and this inhibition is enhanced when irradiation is carried out in the presence of oxygen. Thus, three different effects can be distinguished: (1) promotion of breakage through radiation-chemical reactions, (2) promotion of damage to the reunion mechanism, and (3) increase in the rate of reunion by enhancement of cellular respiration and ATP formation. The first two reactions require that oxygen be present during irradiation; this tends to increase the probability of rearrangement. The third requires that oxygen be present after irradiation; this tends to decrease the probability of rearrangement. Some of these relations are summarized in Fig. 12-3.

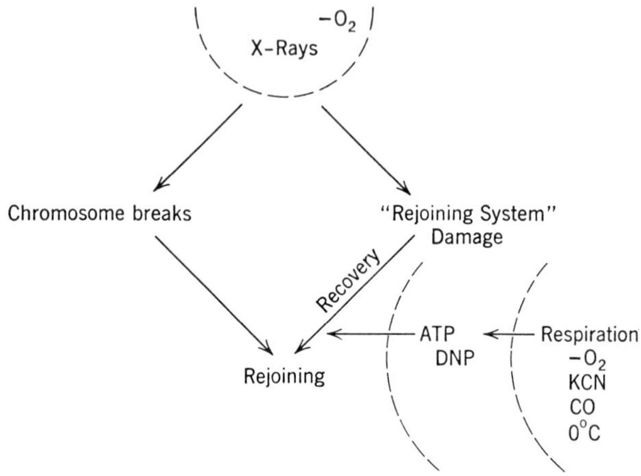

Fig. 12-3. Diagram from Wolff and Luippold (1956) of the influence of radiation and metabolism on chromosome breakage and reunion. (Reproduced by permission of Oliver and Boyd Ltd., London.)

The influence of the reunion time on the number of aberrations depends on several circumstances. A long reunion time will mean more aberrations if the dose is given at low intensity or in widely separated fractions. Change in reunion time may have no effect on the number of aberrations if the dose is given over a relatively short period.

The number and kind of aberrations also vary considerably with the stage at which the cell is irradiated. Some variations with stage seem to depend on whether the chromosomes are single or double at the time of irradiation. A full understanding of other factors has not yet been reached.

Mitotic Delay

Another effect of radiation that has been extensively investigated is mitotic inhibition or delay. Ordinarily, the delay results from a block in the progress of the cell through some particular stage in its life cycle, and the block is usually reversible. Excellent reviews are given by Carlson (1954) and Whiting (1961).

One of the most carefully studied cases is that of the dividing neuroblast cells of the grasshopper embryo. After exposure to about 10 rad, cells in metaphase, anaphase, and telophase complete division on schedule; cells in prophase halt in this stage, and those in interphase proceed to prophase and then stop. Some cells in prophase may even revert to interphase. Later, as recovery occurs, the cells that have accumulated in prophase or late interphase pass through subsequent mitotic stages together. These effects produce a characteristic curve (Fig. 12-4) for the relation of the mitotic index (the ratio of the number of mitoses in the irradiated embryo to the number in the control) to time after irradiation. Similar inhibition of division occurs in mammalian cells at low dose levels, giving rise to curves like those in Fig. 12-4 for a variety of mammalian systems.

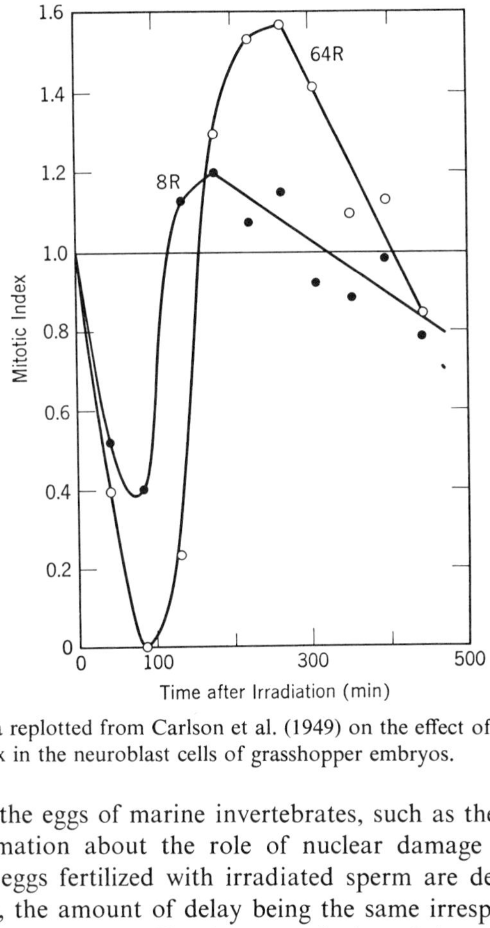

Fig. 12-4. Data replotted from Carlson et al. (1949) on the effect of X-rays (2 R/min) on the mitotic index in the neuroblast cells of grasshopper embryos.

Studies on the eggs of marine invertebrates, such as the sea urchin, have yielded information about the role of nuclear damage in mitotic delay. Unirradiated eggs fertilized with irradiated sperm are delayed in cleavage (cell division), the amount of delay being the same irrespective of the time between irradiation and fertilization. Irradiation of the egg produces similar delay, but the effect decreases with increasing time between irradiation and fertilization. These observations suggest that recovery occurs in the egg but not in the sperm. When either the sperm or egg is irradiated with but moderate doses, only the first cleavage is appreciably delayed, the interval between remaining cleavages being nearly normal; recovery apparently takes place after the first cleavage. The production of cleavage delay by irradiating the sperm, considered with other evidence, strongly suggests that the primary damage is to the nucleus and that recovery from this damage requires cellular metabolism since recovery occurs only in the egg, fertilized or unfertilized.

Inhibition of DNA Synthesis

A problem sometimes thought to be closely related to mitotic delay is inhibition of DNA synthesis. This radiation effect has now been reported by a large number of workers (see review by Ord and Stocken, 1960), and all agree that it can be detected after relatively low doses, in contrast to most other biochemically detectable effects. Synthesis of DNA usually occurs at some time during interphase in the so-called synthesis, or S, period. Obviously, cells blocked at another stage, such as prophase, will fail to reach the next S period at the proper time so that DNA synthesis will appear inhibited. Some workers suggest that this blocking can account for all cases of inhibition of DNA synthesis, at least at low doses. Others think that there is evidence for a more specific block in the onset of the S period itself. With higher doses (several hundred to several thousand rad), progress of the cell through its whole division cycle may be slowed down, including progress through the S period. Thus the finding that DNA synthesis is inhibited or slowed down in various tissues may have a variety of explanations, some of which involve DNA synthesis only secondarily.

Cell Death and Cell Inactivation

Many of the effects of radiation on the whole animal, such as injury to blood-forming organs and the digestive tract, are probably caused by death of cells. Studies on cell death have been made with a great variety of cells, but most of our information comes from microorganisms. Reviews of this work are to be found in the appropriate chapters of the books listed in Section 12-1. The usual method in such work is to spread a known number of cells on agar culture plates and to count the number of macroscopically visible colonies that are formed. It is assumed that each viable cell will produce a colony of daughter cells. This technique has recently been applied successfully to mammalian cells (Puck et al., 1957). Failure to form a visible colony may be the result either of outright cell death or the failure of surviving cells to multiply; consequently, the more neutral expression, *inactivation*, is often used instead of *death*.

Radiation dose-mortality data obtained in such experiments are usually presented in the form of a survival curve. This usage is based on the principle that the number of cells surviving a given increment of dose is a function of the number surviving up to that increment, and that once a cell is inactivated further damage is inconsequential. Mathematically, this relation can be expressed by the differential equation

$$\frac{dN}{dD} = f(N) \qquad (12\text{-}1)$$

where N is the number of survivors and D is the dose. In the simplest case, $f(N)$ equals $-kN$, and the widely used exponential survival curve

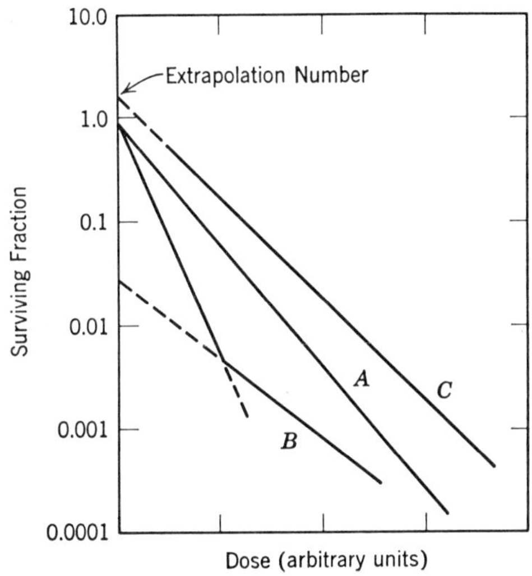

Fig. 12-5. Diagrammatic representation of three forms of survival curves (see text).

results from integration of Eq. (12-1)

$$\frac{N}{N_0} = e^{-kD} \qquad (12\text{-}2)$$

where N_0 is the number of viable cells at zero dose. This relation implies that all cells in the population have an equal chance of being inactivated and that any given cell is either inactivated or unaffected after a given increment of dose. This kind of survival curve, or a reasonable approximation to it, has been found to hold for a variety of microorganisms. A plot of the logarithm of the fraction of survivors yields a straight line (Fig. 12-5, curve A). Examples of two other common forms of survival curves are also given in the same figure. Curve B shows the result of irradiating a mixture of cells of very different sensitivities. Curve C shows the result for cells that are inactivated only after first sustaining sufficient nonlethal damage. The initial shoulder on curve *C* occupies the dose range in which most cells are acquiring sufficient damage so that any additional injury will inactivate. Once this point is reached, the curve essentially straightens out and follows the simple exponential form. Curves such as *C* can be characterized by two numbers, the slope of the linear portion of the curve and the intercept derived by extrapolating the linear portion to the ordinate. The latter number is sometimes called the "hit"-number, or "hitness," from a form of target theory that can be applied to such curves. It seems better, however, to use the more neutral term "extrapolation number," since the intercept can be interpreted in a variety of ways.

The published curves for mammalian cells frequently seem to have an extrapolation number around 2, but this number may vary considerably. In any case, the curves are of a form that suggests that the cell must first receive some nonlethal injury before it is finally inactivated. Reversible injury is also suggested by results of fractionated dose studies (Elkind and Sutton, 1960).

It is customary to express the sensitivity to killing as the dose required to reduce the proportion surviving to 50% (LD_{50}) or to 37% (inactivation dose). The latter is the dose required to give an average of one killing hit per cell on the simple exponential model, thus it is sometimes called the mean lethal dose. Values of the LD_{50} of 100 to 200 rad have been reported for mammalian cells cultured by single-cell techniques.

Chromosome aberrations are potentially an important cause of cell death; that is, the loss of large segments of chromosomes through misdistribution of rearranged fragments at mitosis probably leads to lethal chromosomal imbalance. An attempt has been made (Puck, 1958) to deduce, from the form of the survival curve, that this phenomenon is the main cause of death for mammalian cells in culture. Serious difficulties exist for this interpretation, however (Bender and Wolff, 1961): consequently, the role of chromosome aberrations in causing the observed mortality of mammalian cells is still in doubt even though it is perfectly clear that many types of aberrations could cause death if no other cause supervened.

There are many instances in mammals where cell death occurs before the first postirradiation division; for example, in the very sensitive late type A and intermediate spermatogonia and early oocytes of the mouse (Oakberg, 1960) for which the median lethal doses are quite low (25 and 8 rad, respectively). Such deaths before cell division cannot be caused by loss of chromosomal parts through misdistribution but may be caused by some other kind of injury to the genetic mechanism. There is appreciable evidence from experiments with lower organisms that death after small doses is predominantly caused by nuclear injury (Whiting, 1961).

Other Effects

A variety of other effects of radiation on cells has been reported, such as effects on various enzyme systems and effects on permeability. Results are still quite limited, and widely accepted generalizations of the kind that are available for chromosome aberrations, mutations, division delay, and cell inactivation do not yet exist. Part of the problem of elucidating such effects stems from the difficulty of distinguishing between effects on cell populations and on individual cells.

A priori, it might seem improbable, because of the multiplicity of enzyme molecules present in every cell, that low doses of ionizing radiation could seriously alter the enzyme constitution of the cell by direct destruction of enzymes or enzyme systems. Nonetheless, there have been a number of reports that relatively low doses alter some of the phosphorylations involved in energy metabolism. It is rather generally agreed that over-all protein synthesis is not especially sensitive; however, inhibition of synthesis of specific enzymes has been reported. Inhibition of DNA synthesis has already been discussed in connection with mitotic delay.

Further discussion of metabolic and other effects may be found elsewhere (Hollaender, 1954–1956; Claus, 1958; Bacq and Alexander, 1961; and Errera and Forssberg, 1960).

Protection and Recovery in Cells

Cells may be protected against radiation damage before and during irradiation and helped to recover afterward (see Hollaender, 1960, for detailed coverage of this subject). We have already touched on this problem in the section on chromosome aberrations and have also pointed out that mitotic delay is a reversible phenomenon. Much of the work on protection and recovery has been done with microorganisms.

Many kinds of cells can be protected from inactivation by anoxia during irradiation. Usually the effectiveness of the radiation can be reduced to about one-third by such treatment. A number of substances that protect against radiation probably act by removing oxygen. Others, however, expecially certain organic sulfur compounds, also act in different ways. Lowering the temperature and removing water also afford protection.

In addition to measures that operate only during irradiation, and so presumably affect rapid radiation-chemical reactions, there are others that act after irradiation, the so-called recovery measures. Recovery often, perhaps always, involves the active participation of cellular metabolism. In some instances, inhibition of one or more phases of cellular metabolism after irradiation has led to recovery. Perhaps the cell is given more time to recover if processes such as chromosome duplication or cell division are inhibited for a time. Perhaps some of the initial radiation damage is converted to irrevocable injury through one or more metabolic steps but is dissipated harmlessly if these steps are inhibited. In other instances, temporary support of the irradiated cell by adding nutrients has proved useful; this result suggests that loss of the ability to make certain substances occurs but is reversible. Although it is clearly established that both potentially lethal or hereditary damage to cells can be repaired after irradiation and the time available for such repair varies greatly, most evidence suggests that the interval is rarely, if ever, greater than one cell generation.

12-3 INHERITED EFFECTS

Germ Cell Formation

We turn now to the problem of mutation and its effects on future generations. Mutations can be produced in any cell of the body, but only those in the germ cells are transmitted to future generations. Before discussing mutations, however, we will review briefly the development of germ cells in the mammal and describe their sensitivity to radiation.

In the adult male, immature stem cells, called type A spermatogonia, are scattered throughout the tubules of the testis. These cells give rise by division both to more type A spermatogonia, thus maintaining the stem cell line, and to other cells known as type B spermatogonia. The latter multiply by successive cell divisions and then undergo two meiotic divisions to produce mature germ cells, or sperm.

The development of female germ cells, or ova, is different. According to the best evidence, all early stages of oogonial development have already occurred by the time of birth. The postnatal ovary therefore contains no stem cells but does contain primary oocytes, which remain throughout most of life in an early stage of the first meiotic division. From time to time during reproductive life, a few of these oocytes increase in size and complete the meitotic divisions to produce mature ova at about the time of fertilization. Thus, the ovary, unlike the testis, has only a limited number of nonrenewable germ cells, which are gradually depleted by successive ovulations.

Type B, intermediate, and some type A spermatogonia are very sensitive to radiation and are killed by low doses; whereas mature sperm, intermediate stages in spermatogenesis, and some type A spermatogonia are considerably more radioresistant. Other type A spermatogonia may be quite sensitive. This differential sensitivity accounts for the typical course of events in males exposed to moderate doses: The death of many of the spermatogonia curtails the production of new sperm until regeneration of the germinal epithelium from the more resistant type A cells has had a chance to occur. Thus, first, there is a period of declining fertility, next a sterile period, and then the eventual recovery of fertility. Progeny obtained from irradiated males mated in the presterile period come mainly from cells irradiated as sperm or as other relatively mature stages in spermatogenesis. Progeny obtained after the sterile period, however, come from cells irradiated as type A spermatogonia. Since the total development of mature sperm from type A spermatogonia occupies only a period of a few weeks, the dose accumulated by type A spermatogonia is likely to exceed that accumulated by any other maturation stages.

The situation in the female seems to vary from species to species. In the mouse, the primary oocytes in the earliest follicle stages are very sensitive (LD_{50} about 8 rad), and permanent sterility sets in once the supply of the later, more resistant stages is exhausted. In the human, even doses of several hundred roentgens do not induce permanent sterility. Instead, the fertile-sterile-fertile pattern is found (Peck et al., 1940).

Segregation and Recombination

Most higher organisms have two sets of chromosomes in nearly every cell in the body, one set inherited from each parent. Each chromosome is thus one of a pair differing from all other pairs in substituent genes. In man, there are normally 23 chromosomes per set, or 23 pairs of chromosomes per diploid cell. Although both members of the same pair of chromosomes are alike in many of their hundreds of homologous substituent genes, they usually differ in some genes. Nonidentical homologous genes are called *alleles* of one another. Each allelic form of a gene usually influence the same biological character, or trait, as the other, but in a different way. Allelic differences arose originally by mutation, most of those presently in the population probably arising many generations ago. When a given gene is present in an individual in two different allelic forms (the genes of the pair differ), the individual is said to be *heterozygous* for that gene. When the gene is present in only one allelic form (both genes of the pair are identical), the individual is said to be *homozygous* for that gene.

The meiotic divisions in spermatogenesis and oogenesis reduce the double or diploid number of chromosomes to the single or haploid number in each sperm and ovum. In so doing one member of each chromosome pair is chosen at random to make up one complete set of chromosomes in each gamete. The total number of haploid sets that can be drawn from the 23 pairs of chromosomes in man is obviously very large and it is improbable that any children, except identical twins (derived from the same fertilized egg), ever get precisely the same set of chromosomes from a given parent.

The new individual resulting from the union of egg and sperm thus starts life with one full set of chromosomes taken at random from each parent. The various laws of inheritance, such as Mendel's laws, all follow from this rather simple fact plus certain complications such as the "linkage" of many genes in the same chromosomes and the exchange of genes between homologous chromosomes by crossing over. Such complications, however, need not concern us here. More detailed expositions are presented by Auerbach (1956), Wallace and Dobzhansky (1959), and in elementary texts on genetics.

Mutations

The induction of mutations is studied experimentally in higher organisms by irradiating either the male or female parent and looking for certain kinds of changes in the immediate (F_1) offspring or in later generations. When a mutant allele has a readily detectable effect in the presence of a nonmutant allele, the mutation is said to be *dominant*. When the mutation must be made homozygous or heterozygous with a similar mutant allele before its effects can be easily detected, it is said to be *recessive*; however, there is reason to think that many, perhaps all, so-called recessive mutations have effects even when heterozygous, although these effects are often too small to be readily detected.

In the mouse a number of categories of mutations or mutation-like events have been studied. Among these are: (1) dominant lethals, (2) dominant complete sterility, (3) inherited semisterility, (4) recessive lethals, (5) specific locus mutations, and (6) dominant visibles (W. L. Russell, 1954).

The first three categories seem to be caused by gross chromosomal changes. Dominant lethals (1) kill the fertilized egg, or zygote, very early in its development, usually before its implantation in the uterus. They are believed to be, at least in part, the consequence of chromosome losses occasioned by chromosome aberrations. Dominant complete sterility (2) is manifested by full sterility of the offspring of the irradiated parents, but its cause has not been analyzed. Inherited semisterility (3) results from certain kinds of transmissible chromosome aberrations (reciprocal translocations) that are not in themselves lethal but can cause difficulties in chromosome distribution at meiosis; some of the gametes receive incomplete chromosome sets and produce inviable zygotes.

These changes are found in irradiated oocytes and sperm but rarely in surviving spermatogonia. To avoid the consequences of such changes, therefore, males who have received an acute dose of radiation should refrain from procreation for a few weeks (W. L. Russell et al., 1958). This precaution would also greatly reduce the frequency of other transmissible mutations since these are produced at a lower rate in spermatogonia than in sperm. The bulk of the information on mutation rates in *Drosophila* has been obtained with recessive lethals (4), which are mutations that do not allow survival when present in homozygous condition. Studies of specific locus mutations (5) have yielded information on mutation rates in both the mouse and *Drosophila*. These are mutations at gene loci for which a laboratory mutant is already available. They are produced by irradiating a homozygous dominant animal and crossing it to a mate homozygous for the recessive laboratory mutant allele. If a new mutant (not necessarily identical with the laboratory allele) has been induced, the progeny will show the mutant character. In both (4) and (5), the mutations may be true gene mutations or small chromosome rearrangements, and a distinction is often difficult to make. About half of the mutations in category (5) in the mouse are also lethal when made homozygous.

The data from both the mouse (Fig. 12-6) and *Drosophila* are consistent with the interpretation that mutations increase as a linear function of dose. Therefore, despite the fact that the data are subject to rather large statistical errors and have not been extended to very low doses (25 rad in *Drosophila*), the best inference is that there is a finite amount of mutation produced no matter how low the dose. There are, however, various saturation and selective effects at high dose levels that make the curve flatten or even decline (Fig. 12-6). These are not contradictory to the idea of essentially linear increase at low doses. The influence of LET in mutagenesis is not as clearly established as it is in the production of major chromosome aberrations, but the available evidence suggests that it is significant (United Nations, 1966).

Many early workers believed that there was no dose-rate effect. Indeed, this is true for mutations induced in mature sperm, studies on *Drosophila* sperm being especially responsible for the earlier generalization. Work on the mouse (W. L. Russell et al., 1958, 1960), however, has shown that appreciably fewer mutations are produced at low dose rates than at high ones in irradiated oocytes and spermatogonia (Fig. 12-6). The discovery of this effect of the rate or intensity at which the dose is given is still so recent that only a limited number of doses have been tested. Nonetheless, it would be incorrect to assume that the amount of mutation per roentgen decreases indefinitely with decreasing intensity. It seems probable (W. L. Russell et al., 1960) that below a certain intensity there is no further decrease in effectiveness.

Russell, Russell, and Kelly (1960) have suggested that the intensity effect is based on some form of repair of the initial damage leading to mutation. High-intensity radiation perhaps damages this hypothetical repair mechanism, thereby allowing more of the initial damage to evolve fully into mutation. Evidence for this interpretation is available in studies with plants and microorganisms. It may be recalled from the section on chromosome aberrations

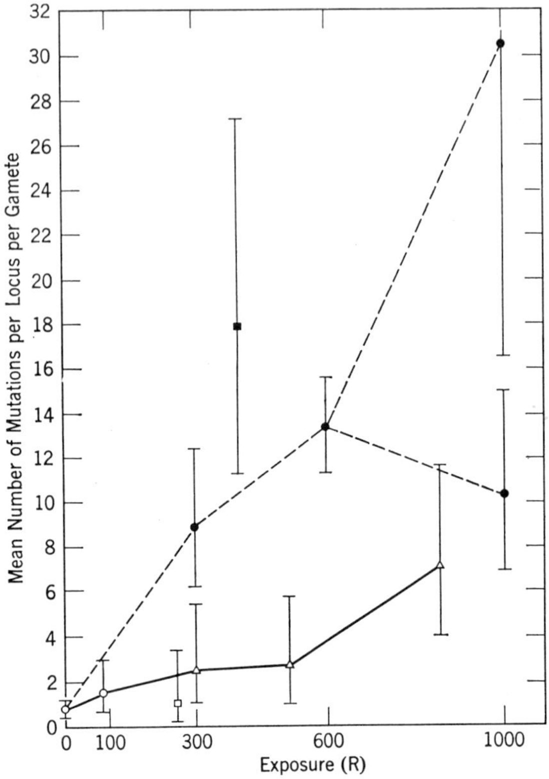

Fig. 12-6. Exposure curves for specific locus mutations in the mouse; 90% confidence intervals shown. Solid points represent results with acute X-rays (80–90 R/min). Open points are chronic γ-ray results (triangles and square, 90 R/week; circle, 10 R/week). Squares are mutation rates in females; all other points, mutation rates in males. The point for zero dose is the sum of all male controls. The top 1000 R point represents results of a single exposure; the lower one, results of successive exposures to 600 and 400 R, respectively, separated by an interval of 15 weeks. (Russell et al., 1960.)

that radiation has been shown to damage the chromosome rejoining or repair mechanism. Studies on mutation induction in microorganisms have shown that processes involving cellular metabolism intervene between the initial radiation damage and the final mutation, which is not established until the chromosomes duplicate (Hollaender and Kimball, 1956; Conger, 1961). The exact nature of these intervening processes is still the subject of disagreement, but their existence is clearly established. Radiation could thus influence the course of these intervening events by modifying the cellular metabolism involved or by changing the time available through delays in chromosome duplication (Kimball et al., 1961). For further discussion of these problems see Sobels (1963).

The work with microorganisms shows that the time available for such intermediate processes is limited to the first cell generation after irradiation. Once the cell has divided the mutation is irreversible except for reverse mutations, which are just as rare as the original event. This is clearly shown for the mouse by the following observation (W. L. Russell et al., 1958): Young males were irradiated, and after the sterile period was over, they were bred to unirradiated females throughout the rest of their lifetimes; the number of mutations detected in offspring conceived during the two halves of the reproductive life-span was not significantly different. Thus, the type A spermatogonia containing mutations must have reproduced at a rate not distinguishably different from the nonmutant ones, and no "cure from mutation" occurred with time.

The long-term consequences of mutation have been the subject of a large number of reviews and discussions. Note especially the appropriate chapters in books edited by Hollaender (1954–56) and Claus (1958), and to the reports of the National Academy of Sciences—National Research Council (1956, 1960), the British Medical Research Council (1956), the World Health Organization (1957), and the United Nations Scientific Committee on the Effects of Atomic Radiation (1962, 1964, 1966). Semipopular books by Auerbach (1956) and Wallace and Dobzhansky (1959) may also be consulted as well as articles by Wright (1948) and Muller (1950). First, it is sometimes useful to distinguish between first generation effects and those involving later generations. Some of the most severe first generation effects, the dominant lethals, have been previously mentioned. They represent a relatively minor hazard in humans because (1) they usually cause death before implantation and would often go completely unnoticed; (2) they are produced in very low frequency in spermatogonia which are the germ cells accumulating most of the dose from chronic irradiation in males; (3) in the case of acute irradiation of the male, the simple expedient of refraining from procreation for a few weeks would avoid the consequence of such changes.

Many mutations that are normally considered recessive for detection purposes nevertheless have appreciable, if small, effects when heterozygous. The aggregate of such small heterozygous effects in the first and later generations may well be of greater importance than their more noticeable homozygous effects. This result is probable, provided the allele is not very common, since the number of individuals carrying only one representative of such mutant alleles would be much larger than the number carrying two. The existence of such small dominant effects in heterozygous mice has been demonstrated (W.L. Russell et al., 1958) for what are conventionally considered recessive mutants. Moreover, in addition to such dominant effects of specific locus mutants, over-all reduction in survival to weaning age and reduction in total life-span have been found in the offspring of matings between irradiated male mice and unirradiated females. These first generation effects, therefore, must be considered in assessing the genetic hazards of radiation for man. Their existence strengthens the view that an appreciable part of the total damage to all generations results from the aggregate of small heterozygous effects. Such deleterious mutations would be expected to accumulate over many generations if the exposure to radiation were permanently increased. It is this gradual accumulation that makes the hazard of small exposures to large populations so potentially important and at the same time so insidious a danger.

In the human population there are many representatives of each gene, two representatives per individual. These allelic representatives of the 10,000 or more different genes are being continuously reshuffled and recombined as a result of sexual reproduction. The number of possible combinations is tremendous, and only a small proportion is realized in any one generation. Since different individuals contribute unequally to the population of the next generation, partly because of the different gene combinations they possess, certain alleles are more likely to be transmitted to the next generation than others. This selection, operating upon the large store of potential variation already in the population, is basically how man and other organisms have evolved. Successes of plant and animal breeding result from artificial control of selection.

Mutation converts one allele into another, but it is especially likely to convert alleles that are most frequent in the population (that is, those favored previously by selection) simply because there are so many more representatives of these alleles. Thus, the general effect of mutation is to decrease the frequency of the commonest alleles and increase the frequency of others. Mutation and selection are therefore balanced forces, selection tending to increase the frequency of certain alleles at the expense of others, mutation tending to bring the frequency back to the distribution expected if all alleles were equally favorable. The equilibrium frequency of individuals with traits determined by less favorable alleles will increase directly with an increase in mutation rate. For example, a permanent doubling of mutation rate will eventually double the frequency at which such individuals occur. But this increase to a new equilibrium level requires a number of generations and depends on the magnitude of the selective disadvantage of such individuals; the greater the disadvantage the more rapid the approach to equilibrium. Likewise, a decrease in mutation rate will result in a decrease in number of such individuals, but the decrease is just as gradual as the increase. Thus, the full effects of increase in mutation rate will not be apparent for some generations, and even an immediate reduction in mutation rate after the full effects are reached would not change the situation very rapidly.

In addition to alleles controlled by a balance between selection and mutation, there are alleles whose frequency depends on a balance between opposing selection pressures. For example, cases are known in which the heterozygote for two alleles has a selective advantage over either homozygous type. Since the heterozygote cannot breed true, it cannot be established as the only type in the population. Equilibrium will be established when the frequency of the two alleles (A and a) yields the optimum possible frequency of the more favorable heterozygote (Aa) in relation to the less favorable homozygotes (AA and aa). If the two homozygotes are equally unfavorable, then the two alleles will be present in equal frequency at equilibrium. If one homozygote is less favorable than the other, one allele will be less frequent than the other. The distribution of such alleles in the population will usually be nearly independent of mutation rate, since the opposing selection pressures usually exert greater influence than the mutation rates. Thus, one of the important and controversial questions is the relative frequency in the human population of the two types of equilibrium. There is evidence (see Morton, 1960) that a large proportion of mankind's genetic disabilities is determined by alleles kept in the population through recurring mutation. Another problem, and perhaps the most important and difficult one, is to estimate the social load imposed on the population by a given rise in mutation rate. This problem is discussed by Wright (1960) and is referred to in greater or lesser detail in some of the other publications cited at the beginning of this section.

Considerable effort also has gone into estimating the increase in mutation rate that would be expected for a given dose of radiation. If all spontaneous mutation was attributable to background radiation, then doubling the present background would double the mutation rate. Many spontaneous mutations probably are the results of other kinds of accidents to the genetic material, however. Thus, the amount of radiation to double the mutation rate must be greater, probably considerably greater, than the background. Several kinds of information are now available for making estimates. There are, for example, the quantitative data for specific locus mutations in the mouse, which yield rates in the order of 10^{-7} mutations per locus per roentgen, varying with the stage in gametogenesis, sex, and rate of irradition. There are the estimates of the spontaneous mutation rate in man, which lie between 10^{-5} and 10^{-6} mutations per locus per generation, although there are some grounds for suspecting that the sample of loci on which this estimate was made has been unduly weighted by those loci with high rates. Finally, there exist a few irradiated human populations on which to make some uncertain, limiting estimates. The largest of these, the Japanese populations irradiated at Hiroshima and Nagasaki, have shown a small but possibly significant decrease (Schull and Neel, 1958) in the frequency of males among the offspring of irradiated mothers, an effect that would be expected from recessive lethal mutations induced in the sex chromosomes of the mothers. The doses involved are uncertain, but on the basis of the best available estimates Schull and Neel (1958) and Morton (1960) conclude that the results are reasonably consistent with the estimates of mutation rate for the mouse. Various point estimates of the doubling dose lie in the range of 30 to 80 rad, but higher or lower values are not at all impossible.

The over-all dose to the whole breeding, or potential breeding, population is the important factor. A small dose to many individuals will have the same effect as a much larger dose given to a few, provided, of course, that the intensity in both cases is low enough so that there is no difference due to dose rate.

It is obvious therefore that there are still great uncertainties about the quantitative aspects of the problem. Both the radiation-induced mutation rate in man and the magnitude of the additional social burden imposed by a given increase in rate are poorly known. Nevertheless, both could be high enough to warrant considerable caution. A misjudgment now cannot be rapidly rectified and may have repercussions for generations. Furthermore, any increase in radiation, no matter how small, will impose some additional social burden. The problem is not to completely eliminate the hazard but to keep it as low as possible while taking the others needs of society into account. These considerations have gone into the various recommendations designed to keep the exposure of the over-all population as low as possible.

12-4 EFFECTS ON ANIMALS

In multicellular animals, effects on individual cells are complicated by an interaction between injured and uninjured cells. The resulting secondary changes are mostly adaptive or reparative processes like those occurring in response to injurious agents other than radiation. Although such changes mask the underlying effects of radiation on single cells, there is no evidence that individual animals cells differ qualitatively from microbial or plant cells in their response to radiation.

Prompt versus Late Effects

Effects of radiation on the various organs and tissues of the body may be grouped into two major categories: (1) those appearing relatively soon (within days or weeks) after exposure, and (2) those developing slowly, becoming manifest only after a relatively long induction period (months or years). In general, the former are associated with extensive killing of cells and acute radiation sickness; the latter arise through undefined mechanisms and resemble the degenerative and neoplastic phenomena of growing old (Upton, 1960).

Variations in Radiosensitivity

Owing to the high radiosensitivity of dividing cells, organs in which cells are being continually replaced are the most vulnerable. These include notably the gonads, blood-forming tissues, lymphatic system, skin, and gastrointestinal tract. Other factors, however, also influence radiosensitivity, such as oxygen tension, metabolic activity, and temperature. Certain nondividing cells (for example, the primary oocyte and the lymphocyte) are, furthermore, unusually radiosensitive for unexplained reasons. Accordingly, the effects of a given dose of radiation may vary widely, depending on the organ irradiated and its physiologic state at the time of exposure.

Another factor is genetic constitution. Although vertebrates in general share the property of radiosensitivity of dividing cells, there are rather marked species variations in the relative sensitivity of various tissues and organs. In fowl, for example, injury of blood vessels is an early and serious complication of whole-body irradiation, an event having no real counterpart in mammals.

Effects on the Mammalian Embryo and Fetus

Throughout embryonic and fetal life, the effects of radiation vary markedly in relation to the developmental stage of the organism at the time of irradiation (L. B. Russel, 1954). According to the end results, three major stages may be recognized: (1) the preimplantation period (between fertilization of the egg and its implantation in the lining of the uterus), (2) the period of major organogenesis (the time of maximal formation of new organs), and (3) the fetal period (characterized predominantly by growth of preformed organs and minimal organogenesis).

During the first period the organism is more sensitive to killing than at any other time of life, the lethal dose being one-third or less than that during the adult period. Death in this case, however, occurs predominantly before birth. Individuals born tend to be relatively normal.

Four trends are evident during the second period (two to six weeks after conception in human embryos): (1) susceptibility to prenatal killing decreases with age; (2) the time of death is delayed; (3) neonatal mortality assumes more importance; and (4) the organism is more susceptible to induction of malformations. Because organs enter critical stages at different times, the type of induced deformity varies markedly with the time of irradiation.

Irradiation during the fetal period generally causes less severe injury than irradiation at earlier stages and may lead to defects that become apparent only later in life. Organs most likely to be affected are those maturing relatively late, such as the brain, retina, gonads, and lens of the eye.

Since radiosensitivity is relatively high throughout embryonal and fetal life, radiation during pregnancy should be avoided as much as possible. The hazard is particularly high during the period of major organogenesis, which begins in man about $1\frac{1}{2}$ weeks after conception (so early that pregnancy may be unsuspected). Russell and Russell (1952) have recommended, therefore, that irradiation of the uterus in women of childbearing age be restricted to the first fortnight after the menses to avoid inadvertent exposure of an embryo in an especially sensitive stage.

Effects on Organs and Systems

In this discussion, the radiation effects described are those characteristic for mammals (Bloom, 1954; Jacobson, 1954). Mammalian organs show different effects of irradiation. The following is a description of some of the changes observed.

Gonads. In both the male and female gonad, the cells most sensitive to radiation death are the immature stages of germ cells (spermatogonia in the testis and early oocytes in the ovary), as noted earlier.

Although relatively small amounts of radiation cause depletion of spermatogonia, thereby curtailing production of spermatozoa, the more mature stages are less radiosensitive. Fertility is therefore not affected until the supply of the more mature cells becomes exhausted, which may require several weeks. Then sterility ensues and persists until the testis is repopulated by proliferation of surviving spermatogonia. The dose required to cause lasting sterility generally exceeds the lethal dose for the organism if applied to the whole body in a single exposure ($>$1000 rad); hence, permanent sterilization usually is not a complication of acute whole-body irradiation in male mammals.

In the mammalian ovary, since the supply of oocytes is not renewable after birth, killing of primary oocytes by radiation irreparably reduces the number of potential gametes. The reproductive period is thus shortened. The dose of radiation required to kill all the oocytes, including the mature stages (that is, to sterilize promptly), is several hundred rad.

Depletion of oocytes in the mouse ovary, whether through natural aging or irradiation, is often followed by the development of ovarian tumors. Because of the high radiosensitivity of early oocyte stages in the mouse, the incidence of ovarian tumors may be greatly increased by only 50 rad of X-rays. As yet, however, there is no evidence of comparable effects in other species.

Of greater biological importance than sterilizing effects are genetic changes induced by irradiation in germ cells, since these may be transmitted through successive generations to countless victims and presumably may be caused by minute amounts of radiation. These effects have been discussed in Section 12-3.

Eye. The part of the eye most easily injured by radiation is the lens. Opacification has been detected after only 15 rad of X-rays in the mouse (200 rad in man) absorbed in a single exposure (see Upton et al., 1957, 1960). Even smaller amounts of neutron radiation are cataractogenic. Furthermore, successive exposures to neutrons are essentially additive in their effects, whereas successive exposures to X-rays and γ-rays decrease in effectiveness with decreasing dose rate.

The basis for the high radiosensitivity of the lens is not known, although cells of the lens epithelium continue to divide at a low rate throughout life. High susceptibility to cataract formation perhaps reflects both the inability of the lens to rid itself of damaged cells and debris, because of its lack of blood vessels and scavenger cells, and the tendency for even minute abnormalities of the lens to have adverse effects on its refractive qualities.

Other structures are only moderately radiosensitive, that is, cornea, conjunctiva, lids, etc. The retina, however, because of its unique photo-amplifying properties, is able to perceive very small amounts of radiation (>1 rad) through photochemical reactions which are per se without harmful effects (Lipetz, 1955).

Bones and Teeth. Developing bone-forming and tooth-forming cells are radiosensitive. Abnormalities of skeletal and dental growth may be caused by less than 500 rad of X-rays applied to regions of active growth. Although mature tissue is relatively radioresistant, large doses of radiation (such as those accumulating from locally deposited radioisotopes) may cause demineralization and tissue death (radiation osteitis), leading to bone fractures, loosening of teeth, and other complications, including bone cancer.

Endocrine Glands. The glands of internal secretion withstand relatively large amounts of radiation without developing significant morphological lesions. On the other hand, there is growing evidence that relatively small doses may produce functional changes, some of which are merely part of a nonspecific reaction to stress. The radiosensitivity of these organs is therefore subject to reappraisal.

Circulatory System. The effects on the circulatory system are described in two parts.

BLOOD-FORMING TISSUES. Because the circulating blood cells have only a short lifetime, they must be continually replaced through multiplication of precursor cells in the bone marrow and lymphoid tissues. The precursor cells, like dividing cells in general, are highly radiosensitive. Later maturation stages and fully mature circulating blood cells are, however, relatively resistant (the lymphocyte is an exception). Consequently, depression of the blood count is generally a gradual change, occurring over a period of days or weeks, depending on the life-span of the preformed blood cell in question. Since immature precursors circulate in the bloodstream and can multiply faster to compensate for blood loss, severe depletion of blood cells does not occur unless the entire body or a major fraction of its blood-forming tissues is heavily irradiated.

The *lymphocyte*, one of the more numerous of the white blood cells, is an unusually radiosensitive cell. It exhibits structural abnormalities in response to as little as 1 rad or less of X-rays (Dobson and Chupp, 1957). Furthermore, circulating lymphocytes decline within hours after whole-body irradiation. They may be almost totally lacking in the blood within one to two days after exposure. These changes are accompanied by lymphocyte destruction in lymphoid tissues throughout the body. The depletion of lymphoid cells is thought to play a role in the depression of immunity resulting from whole-body irradiation. This depression lowers the resistance of the organism to foreign substances; for example, foreign tissues, bacteria, etc. Although the radiosensitivity of rapidly dividing lymphocyte precursor cells is not surprising, the high radiosensitivity of mature lymphocytes is unexplained.

The *granulocyte*, another major type of white blood cell, is relatively radioresistant. Paradoxically, however, the number of granulocytes in the blood may temporarily increase during the first one to two days after irradiation. This results from redistribution of preformed granulocytes from tissue reservoirs into the circulation. In the face of large-scale destruction of granulocyte precursors, the level of granulocytes in the blood eventually declines, reaching its lowest point within five to ten days in lethally irradiated rodents (later in larger animals). The rate of decline varies with the radiation dose. Depletion of these cells is of grave importance; their decline predisposes to infection, a major cause of death in heavily irradiated individuals.

The blood *platelet* is another circulating blood corpuscle. Its production is curtailed by irradiation through injury of precursor cells in the bone marrow. In general, the rate of decline of the platelet count parallels that of the granulocytes. Since an adequate supply of platelets is essential to prevent bleeding, marked depletion leads to hemorrhage, which may be profuse enough to cause exsanguination.

The *red blood cell* count declines only slowly after irradiation in the absence of hemorrhage, since the red cell is radioresistant and has a relatively long life-span (60 days in the mouse, 120 days in man). Consequently, severe depletion of red blood cells (that is, anemia) develops later than the blood changes just described. It may, nevertheless, constitute an important complication of irradiation injury and necessitate blood transfusion.

Recovery of the blood-forming tissues occurs through compensatory multiplication of surviving stem cells in the bone marrow, lymphoid organs, spleen, and thymus. Hence, the survival of the animal depends on prompt regeneration of these cells and their ability to repopulate the depleted organs. This process can be facilitated either by shielding blood-forming tissues from direct radiation or by implanting blood-forming cells to replace those destroyed.

Transplantation, however, is usually soon complicated by severe immunological reactions when the grafted cells and the recipient are not genetically identical. Bone marrow transplantation in humans will be of limited clinical value until means are found to overcome the attendant immunological difficulties.

VASCULAR NETWORK. Apart from transient dilation of blood vessels (such as occurs in erythema of the skin), swelling and death of vascular lining cells—with occlusion or rupture of the affected vessel—may result from radiation doses commonly used in cancer therapy (2–6 Krad). Other alterations include progressive and permanent vascular dilation and sclerosis. The blood supply to a particular tissue may be impaired and cause secondary disturbances of metabolism, growth, and function in the involved tissue. At lower dose levels, vascular changes are less conspicuous or lacking; however, even doses below 500 rad cause severe injury of renal blood vessels in the mouse, and are postulated to accentuate arteriolocapillary sclerosis in aging rats.

Skin. Because of its high radiosensitivity and external location, the skin was the first site of radiation injury recognized and has been most thoroughly investigated. Its susceptibility is generally attributed to the high rate of cell division in its basal, or germinal, layer through which desquamating cells of the superficial layers are continually replaced. Within minutes after irradiation, mitotic inhibition and cytological abnormalities appear in cells of the germinal layer. Within hours a transitory reddening of the skin, or erythema, may ensue. The initial reddening usually subsides within a day or two and is followed by a more pronounced erythema about ten days later. After this reaction, a tertiary erythema may also occasionally develop. The reddening is caused by engorgement of blood vessels in the skin. Radiologists used it as a crude dosimeter in the days before adequate instrumentation. In general, the reaction requires 400–500 rad of X-rays. (The requisite dose varies appreciably, however, depending on the energy of the radiation, dose rate, size of area of skin irradiated, and physiological factors such as blood-flow, complexion, and region of body exposed.) Following erythema, tannish pigmentation of the skin ensues, which may last for weeks or longer.

With even larger doses of radiation, the destruction of cells in the germinal layer may be so extensive that the skin blisters. At still higher dose levels, ulceration and sloughing of the epidermis occur; these, however, are slow in developing. These complications are later associated with thinning of the epidermis and scarring of the underlying tissue ("radiodermatitis"), which predispose eventually to cancer of the skin.

Injury of hair follicles may cause loss of hair, or epilation. This effect appears two to three weeks after irradiation, depending on the dose and species, and may be permanent with large doses. Injury of pigment cells in the hair follicle may cause the newly formed hair to be lacking in pigment (that is, gray). The dose required for epilation or for graying varies, but generally exceeds 200–300 rad.

Gastrointestinal Tract. The lining of the intestinal tract is continually renewed by proliferation of dividing cells that are located in crypts below its surface. A dose of less than 50 rad of X-rays causes significant mitotic inhibition of these cells. Above about 1000 rad, depending on the species, depletion of the dividing cells is so severe that the lining of the intestine cannot be adequately maintained. Denudation of the bowel ensues, along with other disturbances, within several days. Hence, excessive fluid and salts escape into the intestine, and bacteria and toxic materials are allowed to gain entry into the bloodstream. The immediate results include diarrhea, dehydration, infection, and toxemia. These complications usually begin to occur several days or more after irradiation, depending on the species and the radiation dose. They are not to be confused with the transient nausea and vomiting frequently developing in heavily irradiated individuals during the first 24 hours after exposure. Severe diarrhea, associated with inflammation of the upper gastrointestinal tract and mouth, may also follow infection and ulceration arising as secondary effects of bone marrow injury.

Nervous System. Relatively large amounts of radiation (500 rad) are required to cause detectable morphologic changes in adult nervous tissue, although smaller doses have been reported to cause transitory functional disturbances which are as yet of equivocal significance. Well documented, however, is the occurrence of avoidance reactions in rodents, snails, fish, and other organisms exposed to only a few rads at high intensity. Whether such reactions reflect injury of nervous tissue remains to be determined.

Intensive exposure of the brain (1000–10,000 rad, depending on the species) can produce incapacitating effects (nausea, vomiting, convulsions, unconsciousness) within minutes or hours, rapidly leading to death from damage to vital centers.

Urinary System. The kidney and lower urinary tract are generally regarded as radioresistant. Nevertheless, doses in excess of 500–1000 rad (depending on the species) may cause progressive scarring and atrophy of the kidney, leading to fatal impairment of renal function. Similar lesions may also result from chronic irradiation of the kidney by locally deposited radioelements (radon, polonium).

Respiratory System. The lungs and respiratory tract are relatively radioresistant; however, intensive irradiation may result in a chronic, pneumonia-like disease, with progressive scarring of the lung and pulmonary blood vessels. Irradiation of the respiratory tract may also cause cancer of the nasopharyngeal mucosa, bronchial epithelium, and alveolar lining cells.

12-5 ACUTE RADIATION SYNDROME

Intensive irradiation of the entire body severely depletes radiosensitive cells in many organs simultaneously. The combined effects produce "radiation sickness" (acute radiation syndrome). The resulting symptoms (Table 12-1)

Table 12-1 Symptoms of Acute Radiation Syndrome[a]

Time After Exposure	Supralethal Dose (1000 rad)	Median Lethal Dose (500 rad)	Sublethal Dose (200 rad)
	Nausea and vomiting on first day	Nausea and vomiting on first day	
First week	Nausea, vomiting, diarrhea, fever, inflammation of throat, prostration, dehydration, emaciation leading to death		
Second week			
Third week		General malaise, loss of appetite, loss of hair, hemorrhage, pallor, diarrhea, fever, inflammation of throat, emaciation leading to death in 50% of victims	Loss of appetite, loss of hair, inflammation of throat, pallor, hemorrhage, diarrhea
			Recovery begins No deaths in absence of complications

[a] Modified from Alexander (1957).

vary in kind, severity, and time of onset, depending on the species, dose, dose rate, and quality of radiation. Death, when it occurs, is often ascribable primarily to injury of the bone marrow, gastrointestinal tract, or brain (Fig. 12-7). Within the lower-lethal exposure range, depression of functional bone marrow is the predominant cause of death in the mouse, dog, monkey, and some other mammals (including man). At higher exposure levels, intestinal

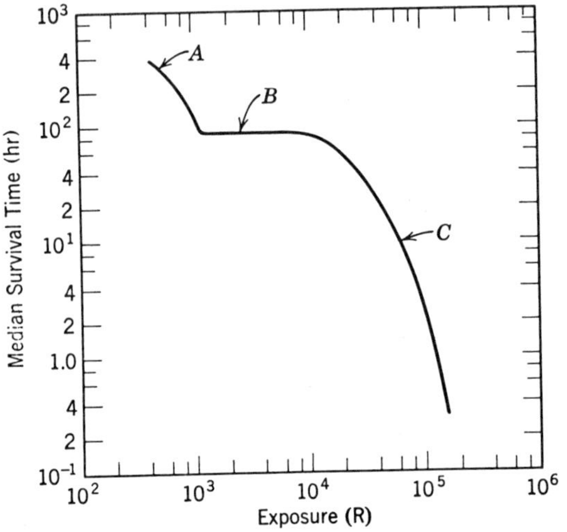

Fig. 12-7. Median survival times of mice, rats, and monkeys after large, rapidly delivered (1000–6500 R/min) exposures of γ-radiation. A, Marrow death range; B, intestinal death; C, brain death. (Modified from Langham et al., 1956.)

injury supervenes, causing earlier mortality; that is, within the first week, as opposed to the second and third weeks after irradiation. After massive supralethal exposure, death may occur within minutes from brain damage.

LD_{50}/*30 Days*

Since death from acute radiation sickness occurs within days or weeks after irradiation (Fig. 12-7), it is customary to analyze mammalian dose-mortality data on a 30-day survival basis (although in man the mortality rate may remain high until 60 days after exposure). Radiation toxicity is often expressed in terms of the median lethal dose—the dose equivalent required to kill 50% of the exposed subjects within 30 days, that is, the LD_{50}/30 days (Fig. 12-8). This value varies for different species (Table 12-2), in general, approximating 500 rem for most mammals. Microorganisms and various invertebrates have been reported to survive thousands, or even hundreds of

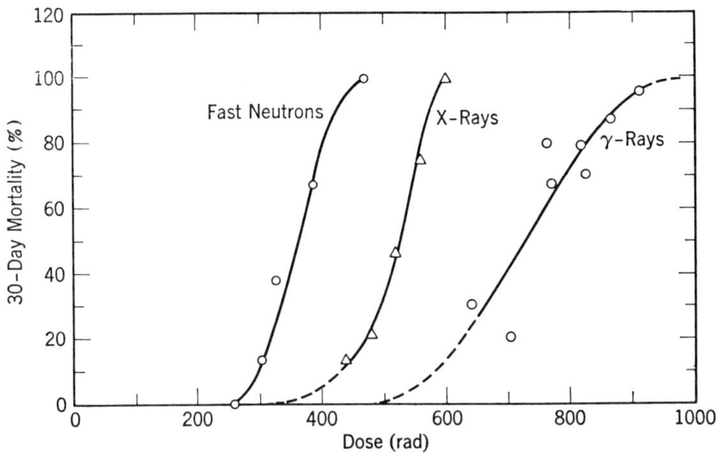

Fig. 12-8. Thirty-day mortality of mice exposed to X-rays, fast neutrons, and γ-rays. The mice were 9–12 weeks old at the time of whole-body irradiation. (From A. C. Upton et al., 1956; reproduced by permission of Academic Press, Inc., New York.)

thousands, of rads. However, the criteria and causes of death for these forms are so different that meaningful comparisons with mammals cannot be made. It is clear, however, that much higher doses are required to wipe out most populations of bacteria than to kill any given vertebrate species.

The LD_{50}/30 days is relatively reproducible in homogeneous populations of inbred laboratory animals. It varies appreciably with age at exposure

Table 12-2 Median Lethal Dose Equivalent of X-Rays for Various Species of Vertebrates

Organism	Median Lethal Dose Equivalent (rem)
Goldfish	750
Mouse	600
Guinea pig	400
Rabbit	700
Hamster	700
Rat	700
Dog	300
Monkey	550
Goat	300
Man	500 (?)

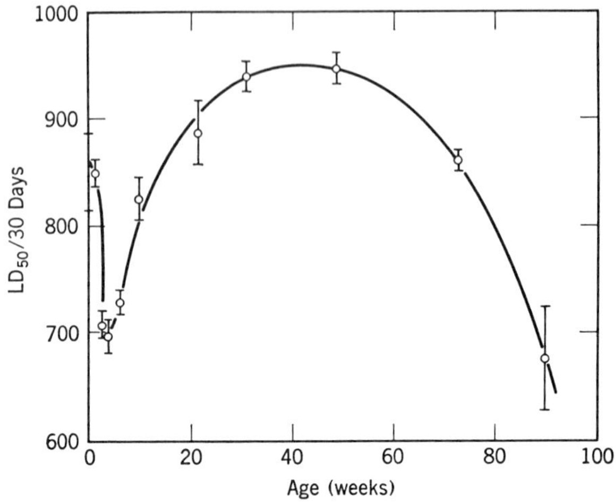

Fig. 12-9. Curve showing $LD_{50}/30$ of mice in relation to age at time of γ-irradiation. (Modified from Crossfill et al., 1959; reproduced by permission of *Nature*.)

(Fig. 12-9) and, to a lesser extent, with other biological variables. Despite extensive killing of radiosensitive cells, animals are able to withstand up to 200 rad with negligible mortality, owing to the functional and cellular reserve capacity of injured tissues.

12-6 DELAYED SOMATIC EFFECTS OF RADIATION

Life-Shortening

Apart from the early mortality associated with acute radiation sickness, radiation injury may cause death months or years later (Fig. 12-10). The extent of life-shortening varies in relation to the absorbed dose and quality (RBE) of the radiation (Fig. 12-11) and the dose rate (Fig. 12-12). However, the relation among these variables is not precisely known. The dose-survival curve may also change with genetic differences and other biological variations. It is not possible, therefore, to extrapolate the limited data on rodents to other species or to man.

Prediction of the effects of low-level irradiation is further complicated because "oversurvival" (that is, increase in the mean survival time) may occur in mice and rats exposed daily to 1 rad or less of γ-rays. This effect is generally ascribed to reduction in mortality early in life rather than to postponement of aging. Since early death in nonirradiated laboratory animals usually results from infection, the practical significance of the effect for man is equivocal.

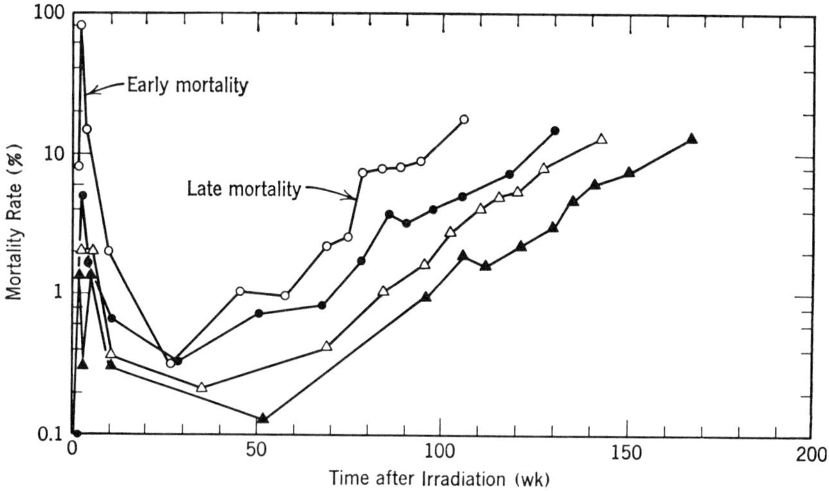

Fig. 12-10. Early and late radiation mortality responses in $(L \times A)F_1$ female mice exposed to atomic bomb γ-rays at 6–12 weeks of age. (Modified from Upton, 1960.) ○, 687 rad; ●, 578 rad; △, 368 rad; and ▲, nonirradiated control. (Reproduced by permission of *Gerontologia.*)

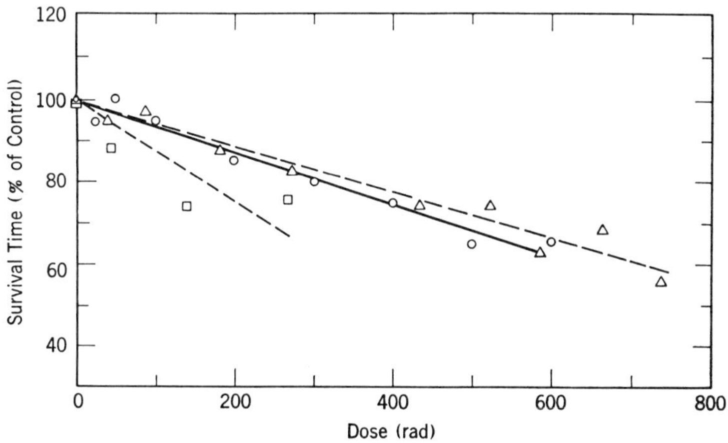

Fig. 12-11. Mean survival time in mice after a single exposure to radiation early in life (animals dying within first 30 days excluded). △, γ-ray data from Lindop and Rotblat, 1959; ○, γ-ray data from A. C. Upton (unpublished); □, neutron data from Upton, 1960 (reproduced by permission of *Gerontologia*).

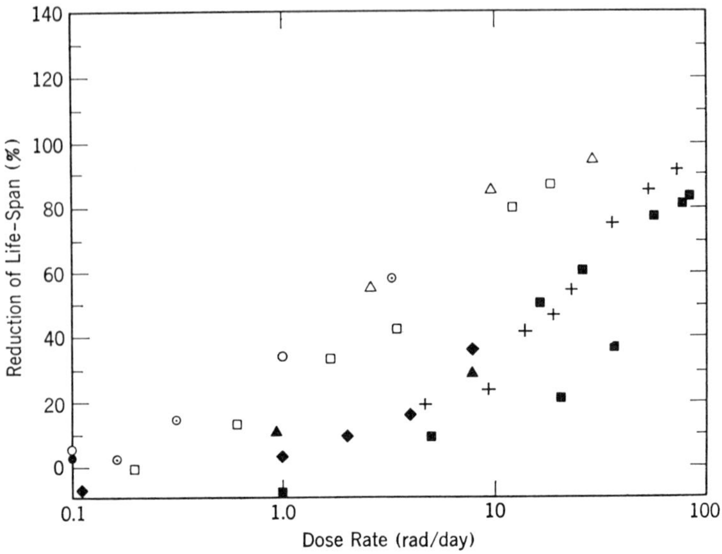

Fig. 12-12. Reduction of mean or median survival time in mice exposed daily for the duration of life (shaded symbols denote data for X- and γ-rays, unshaded symbols denote data for fast neutrons). +, data from Henshaw, 1944; △, data from Henshaw et al., 1947; ⊙, data from Evans, 1948; ◇, data from Lorentz et al., 1954; ○, data from Neary et al., 1957; □, data from A. C. Upton, unpublished (reproduced by permission of *Gerontologia*).

Whether radiation can cause life-shortening in man is not yet known. Although the average age at death in radiologists is about five years less than that in other physicians, this difference can be ascribed to the younger average age of living radiologists. On the other hand, some evidence indicates that the age-specific death rate of radiologists may be abnormally high. Provided that other causative factors can be excluded, this phenomenon (if verified) may denote a life-shortening effect of occupational irradiation (Upton, 1960).

The biological basis of life-shortening is obscure, but the effect seems to be correlated with premature onset of diseases otherwise associated with growing old, including neoplasms as well as nonneoplastic degenerative changes. Whatever the mechanism, the incidence of affected diseases is not altered uniformly; that is, some diseases are increased in frequency and others decreased. Furthermore, not all age-dependent changes are similarly hastened in onset. The sequence of events cannot, therefore, be ascribed to premature aging in the simplest sense (Upton, 1960).

Many theories have been proposed to explain the effects of radiation on longevity (Upton, 1960; Neary, 1960). Some postulate a nonspecific disorganization of the biochemistry of cells (entrophy) through random ionization. Others implicate radiation-induced changes in genes or chromosomes (somatic mutations). Others seek to explain aging in terms of various types of functional impairment. None of the hypotheses adequately accounts for the observed facts. It is doubtful, moreover, that the question will be further elucidated without more adequate information about the radiation changes themselves and about spontaneous aging.

Radiation Cancer

Within only a few years after it was noted that X-rays caused regression of growing tumors, X-rays were also observed to induce tumor formation. Extensive clinical and experimental observations have since shown that nearly any form of cancer may be induced by irradiation under appropriate conditions, and that all ionizing radiations are carcinogenic (Furth and Lorenz, 1954). Little is known, however, about the mechanism of tumor induction or the relation between tumor incidence and dose. Hence, precise assessment of the carcinogenic hazards of radiation, particularly at low dose levels, cannot be made (United Nations, 1964).

The relation between tumor incidence and dose has been the subject of several investigations in rodents, principally in mice. From existing data, however, the only generalization that can be made about the shape of the dose–response curve is that it seems to vary from one type of tumor to another and to be influenced by radiological and biological variables. A linear relation has not been conclusively demonstrated. Instead, the dose–incidence curve tends to be sigmoid, often showing saturation or downward deflection at high dose levels, presumably because of excessive radiation injury (Fig. 12-13). In general, the tumor-inducing effectiveness of X-rays and γ-rays seems to be significantly lower and appreciably more intensity-dependent than that of heavy particles, but the influence of intensity and LET is still largely unexplored (Upton et al., 1960).

At the lowest exposure rate thus far systematically investigated (0.11 R/day), mice exposed daily throughout life to radium γ-rays exhibited a slightly increased incidence of certain types of cancer (Lorenz et al., 1955). Because of statistical limitations, however, the data do not enable confident extrapolation to lower dose levels.

Whether there is a threshold dose for cancer induction likewise cannot be decided from present evidence. Since, however, susceptibility to radiation-induced tumors is influenced by physiological variables, and since cancer presumably arises through the interaction of multiple contributing factors, it is conceivable that a single quantum of radiation may be carcinogenic in an appropriately conditioned individual. Furthermore, since radiation seems in many instances merely to hasten the onset of an otherwise spontaneously

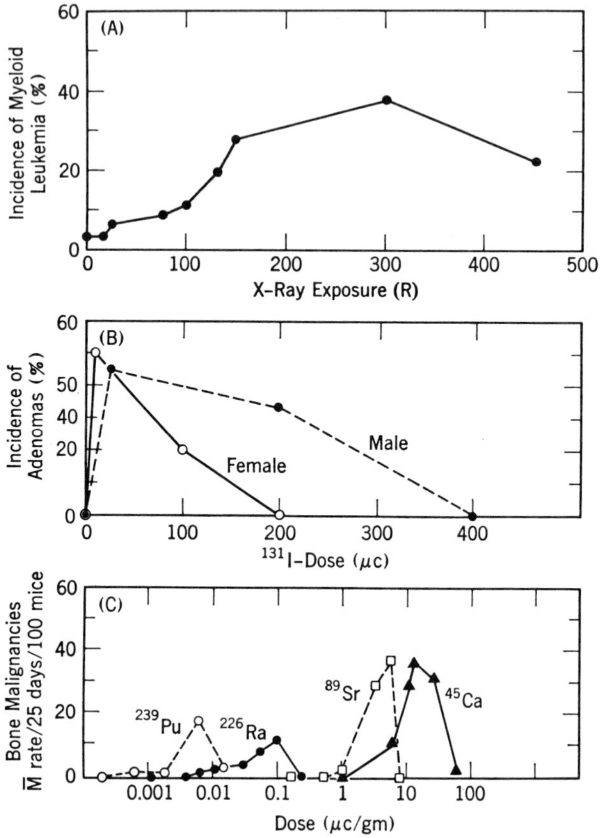

Fig. 12-13. Incidence of various neoplasms in relation to radiation dose. (A) Myeloid leukemia in RF male mice exposed to X-rays (from Upton et al., 1958 and A. C. Upton, unpublished data). (B) Thyroid adenomas in rats injected with ^{131}I (from Lindsay et al., 1957). (C) Osteosarcomas in CF_1 female mice injected with the radionuclides shown (from Finkel, 1956).

occurring cancer, evaluation of the effects of radiation on the incidence of malignant growth is by no means a straightforward statistical or biological problem.

Two of the most plausible hypotheses of the mechanism of tumor formation are the *somatic mutation* theory and the *virus* theory. According to the former, cancer results from one or more mutations in the genetic constitution of a cell that cause it to undergo a transformation toward malignant growth. Theoretically, the role of radiation in this process may be twofold: (1) to cause cancer mutations through mutagenic effects on the potentially cancer-forming cell, or (2) to produce conditions favoring the selective growth of spontaneous cancer mutants. Although the cancer change has all the hallmarks of a genetic alteration of the cell (in some instances cancer cells have actually been found to have chromosomal abnormalities), cancer has never been proved to result from chromosomal changes or mutations. Hence, the somatic mutation hypothesis remains speculative.

The virus theory of tumor formation is based on evidence that radiation exerts carcinogenic effects through changes in the environment of the cell rather than through direct action on the cell itself. For example, lymphoid tumors may be induced in nonirradiated thymus cells by implantation of these cells into a previously irradiated recipient mouse (Kaplan, 1959). Under such conditions, the carcinogenic action of radiation cannot be dependent on its mutagenicity in the classical sense since the thymus is not irradiated. The nature of the carcinogenic effect remains to be disclosed, but preliminary studies suggest that radiation-induced thymic tumors may arise through the intermediate action of a virus. This mechanism, if verified, will call for systematic study of the effects of radiation on the virus-host relation and on the possible role of viruses in other radiation cancers. From the limited information now available, radiation may be postulated to "activate" tumor viruses by one or more of several hypothetical mechanisms (see Upton, 1958). Apart from the possible activation of viruses, however, other effects of radiation on the environment of potentially malignant cells involve manifold disturbances of normal growth-regulating factors. Thus, in any analysis of the cancer-forming action of radiation, the influence of extracellular radiation effects must be assessed along with effects on the irradiated cancer-forming cell itself.

12-7 FACTORS MODIFYING THE RADIATION RESPONSE

Time-Intensity Factors

In general, the influence of these factors is the same in animals as in individual cells, that is, the effectiveness of low LET radiation such as X-rays and γ-rays decreases with decreasing dose rate and increasing interval between successive exposures. Intensity-dependence is especially marked in injuries resulting from killing of cells (for example, those injuries prominent in acute radiation sickness). Hence, the $LD_{50}/30$ in rodents increases appreciably with prolongation of the exposure time. Various theories of recovery have been formulated from this effect (Sacher, 1958). Recovery, however, represents chiefly cellular replacement rather than repair. The intensity-dependence of X-rays and γ-rays for delayed somatic effects (life-shortening, etc.) is less marked. The γ-rays represent effects that are inferred as relatively "irreparable."

For certain effects, a given dose of radiation may be more effective when broken into multiple, appropriately timed exposures than when administered in a single brief treatment. The response, thus, depends on the interrelation of several factors—total dose, number of exposures, dose per exposure, and interval between exposures. Although this interrelation is still poorly understood, it may involve the interaction of injury and recovery processes at various levels of biological organization.

Relative Biological Effectiveness (RBE)

The effectiveness of radiation depends partly on the distribution of ionizations along the paths of the ionizing particles, or on the linear energy transfer (LET) of the radiation (Zirkle, 1954). The RBE of a given radiation is determined by comparing the absorbed dose of that radiation required to cause a particular biological effect with the absorbed dose of a reference radiation (usually X-rays or γ-rays) required for the same effect, that is,

$$\text{RBE neutrons} = \frac{\text{Dose } \gamma\text{-rays (rad)}}{\text{Dose neutrons (rad)}} \tag{12-3}$$

Because the RBE varies with dose rate and with the biological effect in question, it is essential that the radiations under comparison be administered at equivalent intensities and evaluated for the same biological effect. When, on the other hand, different radiations are compared for over-all toxicity, it is preferable to use a term such as *relative toxicity* rather than RBE.

For most effects on animal tissues, the RBE increases with LET up to the LET range of fission neutrons (about 50 keV/μ) (Zirkle, 1954). Because, however, penetration of neutrons and other heavy particles into the deeper tissues of large animals is associated with attenuation, moderation, secondary ionization, and other perturbations of the radiation, determination of the RBE is greatly complicated.

In general, the RBE of high LET radiations increases with decreasing dose rate because of the relatively greater intensity-dependence of low LET radiations noted earlier (Evans, 1952).

Protective Agents

Many common factors influence the response of individual cells and multicellular organisms to irradiation. One factor consistently affecting radiosensitivity to X-rays and γ-rays is oxygen tension. Reduction of the oxygen tension has been observed to increase the X-ray $LD_{50}/30$ by a factor of 2 to 3 and reduce the severity of delayed somatic effects in experimental animals. Radioprotective chemicals may exert their effects by lowering oxygen tension.

Chemical protection is also believed to occur through other mechanisms (Patt, 1953; Doherty, 1960); for example, compounds such as cysteine, cysteamine, and AET are thought to trap radiation-induced radicals. These chemicals, likewise, increase the X-ray $LD_{50}/30$ by a factor of about 2 and reduce the severity of some delayed somatic effects of radiation in mice. Drawbacks to the practical use of these agents, however, are high toxicity, relatively rapid excretion or degradation, and failure to protect when administered after irradiation. An additional hindrance is the requirement that they be present in the body at precisely regulated concentrations throughout the period of exposure. Like hypoxia, they are more effective against X-rays and γ-rays than against heavy particles, which is consistent with the theory that they act through interference with radiation-induced radicals.

Recovery-Promoting Agents

There are few known ways to promote repair of radiation injury in previously irradiated animals. The most notable method is replacing damaged blood-forming cells by transplantation. Infusion of viable, nonirradiated bone marrow cells enables otherwise lethally irradiated animals to escape death from marrow injury. This procedure also protects irradiated mice against the subsequent development of certain types of leukemia. Marrow replacement is not uniformly successful, however, unless the donor and recipient are genetically identical. Hence the clinical use of this measure will be limited until the attendant immunological problems are solved (Congdon, 1957 and Mathe et al., 1965).

12-8 EFFECTS OF INTERNALLY DEPOSITED RADIONUCLIDES

The biological effects of an ionizing particle are the same whatever its source, but certain considerations applying only to internally deposited sources of radiation deserve special mention.

Uptake and Distribution

An internal emitter must enter from the environment, a process involving a series of complex reactions associated with uptake and distribution (Table 12-3). The same concentration of a given radioisotope in the environment may therefore have radically divergent effects on different organisms, depending on the species in question, type of organ or cell involved, cellular metabolism, state of isotope in the environment (physical form, chemical state, concentration), concentration of the element in the organism, route of entry (lungs, skin, digestive tract, other tissue), and additional factors (Wasserman, 1960). The uptake of radioiodine by the mammalian thyroid gland, for example, varies so profoundly with the physiological activity of the gland that the extent of uptake of a test dose of ^{131}I is widely used to gage thyroid function in medicine and physiology.

The prediction of certain elements for specific organs has an important bearing on subsequent biological effects of their radioisotopes. For example,

Table 12-3 Factors Influencing the Entry and Distribution of Radionuclides in Living Organisms and Cells[a]

1. Physicochemical properties of the radionuclide
   a. Physical state
   b. Solubility
   c. Particle size
   d. Tendency to form colloids
   e. Formation of unabsorbable complexes or chelates
2. Radiochemical properties of the radionuclide
   a. Physical half-life
3. Biological considerations
   a. Transfer of the radionuclide in food chain
   b. Metabolic pathway of the radionuclide *in vivo*
   c. Degree and route of absorption of the radionuclide
   d. Nutritional and physiological status of the organism

[a] Modified from R. H. Wasserman (1960).

because iodine is selectively concentrated in the thyroid gland, the effects of radioiodine are confined principally to this and adjacent organs. In the case of bone-seeking elements (such as strontium, calcium, radium, plutonium) the skeleton is the "critical" organ. A rapidly growing body of information has become available for the more abundant radionuclides in man, mammals, certain marine forms, and some plants (Norman, 1960; Robertson, 1960; and A. H. Wolff, 1960). Many other radionuclides are still incompletely characterized.

The energy of the emitted radiations and their geometry also determine the area of injury. Thus, energetic γ-rays may penetrate to the remotest parts of the human body, whereas α-particles affect only adjacent cells. The tendency, however, for many radioelements to be concentrated *in vivo* in irregularly distributed microscopic "hot spots" and to decay with emissions varying in mass, charge, and energy greatly complicates the assessment of the radiation dose delivered to a given tissuc.

Retention

Once incorporated, the effects of a radionuclide depend on its retention inside the organism. These vary widely, depending on the element, the manner of incorporation, and the metabolism of the organism. Retention is usually expressed in terms of biological half-life. (The rate of loss of an element from an organism tends to become exponential with time after an initially more complex situation associated with its uptake, distribution, and equilibration.) The cumulative dose of radiation administered by an absorbed isotope depends, therefore, on the *effective half-life* E, which is the relation between the biological half-life B, and physical half-life P, given by the formula

$$E = \frac{BP}{B + P}$$

The interrelation of the various half-lives of ^{131}I is shown in Fig. 12-14.

Because of their characteristically long effective half-lives, bone-seeking radionuclides tend to be more hazardous biologically than any other single class of isotopes. Certain other radionuclides (for example, thorium and polonium), however, are stably incorporated into soft tissues in the form of biologically inert colloids; they also are sources of considerable biological danger.

Owing to the great diversity of the many radionuclides themselves and the vagaries of their behavior *in vivo*, no precise prediction can be made about the biological effects of a given radioisotope until its metabolism is characterized in the organism in question. For this reason, assessment of the hazards of fission products in man will be seriously limited without better knowledge of their behavior in human beings.

Factors Modifying Uptake and Retention

Certain factors that modify the metabolism of isotopes have importance from the standpoint of radiation protection. Various measures may be taken to inhibit the uptake and retention of potentially injurious internal emitters. Practical measures have not been developed, but encouraging preliminary results deserve mention.

First, it is axiomatic that the uptake and retention of any radioactive isotope is competitively inhibited by the presence of its stable isotope. Greatly increasing the concentration of stable isotope relative to that of the radionuclide generally serves to reduce the biological contamination of a cell or organism. This measure is limited in many instances by chemical toxicity. A related phenomenon is a form of biological discrimination that favors the

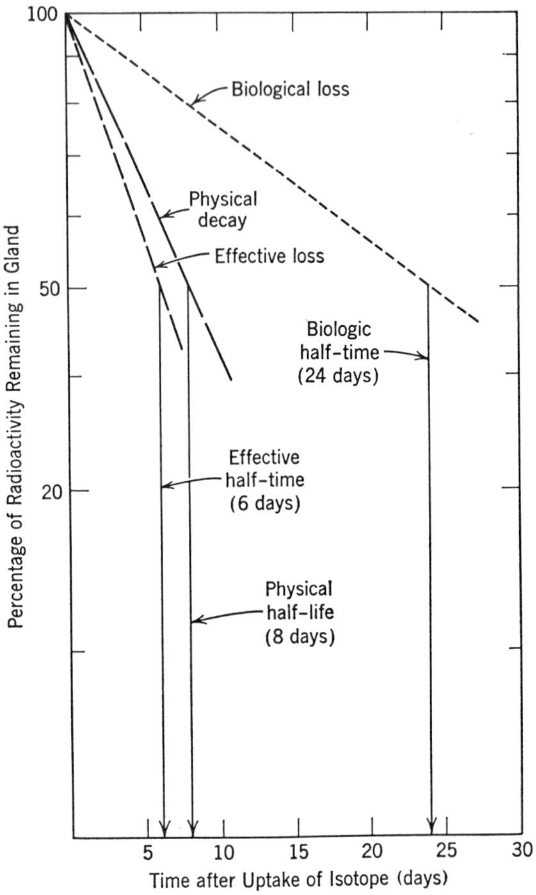

Fig. 12-14. Rate of loss of radioactive iodine ^{131}I after maximal uptake in the human thyroid gland. The rate of biological loss is plotted as exponential, although it is actually somewhat more rapid initially than at later times after uptake. (From G. A. Andrews, 1957, reproduced by permission of The American College of Physicians, Philadelphia.)

selective absorption of one element over a sister element when both are present in equal concentrations; for example, the calcium/strontium uptake ratio in man is about 2:1 to 7:1. Hence, depending on the diet and other factors, the concentration of strontium in man is only about one-tenth that in the soil and plants. It is customary, therefore, to express the maximum permissible concentration of radiostrontium in terms of the Sr/Ca ratio, one strontium unit being equal to 1 $\mu\mu c$ of ^{90}Sr per gram of Ca. A similar consideration applies to the $^{137}Cs/K$ relation. These discriminatory factors are not constant but vary widely with biological and chemical factors (Kornberg,

1960). Hence their ultimate importance in the passage of radionuclides through man's food cycle remains to be elucidated.

A second measure involves the use of chelating agents which complex with the radioelements in question and facilitate their clearance from the body. This attack promises a means for promoting decontamination of the body after a radioelement has already been absorbed. The action of such agents, however, is contingent on chemical and biological requirements that limit the range and effectiveness of the compounds thus far available (Wasserman, 1960).

Transmutation

Internal emitters also may produce chemical effects through transmutation. In general, these effects are not considered as important as those arising through radiobiological mechanisms, but they have not been thoroughly explored. In radionuclides having emissions of low energy, transmutation may be of relatively greater importance, especially if the element is located in a biologically strategic molecule; for example, the mutagenic effects of transmutation from ^{14}C disintegrations in DNA may be as great as or greater than those of the associated β-radiations (Totter et al., 1958).

12-9 CONCLUSION

The alteration of atoms and molecules in living cells by randomly absorbed ionizing radiations may be amplified, through a complex sequence of chemical and physiological reactions, into gross disturbances of function and form affecting whole animals or populations of animals. The nature of the changes underlying this process is still poorly understood, but evidence implicates effects on genes, and possibly other macromolecules, critical to the cell.

Although any cell may be killed if subjected to a high enough dose of radiation, the median lethal dose varies enormously among cells, depending on species, organ, mitotic activity, physiologic state, and environmental factors (nutrition, oxygen tension, temperature, etc.), as well as on radiological variables (dose rate, number of exposures, interval between exposures, LET). Because of these variations, somatic radiation injury in higher animals tends to appear in certain cells and tissues more than in others. The effects of acutely lethal whole-body irradiation are, for example, ascribable primarily to destruction of dividing cells that maintain the number of the circulating blood corpuscles, the lining of the gastrointestinal tract, the skin, and the gonads. Recovery from such effects, moreover, is contingent on proliferation of surviving cells to replace those destroyed.

Not obviously related to the prompt killing of cells are delayed somatic effects which appear months or years after irradiation. These include the induction of cancer and various nontumorous degenerative changes associated with shortening of the life-span. Taken together, these effects suggest precocious senescence, but their relation to natural aging remains to be disclosed. Although the dose–response relation for such effects is not well known, they may be induced by relatively small doses accumulated over long periods of time. Hence they are of great concern in health physics.

Of even greater importance to human safety are the genetic effects of radiation, which may be transmitted to future generations. These vary in severity from dominant lethality to mild or undetectable effects. They are predominantly harmful. The consequences of a small radiation-induced increase in the mutation rate throughout a population cannot be precisely estimated. These mutations should be kept to a minimum because of their potential harm and relative irreversibility. For this reason, and because it is likely that any increase in radiation level will increase the mutation rate, genetic effects constitute a limiting factor in the maximum permissible dose for any species.

REFERENCES

Alexander, P., *Atomic Radiation and Life*, Penguin Books, Baltimore, Maryland, 1957.

Andrews, G. A., "A Few Notions Involved in the Clinical Use of Radioisotopes," *Ann. Internal Med.* **47,** 922–938 (1957).

Auerbach, Charlotte, *Genetics in the Atomic Age*, Essential Books, Fair Lawn, New Jersey, 1956.

Bacq, Z. M., and Peter Alexander, *Fundamentals of Radiobiology*, Second Edition, Academic Press, New York, 1961.

Bender, M. A., and Sheldon Wolff, "X-Ray-Induced Chromosome Aberrations and Reproductive Death in Mammalian Cells," *Am. Naturalist* **95,** 39–52 (1961).

Bloom, W., "Histological Changes After Irradiation," in *Radiation Biology* (A. Hollaender, ed.), McGraw-Hill, New York, 1954, Vol. I, pp. 1091–1143.

British Medical Research Council, *The Hazards to Man of Nuclear and Allied Radiations*, Her Majesty's Stationery Office, London, 1956.

Carlson, J. G., "Immediate Effects on Division, Morphology, and Viability of the Cell," in *Radiation Biology* (A. Hollaender, ed.), McGraw-Hill, New York, 1954, Vol. I, pp. 763–824.

Carlson, J. G., M. L. Snyder, and Alexander Hollaender, "Relation of Gamma-Ray Dosage Rate to Mitotic Effect in the Grasshopper Neuroblast," *J. Cellular Comp. Physiol.* **33,** 365–372 (1949).

Claus, W. D. (ed.), *Radiation Biology and Medicine*, Addison-Wesley, Reading, Massachusetts, 1958.

Congdon, C. C., "Experimental Treatment of Total-Body Irradiation Injury. A Brief Review," *Blood* **12,** 746–754 (1957).

Conger, A. D., "Genetical Effects," in *Radiation Protection and Recovery* (A. Hollaender, ed.), Pergamon Press, New York, 1961.

Crossfill, M. L., P. J. Lindop, and J. Rotblat, "Variation of Sensitivity to Ionizing Radiation with Age," *Nature* **183,** 1729–1730 (1959).

Dobson, R. L., and M. M. Chupp, "Effect of Small Measured Dose of Radiation on Lymphocyte Morphology in Man," *Proc. Soc. Exptl. Biol. Med.* **95,** 360–361 (1957).

Doherty, D. G., "Chemical Protection to Mammals against Ionizing Radiation," in *Radiation Protection and Recovery* (A. Hollaender, ed.), Pergamon Press, New York, 1960, pp. 45–86.

Elkind, M. M., and Harriet Sutton, "Radiation Response of Mammalian Cells Grown in Culture. I. Repair of X-Ray Damage in Surviving Chinese Hamster Cells," *Radiation Res.* **13,** 556–593 (1960).

Errera, M., and A. Forssberg (eds.), *Mechanisms in Radiobiology*, Academic Press, New York, 1960.

Evans, T. C., "The Influence of Quantity and Quality of Radiation on the Biological Effect," in *Symposium on Radiation Biology* (J. J. Nickson, ed.), John Wiley and Sons, New York, 1952, pp. 393–413.

Evans, T. C., "Effects of Small Daily Doses of Fast Neutrons on Mice," *Radiology* **50,** 811–834 (1948).

Finkel, M. P., "Relative Biological Effectiveness of Internal Emitters," *Radiology* **67,** 665–672 (1956).

Furth, J., and E. Lorenz, "Carcinogenesis by Ionizing Radiations," in *Radiation Biology* (A. Hollaender, ed.), McGraw-Hill, New York, 1954, Vol. I, pp. 1145–1201.

Giles, N. H., "Radiation-Induced Chromosome Aberrations in *Tradescantia*," in *Radiation Biology* (A. Hollaender, ed.), McGraw-Hill, New York, 1954, Vol. I, pp. 713–761.

Giles, N. H., "Comparative Studies of the Cytogenetical Effects of Neutrons and X Rays," *Genetics* **28,** 398–418 (1943).

Gray, L. H., "The Influence of Oxygen on the Response of Cells and Tissues to Ionizing Radiation," in *Lectures on the Scientific Basis of Medicine*, Pergamon Press, New York, 1960, Vol. 8, pp. 314–347.

Henshaw, P. S., "Experimental Roentgen Injury. IV. Effects of Repeated Small Doses of X-Rays on Blood Picture, Tissue Morphology, and Life Span in Mice," *J. Natl. Cancer Inst.* **4,** 513–522 (1944).

Henshaw, P. S., E. F. Riley, and G. E. Stapleton, "The Biologic Effects of Pile Radiations," *Radiology*, **49,** 349–360 (1947).

Hollaender, A. (ed.), *Radiation Protection and Recovery*, Pergamon, New York, 1960.

Hollaender, A. (ed.), *Radiation Biology*, McGraw-Hill, New York, 1954–56, Vols. I, II, and III.

Hollaender, A., and R. F. Kimball, "Modification of Radiation-Induced Genetic Damage," *Nature* **177,** 726–730 (1956).

Hutchinson, Franklin, and Josephine Arena, "Destruction of the Activity of Deoxyribonucleic Acid in Irradiated Cells," *Radiation Res.* **13,** 137–147 (1960).

Jacobson, L. O., "The Hematologic Effects of Ionizing Radiation," in *Radiation Biology* (A. Hollaender, ed.), McGraw-Hill, New York, 1954, Vol. I, pp. 1029–1090.

Kaplan, H. S., "Some Implications of Indirect Induction Mechanisms in Carcinogenesis: A Review," *Cancer Res.* **19,** 791–803 (1959).

Kaufmann, B. P., "Chromosome Aberration Induced in Animal Cells by Ionizing Radiations," in *Radiation Biology* (A. Hollaender, ed.), McGraw-Hill, New York, 1954, Vol. I, pp. 627–711.

Kimball, R. F., Nenita Gaither, and Stella W. Perdue, "Metabolic Repair of Premutational Damage in *Paramecium*," *Intern. J. Radiation Biol.*, **3,** 133 (1961).

Kornberg, H. A., "Radioisotopes and Environmental Circumstances: The Passage of Pairs of Elements Through Food Chains," in *Radioisotopes in the Biosphere* (R. S. Caldecott and L. A. Snyder, eds.), Univ. of Minnesota Press, Minneapolis, 1960, pp. 255–268.

Langham, W. H., K. T. Woodward, S. M. Rotherwel, P. S. Harris, C. C. Lusbaugh, and J. B. Storer, "Studies of the Effects of Rapidly Delivered, Massive Doses of Gamma Rays on Mammals," *Radiation Res.* **5,** 404–432 (1956).

Lea, D. E., *Actions of Radiation on Living Cells*, Cambridge University Press, London, 1955, 2nd ed.

Lindop, P. J., and J. Rotblat, "Shortening of Lifespan of Mice as a Function of Age at Irradiation, "*Gerontologia* **3,** 122–127 (1959).

Lindsay, L., G. D. Potter, and I. L. Chaikoff, "Thyroid Neoplasms in the Rat: A Comparison of Naturally Occurring and ^{131}I-Induced Tumors," *Cancer Res.* **17,** 183–189 (1957).

Lipetz, L., "Electrophysiology of the X Ray Phosphene," *Radiation Res.* **2,** 306–329 (1955).

Lorenz, E., J. W. Hollcroft, E. Miller, C. C. Congdon, and R. Schweisthal, "Long-Term Effects of Acute and Chronic Irradiation in Mice. I. Survival and Tumor Incidence Following Chronic Irradiation of 0–11 r Per Day," *J. Natl. Cancer Inst.* **15,** 1049–1058 (1955).

Lorenz, E., L. O. Jacobson, W. E. Heston, M. Shimkin, A. B. Escherbrenner, M. K. Deringer, J. Doviger, and R. Schweisthal, "Effects of Long-Continued Total-Body Gamma-Irradiation on Mice, Guinea Pigs, and Rabbits. III. Effects on Life Span, Weight, Blood Picture, and Carcinogenesis and the Role of the Intensity of Radiation," In *Biological Effects of External X- and Gamma-Radiation* (R. E. Zirkle, ed.), McGraw-Hill, New York, 1954, Part I, pp. 24–168.

Mathe, G., J. L. Amiel, L. Schwarzenberg, A. Cattan, M. Schneider, M. J. de Vries, M. Tubiana, C. Lalanne, J. L. Binet, M. Papiernik, G. Seman, M. Matsukura, A. M. Mery, V. Schwarzmann, and A. Flaisler, "Successful Allogenic Bone Marrow Transplantation in Man: Chimerism, Induced Specific Tolerance and Possible Anti-Leukemic Effects," *Blood* **25,** 179–196 (1965).

Morton, N. E., "The Mutational Load due to Detrimental Genes in Man," *Am. J. Human Genet.* **12,** 348–364 (1960).

Muller, H. J., "Our Load of Mutations," *Am. J. Human Genet.* **2,** 111–176 (1950).

National Academy of Sciences—National Research Council, *The Biological Effects of Atomic Radiation.* Summary Reports, Washington (1956, 1960).

Neary, G. J., "Ageing and Radiation," *Nature* **187,** 10–18 (1960).

Neary, G. J., R. J. Mumson, and R. H. Mole, *Chronic Radiation Hazards, An Experimental Study with Fast Neutrons*, Pergamon, London, 1957.

Norman, A. G., "Radioisotopes in Soils and Plants: A Critical Review." In *Radioisotopes in the Biosphere* (R. S. Caldecott and L. A. Snyder, eds.), Univ. of Minnesota Press, Minneapolis, 1960, pp. 110–122.

Oakberg, E. F., "Degeneration of Spermatogonia of the Mouse Following Exposure to X Rays and Stages in the Mitotic Cycle at which Cell Death Occurs," *J. Morphol.* **97,** 39–54 (1955).

Ord, M. G., and L. A. Stocken, "The Biochemical Lesion *in vivo* and *in vitro*," in *Mechanisms in Radiobiology* (M. Errera and A. Forssberg, eds.), Academic Press, New York, 1960, Vol. 1, pp. 259–331.

Patt, H. M., "Protective Mechanisms in Ionizing Radiation Injury," *Physiol. Rev.* **33,** 35–76 (1953).

Peck, W. S., J. T. McGreer, M. R. Kretzschmar, and W. E. Brown," Castration of the Female by Irradiation," *Radiology* **34,** 176–186 (1940).

Proc. Intern. Conf. Peaceful Uses At. Energy, Geneva, 1955, **11,** United Nations, New York, 1956.

Proc. Intern. Conf. Peaceful Uses At. Energy, Geneva, 1958, **22,** United Nations, New York, 1958.

Puck, T. T., "Action of Radiation on Mammalian Cells. III. Relationship Between Reproductive Death and Induction of Chromosome Anomalies by X Irradiation of Euploid and Human Cells *in vitro,*" *Proc. Natl. Acad. Sci. U.S.,* **44,** 772–780 (1958).

Puck, T. T., D. Morkovan, P. I. Marcus, and S. J. Ciecuir, "Action of X-Ray on Mammalian Cells. II. Survival Curves of Cells from Normal Human Tissues," *J. Exptl. Med.* **106,** 485–500 (1957).

Robertson, J. S., "Radioisotopes in the Skeleton: A Critical Review." In *Radioisotopes in the Biosphere* (R. S. Caldecott and L. A. Snyder, eds.), Univ. of Minnesota Press, Minneapolis, 1960, pp. 423–430.

Russell, L. B., "The Effects of Radiation on Mammalian Prenatal Development," in *Radiation Biology* (A. Hollaender, ed.), McGraw-Hill, New York, 1954, Vol. I, pp. 861–918.

Russell, L. B., and W. L. Russell, "Radiation Hazards to the Embryo and Fetus," *Radiology* **58,** 369–376 (1952).

Russell, W. L., "Genetic Effects of Radiation in Mammals," in *Radiation Biology* (A. Hollaender, ed.), McGraw-Hill, New York, 1954, Vol. I, pp. 825–859.

Russell, W. L., L. B. Russell, and E. M. Kelly, "Dependence of Mutation Rate on Radiation Intensity," In *Symposium on the Immediate and Low Level Effects of Ionizing Radiation* (A. A. Buzzati-Traverso, ed.), Taylor and Francis, London, 1960.

Russell, W. L., L. B. Russell, and E. F. Oakberg, "Radiation Genetics of Mammals," in *Radiation Biology and Medicine* (W. D. Claus, ed.), Addison-Wesley, Reading, Massachusetts, 1958, pp. 189–205.

Sacher, G. A., "Reparable and Irreparable Injury: A Survey of the Position in Experiment and Theory," in *Radiation Biology and Medicine* (W. D. Claus, ed.), Addison-Wesley, Reading, Massachusetts, 1958, pp. 283–313.

Sax, Karl, "Types and Frequencies of Chromosomal Aberrations Induced by X-Rays," *Cold Spring Harbor Symp. Quant. Biol.* **9,** 93–103 (1941).

Schull, W. J., and James V. Neel, "Radiation and the Sex Ratio in Man," *Science* **128,** 343–348 (1958).

Sobels, F. H. (ed.), *Repair from Genetic Radiation Damage*, Pergamon, New York, 1963.

Totter, J. R., M. R. Zelle, and H. Hollister, "Hazard to Man of Carbon-14," *Science* **128,** 1490–1495 (1958).

United Nations Scientific Committee on the Effects of Atomic Radiation. Report: General Assembly, Official Records, Seventeenth Session, Supplement No. 16 (A/5216), 1958; Nineteenth Session, Supplement No. 14 (A/5814), 1964; Twenty-first Session, Supplement No. 14 (A/6314), 1966; United Nations, New York.

Upton, A. C., "Ionizing Radiation and Aging," *Gerontologia* **4,** 162–176 (1960).

Upton, A. C., "The Radiobiology of the Cancer Cell," *Federation Proc.* **17,** 698–713 (1958).

Upton, A. C., A. W. Kimball, J. Furth, K. W. Christenberry, and W. H. Benedict, "Some Delayed Effects of Atom-Bomb Radiations in Mice," *Cancer Res.* **20** (No. 8, Part 2), 1–62 (1960).

Upton, A. C., F. F. Wolff, J. Furth, and A. W. Kimball, "A Comparison of the Induction of Myeloid and Lymphoid Leukemias in X-Radiated RF Mice," *Cancer Res.* **18,** 842–848 (1958).

Upton, A. C., K. W. Christenberry, G. S. Melville, J. Furth, and G. S. Hurst, "The Relative Biological Effectiveness of Neutrons, X-Rays, and Gamma-Rays for the Production of Lens Opacities: Observations in Mice, Rats, Guinea Pigs, and Rabbits," *Radiology* **67,** 686–696 (1957).

Upton, A. C., F. P. Conte, G. S. Hurst, and W. A. Mills, "The Relative Biological Effectiveness of Fast Neutrons, X-Rays, and γ-Rays for Acute Lethality in Mice," *Radiation Res.* **4,** 117–131 (1956).

Wallace, Bruce, and Th. Dobzhansky, *Radiation, Genes, and Man*, Holt, New York, 1959.

Wasserman, R. H., "Radioisotope Absorption and Methods of Elimination: A Critical Review," in *Radioisotopes in the Biosphere* (R. S. Caldecott and L. A. Snyder, eds.), Univ. of Minnesota Press, Minneapolis, 1960, pp. 564–573.

Whiting, Anna R., "Protection and Recovery of the Cell from Radiation Damage," in *Radiation Protection and Recovery* (A. Hollaender, ed.), Pergamon, New York, 1961, pp. 117–156.

Wolff, A. H., "Radioisotopes and Environmental Circumstances: A Critical Review," in *Radioisotopes in the Biosphere* (R. S. Caldecott and L. A. Snyder, eds.), Univ. of Minnesota Press, Minneapolis, 1960, pp. 331–338.

Wolff, Sheldon, "Chromosome Aberrations," in *Radiation Protection and Recovery* (A. Hollaender, ed.), Pergamon, New York, 1961, pp. 157–174.

Wolff, Sheldon, "Radiation Genetics," in *Mechanisms in Radiobiology* (M. Errera and A. Forssberg, eds.), Academic Press, New York, 1960, Vol. 1, pp. 419–475.

Wolff, Sheldon, and H. E. Luippold, "The Biochemical Aspects of Chromosome Rejoining," in *Progress in Radiobiology* (J. S. Mitchell, B. E. Holmes, and C. L. Smith, eds.), Oliver and Boyd, Edinburgh, 1956, pp. 217–221.

World Health Organization, *Effect of Radiation on Human Heredity* (Report of a Study Group and Papers Presented at a Study Group), Geneva, 1957.

Wright, Sewall, "On the Appraisal of Genetic Effects of Radiation on Man," in *The Biological Effects of Atomic Radiation*, Natl. Acad. Sci. U.S.—Natl. Res. Council, Summary Rept. (1960).

Wright, Sewall, "Discussion on Population Genetics and Radiation," *J. Cellular Comp. Physiol.* **35** (Suppl. 1), 187–210 (1948).

Zirkle, R. E., "The Radiobiological Importance of Linear Energy Transfer," in *Radiation Biology* (A. Hollaender, ed.), McGraw-Hill, New York, 1954, Vol. I, pp. 315–350.

QUESTIONS

1. What chemical changes are considered to be most important in radiation injury?

2. What are the principal effects of radiation on chromosomes, and what is their significance in radiation injury?

3. What types of radiation injury are transmitted to the descendants of irradiated individuals, and what is the significance of such effects for exposed human populations?

4. What are the major signs and symptoms of the acute radiation syndrome?

5. What is meant by the term $LD_{50}/30$?

6. What is considered to be the lethal dose of whole-body radiation for man?

7. What are the major late somatic effects of radiation, and what is known of their dose-effect relationships?

8. What factors modify the radiation response?

9. Define and illustrate the term RBE.

10. Discuss the importance of the biological half-life, the physical half-life, and the effective half-life of an internally deposited radionuclide in relation to the hazard of injury from the nuclide.

CHAPTER 13

Evaluation of Human Exposure Data

Niel Wald

A. INTRODUCTION

13-1 PURPOSE

The profession of Health Physics arose to fill a human need in human society. It is not surprising, then, that the ultimate test of the value of the information accumulated in the disciplines basic to Health Physics, such as physics, chemistry, and biology, is whether it is relevant to man and his particular problems. The purpose of this chapter, therefore, is to review and to evaluate the evidence *in man* of (1) the occurrence of the biophysical processes at the molecular level described in Chapter 11 and (2) the forms of radiobiological damage demonstrated in plants and animals which are discussed in Chapter 12. In addition, being human beings, we are concerned with the possibility of prevention and reversal of harmful biophysical processes, and of the repair of radiobiological damage in man. The medical management of radiation injury is considered in limited detail for this reason. Finally, the means for adding to our insufficient supply of useful information about human radiation injury are discussed.

In reviewing a subject which has aroused so much interest and study in recent years, it is clearly impossible to include all relevant information. In general, various aspects have been selected for detailed treatment on the basis of their particular interest and value to the reader who is not specialized in this scientific area. The most complete and recent reviews of each aspect have been cited, and references have been made to relevant material published subsequent to the reviews. The selections are not all-inclusive but, rather, reflect the limitations of space and the interests and capabilities of the writer.

13-2 DEFINITIONS

The terminology used in this chapter is essentially that used by the majority of workers in this field. However, to reduce the ambiguity of certain terms such as "acute" and "chronic," which have been used interchangeably to describe both radiation exposure and its effects, we have adhered to the definitions set up by the Subcommittee on Hematologic Effects of the Committee on Pathologic Effects of Atomic Radiation, National Academy of Sciences-National Research Council[59] and utilized by the United Nations Scientific Committee on the Effects of Atomic Radiation.[100]

Acute describes disease conditions having a short and sharp course as distinguished from those which are *chronic*, that is, having slow progression and long duration. *Short-term exposure* includes (1) total-body exposure to radiation over a short period of time, for example, in nuclear warfare and nuclear reactor or accelerator accidents, and (2) exposure to limited but substantial body areas when the radiation is delivered either as a single dose or fractionated over a few days or weeks; for example, in therapeutic radiation, diagnostic radiology, and tracer or therapeutic use of radioactive isotopes. In considering short-term exposure, a dose of 50 rad or more is defined arbitrarily as a *high dose*, less than 50 rad as a *low dose*. *Long-term exposure* refers to continued or repeated exposure to radiation over long periods of time—months or years, for example, routine occupational exposure, body burdens of radioactive isotopes with relatively long effective half-lives, X-ray examinations repeated frequently over a long time, exposures to cosmic radiation, naturally occurring radioactive isotopes, and fallout.

The *immediate effects* of radiation are those events occurring in tissue during exposure as a result of the interactions of the radiation with matter. These interactions may lead to clinical manifestations or observable alterations in various organs that occur promptly or may appear months or years after irradiation. The development of these clinical findings will depend not only on irradiation but will also be influenced by secondary factors such as hormones. A difference should also be noted between those effects that produce only a cytologic abnormality and those that produce a serious disease. An example of the former is the occurrence of binucleate lymphocytes in the circulating blood, whereas leukemia is an illustration of the latter.

It is not possible to draw a sharp distinction between early and delayed effects since some that occur soon after irradiation may persist. Nevertheless, it is convenient to consider as *early effects* those which are observable within 60 days after the beginning of exposure, while *late effects* are those occurring thereafter.

13-3 SOURCES OF HUMAN DATA

Ethical considerations concerning human rights and dignity properly preclude the deliberate production of experimental injury of unknown severity, duration, and treatment in a fellow man by a scientific investigator. The sources of data concerning human radiation exposure, its effect, and their management, therefore, are the observations made in three population groups: (1) individual human cases of major accidental exposure, (2) patients exposed purposefully to radiation as a form of medical treatment, or (3) general populations exposed to nuclear weapons as an unhappy consequence of military action or of weapons tests. These include workers in nuclear research and industry as well as in other industries using radioactive materials; patients who received whole-body radiation for treatment of widespread cancer or leukemia; patients given nearly whole-body radiation for nonneoplastic diseases; patients receiving localized external exposure to diseased areas of the body; patients treated with internally administered radioisotopes for cancer or noncancerous conditions; and members of the general population exposed to the effects of nuclear weapons in the Japanese cities of Hiroshima and Nagasaki, in the Marshall Islands, and on a Japanese fishing boat.

A number of individuals and committees have reported and reviewed the information concerning human radiation exposure and its effects [Refs. 8, 18, 19, 36, 39, 44, 55, 59, 63, 66, 67, 98, 100, 101, 106, and 109]. Table 13-1 contains brief summaries of human experiences with the acute and the long-term or chronic effects of radiation in man. The tabulation is not all-inclusive, but contains the most useful of the human experiences. The references cited are not necessarily the original sources, but may instead be suitable data compilations and review articles which themselves cite the original sources. Other relevant references are cited when discussed in the subsequent text.

13-4 GENERAL EVALUATION OF HUMAN DATA

In considering the information to be obtained by the study of human experience, it is important to be aware of certain significant differences between such data and those obtained by laboratory experiment. The laboratory investigator can produce at will the condition to be studied in a carefully selected laboratory population. The student of human experience must utilize the tools developed in the field of epidemiology, which MacMahon has defined aptly as "the study of the distribution and determinants of disease prevalence in man."[58]

Descriptive epidemiology is primarily concerned with the *distribution* of disease by such variables as age, sex, and socioeconomic status. Although such information has proved useful, the major contribution to our knowledge of the effects of radiation in man have come from the use of the techniques of analytical epidemiology. This is concerned with the *determinants* of disease frequency or prevalence and makes the fullest possible use of technical knowledge in the fields of medicine and health physics, involving as it does the integration of laboratory, clinical, and epidemiological information. Both retrospective and prospective approaches are employed. In the former, the frequency with which the suspected stimulus (that is, radiation exposure) is found in the history of patients showing the response under study (that is, leukemia, etc.) is determined. In the latter, the incidence of the suspected response is determined in follow-up studies of the population receiving the stimulus of interest.

It is evident that the study of disease or injury, in whatever population it is found, will not produce the same clean-cut answers as the study of carefully quantitated disease or injury produced in a selected population whose characteristics are fully defined in advance. The nonexperimental radiation exposures, the doses of which have been estimated in retrospect in most instances, have consisted of a varied mixture of types of radiation delivered externally and/or internally at varying dose rates, irradiating the bodies of the recipients nonuniformly. The inconsistently selected recipients have been in varied states of health on exposure and have received a wide range of treatments subsequently, in addition to exhibiting the usual nonspecific variations in biological response. The range of doses is incomplete, requiring assumptions concerning dose–response relationships in order to extrapolate to dose levels of interest. Without thereby negating the value of the information produced, it is useful to consider the specific limitations of the various kinds of human exposure situations from which our information of radiation-induced biomedical effects has been derived.

Table 13-2 presents a generalized evaluation of the data obtainable from the three main types of populations in which radiation effects have been studied. It will be noted that each has some advantages and some disadvantages. For example, the dosimetric information readily available in accidental exposure situations is generally quite limited, although it may be reconstructed retrospectively if the information to be obtained warrants the effort. On the other hand the population is generally a healthy one prior to exposure, and baseline information in the form of medical and laboratory examinations on a preemployment and subsequent periodic basis is often available. The examinations after such exposures are generally quite complete and the ability to trace the occupational worker through Workman's Compensation and Social Security records provides good possibilities for long-term follow-up.[62] In addition, the number of significant external

Table 13-1 Sources of Human Data

Population			Exposure			
Type	Size	Location	Type	Date	Variables Studied	References
A. Short-term, external						
1. Occupational	2	Los Alamos Scientific Laboratory	Experimental assembly, criticality	1945	Acute effects	42, 66, 98; 100, pp. 140–144; 106, pp. 49–66
	8	Los Alamos Scientific Laboratory	Experimental assembly, criticality	1946	Acute effects	Same as above
	4	Argonne National Laboratory	Reactor, criticality	1952	Acute effects	43, 66, 98; 100, pp. 140–144
	2	U.S.S.R.	Reactor, criticality	ca. 1953	Acute effects	35, 66, 98; 100, pp. 140–144
	8	Oak Ridge National Laboratory	Chemical processing, criticality	1958	Acute and delayed effects	4, 13, 66, 98; 100, pp. 140–144; 106, pp. 27–48
	6	Kidric Institute, Yugoslavia	Reactor, criticality	1958	Acute and delayed effects	47, 66, 98; 100, pp. 140–144; 106, pp. 67–81; 105–111
	2	Los Alamos Scientific Laboratory	Chemical processing, criticality	1958	Acute effects	66, 81, 98; 100, pp. 140–144
	1	Moscow, USSR	^{137}Cs capsule (self-inflicted), gamma	1960	Acute effects	1
	9	Lockport, New York	Klystron tube, X-Ray	1960	Acute and delayed effects	100, pp. 140–144; 106, pp. 11–26
	1	University of Wisconsin	^{60}Co capsule, gamma	1960	Acute and delayed effects	72
	3	Hanford Atomic Products Operation	Chemical processing, criticality	1962	Acute effects	33

	6	Puerto Rico Nuclear Center	Reactor, gamma	1962	Acute effects	54
	4	Rhode Island	Chemical processing, criticality	1964	Acute effects	50
	1	Mol, Belgium	Reactor, criticality	1965	Acute effects	111, 112
2. Medical						
Metastatic cancer	263	M. D. Anderson Hospital, Houston, Texas	X-Ray, whole-body	1957	Acute effects	55
Cancer of the cervix	71,582	U.S., Europe, etc.	Radium, pelvis	ca. 1950	Leukemia	76, 101, 109
Skin cancer	ca. 300	Japan	Therapeutic		Exposure incidence	101
Radiation dermatitis	380	U.S.	Therapeutic	1930–ca. 1950	Skin cancer	96, 101
Adult leukemia	1600	Denmark	X-Ray, diagnostic and therapeutic	1940–1954	Exposure incidence	101, 109
Adult leukemia	1023	U.K.	X-Ray, diagnostic and therapeutic	1952–1960	Exposure incidence	84, 101
Adult leukemia	590	New Zealand	X-Ray or radioisotope, diagnostic and therapeutic	1958–1961	Exposure incidence	34, 101
Ankylosing spondylitis	13,352	U.K.	X-Ray, full spine	1935–1955	Leukemia	22, 101, 109
Thyroid cancer	ca. 900	U.S. and Japan	X-Ray, head and neck, therapeutic	to 1960	Exposure incidence	101
Neck cancer	906	Japan	X-Ray, neck, therapeutic	to 1960	Exposure incidence	101
Pediatric thymic enlargement, etc.	ca. 11,000	U.S.	X-Ray, head, neck, chest	ca. 1930–1950	Cancer and leukemia	101, 109

Table 13-1 (Continued)

Population			Exposure			
Type	Size	Location	Type	Date	Variables Studied	References
Pediatric leukemia and cancer	ca. 3400	U.K. and U.S.	X-Ray, prenatal, whole-body	ca. 1951–1957	Exposure incidence	101, 109
Prenatal exposure	39,166	U.K.	X-Ray; whole-body	1945–1956	Leukemia	101, 109
Prenatal exposure	734,243	Northeastern U.S.	X-Ray; whole-body	1947–1954	Leukemia and cancer	101
Various diseases		France, Holland, Canada	X-Ray		Genetic effects	61, 71
3. Nuclear weapons						
General	450,000	Hiroshima and Nagasaki	Neutron, γ, fission products	1945	Acute effects	67
General	100,000	Hiroshima and Nagasaki	Neutron, γ, fission products	1946 to present	Delayed and genetic effects	2, 40, 61, 101, 109, 110
General	267	Marshall Islands	β-γ fallout	1954	Acute and delayed effects	24
General	23	Japanese Fishing Boat	β-γ fallout	1954	Acute effects	94, 97
B. Short-term, internal						
1. Medical						
Various diseases		U.S.	Therapeutic amounts of ^{198}Au, ^{32}P, ^{72}Ga, ^{90}Y, ^{131}I		Acute effects	109
Hyperthyroidism	ca. 60,000	U.K., U.S., Canada, Austria	^{131}I	ca. 1940–1960	Leukemia	101, 109
Polycythemia vera	1238	U.S.	^{32}P	1936–1960	Leukemia	101, 109

C. Long-term, external						
1. Occupational						
Radiologists	ca. 1200	U.S. and U.K.	X-Ray and fluoroscope	1929–1961	Leukemia	101, 109
Physicians	16,808	U.S.	X-Ray and fluoroscope	1915–1958	Life-span	88, 89
X-Ray workers	137	Japan	X-Ray	1958	Fertility and genetic effects	49
Radiologists		U.S.	X-Ray		Genetic effects	71
2. Environmental						
General	6	Mexico	^{60}Co Source	1962	Acute effects	53
General	All U.S. cities	U.S.	Cosmic radiation	1949–1951	Leukemia	101, 109
General	Total population	Scotland	Natural sources	1939–1956	Leukemia	101, 109
General	1,242,744	Upstate New York	Natural sources	1948–1955	Cancer and congenital anomalies	38
General	87,184	Northern New England	Natural sources	1952–1961	Congenital anomalies	113
D. Long-term, internal						
1. Occupational						
Watch dial painters and chemists	ca. 600	Illinois, New Jersey, Massachusetts	Ra and Ra-mesothorium	1917–1925	Bone cancer	101
Miners, pitchblende		Schneeburg and Joachimsthal	Airborne radon and daughters	ca. 1921–1930	Lung cancer	101
fluorspar	71	St. Lawrence, Newfoundland	Radon daughters	1952–1961	Lung cancer	101
uranium	3415	Colorado Plateau	Radon daughters	1950–1963	Lung cancer	105
2. Medical						
Radium	ca. 160	Illinois, Massachusetts	Ra and Ra-mesothorium ingestion and injection	1915–1929	Bone cancer	101
Thorium	ca. 1000	U.S.	Thorotrast injection	1932–1947	Cancer	101

Table 13-2 Generalized Evaluation of Human Data

	Dosimetry Data				Biomedical Data					
	Short-Term Exposure		Long-Term Exposure			Short-Term Exposure		Long-Term Exposure		
Exposed Population	External	Internal	External	Internal	Preexposure Baseline	Acute	Chronic	Acute	Chronic	Influence of Intercurrent Variables
Accidental occupational	Poor–fair	Poor	Poor	Poor	Fair–good	Excellent	Excellent	–	Good	Limited
Medical radiation	Good	Good	Good	Fair–good	Variable	Excellent	Fair	–	Fair	Marked
Nuclear weapons	Fair–good	Poor	—	—	None	Poor–fair	Excellent	–	–	Moderate

variables which may invalidate the biomedical follow-up observations is generally limited.

In the populations of patients receiving radiation therapy, on the other hand, the dosimetric information is good since the radiation exposure is planned in advance. The only limitations here are the fundamental dosimetric problems involving local internal dose estimations both from external sources and, more particularly, from internal sources of radiation. The difficulties in the use of this type of study population lie mainly in the biomedical area because the very selection of such individuals for radiation exposure implies that serious medical abnormalities are present even before such treatment. The setting of a good biomedical baseline, therefore, as a reference point from which to judge the magnitude of any changes related to the radiation exposure becomes a difficult problem. In addition, there is a definite limitation to the possibility of good, long-term, follow-up studies in such a population since the postirradiation survival of such patients, of whom many have various forms of cancer, may be quite limited.

The third major category, nuclear weapons survivors, has provided populations whose health status is average but for whom there is little specific baseline information available. Although the nature of the exposure events has generally precluded good observations of acute effects, the long-term follow-up studies have been excellent. It is the dosimetric data which present serious problems in such population studies. Retrospective reconstructions of the exposure events probably give a fairly good estimate of the exposure. Because of the careful follow-up of such populations on a long-term basis, the influence of intercurrent variables, while it cannot be controlled, can at least be evaluated over time.

With these shortcomings and strengths of the various available human populations in mind, a review of the effects of radiation exposure in man, based on these population studies, can be carried out.

B. EFFECTS OF RADIATION OVEREXPOSURE IN MAN

13-5 IMMEDIATE SOMATIC EFFECTS

Mortality

Death due to radiation overexposure has been recognized in all of the three general population catagories considered here. The determination of the LD_{50} dose (that dose of radiation that will result in death within a stated period of time in 50% of a population of human beings exposed) would seem to be a simple matter, since the biomedical endpoint is a generally undisputed one. Unfortunately, however, there are unresolved problems in each of the various appropriate study populations which preclude the setting of a precise $LD_{50}/60$ days which would be accurate for man.

The Japanese exposed to the wartime atomic bombings comprise a large population from which to determine this information. The retrospective dosimetric data are, however, based on a knowledge of the exact location of each individual under study. In the case of the nonsurvivors, this information is either unreliable or unavailable. Furthermore, the superimposed effects of blast and thermal injury and the stresses of the postbombing environment cannot be measured. A crude first estimation led to a relatively high probable LD_{50} of the order of 700–800 rad.[7] In the light of recent refinements in the dosimetric data developed by Auxier et al.[5] this estimate appears to have been high by a factor of about 2. Revision of tentative dose estimates for the individual Japanese cases using the modified air dose data is now in progress.

In the case of the Marshall Island population there were no fatalities. The fallout γ-radiation dose was increased by the presence of β-emitting radioactive material which also produced skin lesions in a high percentage of the individuals. Estimation of the LD_{50} in this group, based on the magnitude of hematologic damage, has ranged from 350 to over 400 rad.[19]

In the radiation accident population no fatalities occurred in any individuals whose dose was less than 500 rad. However, the patients were all subsequently under various forms of treatments, which may well have altered the outcome. In the radiation therapy population, severe hematologic damage occurred at exposures of 200 R midline, but the subjects were not normal before exposure.

Consideration of experimental animal studies, the results of which are presented in Table 12-2, suggests that LD_{50} is somewhat lower, although somewhat inconsistently so, in large animals as compared to the smaller ones. Bond and Robertson[12] have proposed that the absorbed dose in rad at the midcenter of the animals is a more appropriate measure than the air exposure dose. This is a convenient measure representing the approximate dose that all tissues received. When it is used there appears to be a more consistent pattern in which the LD_{50} decreases as the size of the animal species increases. They therefore suggest that for uniform whole-body exposure, the LD_{50} value is approximately 300–400 R for man, expressed as midcenter tissue dose, rather than the usual estimate of 450–500 rad which has been used in many publications.

It is evident that the resolution of this question cannot be reached without additional human data.

Morbidity

Acute Radiation Syndrome. When man is exposed to a large, single, short-term whole body dose of ionizing radiation, the resultant injury is expressed as a complex of clinical symptoms (abnormalities of function, appearance, or sensation noted by the patient), signs (abnormalities of function or appearance noted by the examiner), and laboratory findings, which are collectively termed the Acute Radiation Syndrome. As indicated in Chapters 11 and 12, this syndrome represents the clinical expression of the damage to many important organs simultaneously by depletion of radiosensitive cells resulting from their inability to reproduce themselves because of radiation-induced disturbances at the molecular level. The nature, frequency, time of onset, and severity of the elements of this syndrome are functions of the magnitude and distribution of the exposure, and of the particular radiosensitivity of the individual. The clinical and laboratory findings of the Acute Radiation Syndrome in man have been reported and reviewed in detail by various observers studying nuclear weapons survivors, radiation therapy patients, and occupational accident victims (see Table 13-1). Although the outlines of a clear-cut symptom complex are confused by variations in radiation exposure, in the general health of the recipient, and in the feasibility of valid controlled clinical observations, certain landmarks do emerge.

The typical chronologic sequence of events following a large whole-body radiation exposure can be divided into four clinical stages:

I. The initial or prodromal stage—0 to 48 hours postexposure
II. The latent stage —48 hours to 2–3 weeks postexposure
III. The manifest illness stage —2–3 weeks to 6–8 weeks postexposure
IV. The recovery stage —6–8 weeks to several months postexposure

The *prodromal* clinical findings include: anorexia (loss of appetite), nausea, extreme sweating, fatigue, and prostration. These changes remit in the *latent* period, which begins about two days later. After about two to three weeks of well-being a number of developments begin within a short time of one another.

These symptoms of the *manifest illness* stage may include: fever, overt or occult infections, scalp hyperesthesia (abnormal sensitivity), epilation (loss of hair), purpura and hemorrhage (bruises and bleeding), diarrhea, ileus (cessation of intestinal movements), cardiovascular collapse, severe lethargy, and changes in sensorium (disturbances in consciousness and perception).

By about the end of the sixth week after exposure the situation has usually been resolved and the clinical improvement of the surviving patients is rapid, although fatigue may continue for some time.

About 15% of patients exposed to a dose of 100 rad may be expected to show some of the signs and symptoms of the syndrome. The frequency increases sharply up to the level of approximately 200 rad, at which level most of those exposed would be expected to exhibit some clinical symptomatology. The occurrence of fatalities has been considered generally to begin at about 200 rad, increasing to 50% at about 450 to 500 rad. However, with optimal therapy, fatalities may not occur below 500 rad. A fatal outcome is probably inevitable at doses greater than 800–1000 rad despite current methods of treatment.

An analysis of the clinical signs and symptoms and laboratory findings in the victims of nuclear criticality accidents was reported recently.[66,98] The data from these irradiated previously healthy individuals indicate that there are five well-defined clinical injury patterns into which patients can be grouped. The pertinent characteristics of each of these injury groups are summarized as follows.

Group I: Most of these patients are asymptomatic; a few may have minimal prodromal symptoms such as nausea and anorexia. Little therapy is required.

Group II: These patients develop the acute radiation syndrome in a mild form. After transient prodromal nausea and vomiting, laboratory and mild clinical evidence of hematologic derangement dominates the picture. Careful observation but conservative therapy of symptoms is indicated.

Group III: Severe manifestations occur in these patients. Complications of hematologic malfunction, including hemorrhage and infections, are pronounced. In the sicker members of the group, some evidence of gastrointestinal damage may also be present. Survival of these patients may depend on wise clinical management.

Group IV: An accelerated version of the Acute Radiation Syndrome takes place. Complications of gastrointestinal injury, such as diarrhea and ileus, dominate the clinical picture. The severity of hematopoietic disturbances is related to the length of survival time following exposure. Recovery is not likely.

Group V: A fulminating course occurs with marked central nervous system impairment including paresthesias, excitement, and coma, as well as cardiovascular complications, such as severe hypotension. A fatal outcome may be expected.

In the Acute Radiation Syndrome, the very radiosensitive hematologic system is the most prone to manifest evidence of injury. It is only when injury is very severe that gastrointestinal symptoms dominate the picture as it is seen in Group IV. The relatively radioresistant central nervous system and the cardiovascular system do not manifest clinical evidence of damage, such as that in Group V, unless the injury is extreme. The predominance of gastrointestinal symptomatology in Group IV cases and central nervous system and cardiovascular signs in Group V cases does not indicate that at these levels severe hematologic damage has not taken place. These patients do not survive long enough to display the full manifestation of their hematologic injury.

The foregoing information is based on cases of external radiation exposure. Similar manifestations of the Acute Radiation Syndrome may accompany the administration of internal radiation such as ^{198}Au or ^{32}P when given in very high doses in cancer therapy. In general, however, the Acute Radiation Syndrome is not seen after exposure to internal emitters. The effects of internal emitters are usually chronic after long-term, lower level exposure.

Local Radiation Injury. The immediate effects of excessive external radiation exposure to a discrete part of the body are seen most often as changes in the skin. The information concerning this type of injury is based on observations of patients treated with radiation, as well as those inadvertently exposed to fission products, accelerator beams, and targets.[60]

Erythema appears on exposure to more than 200 or 300 rad. It is equivalent to a mild sunburn or a thermal burn of the first degree. A transient first wave of erythema may be present within hours after exposure, associated with a sensation of warmth or itching. The major redness appears two or three weeks later, the interval depending on the dose. In the lower dose range no further changes may occur and medical care is not necessary.

A counterpart to this reaction has been described in the conjunctiva, where inflammatory changes have been observed promptly after exposure. An inflammatory response also has been reported in the anterior chamber inside the eye within 30 minutes after irradiation, reaching a peak in one day and ending in one week.[25]

Epilation or loss of hair may occur after exposure to any form of radiation in doses of about 200 rad or more. Regardless of the dosage it generally becomes apparent only two to three weeks after exposure. Associated skin or scalp tenderness may occur one or two days preceding the actual hair loss. With doses greater than about 500 rad epilation may be complete. If the exposure is much greater than 600 rad, hair may not regrow.

Transepidermal injury (dry or wet dermatitis) is equivalent to a thermal burn of the second degree. Blisters form and break open leaving raw, painful wounds which are vulnerable to infection. These may occur within one to two weeks after exposure, depending on the dose. They are preceded or accompanied by erythema or epilation. Recognizable injury of this grade requires skin doses in excess of 1000 rad or 200 $\mu c/hr/cm^2$. The need for medical care depends on the size and severity of the burn.

With higher doses, probably on the order of 5000 rad, a more serious version of transepidermal injury occurs in which the lesion resembles a scalding

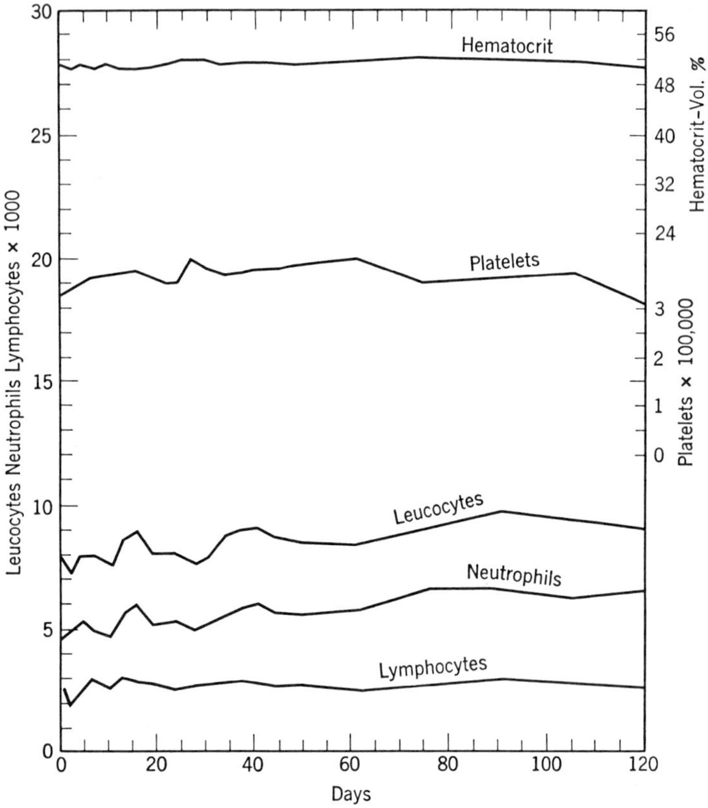

Fig. 13-1. Blood counts—injury group I.

or chemical burn. Pain occurs promptly and is intense. The raw areas may be very slow in healing or may not heal until surgical resection and skin grafting is performed. Epilation is permanent.

Cellular Damage. Although damage at the cellular level after radiation exposure can be demonstrated by many clinical and research laboratory techniques, only those utilized conveniently in the evaluation of human injury will be considered here. Two methods have been widely applied. One involves the enumeration of the number of cells of the various blood elements which are present in the circulating blood, that is, erythrocytes (red blood corpuscles); leukocytes (white blood cells), which occur as the following varieties; lymphocytes, monocytes, polymorphonuclear neutrophils (granulocytes), eosinophils, and basophils; and platelets. The fluctuations in the count are the resultant of the rates of production and destruction as well as the normal life-span of the type of element.[109] The changes in these levels

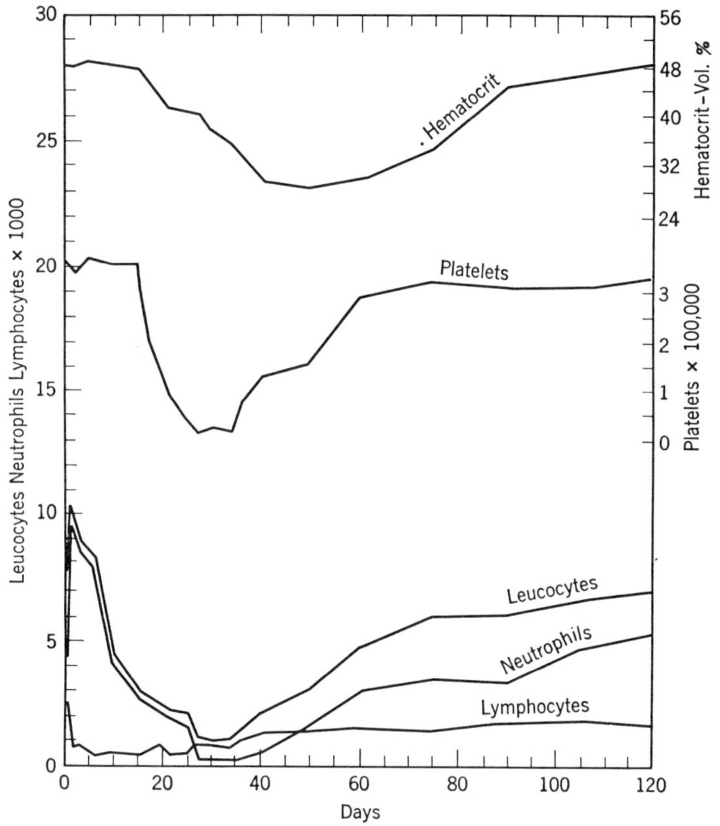

Fig. 13-2. Blood counts—injury group II.

associated with the various levels of radiation injury already described as Injury Groups I through V are portrayed graphically in Figs. 13-1, 13-2, and 13-3. These changes are of themselves nonspecific. However, with other historical and clinical information they are useful indicators of radiation overexposure.

The second method for measuring cell damage is the cytogenetic examination of the chromosomes of individual cells. Recent developments in cytogenetic techniques have made it possible to study the chromosomes of dividing cells from the bone marrow or circulating blood of man.[104] In view of the relationship between radiation and chromosome aberrations, as described in Chapter 12, this methodology has been applied to the study of individuals immediately after radiation exposure. Fliedner et al.[32] have noted

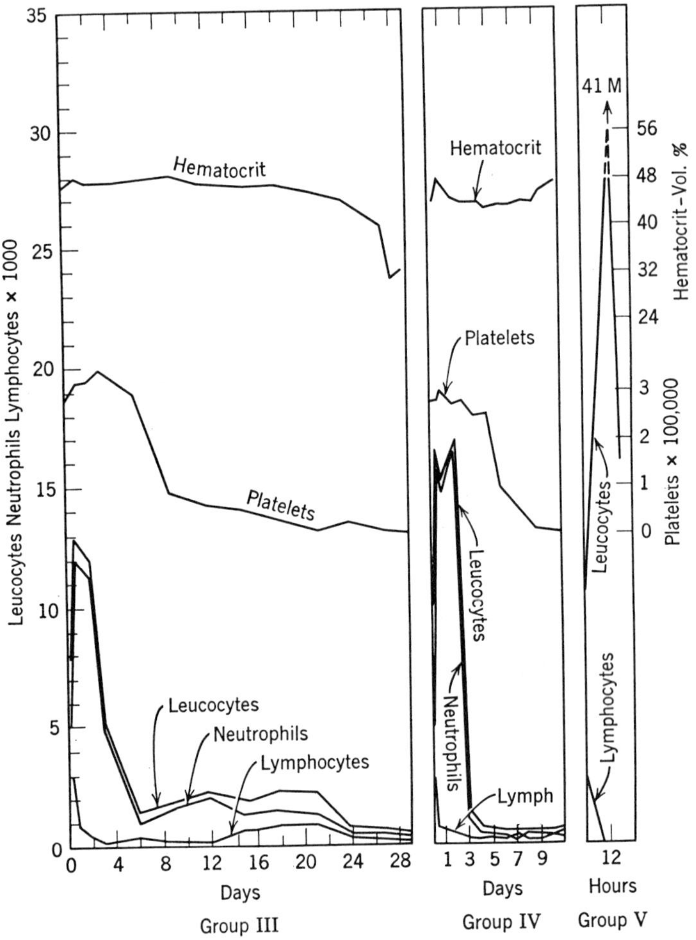

Fig. 13-3. Blood counts—injury groups III, IV, and V.

chromosomal stickiness, fragmentation, clumping, and bridging in bone marrow specimens taken early after a radiation criticality accident. Tough et al.[91] observed such fragmentation in cultured circulating blood cells as well as a fall in the percentage of cells bearing the normal human chromosome complement of 46 within 24 hours after therapeutic radiation of 250 rad to the skin over the spinal column. Buckton et al.[11] and Norman et al.[65]

confirmed and extended these findings in radiation therapy patients. Similar observations have been made in patients receiving diagnostic X-ray examinations such as fluoroscopic studies of the gastrointestinal tract as well as in patients receiving therapeutic doses of radioiodine.[48] Approximate doses with which these findings have been associated have been as low as 1870 mrad skin dose[87] and 0.8 rad infant whole-body dose[20] of diagnostic X-rays, 3 rad from fluoroscopy,[15] and 6–11 mc ^{131}I administered orally in the case of radioiodine exposure.[6] [The estimated dose to hematopoietic tissue is about 0.5 to 1 rad/mc ^{131}I.[80]] These data have recently been summarized in tabular form.[48] As yet, however, the dose–response relationship for such changes is not completely known.

13-6 LATE SOMATIC EFFECTS

Gradually, about a decade after the discovery of the X-ray and of natural radioactivity, the realization developed that the somatic effects of an overexposure to ionizing radiation might not be evident until years or decades following the event. It was based primarily on the clear progression of acute radiation burns and dermatitis to cancer of the skin. That a late radiation effect might occur in the absence of any striking initial changes or continuing evidence of abnormality was a much more difficult concept. A major obstacle to this realization was the fact that none of the conspicuous late effects of radiation exposure are uniquely induced by radiation alone. Various forms of cancer, cataracts, and reductions of life-span, fertility, growth, and development are all known to occur in man in the absence of significant radiation overexposure. Their relationship to radiation overexposure in man can only be deduced from the demonstration of an increased frequency of such changes which occurs in a logical relationship to the time of occurrence and the magnitude of the radiation overexposure. Such a relationship cannot be shown conclusively in the individual case but can be demonstrated in properly defined populations, using the techniques of epidemiology. Supporting evidence from laboratory animal experiments is used to supplement and complement the statistical information obtained from such population studies.

Carcinogenesis

One of the first late effects of radiation overexposure to be recognized was the increased likelihood of the development of cancer. Among the organ systems of man in which there is some evidence for this are skin, blood-forming tissue, bone, lung, thyroid gland, and connective tissue.

Skin. The first description of the progression of acute radiation dermatitis to an epidermoid carcinoma of the skin in a radiologist appeared in 1902, just seven years after the discovery of X-rays. By 1907, seven fatal cases were recorded in early radium and X-ray workers and the number increased rapidly to several hundred in the next two decades. The subject has been reviewed in detail by Traenkle.[96]

In therapeutically irradiated patients who developed chronic radiation dermatitis, the incidence of skin cancer in the two largest studies was 10%[83] and 28%[93], respectively. In a recent survey of medical exposure[92] there appeared to be a correlation between magnitude of dose and increase in relative risk of skin cancer. The over-all risk was 6.0 with limits of 16.3 and 2.2. With an average excess radiation of 240 R in the cancer group, the relative risk per roentgen was 0.026. Since radiation therapy usually involves fractionated exposures, these risk estimates may not apply to single exposures.

No increase in the incidence of skin cancer has been noted as yet among the survivors of nuclear weapons exposure. Clinical and occupational experience also suggests that the risk of cancer of the skin is low in the absence of acute radiation-induced skin damage. It may be, however, that the absence of such acute damage may, in inadequately designed surveys, preclude the recognition of an association of skin cancer with radiation exposure.

Blood-Forming Tissue. The first cases of leukemia or cancer of the blood-forming tissue in individuals occupationally exposed to X-radiation were reported by von Jagie and others in 1911.[102] Since then there has been a continuing series of case reports, some epidemiologic studies and many reviews suggesting a cause and effect relationship between chronic low-level radiation exposure and leukemia in man. More recently, the studies of the Atomic Bomb Casualty Commission of the survivors of the Hiroshima and Nagasaki bombings have supplied evidence for a similar relationship to a single high-level exposure. More recently still, various patient populations exposed to therapeutic or diagnostic radiation in the course of their medical management have come under observation and analysis in epidemiologic studies of leukemia. Environmental exposure situations, both internal and external, have been examined as well. Also, over the past 30 years since the early work of Krebs and others, animal experiments under a variety of exposure conditions have shown the leukemogenic property of radiation exposure. Some of these have been discussed in Chapters 11 and 12.

This general field has been of great interest to students of hematology, oncology, and radiobiology, and numerous reviews of the general subject matter have been published. Among the most recent and complete of these are the three reports of the United Nations Scientific Committee on the Effects of Atomic Radiation,[99–101] particularly the last one; a detailed review article[109]; and the reports of two subcommittees of the National

Academy of Sciences—National Research Council Committee on Pathologic Effects of Atomic Radiation,[59,64] the latter of which includes an extensive bibliography. The reader is urged to consult these reviews and the original sources for details since only a brief summary of the highlights of the many studies can be attempted here.

OCCUPATIONAL EXPOSURE. About a decade after the initial utilization of radium and X-rays in medicine, case reports of leukemia in radiologists and their associates began to appear in the scientific literature. Since 1944 a number of investigations have applied the retrospective epidemiological study technique to the radiologist population. Initial studies revealed that death from leukemia occurred about ten times more frequently in radiologists than in other physicians, while death by leukemia in all physicians occurred about two and one-half times more frequently than it did in the general population. As the study population neared the present decade, this ratio had fallen to 3.6 for radiologists to 1.0 for all physicians to 0.4 for the general population. In a recent study of radiologists in California,[51] in addition to the threefold increase incidence of leukemia, five deaths from multiple myeloma, another bone marrow neoplasm, were found where one was expected.

The major defect in the study of radiologists is the absence of any data pertaining to the radiation dose received by those who developed leukemia or other blood dyscrasia. Braestrup[14] has estimated that as high as 100 R/year may have been received by the workers before 1930. In any event the evidence seems good for the qualitative conclusion that long-term intermittent radiation exposure increases the incidence of leukemia in man.

NUCLEAR WEAPONS EXPOSURE. The surviving population of Hiroshima and Nagasaki, Japan, contains about 60,000 people, still living there in 1950, who are presumed to have received some radiation exposure. This includes some 7000 persons who underwent major symptoms of the Acute radiation Syndrome. This population provides the most informative data available on the relationship between radiation exposure and leukemia in man. In view of the tremendous value of a population of this size, a major effort has been made to establish in retrospect the specific radiation dose received by each individual survivor. Arakawa[3] has reported the current status of this effort, stemming from the joint program of the Health Physics Division of Oak Ridge National Laboratory and the Atomic Bomb Casualty Commission. Based on theoretical and experimental information concerning the air dose from the weapons, the location and shielding history of the individual, and the attenuation factors for the particular shielding, a tentative dose assignment has been made for each survivor. There is obviously a large uncertainty about the absolute exposure dose, although recent refinements[5] are reducing and quantitating the error estimate. In any event, the relative exposure dose of one individual as compared to that of others is reasonably accurate.

In Hiroshima, 127 cases of leukemia with diagnosis confirmed by the Atomic Bomb Casualty Commission were observed through 1958, whereas in Nagasaki the total was 82 cases. This number represents a highly significant increase in the incidence of leukemia. The total number of cases observed since 1945 represents a several-fold increase over the total number to be expected over the total life-span of the group on the basis of the current Japanese leukemia life-table statistics. Thus the increase represents more than a simple acceleration of the time of occurrence of leukemia among those susceptibles who would have developed it during a normal lifetime. In the range between 100 and 900 rad, the relationship of dose to annual leukemia incidence can be described by a straight line. The average rate of increase in incidence with dose was 1.1 cases/10^6/year/rad in Hiroshima and 1.6 cases/10^6/year/rad in Nagasaki, based on previous air-dose estimates.[3] The revised estimates[5] may reduce these rate estimates by a factor of 2. Chronic granulocytic and the acute leukemias were the types of most greatly increased incidence observed in the survivors within the exposure range. The incidence of leukemia of the various hematologic types has varied systematically in relation to age at the time of irradiation.

MEDICAL RADIATION EXPOSURE. Information concerning the populations receiving medically administered therapeutic and diagnostic radiation has been accumulating since about 1955. Groups receiving therapeutic radiation exposure include patients treated for ankylosing spondylitis, gynecologic neoplasms, and other malignant and benign conditions, as well as children treated for real or potential thymic enlargement. Groups receiving diagnostic radiation exposure have included adults and children, both after birth and *in utero*.

Anklylosing spondylitis, a form of rheumatoid arthritis of the spine, is a painful, chronic but nonfatal disease seen preponderantly in males of the third decade of life and older. In Great Britain between 1935 and 1955, it was treated by radiotherapy, using wide-field X-ray administration to the full spine in dosages ranging from 121 to over 3000 R in from one to eight courses of treatment.[22] Among some 13,000 patients so treated, 33 proved and 5 probable cases of leukemia were found while the expected number of cases in this population was 2.9.

The highly significant excess of death due to leukemia as compared to the general British male population is most probably due to the radiation exposure. However, certain considerations suggest that the increase in incidence might be greater than that which one would find after the same exposure of an initially normal population.[109] An association between leukemia and rheumatic diseases had been observed and there is evidence for the action of hereditary factors in the rheumatic diseases. In addition, virtually all of these patients were also exposed to other forms of therapy for their chronic and painful disease, including medicines which have been known to depress bone marrow cellular activity.

The relationship of the incidence of leukemia to radiation dose is not simple in this population because the exposures were protracted and in many instances involved multiple courses of treatment over a period of years. Using various estimates and assumptions to overcome the difficulties, a regression line, fitted through the incidences observed at exposures in the range of approximately 300 and 1500 R to the spine, has a slope of $0.5/10^6$/year/R. The similarity of this slope to that drawn in the study of the Hiroshima survivors suggests that a common risk estimate may apply to both populations.

Enlargement of the thymus gland located in the lower part of the neck and upper mid chest, was a relatively frequent indication for radiation therapy in the pediatric population in the early decades of this century. Since the gland is composed of radiation-sensitive lymphoid tissue, it does indeed diminish in size under radiation exposure. Whether the "enlargement" of this gland was real or was the result of misconceptions concerning its normal size at a given age, or the seeming enlargement was due to nonrecognition of the effects of exposure geometry in producing distortions of organ size on diagnostic X-ray examination is not yet certain. Possibly all three factors are involved.

In 1955 Simpson and Hemplemann[77] initiated an investigation of children who had been treated in this manner. This study, which has been expanded over the years by Hemplemann and his colleagues,[70] is the largest and the only one which has shown an increased incidence of leukemia under these circumstances. Other investigations[95] of these children, or children who received smaller doses of radiation for other benign conditions, have not shown this increase or have been inconclusive because of small population size. One study,[29] involving children who received radiation exposure in this body region prophylactically without evidence of disease, has shown no increased incidence of leukemia. The doses employed in the latter series, however, may well have been of smaller magnitude. In general, these data are inadequate to distinguish between the influence of the irradiation and that of the condition, so-called "thymic enlargement," or the underlying basis for the diagnosis which served as the indication for therapy.

Therapeutic and diagnostic radiation exposure in adults has also been investigated for its possible effects on the incidence of leukemia. Two approaches have been used: (1) follow-up information has been sought in populations selected because of their history of radiation therapy, and (2) retrospective radiation exposure information has been sought in populations selected because they were suffering from leukemia. An example of the first approach was the study of Simon et al.[76] of women treated for carcinoma of the cervix and of the uterus in large treatment centers throughout the world. The incidence of leukemia in over 70,000 patients was not increased over that of women in the appropriate age and national groups. However, since 25% of the patients did not survive one year and 60% did not survive five years, the population at risk was greatly reduced before the maximum incidence of radiation-related leukemia, if it occurred, would have been anticipated.

In the other approach, studies by Faber,[109] Stewart et al.,[84] and Gunz[34] have suggested that prior radiation therapy was found significantly more frequently in the histories of leukemia patients than in those of other populations. The evidence for a similar role for diagnostic radiation exposure is more ambiguous. It would appear that while therapeutic radiation in adults may give rise to a detectable increase of risk of leukemia, that more intensive investigation and dosimetry are required before meaningful estimates of risk from diagnostic radiation of adults can be obtained.

Prenatal diagnostic radiation exposure has been studies in both retrospective and prospective investigations. In the former, when the radiation history of leukemic children was compared to that of hopefully appropriate control children, Stewart[90] found that there was a relative risk of 1.83 for leukemia in the children who were exposed prenatally to maternal abdominal diagnostic radiation. The incidence of early postnatal diagnostic or therapeutic exposure was almost twice as high in the leukemic children as in the control group for the first three years of life. Other similar studies show more ambiguous results, probably related to the variation in selection of the controls in the retrospective studies.

An extensive prospective study was carried out by MacMahon[52] of over 700,000 children born in northeastern United States in the years 1947 to 1954. The frequency of intrauterine X-ray exposure in the population was estimated by a record search of a 1% systematic sample. It was then determined on the basis of a prospective follow-up that the leukemia mortality was about 40% higher in the X-rayed than in the non-X-rayed members of the study population, giving a relative risk of 1.4 with 95% confidence limits of 1.0 and 2.0. Using the total of 11 pertinent surveys, the joint maximum likelihood estimate of the relative risk was found to be 1.4 with 95% confidence limits of 1.4 and 1.6. This strongly suggests that irradiation of fetal tissue gives rise to a greater risk per unit dose than does the exposure of tissue in children or adults, possibly by a factor as high as five.

Medical Internal Radiation Exposure. Iodine-131 has been used for the treatment of hyperthyroidism for over 20 years. Pochin[69] studied an estimated 59,000 patients treated in various countries and determined from the appropriate national leukemia rates that the expected number of cases was not exceeded in this population. It is noteworthy, however, that 70% of the cases which have occurred have been in males, although only 20% of the patients so treated were males. In addition, significantly more cases of acute leukemia have been observed, as well as a significant deficit in chronic cases. The basis for these atypical features is not as yet understood.

Phosphorus-32 has been used as the treatment of choice in cases of polycythemia vera for over 20 years. A review[109] of over 1200 cases reveals that 3.3% of the patients died of acute leukemia. Since nearly 75% of the total number of cases were still alive at the time that it was made, this represents an incomplete ascertainment of leukemia incidence.

In a more recent review[68] of the causes of death of polycythemia vera patients, 11 to 12% were found to be due to acute leukemia in the patients who received irradiation therapy with ^{32}P and/or X-ray. This contrasted strongly with the finding of no deaths due to acute leukemia in 196 comparable patients who did not receive any form of radiation exposure as part of their treatment. Since the same extension of life expectancy from a mean of about 2 years in the untreated disease to about 13 years following treatment with ^{32}P can also be accomplished by the use of myelosuppressive medicines, these are recommended as the treatment of choice.

Other internal emitting radioisotopes, including radium, thorium, and strontium-90, have not been shown to be leukemogenic in man.

Bone. Bone tumors have been reported after radiation therapy in a relatively small number of instances. Bone sarcoma has generally occurred in these areas which initially contained either benign tumor or chronic osteomyelitic infection. Most cases involved exposures between 3000 and 7500 R. An increased incidence of a benign tumor, the osteochondroma, has been reported by Pifer et al.[70] in the population of children treated for thymic enlargement which was discussed above. A sevenfold increase in the risk of development of this tumor was present in the children treated with exposures on the order of 120 to 330 R.

Bone tumors arising as a result of the presence of internal emitters have been studied in the population which received exposure to radium or radium-mesothorium mixtures in the course of their occupations as watch-dial painters, chemists, and physicists or as patients treated with such solutions by injection. Hasterlik[41] reported the occurrence of 15 bone sarcomas and 11 tumors of cranial structures, seven of which involved the mastoid bone, in 264 such persons in the Chicago area. For 30 to 40 years, the patients developing tumors had carried radium body burdens of 0.6 to 10 μc as currently measured, although the initial body burdens were probably 10 to 100 times higher. No tumors developed in patients bearing less than six times the present occupational permissible level of radium at the time of measurement.

The relationship between the incidence of bone cancer and the radium body burden is a difficult one to ascertain because of the necessity to extrapolate present body burdens back to the initial quantities. Other problems concern the effect of the biological localization of the radium into "hot spots" distributed through the bone in a nonuniform manner. A tentative and very rough estimate of the relationship suggests that the risk of development of bone tumor is roughly 22 to 33×10^{-6} cases/year/μc intake. Making further presently unsupportable assumptions concerning which cells give rise to the malignancies, a crude estimate of risk per unit dose comes out to be about 4 cases/10^6/year/rad. Similar studies carried out by Evans et al.[31] in the Boston area and Barrer et al.[101] in New Jersey have shown 23 cases of osteogenic sarcoma with body burdens ranging from 0.6 to 24.0 μc and three cases of such sarcomas associated with body burdens of 0.6 to 1.67 μc, respectively. In the Evans study eight cases of cancer of the paranasal sinus or the mastoid bone were also found associated with body burdens of 1.0 to 10.0 μc. In general, it appears that with residual radium body burdens beginning in the neighborhood of 1 μc and extending to about 25 μc, the fraction of the total number of cases which involve the various forms of cancer of the bone amounts to roughly 25%.

Lung. A relationship between the inhalation of radioactive materials and an increased incidence of cancer of the lung in man has been suggested by several studies involving workers in the mining industry. Suspicions were first aroused by the finding in the third decade of this century that about 50% of the miners in the Schneeberg region of Saxony and the Joachimsthal mines of Bohemia were dying of carcinoma of the lung. Although these mines had been worked since the fifteenth century for a variety of metals, the common denominator was pitchblende and its associated air-borne radon. The concentration of radon has been estimated at about 2×10^{-8} liter.[101]

More recently DeVilliers and Windish[30] reported that 45% of 51 deaths occurring between 1952 and 1961 among miners in a fluorspar mining community in St Lawrence, Newfoundland were due to primary lung cancer. Here also, the concentrations of radon and its daughters in air were well in excess of suggested MPC levels.

A study[105] of United States uranium miners and millers in the Colorado plateau area showed significantly increased mortality. Further analysis revealed that three factors contributed to this excess: (1) a tenfold increase in respiratory neoplasms in miners with five or more years of underground experience; (2) other miners' diseases such as pulmonary fibrosis and its complications; and (3) accidents, particularly in the mines.

All of these data strongly suggest that cancer of the lung can be induced by inhaled radon and its daughters.

Thyroid. Thyroid neoplasms have been sought in the populations exposed to nuclear weapons radiation in Hiroshima and Nagasaki, and in the Marshall Islands. Twenty-one patients with histologically confirmed thyroid carcinomas were detected in the course of routine medical examinations in Japan over a three-year period.[78] Although the over-all incidence of thyroid cancer was not greatly different than that cited for Japan in general, the age distribution differed significantly in that the patients among the atomic bomb survivors were younger. The shift was particularly noticeable in those exposed within 1400 meters of the hypocenter. If these findings were representative of the incidence in the total exposed and nonexposed populations, the extrapolated incidence of thyroid carcinoma in the exposed would be 0.13% and in the nonexposed 0.04%.

In a review of thyroid lesions found in autopsy and surgical pathologic specimens in Hiroshima atomic bomb survivors, the incidence in the group exposed within 1400 meters was about twice that in the nonexposed group.[110] The incidence of thyroid carcinoma in surgical specimens from the group exposed within 1400 meters was 3.75 times that in the 1400- to 1999-meter group.[2] These surveys both suggest that the incidence of thyroid carcinoma has been increased in the irradiated population of Hiroshima and Nagasaki and that the incidence varies inversely with distance from the hypocenter. The incidence of benign thyroid tumors was also more than doubled in the most proximally exposed group. More recently, follow-up studies[24] in the Marshall Island population have revealed six instances of benign thyroid nodules in the Rongelap Islanders, who received the heaviest exposure to β- and γ-fallout. Five of these occurred in children. No thryoid nodules were detected in 75 comparable but unexposed children.

The pediatric population receiving radiation therapy prophylactically or for various benign conditions has been discussed in connection with radiation leukemogenesis above. The difficulties in distinguishing between effects of radiation and effects of conditions which prompted the radiation have already been pointed out. In many prospective surveys and in numerous retrospective surveys, therapeutic irradiation of children has been associated with a subsequent increase in the incidence of thyroid tumors. In a major retrospective survey of 286 cases of thyroid cancer, Winship and Rosvoll[108] found that 80% of these cases had received therapeutic irradiation to head or neck during infancy or childhood, mostly for enlarged thymus. The exposures ranged from 180 to 6000 R with an average of 600 R. In the prospective survey of Pifer et al.,[70] nine cases of thyroid carcinoma were found in the pediatric population receiving exposures ranging from 156 to 1092 R with a mean of 598 R. Twenty-one cases of thyroid adenoma were associated with exposures ranging from 144 to 756 R with a mean of 372 R. The incidence of thyroid tumors was considerably higher when the dose was given half anteriorly and half posteriorly as compared to that in patients given anterior radiation only. In the prospective study of Saenger et al.,[86] eleven cases of thyroid cancer were found among previously irradiated patients who were under 23 years of age at the time of cancer development, although 0.12 case was expected; these data give a relative risk of about 91.

Although the individual surveys on the incidence of thyroid cancer following irradiation preclude dose–effect regression determinations because of the small number of cases, a crude risk estimate can be obtained by determining the slope of the straight line connecting the origin of points obtained when the observed incidences are plotted against exposures.[101] If a single straight line is fitted through the incidences observed in these surveys, its slope indicates the rate of incidence with exposure is 0.9 cases/10^6/yr/R. Because of the small numbers, the statistical uncertainty of the base, and the even larger uncertainties of dosimetry, it appears that the joint risk estimate may be between 0.5 and 1.5/10^6/yr/R. This estimate is based on an average follow-up time of about 16 years, which may be inadequate because of the long latent period for thyroid cancers and adenomas. Also, it can only be considered of possible validity for short-term exposure of children in an estimated range of 100 to 300 R.

Internal exposure of the thyroid gland to ^{131}I irradiation and a possible relationship to thyroid tumor formation has been studied by Sheline et al.[82] in 256 patients with diffuse hyperthyroidism treated with ^{131}I, eight patients (3.12%) developed multiple nodules 5 to 11 years after therapy. Six of these patients were under 20 years of age at the time of treatment, representing 33% of the total number of patients in the age group. The other two patients represented 0.84% of the 238 adult patients so treated. This again suggests the greater risk of thyroid tumor development after radiation exposure in the young.

Nonspecific Sites. Among the Japanese atomic bomb survivors, data from tumor registry studies of cancer incidence have shown an excess of malignancies in the most proximally exposed group in Hiroshima with a gradient of incidence related significantly to exposure distance. In a mortality study during the years 1950 to 1959, 36 more deaths from leukemia occurred in the population within 1400 meters of the hypocenter in Hiroshima and Nagasaki than was to be expected from the mortality of those exposed to lower doses of radiation between 1400 and 2000 meters. During the same period 30 more deaths from other forms of malignant disease (standard error 14.6) occurred in the former group than were expected on the basis of mortality in the latter group. It would seem evident that the increased incidence of all other forms of malignant disease is much smaller than the increased incidence of leukemia.

In the pediatric population exposed to diagnostic radiation *in utero*, an increased frequency of irradiation in the history of subsequent cancer cases was also observed. In the series of Stewart et al.[90] the ratio was 1.7 for cancer deaths up to the age of four years and 2.5 for deaths from 5 to 7 years. In MacMahon's series,[52] cancer mortality was about 40% higher in the X-rayed than in the non-X-rayed members of the study population. The excess cancer mortality was most marked at ages 5 to 7 years, at which time the relative risk was 2.0, but this excess risk appeared to end by age 8 years. More recently MacMahon and Hutchison[56] reviewed 11 published studies concerning the relative risk of malignant disease following *in utero* exposure. In five studies the relative risk was less than one, but all of these included small samples with large sampling variability. In six studies, the risk was greater than one. By weighting the relative risk observations of the 11 studies, they obtained a maximum likelihood estimate of risk of 1.40, and the true value may be expected to be within the range of 1.20 to 1.64 ($P = 0.05$).

An internal emitter associated with a large number of subsequent malignant tumors at various sites is Thorotrast. This substance is thorium dioxide in a colloidal suspension and was widely used in diagnostic radiology between 1928 and 1945 primarily for visualization of the liver, spleen, cerebral arteries, and the cavities of the body. The thorium particles, usually about 25% by weight, may remain at the site of injection or may be distributed in the phagocytes of the reticuloendothelial system and thereby concentrated in the liver and spleen. Approximately 74 cases of malignant tumors following the injection of this material have been collected in the scientific literature.[101] A variety of organs have been involved after a long latent period of between 15 and 20 years. There is no good population base to utilize in determining the incidence of cancer. Also, it is not feasible to arrive at any reasonable approximation of the radiation exposure involved, since the distribution of the material is extremely nonuniform and the radiation is primarily of the α-type and therefore highly localized.

Degenerative Conditions

Most of the information concerning the various degenerative changes seen as late sequelae of radiation overexposure in man has been developed from observations in human beings who received local or partial body irradiation as a result of occupational or therapeutic procedures. The doses involved are generally higher than those which would be compatible with survival if delivered to the whole body. In general, their effect is to reduce the capacity of the exposed tissue to regenerate. The failure to produce the normal number of working cells in a given organ is often associated with a relative increase in supporting tissue and changes in the blood vessels supplying the organ. Fibrous connective tissue is not able to carry out the functions of the specialized cells of the particular organ and the resultant inadequately functioning organ produces a clinical disease state appropriate to its particular activities.

Skin. Injuries of the skin were among the first recognized late radiation effects in man. Permanent hair loss has been induced by exposures of 500 to 700 rad. Thinning of the skin and loss of sweating are other late effects. Repeated destruction of the intact outer layer and resultant infections are an expression of the damage to the blood vessels supplying the skin. Other visible effects include changes in pigmentation and the production of permanent dilation of capillaries in the irradiated areas.

Kidney. Damage to the kidney is a well-known complication of radiation overexposure. Permanent hypertension may be induced in man within a few years after a single localized X-ray of about 3000 to 5000 rad or by smaller fractionated doses. Thickening or sclerosis of the kidney's fine vasculature is an early change revealed on histopathological examination. All of the consequences of the nephrosclerosis and its related hypertension which follow are the same as those which occur spontaneously in generalized arteriosclerotic disease.

Central Nervous System. Irradiation of human brain or spinal cord with several thousand rad given over a short time may injure blood vessels, causing damage to the tissues due to inadequate blood and oxygen supply. The brain or spinal cord then undergoes secondary degeneration and loss of function.

Bone. Radiation osteitis is a late degenerative effect of intensive radiation of several thousand rad to bone. Damage to bone cells and their blood vessels is revealed slowly and after many years leads to breakdown of the tissue, including spontaneous fractures. The bone marrow cavity may be replaced by fibrous tissue so that the blood-forming cells are decreased in number because of inadequate space and blood supply. Secondary anemia has been associated with this "myelofibrosis."

Gastrointestinal Tract. The gastrointestinal tract has shown some permanent late effects after fractionated doses of several thousand rad. Increased fibrous connective tissue and thinning of the lining with discontinuities and ulceration have been reported. Permanent reduction of the secretion of acid and pepsin by the stomach lining has been observed. The latter effects have also been produced intentionally as a form of treatment for acid hypersecretion.

Lungs. High-dose exposure of lung tissue produces progressive fibrosis with blood vessel damage. This is usually a slow process characterized clinically by cough and shortness of breath as the functioning lung tissue volume is gradually reduced. Similar changes have been produced by the inhalation of radionuclides which then reside in the lung.

Life-Span Reduction

Although animal experimentation has demonstrated life-shortening in the radiation-exposed rodent population, it is difficult to extrapolate these findings to man. It is also difficult to obtain such information in human populations. Such populations must be of sufficiently large size and well characterized in order that suitable control populations can serve as a basis for comparison. The Japanese atomic bomb survivors and the radiologist populations in the United States and Great Britian have been examined for this purpose.

Studies of the mortality of atomic bomb survivors[46] have thus far shown no convincing evidence of a higher general mortality in the more heavily radiated groups which can be reasonably attributed to the radiation exposure. A major difficulty in this study is the selection of an appropriate nonexposed group to serve as a control. A survey of the mortality of British radiologists between 1897 and 1957 in comparison with other physicians and men of comparable social class revealed no evidence of life-shortening according to Court-Brown and Doll.[23] The early adoption of effective safety measures was considered responsible for this finding.

In studies of medical specialists in the United States, Dublin and Spiegelman[28] found that mortality of radiologists and dermatologists was 16% and 25%, respectively, above that of all specialists combined. They estimated that the difference between radiologists and dermatologists and the other specialists was one to three years of decreased life span. If one attempts to apply this work to radiologists specifically, the small sample size precludes valid results. In a larger study Warren[103] used 82,000 obituaries of physicians to compare the mean age at death of radiologists with that of other physicians. He concluded that radiologists died 5.2 years earlier than physicians not occupationally exposed to radiation. However, Seltser and Sartwell[85] pointed out that although the conclusion might be correct, the difference could be accounted for simply by differences in age composition between the two groups under study.

More recently these investigators have pursued this problem by studying three medical specialties chosen on the basis of the amount of presumed average exposure of their members to X-rays in the course of their professional activities.[88,89] These were the radiologists, internists, and ophthalmologists-otorhinolaryngologists. The first group uses radiation continually, the second uses radiation much less frequently, and the third uses it rarely. The death rates were studied for three calendar periods; 1935–1944, 1945–1954, and 1955–1958. The median age at death for radiologists in these three time periods was 71.4 years, 72.0 years and 73.5 years. For internists the median age at death was 73.4 years, 74.8 years, and 76.0 years.

For eye, ear, nose, and throat specialists, the median ages at death were 76.2 years, 76.0 years, and 76.4 years. Using the nonexposed specialty as a baseline it would appear that the radiologists of three decades ago showed a deficit of 4.8 years of life-span and that this deficit has gradually decreased to 2.9 years in the latter half of the last decade. The internists, a moderate exposure group, showed a smaller deficit of 2.8 years initially and now have virtually no difference in life expectancy from that of the baseline group.

Thus, it would appear that radiation did indeed reduce the life expectancy in consonance with the amount of exposure. It would further appear that with the known improvements in radiation safety practices in more recent years that this deficit in life-span is decreasing. A breakdown of the causes of death demonstrates the same correlation with magnitude of exposure for all causes; cancer, cardiovascular-renal disease, and all other causes. For leukemia alone, it was more marked than for all malignancies.

Growth and Development

With the use of intensive local radiation as a form of therapy in growing patients, it has been observed that damage to the growing ends of long bones will result in cessation of further growth. Because of the particular sensitivity of fetal tissues to ionizing radiation, a human population of special interest has been the children who were *in utero* at a time of radiation exposure. In 1929, Goldstein and Murphy[37] reported finding defective children among the offspring of 106 women who had received therapeutic doses of radiation during pregnancy. Among the atomic bomb survivors of Hiroshima and Nagasaki, a greater frequency of microcephaly and mental retardation was found in children who were between the seventh and fifteenth weeks of gestation and had received a significant dose of radiation as judged by proximity to the hypocenter or by the occurrence of the Acute Radiation Syndrome in the mother.

A further intensive study was initiated in a group of 286 survivors of *in utero* exposure in Nagasaki to detect differences between exposed and control groups that might become apparent during the time of puberty.[10] The groups compared were: (1) children whose mothers were within 2000 meters of the hypocenter; (2) children whose mothers were located between 3000 and 4999 meters; and (3) children whose mothers were not in the city at the time of the bomb. Measurements were made of head circumference, weight, chest circumference, the age at menarche, and the degree of epiphyseal closure in the wrists. Differences were found which were consistent with radiation effect. This was most significant in head circumference measurements.

Although it is difficult to rule out factors other than radiation, such as malnutrition, the selection of controls who lived under the same circumstances suggests that this is a radiation effect. Parallel studies in laboratory animals with similar results help to substantiate the conclusion.

Cataracts

About 200 cases of radiation cataracts due to X- or γ-radiation exposure in man have been reported.[64] In addition, mixed neutron and γ-exposures from cyclotrons, criticality accidents, and nuclear weapons have provided additional human cases.

The characteristic features of a radiation cataract, an opacity usually at the posterior pole of the lens with small granules and vacuoles around it, may no longer be recognizable as the cataract develops, giving it a nonspecific appearance indistinguishable from that of cataracts due to other causes. The dosimetry in many of the instances reported is extremely uncertain. There are, therefore, many ambiguities in the dose–response relationship for this late radiation effect.

Although changes in the lens have been detected following exposures as low as 200 R, the minimal effects of X-ray dose for the production of clinically significant cataracts is thought to be 600 to 1000 rad at 200 kvp. Neutrons are relatively more effective in cataract production than X-ray by a factor of 5 to 10. The latent period for the production of cataracts is usually several years after exposure.

In the Japanese atomic bomb survivors, lens opacities which did not interefer with vision were found in about 47% of 425 survivors in Nagasaki who had been between 1400 and 1800 meters from the hypocenter. However, fully developed cataracts were rare, ten cases being found among 8000 survivors of Hiroshima and Nagasaki by 1956, eleven years post-exposure.[79] The relationship of the incidence of opacities to the magnitude of exposure has been documented in Hiroshima.[45]

Fertility

On the basis of laboratory animal studies it is known that the late pathological effects of radiation of the gonads are chiefly a hastening of the involution associated with advancing age. Higher doses produced histological sterility in man. Only fragmentary data are available based on observations following clinical therapy and some accident cases.

Certain generalizations may be made on the basis of these data.[64] A single dose of about 150 rad to the gonads may induce brief subfertility in many men and women. A single dose of about 250 rad may induce temporary sterility for one or two years. A dose of 500 to 600 rad will probably produce permanent sterility in many people. Single doses of 800 rad or more will probably produce permanent sterility in virtually all human beings. Limited experience with the Marshallese, exposed Japanese, and certain accident cases suggest that a substantial fraction of the midlethal dose for man (400 to 600 rad) does not have a serious, permanent effect on fertility, but the gonadal doses are not known with certainty. Few people have been studied for this purpose for a prolonged postexposure period.

Radiation-induced sterility in males does not induce permanent changes in hormone balance or libido. In females, greater physiological changes occur since ovarian production of sex hormones is intimately related to the development and discharge of ova. Termination of this process produces an artificial menopause with its associated symptoms. These include amenorrhea and various somatic and occasional psychic manifestations.

Cellular Damage

In addition to the immediate evidences of chromosome aberrations described in Section 13-5, several human population studies have been attempted to determine whether a persistently increased incidence of chromosome aberrations is present as a late sequel to radiation exposure.

The aberrations found in the chromosomes of somatic cells in the immediate post radiation exposure period, including aneuploidy, breaks, and translocations have already been mentioned. Additional observations have been made in certain groups of human beings in the period after the clinical effects of the exposure had subsided.

Buckton et al.[11] did a follow-up of the patients receiving X-ray therapy for ankylosing spondylitis. They examined 85 blood cultures from 58 patients over a time period ranging up to 20 years after the X-ray therapy had been completed. They found both stable and unstable morphological abnormalities of the chromosomes. The proportion of cells with unstable abnormalities rose to a peak value about two to three weeks after the end of treatment and then fell, although some of these cells persisted five years and more after the exposure. The proportion of cells with stable abnormalities such as dicentrics and tricentrics rose to a significant level shortly after the end of treatment and then did not change materially over the years studied. Similarly, Bender and Gooch[9] found a significant incidence of aneuploidy and morphological aberrations 29 and 42 months after a nuclear criticality accident produced significant irradiation of eight men. Warren and Meisner[107] studied a population of cancer patients who had received major radiation therapy to the spine or trunk. Although aneuploidy tended to recede with time after exposure to about normal levels in a year, about 5% of the cells were seen to contain atypical or ring chromosomes as long as 30 years afterwards. Corroborative results in atomic bomb survivors and cancer therapy patients have been reported by Doida et al.[29]

In a somewhat different exposure situation, a group of 33 radiation workers receiving chronic low-level exposure at a median dose rate of 1.4 rad/yr with lifetime doses in excess of 10 rad were reported by Norman et al.[65] to have a significantly increased frequency of aneuploidy and other cytogenetic abnormalities. An increase in chromosome aberrations was also observed by Court-Brown et al.[21] in 34 men occupationally exposed to radiation not exceeding the permissible limits.

It would appear that an increased frequency of aneuploidy, and more particularly of morphological aberrations, may be a persistent sequel to radiation exposure which itself terminated many years before. There is no evidence at this time to relate such aberrations to any specific disease states which might occur as late somatic effects of the overexposure.

13-7 GENETIC EFFECTS

There is a large body of information concerning the genetic effects of radiation in plant, insect, and laboratory animal studies. The opportunities for such observations in as genetically heterogeneous, long-lived, and environmentally uncontrolled experimental animals as man are very limited. Direct observations of the genetic consequences to man of exposure to ionizing radiation are thus limited to studies of first-generation offspring. Only autosomal dominant or sex-linked gene mutations, and chromosomal aberrations can be expected from such studies. Further difficulties are those of estimating the dose to the gonads of the parents. In view of the small absolute and relative increases to be expected in the frequency of traits determined by such mutations, there is a need for large populations for this type of survey. The end points used in such studies are usually the incidence of abortions, stillbirths, neonatal deaths, congenital malformations, and shifts in the sex ratio of the progeny.[61,71]

The most extensive survey of this type was carried out in the Japanese cities of Hiroshima and Nagasaki following the atomic bombings.[61] Data were collected on more than 30,000 offspring of irradiated parents and on a comparable control group. Observations made of stillbirths, neonatal deaths, birth weight, and congenital malformations did not reveal a significant effect of radiation. However, there was a significant shift of sex ratio of immediate progeny. Genetic theory suggests that the production of lethal mutations in the X chromosome will affect the sex ratio in different ways according to the sex of the parent exposed to radiation. The net effects of mutations induced by irradiation of the male would be to increase the production of male births, while the effect of such induction in females would be to reduce the production of male births. The observed changes were in accordance with genetic theory.

Similar studies performed on the progency of French, Dutch, and Canadian patients, American radiologists, and Japanese X-ray techniques have recently been reviewed.[49,61,71] In general, these studies suffer from drawbacks in that (1) the doses of radiation are more poorly defined than they are in the Japanese atomic bomb survivor study; (2) reliance is placed on questionnaires and there is a high proportion of nonresponders which might have biased the results; and (3) the studies are retrospective with the possibilities of motivation for recall on the part of the irradiated population differing from that of the control population. With these reservations in mind it is of interest that similar changes in concordance with genetic theory are present in most of these studies.

Thus, it does appear that there is suggestive evidence in human populations that the production of mutations by radiation occurs in a similar fashion to that already observed in experimental plant and animal studies. It is also evident, based on the absence of findings using other points than the sex ratio, that there is no markedly greater sensitivity for the induction of mutations by radiation in man than in other mammalian species.

C. MEDICAL MANAGEMENT OF RADIATION OVEREXPOSURE

This section summarizes past experience in the medical management of the human radiation everexposure cases already discussed. It is not intended to serve as a basis for actual clinical treatment, but rather to provide the nonspecialized reader with a general idea of what takes place in dealing with the biomedical problems presented in such situations.

13-8 IMMEDIATE SOMATIC EFFECTS

First Aid

Since human radiation overexposure is usually the result of a laboratory or industrial accident, or of a nuclear weapon detonation, there is a great likelihood that conventional first aid will be needed to deal with life-threatening injury. The execution of this first responsibility of those at the scene may be complicated by the presence of radioactive contamination. From the standpoint of both the patients and of their attendants, it is obvious that the prime first aid requirement relating to acute radiation injury is removal of the patient from any further exposure sources. Guidelines to the management of this problem have been provided in a report of the National Committee on Radiation Protection and Measurements,[60] and in a handbook of the U.S. Atomic Energy Commission.[74]

Other than the prevention of further radiation exposure, there is no present method for limiting or reversing the chain of events initiated by the biophysical processes at the molecular level within the cells of the body.

Diagnostic Procedures

The first objective of the diagnostic process is to determine whether biologically significant radiation overexposure has taken place. If so, the next objective is to determine the extent of the damage either immediately present or to be anticipated. This is necessary for planning suitable clinical management as well as for providing a basis for the proper handling of the social and economic consequences of such episodes. A good history of the events leading to the suspicion of radiation overexposure, a careful physical examination, and the performance of some simple laboratory procedures are generally sufficient to distinguish between the individuals in whom the Acute Radiation Syndrome may be evolving, those in whom local radiation injury has taken place, and those in whom local traumatic injury accompanied by internal radionuclide contamination has occurred without any evidence of clinical injury.

Acute Radiation Syndrome. Acute Radiation Syndrome may be identified by the history of possible exposure and the observation of the time of occurrence of various signs and symptoms as well as the recognition of certain abnormalities developing in the conventional peripheral blood cell count. Figure 13-4 presents a flow chart which provides the basis for classifying the individual patient into one of the five clinical injury groups which have already been described. Knowing the general long-term course associated with the severity of the early signs of radiation injury, it becomes possible to provide for proper clinical management and to estimate the prognosis in the individual cases.

Local Radiation Injury. This form of injury is also recognized by carefully elicited information concerning the exposure situation and the clinical phenomena which ensued. The time sequence of the various symptoms and signs is the best basis for the association of the local injury with radiation exposure. In the absence of either of these two elements, it may not be possible to identify the nature of the injury because agents other than radiation may produce the same response in various tissues.

Local Traumatic Injury with Radionuclide Contamination. In this clinical situation the conventional medical diagnostic techniques are used in the evaluation of the traumatic injury. The special problem of contamination of the wound is dealt with by such methods as the use of wound probes, that is, extremely small radiation detectors, and the measurement of radioactivity of cotton swabs which have been introduced into various parts of the wound.

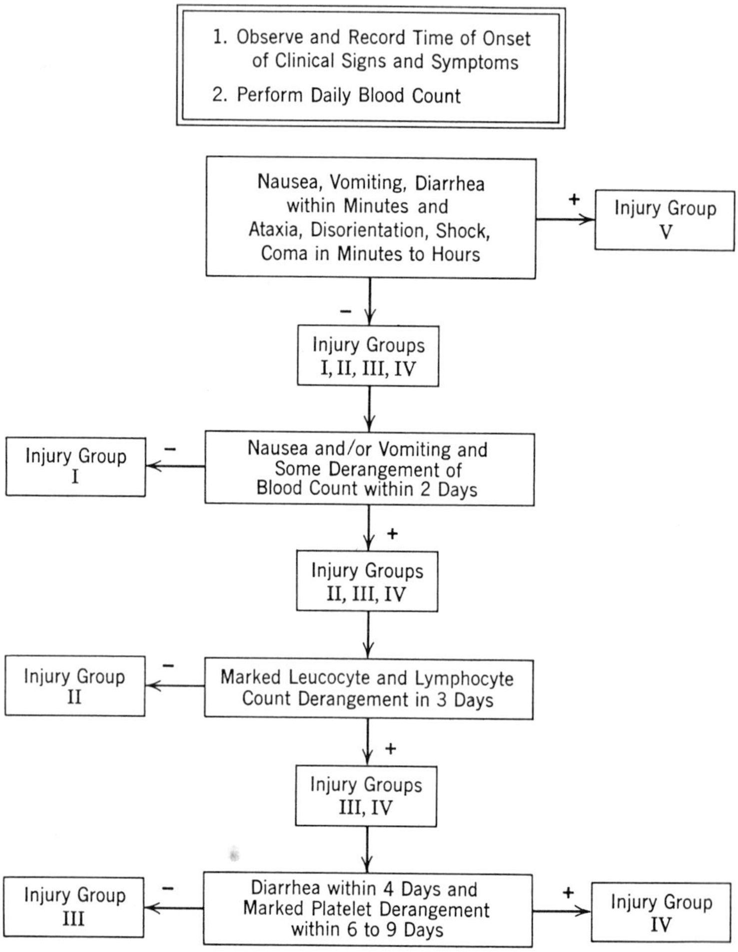

Fig. 13-4. Preliminary evaluation of radiation injury.

A less direct diagnostic procedure for contamination detection is the measurement of various body fluids and excretions such as urine, feces, blood, and saliva for radioactivity. Appropriate sampling intervals are necessary in order to avoid missing excreted radioactivity because of a lag in the time of its appearance.

Internal Contamination without Evident Injury. This may be recognized from an approximate exposure history and the identification of radioactive material within the body or in its excreta. External whole-body, low-level radiation counting and the assay for radioactivity of appropriate body fluids and excreta over a sufficient period of time are necessary to establish the diagnosis.

Clinical Management

Acute Radiation Syndrome. The choice of medical procedures to employ in the clinical management of radiation injury is affected by three types of information. These are the clinical judgement applied to any problem, an understanding of the pathophysiology of the Acute Radiation Syndrome in man, and a knowledge of the magnitude and type of radiation to which the patient has been exposed. Although the last of these would appear to be crucial information on which to base appropriate plans, it may not necessarily be helpful in the individual case for two reasons. First, the dose, if it can be determined, is not usually available as soon as needed. Second, no given individual will necessarily respond exactly as all others to a given dose. Therefore, the diagnostic procedures above are planned to establish the general clinical pattern of injury in the individual under study. With this information available, the planning of clinical management can proceed even in the absence of dosimetric information. It is highly desirable, of course, to obtain physical dosimetry data as soon as possible to supplement the clinical estimate of injury. Subsequent procedures performed in the clinical management of the case serve four purposes: (1) to validate the original diagnostic and prognostic conclusions; (2) to identify promptly the possibility of clinical complications in the patient's course; (3) to provide more data for improved clinical management of future cases; and (4) to supply more information concerning radiation injury.

The recommended clinical procedures to be performed in managing the five injury groups, as well as the time and frequency of performance, are presented in Table 13-3. More specific therapy is reserved for symptomatic indications.

All cases with the probability of acute injury should be hospitalized for close observation for about 72 hours. Those cases falling into injury groups II to V can be recognized by using the diagnostic procedures indicated in the preceding section. These patients should be kept hospitalized. The others can be reassured of their favorable prognosis and released from inpatient care when asymptomatic. More specific treatment considerations for each injury group are as follows:

GROUP I. These patients can be cared for as outpatients using the tests suggested in Table 3-13. When dosimetric information, clinical symptoms

Table 13-3 Recommended Diagnostic Procedures for Clinical Management of Radiation Injury

	Group										
	I–II–III–IV			I	II			III		IV	V
Time (d)	1	2	3	STT	STT	18–48	STT	4–48	STT	4	1
Type A Procedures											
History											
Symptoms } onset	x	x	x	x	x	D	x	D	x	D	D
Signs } duration	x	x	x	x	x	D	x	D	x	D	D
Past medical	x										
Physical examination											
General						d21	3mo$^+$	d15–30	6mo$^+$	d6	D
Body weight	x			x	x	D	x	D	x	D	D
Urinary output	x	x	x		x	D		D		D	6 hr
Laboratory tests											
Hematology											
Hematocrit	x	x	x	x	x	D	x	STT	x	D	6 hr
Leucocytes	x	x	x	x	x	D	x	D	x	D	6 hr
Differential count	x	x	x	x	x	D	x	D	x	D	6 hr
Calculation of total neut. and lymph.	x	x	x	x	x	D	x	D	x	D	6 hr
Platelets	x	x	x	x	x	D	x	STT	x	D	6 hr
Bone marrow aspiration				d30	14d	14d	mo6	14d	mo6	7d	d1
Radioassay											
Blood Na^{24}	x	x									
Whole-body counting	x	x									
Type B Procedures											
Laboratory tests											
Hematology											
Sedimentation rate	x	x	x	x	x	D	x	D	x	D	6 hr
Reticulocytes	x	x	x	x	x	D	x	STT	x	D	6 hr

Bleeding times	x					STT		STT	−75d	3d	6 hr
Clotting times											
Biochemistry											
Blood										STT	6 hr
NPN	x	prn	prn					prn		prn	d1
Sodium	x	prn	prn					prn		prn	d1
Chloride	x	prn	prn					prn		prn	d1
Potassium	x	prn	prn					prn		prn	d1
pH or CO_2	x	prn	prn					prn			
Urine											
Routine analysis	x	x	x		x	D	x	D	x	D	6 hr
Stool											
Occult blood	x					D		d12⁺		D	All
Ophthalmology											
Slit lamp			x	6mo⁺			6mo⁺		6mo⁺		
Type C Procednres											
Biochemistry											
Serum bilirubin	x	x	x		x	STT		STT	−30d ;	D	6 hr
Urine BAIBA	x			−30d	x	D		D		D	D

Recommended frequency of time of performance, key:

STT = Standard Testing Times: 6, 9, 12, 15, 18, 21, 24, 27, 30, 33, 36, 40, 44, 48, 60, 90, 105, 120 days; 6 months, 1 year, and annually.
x = at times indicated in column heading.
d = day(s).
D = daily during time indicated in column heading.
nD = frequency in days.
−Dn = up to and including day at times indicated in column heading.
Dn^+ = on and after day at times indicated in column heading.
Dn = specific day recommended.
n = all time after day specified.
prn = as indicated by clinical course.

and findings, and laboratory data have clearly established the low level of overexposure, there is no specific medical contraindication to a return to work. However, considerations of public opinion, medicolegal sequelae and the like have generally led to the more conservative prescription of a rest period of varying lengths for such patients. Subsequent work should be in an environment in which further radiation overexposure is unlikely.

GROUP II. The prodromal transient nausea and possible vomiting of these patients probably will require at most the administration of the antiemetic agents. The radiation-induced epilation which may occur at this injury level is not permanent, and the patient should be so reassured. Significant complications of the bone marrow damage found in this group, such as anemia, infection, and bleeding, have been infrequent. It is therefore appropriate to maintain a conservative attitude, holding in reserve such definitive treatment as blood and platelet transfusions and antibiotics, pending demonstrated evidence of their need. The maintenance of good personal hygiene and the use of aseptic isolation methods are of value during the period when defenses against infection are reduced. The weakness and fatigue observed in these cases is not understood and therefore only rest on a symptomatic basis can be prescribed.

GROUP III. Treatment in this group may determine whether the outcome of the radiation injury is survival or death. Major complications of serious bone marrow damage, some gastrointestinal damage, and secondary renal and cardiovascular problems may be anticipated. When the diagnostic procedures suggest the stormy clinical course of Group III, prompt bed rest, a bland diet, superior nursing care, an aseptic environment, and the presentation of the realistic but generally reassuring appraisal of the forthcoming medical course to the patient are indicated. The latter is to allay anxiety and to assure the necessary cooperation in the patient's care. Antiemetics are helpful in dealing with the severe and prolonged nausea and vomiting which may accompany the initial phase and recur during the manifest illness stage. Intraveneous feedings may be necessary to maintain fluid and caloric balance.

As the complications of depressed bone marrow functions ensue, aseptic precautions to minimize bacterial exposure are indicated. Although routine prophylactic use of antibiotics is not advisable, evidences of infection should be treated rapidly and vigorously with broad-spectrum antibiotics and fungicides. Bleeding should be treated by fresh blood transfusions and, if the hemorrhage is unresponsive, platelet transfusions may be tried as well. If these measures are unavailing, an attempt may be made to transplant bone marrow from normal related donors in an effort to produce a temporary graft of blood-forming tissue which might tide the patient over until his own bone marrow regenerates. This treatment is not without risk since severe and possibly fatal sequelae may develop during the recovery period following a "successful" take of foreign bone marrow in the radiated host.

GROUP IV. In these patients gastrointestinal difficulties require major therapeutic attention even before hematologic disturbances need treatment. The use of antiemetics, fluid replacement with correction of electrolyte deficit, tubal decompression of the nonfunctioning gastrointestinal tract, antihypotensive agents, antibiotics, and bone marrow transplantations are advocated in this desperate situation. Hemodialysis may be utilized to counteract the secondary effects of cardiovascular and renal failure.

GROUP V. Although experimental animal studies have suggested that the primary disturbances in this category involve the function of the central nervous system, actual human exposures have been complicated by major cardiovascular malfunctions as well. Only general supportive therapy of the type already described for Groups III and IV can be offered at this time.

Local Radiation Injury. The treatment of local radiation injury is similar to that of thermal burns. The surgical removal of dead tissue, the development of suitable temporary covering of the denuded area to minimize fluid loss and prevent bacterial infection, and the use of skin grafts to develop a protective skin layer over the healing deeper tissues are the general procedures employed.

Local Traumatic Injury with Radionuclide Contamination. The twofold clinical problems which must be solved in this situation are the removal of the deposited radionuclides and the routine treatment of the then uncontaminated traumatic wound. The first step generally involves scrubbing of the wound and possibly surgical debridement as well. Wound probes, that is, small volume radiation detectors, wipe samples, and the use of soft-tissue diagnostic X-ray examinations for radionuclides in metallic form aid in determining the extent of the tissue which must be surgically removed. Appropriate consideration is also given to the estimated dose and the relationship to the body burden and to the federal radiation guide numbers. Where warranted by the body-burden estimate, beyond the amount which must be removed locally, attempts may be made at systemic decontamination, as described in the next paragraph.

Internal Contamination without Evident Injury. When the diagnostic procedures outlined above, including external whole-body radiation measurements and radiation bioassays of urine and fecal specimens, suggest internal contamination with a larger body burden of a given nuclide than seems desirable, various techniques may accelerate the excretion of the material. Laxatives, such as magnesium sulfate, will enhance the passage and excretion of nuclides in the gastrointestinal tract.[75] Chelating agents such as EDTA, zirconium citrate, and more recently, DTPA and BAETA have been used with some success in enhancing the excretion of plutonium, yttrium, lanthanum, uranium, and mixed fission products.[17] For strontium body burdens, high calcium intake and acidification of blood pH may reduce bone absorption of the material, as well as increase its excretion.[16]

13-9 LATE SOMATIC EFFECTS

Diagnostic Procedures

As has been previously indicated, the late or chronic effects of radiation overexposure are relatively nonspecific in appearance. Their recognition is generally a matter of increased incidence in an exposed population rather than special characteristics which can be identified in a single individual. For this reason there is no certain way to identify a late effect as being due to radiation exposure. Certain situations may make such as association more likely, as when for example, cataracts appear in the lens of the eye of an otherwise healthy young person with a history of neutron overexposure. Another approach to making this association in the individual case has been attempted in situations where leukemia has developed in an individual with a measured occupational overexposure. Based on population studies performed in the nuclear weapons survivors, the probability factor can be estimated that the disease was initiated by the exposure, within rough limits.[59]

Clinical Management

The main clinical procedure to be applied in connection with late effects of a known radiation overexposure is the early recognition of any such sequelae. Periodic ophthalmoscopic and slit-lamp examinations of the eye for potential cataracts, as well as physical, blood-test and X-ray examinations for the various forms of cancer are examples of these procedures. The actual therapy of any late effects which do develop does not differ from the therapy of such conditions in the absence of a history of radiation exposure. There is no specific therapy for such late effects as life-span reduction and diminished growth, development, and fertility which may follow a serious overexposure.

13-10 GENETIC EFFECTS

There is no clinical method for recognizing or directly altering changes induced by radiation in the germ cells. The passage of time will permit the expulsion of the exposed mature sperm cells but the viable ova and precursor cells for spermatogenesis remain as potential carriers of radiation damage. However, there are repair mechanisms demonstrable for rodent germ cells.[73] It might therefore, be wise to delay procreation until six to twelve months after a major radiation exposure. In addition, the effective use of genetic counseling may be of great value in allaying the anxieties of individuals subjected to radiation overexposure in regard to effects on their future progency. It is important to educate such individuals concerning both the low theoretical likelihood of any recognizable genetic abnormalities and the actual absence of any significant visible effects on the progeny of the Japanese atomic bomb survivors.

D. CONCLUSION

It is evident that a study of the effect of radiation exposure on the various human populations gives qualitative evidence that the same changes which are seen in lower species and, indeed, on the molecular level are found in man himself. It is for the determination of the quantitative probability of such effects that the human data are demonstrably inadequate. Since such quantitation is essential to the professional health physicist, and to society in general, for the appropriate setting of the risk factor in risk-versus-benefit equations, it is evident that further human information is desirable.

In order to obtain additional human data within ethical limitations, better utilization of the situations in which individuals and populations can provide such information is necessary. In all accidental overexposure situations, precise dosimetric information is indicated to complement the usually adequate biomedical information that is accumulated in the immediate post-exposure period. More frequent long-term biomedical observations of such individuals are needed. For the populations already exposed to radiation, the best possible retrospective dosimetric information must be obtained. In addition, there must be whole-hearted support and participation in the long-term follow-up studies of such populations, despite the lack of immediate new results to be disseminated. In the case of populations, both past and future, which are exposed to medically indicated irradiation, the best opportunity exists for careful dosimetric and biomedical observations. Any extra effort and expense will be repaid by the reduction in the operational and the psychological costs of living and working in the nuclear age that will be made possible by our more precise quantitative understanding of the risk of radiation exposure.

The information summarized in this chapter concerning the medical management of human radiation injury is rather reassuring. It is encouraging to note that the outcome of external local as well as even moderately severe whole-body radiation injury, such as that involving the blood-forming tissues, can be influenced favorably by appropriate medical management. Excretion of internally absorbed radionuclides can also be enhanced. On the other hand, it is clear that further biomedical research is necessary in order to provide definitive therapy for the more severe types of injury involving the gastrointestinal, the cardiovascular, and the central nervous systems.

REFERENCES

1. Alekseyeva, O. G., A. F. Bibikova, N. A. Vyalove, A. Ye. Ivanov, N. A. Krayevskiy, N. A. Kurshakov, N. V. Paramonova, V. N. Petushkov, V. V. Snegireva, L. A. Studenikina, Yu. M. Shtukkenberg, and A. Ya. Shulyatikova, "Sluchay Ostroy Lucnevoy Bolezni U Cheloveka (A Case of Acute Radiation Sickness in man)," *Gosudarstvennoye Izdatel'stvo Meditsinskoy Literatury*, Moskva, USSR, 150 pp., 1962. Translated by Foreign Technology Division, U.S. Air Force Systems Command, Wright-Patterson Air Force Base, Ohio, FTD-TT-63-558, 1963.

2. Angevine, D. M., and S. Jablon, *Ann. N.Y. Acad. Sci.* **114,** 823 (1964).

3. Arakawa, E. T., *New Engl. J. Med.* **263** (1960).

4. Andrews, G. A., B. W. Sitterson, A. L. Kretchmar, and M. Brucer, *Health Phys.* **2,** 139 (1959).

5. Auxier, J. A., J. S. Cheka, H. H. Hubbell, Jr., D. R. Johnson, T. D. Jones, P. T. Perdue, and J. H. Thorngate, "Health Phys. Div. Ann. Progr. Rept., July 21, 1965," *Oak Ridge Natl. Lab. Rept.*, ORNL-3849, pp. 153–159.

6. Boyd, E., W. W. Buchanan, and B. Lennox, *Lancet* **1,** 977 (1961).

7. Bond, V. P., E. P. Cronkite, and R. A. Conard, in *Atomic Medicine* (C. F. Behrens and E. R. King, eds.), Williams and Wilkins Company, Baltimore, Maryland, 1964, 4th ed., Chap. 9, pp. 206–207.

8. Bond, V. P., T. M. Fliedner, and E. P. Cronkite, *J. Nucl. Med.* **1,** 221 (1960).

9. Bender, M. A., and P. C. Gooch, *Radiation Res.* **18,** 389 (1963).

10. Burrow, G. N., H. B. Hamilton, and Z. Hrubec, *JAMA* **192,** 357 (1965).

11. Buckton, K. E., P. A. Jacobs, P. A., W. M. Court-Brown, and R. Doll, *Lancet* **2,** 676 (1962).

12. Bond, V. P., and J. S. Robertson, *Ann. Rev. Nucl. Sci.* **7,** 135 (1957).

13. Brucer, M., *The Acute Radiation Syndrome. A Medical Report on the Y-12 Accident*, U.S. Atomic Energy Commission Report ORINS-25, Washington, 1959.

14. Braestrup, C. B., *Am. J. Roentgenol.* **78,** 988 (1957).

15. Bloom, D., and J. H. Tjio, *New Engl. J. Med.* **270,** 1341 (1964).

16. Bradley, F. J., N. Wald, and R. L. Wechsler, "Human Internal Contamination with Strontium-90 Titanate," in *Assessment of Radioactivity in Man*, International Atomic Energy Agency, Vienna, 1964, Vol. II, p. 401.

17. Catsch, A., *Radioactive Metal Mobilization in Medicine*, Charles C Thomas, Springfield, Illinois, 1964.

18. Cronkite, E. P., V. P. Bond, and C. L. Dunham, *Some Effects of Ionizing Radiations on Human Beings*, U.S. Atomic Energy Commission Document TID-5358, Government Printing Office, Washington, 1956.

19. Cronkite, E. P., and V. P. Bond, *Radiation Injury in Man*, Charles C. Thomas, Springfield, Illinois, 1960.

20. Conen, P. E., A. G. Bell, and N. Aspin, *Pediatrics* **31,** 72 (1963).

21. Court-Brown, W. M., K. E. Buckton, and A. S. McLean, *Lancet* **1,** 1239 (1965).

22. Court-Brown, W. M., and R. Doll, *Leukemia and Aplastic Anemia in Patients Irradiated for Ankylosing Spondylitis*, Medical Research Council Special Report No. 295, H.M. Stationery Office, London, 1957.

23. Court-Brown, W. M., and R. Doll, *Brit. Med. J.* **2,** 181 (1958).

24. Conard, R. A., and A. Hicking, *JAMA*, **192,** 457 (1965).

25. Cogan, D. G., in *The Delayed Effects of Whole Body Radiation* (B. B. Watson, ed.), Johns Hopkins, Baltimore, 1960, pp. 54–56.

26. Conen, P. E., *Lancet* **2,** 47 (1961).

27. Conti, E. A., G. D. Patton, J. E. Conti, and L. H. Hempelmann, *Radiology* **74,** 386 (1960).

28. Dublin, L., and M. Spiegelman, *JAMA* **137,** 1519 (1948).

29. Doida, Y., T, Sugahara, and M. Horikawa, *Radiation Res.* **26,** 69 (1965).

30. DeVilliers, A. J., and J. P. Windish, *Brit. J. Ind. Med.* **21,** 94 (1964).

31. Evans, R. D., and C. Goodman, *J. Ind. Hyg. Toxicol.* **23,** 89 (1940).

32. Fliedner, T. M., E. P. Cronkite, V. P. Bond, J. R. Rubini, and G. Andrews, *Acta Haematol.* **22,** 65 (1959).

33. Fuqua, P. A., W. D. Norwood, and S. Marks, *J. Occupation Med.* **7,** 85 (1965).

34. Gunz, F. W., and H. R. Atkinson, *Brit. Med. J.* **1,** 389 (1964).

35. Guskova, A. K., and G. D. Baisogolov, Proc. Intern. Conf. *Peaceful Uses At. Energy, Geneva*, **11,** 35, United Nations, New York, 1956.

36. Gerstner, H. B., *U.S. Armed Forces Med. J.* **9,** 313 (1958).

37. Goldstein, L., and D. P. Murphy, *Am. J. Roentgenol. Radium Therapy Nucl. Med.* **22,** 322 (1929).

38. Gentry, J., E. Parkhurst, and G. Bulin, *Am. J. Public Health* **49,** 497 (1959).

39. Gerstner, H. B., *JAMA* **168,** 381 (1958).

40. Hollingsworth, J. W., G. W. Beebe, M. Ishida, and A. B. Brill, "Medical Findings and Methodology of Studies by the Atomic Bomb Casualty Commission on Atomic Bomb Survivors in Hiroshima and Nagasaki," in *The Use of Vital and Health Statistics for Genetic and Radiation Studies—Proceedings of a Seminar sponsored by the United Nations and the World Health Organization in Geneva, Sept. 5–9, 1960*, A/AC.82/Seminar. United Nations, New York, 1962, pp. 77–100.

41. Hasterlik, R. J., A. J. Finkel, and C. E. Miller, *Ann. N.Y. Acad. Sci.* **114,** 832 (1964).

42. Hempelmann, L. H., H. Lisco, and J. G. Hoffman, *Ann. Internal. Med.* **36,** 279 (1952).

43. Hasterlik, R. J., and L. D. Marinelli, Proc. Intern. Conf. Peaceful Uses At. Energy, Geneva **11,** 25, United Nations, New York, 1956.

44. "Diagnosis and Treatment of Radioactive Poisoning," *Proc. Intern. At. Energy Agency Conf. IAEA, Vienna* (1962.)

45. Inada, G. J., and C. W. Hall, *Am. J. Ophthalmal.* **59,** 216 (1965).

46. Jablon, S., M. Ishida, and M. Yamasaki, *Radiation Res.* **25,** 25 (1965).

47. Jammet, H., G. Mathe, B. Pendic, J. F. Duplan, B. Maupin, R. Latarjet, D. Kalic, L. Schwarzenberg, Z. Djukic, and J. Vigne, *Rev. Franc. Etudes Clin. Biol.* **4,** 210 (1959).

48. Kelly, S., and C. D. Brown, *Am. J. Public Health* **55,** 1419 (1965).

49. Kitabatake, T., *Ind. Med. Surg.* **31,** 347 (1962).

50. Karas, J. S., and J. B. Stanbury, *New Engl. J. Med.* **272,** 755 (1965).

51. Lewis, E. B., *Science* **142,** 1492 (1963).

52. MacMahon, B., *J. Natl. Cancer Inst.* **28,** 1173 (1962).

53. Martinez, R., G. Cassab, G. Ganem, E. Guttman, M. Lieberman, L. Vater, M. Linares, and H. Rodriquez, *Rev. Med. Inst. Mex. Seguro Social* **3,** Suppl. 1, 14 (1964).

54. McCandless, J. B., *JAMA* **193,** 185 (1965).

55. Miller, L. S., G. H. Fletcher, and H. B. Gerstner, *Radiation Res.* **4,** 150 (1958).

56. MacMahon, B., and G. B. Hutchison, *Acta Unio Intern. Contra Cancrum* **20,** 1172 (1964).

57. Miller, R. W., *Pediatrics* **18,** 1 (1956).

58. MacMahon, T. F. Pugh, and J. Ipsen, *Epidemiologic Methods*, Little, Brown, Boston, 1960.

59. "Effects of Ionizing Radiation on the Human Hematopoietic System," *Natl. Acad. Sci. Natl. Res. Council, Publ.* **875** (1961).

60. *Exposure to Radiation in an Emergency*, National Committee on Radiation Protection and Measurements Report No. 29, University of Chicago (1962).

61. Neel, J. V., *Changing Perspectives on the Genetic Effects of Radiation*, Charles C Thomas, Springfield, Illinois, 1963.

62. Proc. Conf. Epidemiologic Res. Occupational Health, Sponsored by Environmental Sciences and Engineering Study Section, N. I. H. *J. Occupational Med.* **4,** 561 (1962).

63. Neel, J. V., and W. J. Schull, "The Effect of Exposure to the Atomic Bombs on Pregnancy Termination in Hiroshima and Nagasaki, *Natl. Acad. Sci—Natl. Res. Council Publ.* **461** (1958).

64. "Long-Term Effects of Ionizing Radiations from External Sources," *Natl. Acad. Sci—Natl. Res. Council Publ.* **849** (1961).

65. Norman, A., M. Sasaki, R. E. Ottoman, and A. G. Fingerhut, *Science* **147,** 745 (1965).

66. Wald, N., and G. E. Thoma, Jr., "Radiation Accidents: Medical Aspects of Neutron and Gamma Ray Exposure," *Rept. ORNL-2748, Part B.*, Techn. Inform. Serv. Oak Ridge, Tennessee, 1961.

67. Oughterson, A. W., and S. Warren, *Medical Effects of the Atomic Bomb in Japan*, McGraw-Hill, New York, 1956.

68. Perkins, J., M. C. G. Isreals, and J. F. Wilkinson, *Quart. J. Med.* **33,** 499 (1964).

69. Pochin, E. E., N. B. Myant, and B. D. Corbett, *Brit. J. Radiol.* **29,** 31 (1956).

70. Pifer, J. W., E. T. Toyooka, R. W. Murray, W. R. Ames, and L. H. Hempelmann, *J. Natl. Cancer Inst.* **31,** 1333 (1963).

71. Purdom, C. E., *Genetic Effects of Radiations*, Academic Press, New York, 1963.

72. Rossi, E. C., A. A. Thorngate, and F. C. Larson, *J. Lab. Clin. Med.* **59,** 655 (1962).

73. Russell, W. L., *Nucleonics* **23,** 53 (1965).

74. Saenger, E. L., ed., *Medical Aspects of Radiation Accidents: A Handbook for Physicians, Health Physicists and Industrial Hygienists*, U.S. At. Energy Comm. Washington, 1963.

75. Saenger, E. L., Personal communication, 1964.

76. Simon, N., M. Brucer, and R. Hayes, *Radiology* **74,** 905 (1960).

77. Simpson, C. L., L. H. Hemplemann, and L. M. Fuller, *Radiology* **64,** 840 (1955).

78. Socolow, E. L., A. Hashizume, S. Neriishi, and R. Niitani, *New Engl. J. Med.* **268,** 406 (1963).

79. Sinskey, R. M., *Am. J. Ophthalmol.* **39,** 285 (1955).

80. Sinclair, W. K., *Proc. Conf. Radioiodine, Nov. 5 and 6, 1956, Argonne Cancer Research Hospital, Chicago*, The University of Chicago Press, Chicago, Illinois, 1957, p. 9.

81. Shipman, T. L., C. C. Lushbaugh, D. F. Peterson, W. Langham, P. S. Harris, and M. S. Lawrence, *J. Occupational Med.* **3,** 145 (1961).

82. Sheline, G. E., S. Lindsay, K. B. McCormak, and M. Galante, *J. Clin. Endocrinol. Metab.* **22,** 8 (1962).

83. Saunders, T. S., and H. Montgomery, *JAMA* **110,** 23 (1938).
84. Stewart, A., W. Pennybacker, and R. Barber, *Brit. Med. J.* **2,** 882 (1962).
85. Seltser, R., and P. E. Sartwell, *JAMA* **166,** 585 (1958).
86. Saenger, E. L., F. N. Silverman, T. B. Sterling, and M. E. Turner, *Radiology* **74,** 889 (1960).
87. Stewart, J. S. S., and A. R. Sanderson, *Lancet* **1,** 978 (1961).
88. Seltser, R., and P. E. Sartwell, *JAMA* **190,** 1046 (1964).
89. Seltser, R., and P. E. Sartwell, *Am. J. Epidemiol,* **81,** 2 (1965).
90. Stewart, A., J. Webb, and E. Hewitt, *Brit. Med. J.* **1,** 1495 (1958).
91. Tough, I. M., K. E. Buckton, A. G. Baikie, and W. M. Court-Brown, *Lancet* **2,** 849 (1960).
92. Takahashi, S., T. Kitabatake, and M. Wakabyashi, *Nippon Acta Radiol.* **23,** 1510 (1965).
93. Teloh, H. A., M. L. Mason, and M. C. Wheelock, *Surg. Gyneol, Obstet.* **90,** 335 (1950).
94. Tsunge, Y., and T. Ohi, "Early Symptoms of the Cases Affected by the Hydrogen Bomb Explosion at Bikini," in *Research in the Effects and Influences of the Nuclear Bomb Test Explosions*, Promotion of Sci., Ueno, Tokyo, 1956, pp. 1309–1311.
95. Toyooka, E. T., J. W. Pifer, S. L. Crump, A. M. Dulton, and L. H. Hempelmann, *J. Natl. Cancer Inst.* **31,** 1357 (1963).
96. Traenkle, H. L., "X-Ray-Induced Skin Cancer," *Natl. Cancer Inst. Monograph* **10,** 423 (1963).
97. Tsuzuki, M., *Medical Consideration on Radiation Injury due to Bikini-fallout.* (A General Review) Japanese Soc. Promotion Sci., Ueno, Tokyo, 1956, pp. 1287–1304.
98. Thoma, G. E., Jr., and N. Wald, *J. Occupational Med.* **1,** 421 (1959).
99. United Nations Scientific Committee on the Effects of Atomic Radiation, Report: General Assembly, Official Records: Thirteenth Session; Supplement No. 17 (A/3838). United Nations, New York, 1958.
100. United Nations Scientific Committee on the Effects of Atomic Radiation, Report: General Assembly, Official Records: Seventeenth Session; Supplement No. 16 (A/5216). Annex D: "Somatic Effects of Radiation," United Nations, New York, 1962.
101. United Nations Scientific Committee on the Effects of Atomic Radiation, Report: General Assembly, Official Records: Nineteenth Session: Supplement No. 14 (A/5814). Annex B: "Radiation Carcinogenesis in Man," United Nations, New York, 1964, p. 81–110.
102. von Jagie, N., G. Dchwartz, and L. Siebenbach, *Klin. Wochschr.* **48,** 122 (1911).
103. Warren, S., *JAMA* **162,** 464 (1956).
104. Wald, N., *Arch. Environ. Health* **5,** 217 (1962).
105. Wagoner, J. K., V. E. Archer, F. E. Lundin, Jr., D. A. Holaday, and J. W. Lloyd, *New Engl. J. Med.* **273,** 181 (1965).
106. *Diagnosis and Treatment of Acute Radiation Injury*, Proceedings of a Scientific Meeting jointly sponsored by the International Atomic Energy Agency and the World Health Organization, Geneva, October 17–21, 1960, World Health Organization.
107. Warren, S., and L. Meisner, *JAMA* **193,** 351 (1965).
108. Winship, T., and R. V. Rosvoll, *Am. J. Surg.* **102,** 747 (1961).
109. Wald, N., G. E. Thoma, Jr., and G. Broun, Jr., *Progr. Hematol.* **3,** 1 (1962).
110. Zeldis, L. J., S. Jablon, and M. Ishida, *Ann. N.Y. Acad. Sci.* **114,** 225 (1964).
111. Parmentier, N. C., R. Boulenger, and G. Portal, "Problems de Dosimetrie lors de l'Accident de Criticite Survenu au Reacteur Venus à Mol, en date du 30/12/65." Presented at the 1st International Congress of the International Radiation Protection Association, Rome, 5–10 September 1966.

112. Jammet, H. P., R. Gongora, R. Le Go, G. Marble, and M. Faes, "Observation Clinique et Traitement d'un Cas d'Irradiation Globale Accidentelle." Presented at the 1st International Congress of the International Radiation Protection Association, Rome, 5–10 September 1966.

113. Segall, A., B. MacMahon, and M. Hannigan, *J. Chron. Dis.* **17**, 915 (1964).

QUESTIONS

1. What three general categories of exposed human beings supply the available data concerning the biological effects of radiation in man? Give a specific example of a population under study in each general category.

2. What is the approximate $LD_{50}/30$ or 60 dose range for man? Discuss some of the limitations which make it difficult to determine a precise human $LD_{50}/30$ or 60.

3. What are the four typical clinical stages of the human acute radiation syndrome and the approximate duration of each following a major whole-body radiation exposure?

4. Give some pertinent clinical and laboratory test characteristics which distinguish the five levels of injury into which patients with the acute radiation syndrome can be grouped. Which of these injury levels requires medical treatment and in which may such treatment be expected to be a significant factor in determining survival?

5. What characteristics of the late somatic effects of radiation make their association with the initial exposure hard to recognize and what study techniques are therefore utilized to demonstrate the relationships?

6. Discuss the evidence linking external penetrating radiation exposure to leukemia occurring after (*a*) instantaneous exposure from a nuclear weapon, (*b*) intermittent occupational exposure, and (*c*) intrauterine X-ray exposure.

7. Discuss the population studies which provide evidence associating various forms of cancer with exposure to radium and its daughters.

8. What effect of radiation, if any, has been demonstrated on human life-span? Give examples of relevant human studies and their limitations.

9. What evidence in human experiences is there, if any, for a genetic effect of radiation exposure?

10. What procedures are available for the diagnosis and treatment of internal radionuclide contamination without evident injury?

CHAPTER 14

Maximum Permissible Exposure Levels—External and Internal

K. Z. Morgan

14-1 EARLY ROLE OF THE ICRU, ICRP, AND NCRP IN ESTABLISHING EXPOSURE LEVELS

The International Congress of Radiology (ICR) established two important organizations which have set radiation standards on an international level. These organizations are the International Commission on Radiological Units, ICRU, and the International Commission on Radiological Protection, ICRP.* It was in 1925 that the first ICR established the ICRU, and in this same year (1925) two scientists, A. Mutscheller and R. M. Sievert, recommended a maximum permissible dose from X-ray and radium of 0.1 of an erythema dose per year.[24] This would correspond to an exposure of about 30 and 70 R/yr from X-rays produced by 100-kv and 200-kv potentials, respectively, applied across the X-ray tube. At the second ICR meeting in 1928, the ICRP was formed under the auspices of ICR. Dr. L. S. Taylor of the National Bureau of Standards was the principal representative from the United States. In this same year (1928) the ICRU made an important step forward in adopting the roentgen as the unit of exposure, and the ICRP published the first set of international recommendations for protection from ionizing radiation. These recommendations were patterned after the proposals of the British X-Ray and Radium Protection Committee.

Stimulated by the formation of the ICRP and as a result of the initiative and guidance of L. S. Taylor, the National Council on Radiation Protection and

* Initially the ICRP was known as the International X-Ray and Radium Protection Committee.

Measurements, NCRP,* was formed in 1929 under the auspices of the National Bureau of Standards of the United States. From its inception the NCRP has been intimately related to and associated with the ICRP. The ICRP has been the only recognized authority to set the values of maximum permissible exposure to ionizing radiation on an international level, and the NCRP has served a similar role in the United States on a national level. In 1934, the ICRP adopted a maximum permissible dose rate of 0.2 R/day and in the same year the NCRP published *Handbook 18*, in which the maximum permissible dose rate was given as 0.1 R/day.[14] For almost 20 years of their history, the ICRP and NCRP were concerned primarily with X-ray and radium protection. In 1946 the NCRP was reorganized and enlarged and in 1950 the ICRP made similar organizational changes so that the many new radiation problems introduced with the operation of the first man-made nuclear reactor on December 2, 1942 could be given appropriate consideration. For more details regarding the history of ICRU, ICRP, and NCRP, the reader is referred to the historical reviews by Taylor.[25–27]

During the past 35 years, there have been many changes in the maximum permissible exposure levels recommended by ICRP and NCRP and, as indicated in Table 14-1, each of these changes has resulted in a decrease in the occupational exposure values. It was some years later that limits of dose to ionizing radiation were suggested—not recommended—for the population-at-large but, as indicated in Table 14-2, each succeeding new value in this case also has represented a decrease in the previous value. The actual amount of decrease intended is not clearly evident, however, from the figures given in Tables 14-1 and 14-2 because the early reports do not make it clear whether the maximum permissible exposure values include or exclude the radiation scattered by the body. The final values given in Tables 14-1 and 14-2 include body scattering.

14-2 REASONS FOR REDUCTIONS IN EXPOSURE LEVELS

It would be impossible to explain just how, beginning in 1934, each of the reductions in maximum permissible dose has come about. In trying to explain the paramount considerations that led to each of these reductions, it must be kept in mind that the ICRP is composed of individuals from various countries; that the NCRP is composed of representatives of many

* Initially the NCRP was known as the Advisory Committee on X-Ray and Radium Protection. The name was later changed to the National Committee on Radiation Protection and Measurements, and then to its present name, the National Council on Radiation Protection and Measurements, an organization created by Act of Congress and signed into law by President Johnson on July 14, 1964.

Table 14-1 Changes in the Maximum Permissible Occupational Exposure[a] of the Total Body to Ionizing Radiation

Recommended Rate[a]	Comments
0.1 Erythema dose/yr ($\approx$0.5 R/wk for 100 kv X-rays) and 1 R/wk for 200 kv X-rays	Recommended by A. Mutscheller and R. M. Sievert in 1925[24]
0.2 R/day (or 1 R/wk)	Recommended by ICRP in 1934[27] and continued in world-wide use until 1950
0.1 R/day (or 0.5 R/wk)	Recommended by NCRP on March 17, 1934[14] and continued in use in United States until 1949
0.3 rem/wk	Recommended by NCRP on March 7, 1949[15] and ICRP in July 1950[16] and continued in use until 1956
5 rem/yr ($\approx$0.1 rem/wk)	Recommended by ICRP in April 1956[4] and NCRP on January 8, 1957[19]

[a] Values given are in addition to doses from medical and from background radiation. Values in parentheses are for a 5-day work week.

Table 14-2 Suggested Limit of Dose Equivalent to the Total Body of Population-at-Large From Man-Made Sources of Ionizing Radiation

Total-Body Rate[a]	Comments
0.03 rem/wk	Suggested by NCRP September 1952[17] as maximum exposure to any organ
0.5 rem/yr (or 0.01 rem/wk)	Suggested by NCRP on April 15, 1958[20] and by ICRP in July 1959[6] as maximum exposure to gonads (or total body)
5 rem/30 yr (or 0.003 rem/wk)	Suggested by ICRP on September 9, 1958[5] as average exposure to gonads (or total body)
0.1 rem/yr (or 0.002 rem/wk)	Suggested by Ad Hoc Committee on May 6, 1959[22] for consideration by the NCRP for average exposure to any body organ

[a] Values given are in addition to doses from medical and from background exposure.

organizations but that each member, in the course of the committee deliberations and decisions, speaks as an individual scientist—not as a representative of a particular organization or of a country. Thus there is frequently a divergence of opinion—as would be expected among individuals—and the final recommendations usually represent some compromises or a majority opinion. However, it is probably safe to say that each of the reductions shown in Tables 14-1 and 14-2 represented an increased awareness of the hazards of ionizing radiation as more people were exposed to it and as more information was accumulated regarding its hazards. It should be emphasized also that with the exception of the early estimate by Rollins in 1902 of tolerance exposure at a level corresponding to about 10 R/day, none of the levels shown in Table 1-16 or in Tables 14-1 and 14-2 were changed because the previous level had been proved to be too high. The statement of NCRP[20] in April 1958 explains how these changes came about:

"The changes in the accumulated MPD (Maximum Permissible Dose) are not the result of positive evidence of damage due to use of the earlier permissible dose levels, but rather are based on the desire to bring the MPD into accord with the trends of scientific opinion; it is recognized that there are still many uncertainties in the available data and information. Consideration has been given also to the probability of a large future increase in radiation uses."

Although positive evidence of radiation damage has been lacking at the earlier and higher levels of permissible exposure, accumulating evidence has failed to prove that there is a threshold dose below which *no* genetic or somatic damage will result. Thus, as work with man-made sources of ionizing radiation increases and as larger numbers of people are exposed, the probability of damage increases. A consequent rise in the number of individuals in which such damage becomes manifested could be expected unless the prudent course of reducing the levels of maximum permissible exposure is followed. At the present time it appears likely that certain types of damage such as genetic changes, leukemic incidence, bone tumors, and shortening of the life-span increase monotonically with dose equivalent—that is, no matter how small the dose rate or accumulated dose, each added increment increases the amount of damage or the probability of detecting some symptoms of damage. If certain types of resulting damage are proportional to the total radiation dose received by the population—a fact neither proved nor disproved—and if one of the objectives is not to permit an increase in radiation damage, then the only recourse has been to reduce the levels of maximum permissible exposure. In a sense, all ionizing radiation is damaging to man and in reality the basic standard of NCRP and ICRP has been to permit no unnecessary exposure and to allow exposure only when it is counterbalanced by equivalent benefits. It is on this basis that the necessity of relatively large medical exposures, both diagnostic and therapeutic, is recognized by the ICRP and the NCRP and that recommended occupational exposures may be from 30 to 100 times those permitted to the average member of the population-at-large, who may receive little or no direct benefit as a result of such exposure.

14-3 MEANINGS OF EXPRESSIONS: TOLERANCE DOSE, MAXIMUM PERMISSIBLE EXPOSURE, AND RADIATION PROTECTION GUIDES

There have been many expressions used to define acceptable levels of exposure to ionizing radiation, for example, tolerance dose, maximum permissible exposure, and, now, radiation protection guide. It would not be meaningful to attempt to give the "correct" or "official" definitions of these expressions, even by quoting from documents of ICRP and NCRP, because these could be found only in the minds of those individuals who first used them and among the thousands of persons that subsequently applied them to everyday health physics problems. Although not explicitly stated in many of the early writings of members of NCRP and ICRP or in the publications of medical physicists, radiologists, biophysicists, and forerunners of health physicists, the general impression seems to have been most prevalent that *tolerance dose* referred to a level set safely below a threshold dose. The threshold dose in turn was a dose below which no damage would accrue. This concept as applied to tolerance dose seems to have evolved[5] from the early radiological experience that indicated substantial recovery of skin damage within a few months following a moderate therapeutic dose and from the fact that some long-term effects like skin cancer developed with a latent period which increased as the dose rate decreased. It was thus reasoned that if the dose rate were low enough, the latent period could be made greater than the remaining life-span of the individual, and the radiation damage would not make itself manifest except for those who lived to be octogenarians.

Since it is rather well established today that there is no threshold for gene mutations (this seems to be better established for the male than for the female in the case of the mammal), and because present studies have not proved that there is a threshold for certain types of somatic damage, there is no tolerance or completely safe dose when all effects on the individual and on future generations are taken into consideration. In more recent times—in particular, after health physics began as a profession in 1942—the term tolerance dose was used to mean a dose that could be expected to produce only tolerable deleterious effects. Because of the ambiguity of this expression and because of increasing evidence that there may not be a true threshold dose either for genetic or somatic types of radiation damage, the term tolerance dose was largely replaced, beginning about 1944, by the term maximum permissible dose (MPD) or maximum permissible exposure (MPE). The MPD was defined by NCRP in 1954[18] as, "The dose of ionizing radiation that, in the light of present knowledge, is not expected to cause appreciable bodily injury to a person at any time during his lifetime." In this case, appreciable bodily injury was defined as "any bodily injury or effect that the average person would regard as being objectionable and/or competent medical authorities would regard as being deleterious to the health and well being of the individual."

From this definition it must be kept in mind that although the MPD "is not expected" to cause bodily injury, it is not stated that there is no possibility of injury to the exposed individual or to future generations. In fact, although the chance is very small, there is nevertheless a low but definite probability in the case of exposed persons in the childbearing age that a radiation dose, no matter how small, will cause a genetic death or produce serious congenital defects in future generations. Furthermore, if as now seems very likely, the probability of occurrence of certain types of somatic damage such as leukemia, bone tumors, and shortening of life-span is a monotonic function of radiation dose, there is the finite, though very small, possibility of serious damage to the individual as a result of exposure at or below the MPD level. Because of the large biological variability and because of the randomness with which serious types of damage may originate within a single cell in the body, a few persons may be damaged at the MPD where no effects can be detected among the vast majority of the exposed individuals.

Since there is no way known at present of detecting in advance the persons most susceptible to radiation damage or to identify which cells of the body may survive radiation damage only to become the precursor of a malignancy, the problem is one of chance. Thus the ICRP and NCRP have attempted to limit the MPD to that which involves a risk or gamble that is considered to be acceptable to the individual and to the population-at-large. The occupational risk must be so small that it is comparable to or less than other commonly accepted industrial risks that one must tolerate in gainful employment in our safest and best managed industries. The NCRP in 1959[21] stated, "In the light of present knowledge, occupational exposure for the working life of an individual at the maximum permissible values recommended in this report[21] is not expected to entail appreciable risk to the individual or to present a hazard more severe than those commonly accepted in other present day industries."

Moreover, the risk of the MPD assigned to the population-at-large must be no greater than that commonly and willingly accepted from other environmental hazards and on an individual basis must be considerably less than that permitted to those who receive occupational exposure and derive more direct benefits from such employment. In any case, all MPD levels must be set such that in the final inventories and assessments the hazards do not exceed the benefits. Although all unnecessary exposure should be avoided, the MPD values must not be set too low, for otherwise man would be deprived of many of the beneficial uses of ionizing radiation, nuclear energy, and its products. In 1958 the ICRP gave a more applicable and specific definition of the MPD[5] than that of the NCRP quoted above. The ICRP states:

"The permissible dose for an individual is that dose, accumulated over a long period of time or resulting from a single exposure which in the light of present knowledge, carries a negligible probability of severe somatic or genetic injuries; furthermore, it is such a dose that any effects that ensue more frequently are limited to those of a minor nature that would not be considered unacceptable by the exposed individual and by competent medical authorities. Any severe somatic injuries (e.g., leukemia) that might result from exposure of individuals to the permissible dose would be limited to an exceedingly small fraction of the exposed group; effects such as shortening of life span, which might be expected to occur more frequently, would be very slight and would likely be hidden by normal biological variations. The permissible doses can therefore be expected to produce effects that could be detected only by statistical methods applied to large groups."

Like all definitions, the preceding one raises a number of questions. Here, we may ask what is a "negligible probability" of severe injuries or what is "an exceedingly small fraction of the exposed group"? Again these questions can only be answered when balancing the risks against the benefits and this ostensibly is what the ICRP and NCRP did when the basic values of MPD were chosen. At the present time, as in the past, the MPD values have been assigned after a careful consideration of the data available. However, it must be kept in mind that in some cases these data have been very scant and from time to time in the future we can expect further modifications in the MPD values as more applicable data become available.

The preceding definitions by NCRP and ICRP refer to "doses that are not expected to cause appreciable injury and to effects of a minor nature that would not be considered unacceptable by the exposed individual." There is little doubt that very small doses—doses that are even smaller than the MPD—produce biological changes in man (Chapter 1). Some of these changes may not constitute damage because of repair or because these are considered to be neither harmful nor beneficial. Some such changes at the MPD rate that have been reported—but not necessarily confirmed—are alterations in a number of the various formed elements of the blood, an increase in the number of bilobed nuclei in the lymphocytes, an increase in the number of chromosomal aberrations, an increase in the number of areas of greater (and of less) density of the bone, an effect on the nervous system, slight decrease in fertility, changes in the urine chemistry, and so on. Perhaps some day certain of these effects of exposure at the MPD level can be developed as reliable indices of human exposure.

In 1959 the Federal Radiation Council (FRC) was formed in the United States to provide a Federal policy on human exposure to ionizing radiation. The major function of the FRC is to "advise the President with respect to radiation matters directly or indirectly affecting health . . . ." From its inception the FRC has recognized the authority and preeminence of the NCRP and the ICRP in the field of radiation protection. In fact, when the FRC was authorized by Congress it was to "consult qualified scientists and experts in radiation matters, including the President of the National Academy of Sciences, the Chairman of the National Committee on Radiation Protection and Measurements and qualified experts in the field of biology and medicine and in the field of health physics. . . ." Although in the first two publications of the FRC[1,2] the impression may be given that there was independent consideration of each of the factors leading to the selection of exposure values, the fact that all values chosen in the first report were identical to those recommended by the NCRP and ICRP serves as proof that there was no effort to dispute the authority of the NCRP and ICRP or to disregard their recommendations. It is to the credit of the FRC that it has supported and endorsed the recommendations of the NCRP and ICRP since it could hardly be expected that an impartial review by a new group of experts of the same data of dose–effect relationships would yield significantly different recommendations.

As explained above, the early expression "tolerance dose" was replaced after about 1944 by "maximum permissible dose" because tolerance seemed to imply to some people a threshold dose below which there would be no radiation damage. When the FRC considered this terminology, it stated that "the words maximum and permissible both have unfortunate connotations not intended by either the NCRP or the ICRP. There can be no single permissible or acceptable level of exposure without regard to the reason for permitting the exposure." Although these observations by the FRC are true, the problems were not solved by their invention and use of the term "radiation protection guides" as a replacement for "maximum permissible exposure." Regardless of the terminology there must be different levels set for the various applications and the goal in any case should be to maintain exposures to the lowest practical level consistent with good practice and without unduly depriving man of the benefits from the use of ionizing radiation.

Although there may be some merit in the use of the newly suggested terminology, it is evident that "radiation protection guides" likewise suggests meanings not intended by the FRC. In particular, the word "guides" suggests to some people that no limit is set for any operation or application and that only guides have been set, guides which may be exceeded with the least provocation. Similar criticisms may be directed toward the term "radioactivity concentration guides" suggested by the FRC as a replacement for the term "maximum permissible concenrtation" as used by the NCRP and the ICRP. Although, at best, any levels that are set must serve as guides, we do not wish to lose sight of the fact that for safe operations, maximum permissible exposures and concentrations must be established for general operations and applications, and employees must regard these levels as limits that must not be exceeded except for extreme emergencies. The exposure levels, such as 3 rem/quarter to the total body or an average concentration of 1 MPC in a water supply as measured over a period of one year, must be respected as maximum permissible limits by all persons and at all times except during emergency situations.

Actually, when the definitions of the maximum permissible dose and radiation protection guide are examined critically, there appear to be few, if any, significant differences in the actual meaning of these terms. The FRC[3] stated in 1965:

"Although radiation doses numerically equal to the RPG's may impose a risk so small that they can be accepted each year for a lifetime if there is significant benefit from the programs causing the exposure, they do not and cannot establish a line that is safe on one side and unsafe on the other. Rather, some risk of injury may exist at any level of dose and the risk continuously increases with dose."

The FRC made it clear that exceeding the radiation protection guides was not to be taken very lightly. It stated[2]:

". . . the establishment of radiation protection standards involves a balancing of the benefits to be derived from the controlled use of radiation and atomic energy against the risk of radiation exposure. This principle is based upon the position adopted by the Federal Radiation Council that any radiation exposure of the population involves some risk; the magnitude of which increases with the exposure. As stated in 'Radiation Protection Guidance for Federal Agencies,' approved by the President, May 13, 1960, 'There should not be any man-made radiation exposure without the expectation of benefit resulting from such exposure.' In recommending use of the term, 'Radiation Protection Guide' it was stated that 'This term is defined as the radiation dose which should not be exceeded without careful consideration of the reasons for doing so; every effort should be made to encourage the maintenance of radiation doses as far below this guide as practicable.' Consistent with these principles, no exposure to radiation should be permitted unless it satisfies two criteria:

"(1) The various benefits to be expected as a result of the exposure, as evaluated by the appropriate responsible group, must outweigh the potential hazard or risk, and

"(2) the reasons for accepting or permitting a particular level of exposure rather than reducing the exposure to a lower level must outweigh the decrease in risk to be expected from reducing the exposure.

". . . In view of the considerations discussed above, ideally, an individual radiation protection guide should be developed for each activity or set of circumstances involving exposure to radiation. Recognizing the impracticability of establishing an individual guide for each application, the Council, in its Report No. 1, pointed out the need for a compromise between this ideal and the application of a single guide to widely differing sets of conditions. The following is taken from the Council's recommendations approved by the President:

> 'There can be no single permissible or acceptable level of exposure without regard to the reasons for permitting the exposure . . . It is basic that exposure to radiation should result from a real determination of its necessity.
>
> 'There can be different Radiation Protection Guides with different numerical values, depending upon the circumstances. The guides recommended herein are appropriate for normal peacetime operations.'

". . . On the basis of extensive consultation, the Council has recommended to the President a set of Radiation Protection Guides which represent a generalized balance between the considerations discussed above. Despite wide differences in the assignment of relative values to the various factors involved, the Council believes that the overall benefits from useful activities involving exposures to radiation at levels within those specified in these guides will outweigh the risks associated with such exposures. There is also sufficient experience in limiting radiation exposures to levels similar to these to demonstrate the general feasibility, with few exceptions, of operating at or below the levels specified in these guides in normal peacetime operations.

". . . The Federal agencies, when applying these Radiation Protection Guides should recognize that they represent generalized and not specific guidance. Because the reasons for accepting or permitting exposure to radiation vary so widely from one situation to another, the guides cannot represent the most appropriate ones for some situations without being inappropriately high or low for others. Each agency should determine, in any specific application, the extent to which the generalized guides apply in the specific situations. For example, certain applications may be able to be conducted at a guide specifying a lower dose than the RPG recommended by the Council. On the other hand, some applications which are not practicable under existing guides and for which the needs are very great

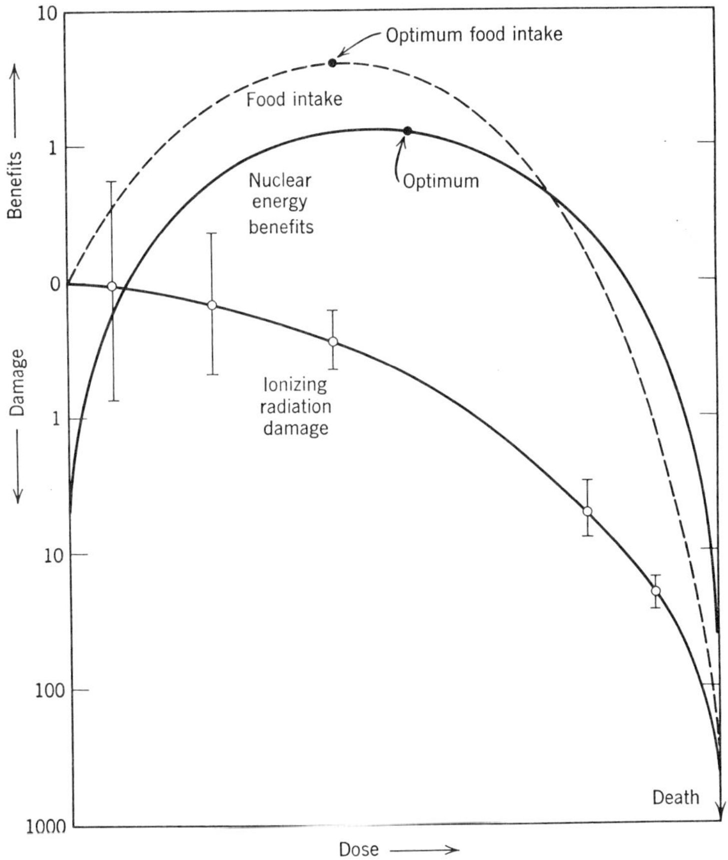

Fig. 14-1. Schematic balancing of benefits and damage with dose.

may require a guide specifying a higher dose. The possibility of certain situations, such as accidents, may require the development of guides to be used when an agency considers such drastic actions as exclusion of persons from a specified area, evacuation, or condemnation of supplies of food.

"... 'Radiation Protection Guidance for Federal Agencies' published in the Federal Register May 18, 1960, recognized that in certain instances the balance of benefit versus risk might necessitate an RPG higher than specified for normal peacetime operations. This was expressed in the following language:

> 'The guides may be exceeded only after the Federal agency having jurisdiction over the matter has carefully considered the reason for doing so. . . .' "

Ideally, the health physicist who sets the exposure limits would like to balance the risks against the benefits as shown in Fig. 14-1. Here it is assumed that all dose from ionizing radiation is harmful to man and at high doses the magnitude of the damage to the individual becomes more certain, that is, the probable errors are smaller, and eventually the damage approaches negative infinity or death. On the other hand, the curve of nuclear energy benefits has a maximum. At low doses, the benefits become negative (or damaging) because of the deprivation of nuclear energy benefits and at high doses the benefits again become negative (or damaging) because of the excessive radiation damage that is uncompensated by benefits. The curve of benefits from nuclear energy is similar to the curve of benefits from intake of food, except that a small intake of food probably does not produce a negative benefit (or damage). Perhaps the most important responsibility of the FRC is the balancing of the benefits resulting from the proper use of ionizing radiation against the hazard and damage it can bring to man.

14-4 APPLICATIONS OF MAXIMUM PERMISSIBLE DOSE OF IONIZING RADIATION FOR OCCUPATIONAL EXPOSURE

Values of maximum permissible dose are applied to external exposure, internal exposure, and, in general, to combinations of external and internal exposure. The calculation of dose, its attenuation in passing through matter, and its decrease with distance from sources of various energies and configurations are discussed in other chapters. In general, discussions of dose in this text refer primarily to absorbed dose (given in units of rad of soft tissue) and to dose equivalent (given in rem units). Unless otherwise specified for purposes of personnel monitoring values of maximum permissible dose equivalent or absorbed dose from total-body human exposure to external sources of X-, γ-, and neutron radiation are given in terms of that measured at the surface of the body at about chest height. In practice, this measurement usually is made with personnel monitoring meters worn by the individuals, and in the case of X-, γ-, and neutron radiation may include a contribution to dose from body scattering. Although a number of general health physics surveys have indicated that for many—and perhaps most—industrial operations in which employees are working in fields of X-, γ-, and neutron radiation, the maximum dose for the average person is received in the region of the chest, very frequently much larger doses are delivered to other parts of the body. For example, a worker may be located most of the day with his back turned toward a well-collimated source of fast neutrons; he may be working with a source of β-radiation located so that most of the dose is delivered to the hands; or a person entering a weapons fallout area may receive most of the dose to the feet and ankles. Appropriate personnel monitoring instruments therefore should be located on the portions of the body surface that are likely to receive the largest dose.

For neutrons or protons, the maximum permissible dose limits are usually given in terms of the flux incident on the surface of the body. These values of maximum permissible flux are calculated to correspond to a maximum permissible dose equivalent rate or absorbed dose rate at the point in the body where the total multiple-collision dose rate is a maximum. Survey measurements of dose rate to determine the working time in a given area corresponding to the maximum permissible dose usually are given in terms of the first-collision dose rate but the measurement often includes a small contribution to the total dose from scatter of the radiation by the body of the observer. With X- or γ-radiation, this scattering contribution to the dose from the body of the surveyor is usually less than 5%; it is ordinarily negligible for fast neutrons but may be greater than 20% for thermal neutrons. The survey instruments are usually calibrated to read directly in terms of first-collision dose of X-, γ-, and β-radiation, in terms of flux of neutrons (n/cm^2/sec) and in terms of disintegrations per minute per 100 cm^2 for α-radiation. Values of maximum permissible internal dose are usually given in terms of the average dose equivalent to a specified body organ, although sometimes the values are given in terms of the maximum dose equivalent to some portion of the organ. Ordinarily, values of maximum permissible internal dose are related to the maximum permissible organ burden qf_2 and/or to the maximum permissible total body burden q (Chapter 10).

As explained earlier, the values of maximum permissible dose rate for external X-ray exposure of occupational workers as recommended by the NCRP and the ICRP have been decreased by an order of magnitude since they were first established by the ICRP in 1934. Unlike external exposure, the basic standard of maximum permissible internal exposure—the body burden of 0.1 μg of ^{226}Ra—has continued in use unchanged since it was first established in 1941 by an Advisory Committee on the Safe Handling of Radioactive Luminous Compounds.[13]

14-5 PRESENT (1967) VALUES OF MAXIMUM PERMISSIBLE DOSE OF IONIZING RADIATION AS APPLIED TO OCCUPATIONAL EXPOSURE

Table 14-3 summarizes the values of maximum permissible dose of ionizing radiation as applied to occupational exposure and as recommended by the ICRP, the NCRP, and the FRC. Detailed discussions of the history of these

Table 14-3 Presently (1967) Recommended Permissible Dose Equivalent[a] to Body Organs of Occupational Workers Exposed to Ionizing Radiation[12]

Body Organ	Maximum Dose Equivalent in Any 13 Weeks (rem/13 wk)	Annual Permissible Dose Equivalent, R_{50} (rem/yr)	Accumulated Dose Equivalent to Age N (rem)[c]
Red bone marrow[b]	3[1,5,20]	5[5,19,20]	$5(N\text{-}18)$[1,5,19,20]
Total body	3[1,9,19–21]	5[5,9,19,21]	$5(N\text{-}18)$[1,9,19–21]
Head and trunk	3[1,20]	5[1,5]	$5(N\text{-}18)$[1,19,20]
Gonads	3[1,5,9,20,21]	5[1,5,9,19–21]	$5(N\text{-}18)$[1,5,9,19–21]
Lenses of eyes	3[1,5,6,9,10,20,21]; 4[7]; 8[8]	5[1,5,6,9,10,19–21]; 15[8]	$5(N\text{-}18)$[1,5,6,9,10,19–21]
Skin	8[5,10,21]; 10[1,21,23]; 15[8]	30[1,5,9,17,20,21,23]	
Thyroid	8[5,10,21]; 10[1]; 15[8]	30[1,5,9,20,21]	
Bone	15[8]	30[9,21]	
Feet, ankles, hands, forearms	20[5]; 25[1,20,21]; 38[8]	75[1,5,17,20,21]	
Other single organs	4[5,9]; 5[1]; 8[8]	15[1,5,9,17,20,21]	

[a] The recommended permissible dose equivalent for the occupational worker is in addition to doses from medical and background exposure.
[b] Referred to in earlier reports as active *blood-forming organs.*
[c] This equation may be considered as an upper limit to the allowable 13-week dose but not an encouragement to accumulate this much dose.

values and the reasons for their choice are not given here. Instead, the reader who wishes more detailed information is referred to publications already cited. Columns 3 and 4 of Table 14-3 indicate that although in some cases specific values have not been considered by one or more of the agencies (NCRP, ICRP, and FRC) for the annual and cumulative occupational dose equivalent, there are no differences in any of these recommended values except for the lenses of the eyes. In the case of maximum dose equivalent for a quarter of a year, however, there are a number of differences. These differences are only minor and do not indicate disagreement among these three agencies. The differences came about because the values were set at different times by separate groups of people making use of various sets of data. The principal differences resulted from the 1960 decisions of the NCRP and the FRC[1] to set the maximum permissible quarterly values at one-third the annual permissible dose equivalent (that is, $R_{50}/3$ rem) and of ICRP in 1966[8] to set the maximum permissible quarterly values at one-half the annual permissible dose equivalent (that is, $R_{50}/2$ rem). The values in the fourth column of Table 14-3 are the maximum permissible dose equivalent values for occupational workers of age N above 18. In the United States, occupational exposure to ionizing radiation at any age less than 18 is forbidden. For those countries where such occupational exposures are permitted, the ICRP has provided that the dose equivalent shall be limited to not more than 5 rem/yr and that the total accumulated dose equivalent to age 30 shall not exceed 60 rem for the group of critical organs listed in Table 14-3 for which the formula $5(N - 18)$ is applicable.

From Table 14-3, it is apparent that there are at present (1967) three principal ICRP regulations that limit exposure to ionizing radiation: (1) no more than one-half the annual permissible dose equivalent or $R_{50}/2$ rem (taken for the purpose of this recommendation to the next highest whole number, that is, $\frac{1}{2} \times 5 \rightarrow 3$, $\frac{1}{2} \times 15 \rightarrow 8$, $\frac{1}{2} \times 30 = 15$, and $\frac{1}{2} \times 75 \rightarrow 38$) to any body organ in a single quarter; (2) no more than $5(N - 18)$ rem to the organs for which the formula is applicable; and (3) no more than the annual permissible dose equivalent (that is, values of R_{50} listed under column 3 of Table 14-3) for the other organs. Thus it is permissible (but should be avoided, if possible) to receive dose equivalents of 12 rem/yr to the gonads, total body, red bone marrow, etc., as long as the limit $5(N - 18)$ is not exceeded. In the other organs, however, one may receive a dose equivalent of $R_{50}/2$ rem in a quarter but no more than R_{50} rem per year except for planned special exposures or emergency exposures (as discussed later in this chapter). Although from a first consideration these three regulations may seem to be very restrictive and conservative, they are probably on the high side for those who may be occupationally exposed for 50 years. Such persons might accumulate a dose equivalent to 250 rem to the gonads or total body, 750 rem to most organs of the body (that is, muscle, GI tract, liver, kidney, brain, etc.), 1500 rem to the skin, thyroid, and bone, and 3750 rem to the limbs. The only justification for such high accumulated doses is that few people are expected to work for 50 years at the maximum permissible exposure limits. As a man gets older, there is less risk of genetic damage and less of the life-span remaining during which delayed types of radiation damage can manifest themselves; on the other hand, there is evidence that as an adult increases in age, less exposure from ionizing radiation is required to produce a given type of damage. In other words, it appears that there is a minimum in the radiosensitivity curve plotted as a function of age, that is, the most radiosensitive members of the population are the very young and the very old.

The ICRP in its latest recommendation[8] has emphasized the special risks associated with allowing a rapid buildup in the body of radionuclides having a long effective half-life (the effective half-life T is given in Chapter 10 by the equation $T = T_r T_b/(T_r + T_b)$ in which T_r = radioactive half-life and T_b = biological half-life). For example, if a person built up a bone burden of a radionuclide with a very large value of T during 13 weeks or a year, so that the dose equivalent rates listed in columns 2 or 3 of Table 14-3 were reached in 13 weeks or a year, respectively, these rates of bone exposure would be continued with very little abatement for many years thereafter. Thus the person, although receiving 30 rem/yr to the bone from a one-year occupational exposure, would be committed to this same annual bone exposure for many years to follow and would not be permitted additional occupational exposure. To avoid this difficulty, the ICRP has introduced the concept of "dose commitment" and stated that in a period of a quarter of a year up to one-half of the maximum permissible dose commitment may be accumulated during occupational exposure. The maximum permissible dose commitment is defined as the body intake of a radionuclide by a person occupationally exposed at the maximum permissible concentration, MPC, of a radionuclide for one year. It can be shown that from a maximum permissible dose commitment the resulting critical body organ dose integrated over 50 years (actually integrated from the beginning of intake to a time 50 years beyond the midpoint of the intake period—thus, if the intake period is τ years and the intake began at a time $t = 0$, the period of integration should be from $t = 0$ to $t = 50 + \tau/2$) would be numerically equal to the annual permissible dose equivalent, R_{50}, values of which are listed in column 3 of Table 14-3 (that is, 5 rem for total body, gonads, or red bone marrow; 30 rem for skin, thyroid, or bone; and 15 rem for most of the body organs).

As an example of a maximum permissible quarterly exposure or single exposure, a person receiving occupational exposure only from ^{24}Na (T = 0.6 days) would be permitted to receive an exposure to ^{24}Na of any duration during a 13-week period if it corresponded to not more than the body intake from an exposure for 13 weeks (40 hr/wk) to air, water, or food contaminated at a level of twice the MPC (based on occupational exposure 40 hr/wk) or it could correspond to the body intake at the MPC for half a year. If, on the other hand, during a 13-week period he received from external exposure one-fourth the annual permissible dose equivalent (that is, $R_{50}/4$ or 5/4 rem) to the total body, his intake of ^{24}Na should be limited to an amount corresponding to exposure at the MPC for 13 weeks. In either case, his total body dose equivalent in the 13-week period would be $R_{50}/2$ or 2.5 rem. Also, since ^{24}Na is distributed rather uniformly in the body and has a short effective half-life T, essentially all the dose equivalent $R_{50}/2$ would be received by the total body during the period of 13 weeks. If, instead of ^{24}Na, a radionuclide with a large value of T had been taken into the body in the 13 weeks, much of the dose equivalent of $R_{50}/2$ would be delivered over a prolonged period during the following 50 years.

It is apparent from the values given in column 4 of Table 14-3 that several persons of the same age may work side-by-side in the same facility but be restricted by different dose limits. For example, three men, all at age 40, may start work at the same time at the same radiation exposure facility. If one of the men has received no previous exposures, he may be permitted to receive a total body dose equivalent of 12 rem/yr until he is age 55, 10 rem the next year, and thereafter be limited to an average of 5 rem/yr; the second man who has already accumulated 102 rem to age 40 may receive 8 rem the first year and thereafter be limited to an average of 5 rem/yr; the third man who has already accumulated 110 rem may receive no exposure until he is 41 and thereafter he may average an exposure of 5 rem/yr. If the person began work with a new employer upon reaching age 40 without any available exposure record and he had worked since age 18 with other employers where occupational exposure to ionizing radiation was possible, regulations specify that his new employer must assume he begins work with an accumulated total body dose equivalent of 5 rem/yr since age 18, or a total of 110 rem.

With skeletal exposure, it is to be noted that the values of annual permissible dose equivalents of 30 rem/yr and 15 rem/13 weeks in Table 14-3 include the relative hazard factor n defined in Chapter 10. This factor is equal to 5 for particulate radiation (that is, α- or β-radiation) from internal emitters in the bone with the exception of radiation from radium and its daughter products when an isotope of radium is the parent element, in which case $n = 1$. In the event of external exposure, for X- or γ-radiation (both external and internal), and for all organs other than bone, $n = 1$. For the lenses of the eyes, the permissible dose equivalents of 15 rem/yr and 8 rem/quarter are applicable only if an additional lens modifying factor of 3 is applied to the quality factor for radiation of high linear energy transfer. Thus, for example, the permissible absorbed dose for the lenses of the eyes for fast neutrons or protons is 0.5 rad/yr and the dose equivalent is $0.5 \times 10 \times 3 = 15$ rem/yr when the quality factor is equal to 10.

Table 14-4 Dose Equivalent Levels[a] Recommended in Early Reports of ICRP[5,7] and Factors to be Applied to Occupational Values

Exposed Group	Dose Equivalent to Gonads[h] or Total Body		Dose Equivalent to Most Other Organs	
	Dose Rate[j]	Factor[e]	Dose Rate[c,j]	Factor[e]
A—Occupational worker	3 rem/qtr 12 rem/yr or 5 rem/yr (av)	1	8 rem/qtr or 15 rem/yr (av)	1
B_a—Work in vicinity of controlled area	1.5 rem/yr	$\frac{3}{10}$	1.5 rem/yr	$\frac{1}{10}$
B_b—Visit area occupationally[g]	1.5 rem/yr	$\frac{3}{10}$	1.5 rem/yr	$\frac{1}{10}$
C—Whole population[d,g]	0.5 rem/yr 5 rem (av) to age 30 or 0.17 rem/yr[b] (av) 0.05 rem/yr[f] (av)	$\frac{1}{10}$ $\frac{1}{30}$ $\frac{1}{100}$[d,f]	1.5 rem/yr or 15 rem (av) to age 30 or 0.5 rem/yr[b] (av)	$\frac{1}{10}$ $\frac{1}{30}$

[a] Values given are in addition to dose from medical and from background exposure. They apply to the sum of external and internal dose.

[b] These average values for whole populations were obtained by taking $\frac{1}{3}$ the maximum value.[5] More recently[8] the ICRP does not specify a specific factor such as $\frac{1}{3}$ but states that those responsible for exposure to whole populations must make certain that appropriate factors are applied such that members of the critically exposed group do not exceed the maximum permissible dose equivalent for members of the whole population.

[c] These values are multiplied by 2 when applied to skin or bone and 5 when applied to hands, forearms, feet, or ankles. These values are multiplied by 2 when applied to thyroid for group A and multiplied by 1 when applied to thyroid for groups B_a, B_b, and C.

[d] The average values for whole population were "suggested" and not "recommended" by the ICRP.[7] The latest report of ICRP[8] does not suggest any specific values for whole population exposure except $\frac{1}{10}$ of values in Group A to be applied to the critical segment of the population. An Ad Hoc Committee of NCRP[22] "suggested" an average rate in the neighborhood of background (or 0.1 rem/yr), but this "suggestion" has not been confirmed by NCRP.

[e] This factor is applied to the average permissible dose equivalent listed under Group A. The factors are applied as a first approximation to the MPC's for 40 hr/wk occupational exposure in the case of Groups B_a and B_b. They are applied as a first approximation to the MPC's for 168 hr/wk occupational exposure in the case of Group C.

[f] This value applies only to the internal dose component.

[g] Originally the ICRP[5] included a special exposed group, B_c, that is, those living in the vicinity of the controlled area. In ICRP *Publication 6*,[7] this group B_c has been considered a part of group C, that is, the whole population.

14-6 RECOMMENDED LEVELS OF EXPOSURE TO VARIOUS SPECIFIED GROUPS

Table 14-4 lists the exposure levels for various population groups as recommended by the ICRP in several of its earlier reports.[5,7] The recent (1966) ICRP report[8] no longer retains groups B_a and B_b as separate categories, but they are listed here because they have been and may continue to be valuable for planning purposes. It should also be emphasized that group C (whole population) is now simply limited to $\frac{1}{10}$ the values listed in column 3 of Table 14-3 as applied to the critical segment of the population. Further reduction or parceling out of this maximum permissible population dose of $R_{50}/10$ is now considered a matter that is left up to the individual countries. Again, these earlier recommendations may provide some useful guidance. In any event, of the levels shown, those for the occupational workers are the highest. The reasons for this were given by the ad hoc committee of the NCRP[22] as follows:

1. These other groups are larger, hence contain a correspondingly larger number of individuals with injurious effects per rem of dose equivalent.
2. Employment involving occupational hazard from radiation exposure is voluntary and, in principle, the hazard can be foreseen by the individual accepting the risk that may be involved.
3. Industrial workers are screened so those least able to meet any particular hazard may be channeled into other activities.
4. In industry there can be evaluation and control of the hazards by radiation monitoring.
5. Population groups other than occupational are more likely to contain children and embryos that are more sensitive to radiation damage and also they may contain older adults who may be more susceptible to radiation damage.
6. The number of years of occupational exposure will be less than the number of years of environmental exposure.
7. Each industry cannot be permitted to impose a full measure of its peculiar occupational hazards on the environment.

If the last condition were not realized, man could not survive long in an industrial area where the hazard would be the sum of every maximum

---

[h] Applies also to blood-forming organs and to mixtures of radionuclides that are taken up by several organs such that the dose equivalents in the several organs are of comparable magnitude and constitute essentially a total-body exposure.[7]

[j] All dose rates are maximum levels except in the cases where average values are indicated.

occupational hazard: we must not all run the same risk of falling from a scaffold as the construction worker, the same risk of lead poisoning as the lead burner, or the same risk of getting burned from a stove as the housewife. It might be added also that if genetic damage relates linearly to accumulated dose at the very low dose rates, $\frac{1}{2}$ mrad/yr to each person in the United States population of 2×10^8 produces as much genetic damage as 1 rad/yr to each of 10^5 occupational workers. In fact, the $\frac{1}{2}$ mrad/yr to each of the 2×10^8 population is probably worse than the 1 rad/yr to each of the 10^5 occupational workers because of the less favorable age distribution of the larger population group.

Values in column 2 of Table 14-4 apply to the gonads, red bone marrow (or blood-forming organs), total body, and to certain exposures from mixtures of radionuclides that essentially constitute whole-body exposure. Values in column 4 apply to most organs of the body, including the lenses of the eyes; when multiplied by 2, the values in column 4 apply to skin or bone, and when multiplied by 5 they apply to the body limbs. Values of column 4 apply to the thyroid for Groups B_a, B_b, and C and, when multiplied by 2, they apply to thyroid for Group A. The values in columns 3 and 5 are the factors by which the average occupational dose equivalent values are multiplied in order to obtain the dose equivalent values for the corresponding exposure groups. The factor of $\frac{1}{10}$ by which the average permissible dose equivalent for occupational exposure must be multiplied to obtain the maximum permissible dose equivalent for members of the whole population applies to any and all body organ exposures and is a factor currently (1967) recommended by the ICRP, the NCRP, and the FRC.

The additional factor of $\frac{1}{3}$ (that is, $\frac{1}{10} \times \frac{1}{3} = \frac{1}{30}$) by which this maximum permissible dose equivalent for members of the whole population must be multiplied to obtain the average permissible dose equivalent for the whole population has been recommended by the ICRP[5] and by the FRC[1] as a first approximation of the safety factor necessary to account for the human variation in dose equivalent for members of the critical exposure group—a variation due to metabolic, physiological, and other factors. The latest reports of the ICRP[8,28] do not give this factor $\frac{1}{3}$ but instead caution that those responsible for exposure of whole populations must make certain that appropriate safety factors are applied so that members of the critical exposure group do not exceed the maximum permissible dose equivalent. In addition, when selecting the critical population exposure group, many factors such as age, sex, personal habits, environmental conditions, etc., must be taken into account in order to narrow the critical population exposure group and ensure that it is relatively homogeneous with respect to the dose equivalent its members receive. In fact, this critical population exposure group is the one that can be expected to receive the most significant exposure from the standpoint of magnitude of dose and of potential radiation damage.[11]

In general, in applying the values in Tables 14-3 and 14-4, if the fraction f of these dose values is received by a body organ from external exposure, only the fraction $1 - f$ may be received from internal exposure. The last value in column 2 of Table 14-4 (that is, 0.05 rem/yr) is an exception to this rule in that it is a suggested allocation specifically for internal exposure. In the case of simultaneous external exposure and exposure to long-lived bone-seeking radionuclides, the ICRP[7,8] has provided a special set of rules applicable to occupational workers in order to assure them adequate protection while at the same time simplifying personnel monitoring procedures. These rules may be summarized as follows. (1) If the estimated body burden of a long-lived bone-seeking radionuclide is less than one-half of the maximum permissible, no consequential restriction of external radiation exposure need be applied. (2) If the estimated body burden is greater than one-half but less than the maximum permissible, the total body exposure to penetrating external radiation shall be limited to not more than 1.5 rem in a year. (3) If the estimated body burden equals or exceeds the maximum permissible, no occupational exposure to penetrating external radiation is permissible.

One question that has always arisen when considering Table 14-4 is, "What is an acceptable size for each of these groups?" The ICRP[5] has suggested that the size be determined by the equation

$$S = \frac{100K}{30I} \tag{14-1}$$

where S = percent of total population

K = average dose equivalent per 30 years to members of the total population as a result of exposure to the group

I = average dose equivalent per year to individual members of the group

Values of I for groups A and C are given in Table 14-4 as 5 and 0.17 rem/yr, respectively. No value of I is given by the ICRP for group B but for purposes of calculation, it has used the maximum value of 1.5 rem/yr given in Table 14-4. Firm values of K have not been recommended, but Table 14-5 lists the values "suggested" by the ICRP for genetic dose in its earlier reports.[5,7] These values were not reprinted in the latest ICRP report[8] because it is now felt best to leave it up to the various countries to apportion the 5 rem per 30 years to the population. In any case, the goal still remains to make sure that the per capita gonad dose equivalent does not exceed the 6–10 rem from conception to age 30 from all man-made sources. This is recommended by different national and international scientific bodies. Obviously, if medical exposure could be reduced, this might conceivably justify an increase in the other man-made exposures. As indicated in Chapter 1, there may be an order of magnitude difference in the medical dose received by the gonads of people living in one of the advanced countries compared with people living in another advanced country.

Applying values of K from this table and the values of I as listed above in Eq. (14-1), we find that the sizes of groups A, B, and C are 0.7%, 1%, and 100%, respectively, of the total population. Originally, when ICRP first published the values listed in Tables 14-4 and 14-5,[5] group B or the special

Table 14-5 Permissible Genetic Dose to the Population-at-Large Suggested in the Early Reports of the ICRP to Serve as a Guide
(Dose Equivalent in rem to Age 30)

Medical[a]			4.5 rem
Background[a]			3.0
Other sources			
Population-at-large			
Internal	1.5		
External	0.5		
Subtotal		2.0	
Other groups			
Occupational	1.0		
Special groups	0.5		
Reserve	1.5		
Subtotal		3.0	
Total other sources			5.0
Total from all sources			12.5 rem

[a] The medical and background values are probable average values in the United States (see Chapter 1) and are not values recommended by the ICRP.

groups were comprised mostly of the subgroup B_c which included those people living in the vicinity of the controlled area. From this it was apparent that nuclear facilities with a large ionizing radiation exposure potential should not be located near large population centers. For example, if it is assumed that the ratio of the genetic population to the total population is the same for this group B in the New York area as for the entire United States and that the maximum allowable for this group B is 1% of the United States population as given by Eq. (14-1), then group B would be limited to only 2×10^6 people or considerably less than the population of New York City. Thus, if one-fourth of this one city were included in group B because of the close proximity of a large potential source of ionizing radiation, this would exceed the size of group B suggested in the early publications of ICRP for an entire country the size of the United States. Care must be taken also in choosing the size of the total genetic population. When determining the size of the exposure groups in the United States, it may be reasonable to take the genetic population as proportional to the population of the entire United States. This genetic population is defined approximately by the equation

$$G = \tfrac{1}{2} \sum_i N_i n_i \tag{14-2}$$

where G = number of persons in the genetic population
N_i = number of persons in this population of age i
n_i = average number of expected children of persons of age i

In some parts of the world, the total genetic population may be much larger or smaller than the population of a single country and in many cases the ratio of the genetic population to the total population in a local group may differ considerably from the average of this ratio for the total population of a country or group of genetically associated countries.

14-7 VALUES OF MAXIMUM PERMISSIBLE SINGLE OR QUARTERLY EXPOSURE VERSUS MAXIMUM PERMISSIBLE CONTINUOUS EXPOSURE

As explained previously, the maximum permissible occupational exposure is limited to no more than that calculated by the age prorated equation, $5(N - 18)$ as given in column 4 of Table 14-3 for the first group of organs or no more than the annual permissible dose equivalent R_{50} given in column 3 of Table 14-3 for the other organs. This limitation is applied separately to the dose equivalent accumulated by each organ of the body. It applies both to single and continuous exposures as well as to external and internal exposure and to any combination of these.[9] In instances in which several radionuclides are taken into the body and the resultant tissue doses in several organs are of comparable magnitude, the combined exposure must be considered to constitute whole-body exposure and the dose equivalents to each of the organs must be restricted to not more than 3 rem/13 weeks or an accumulated dose equivalent of $5(N - 18)$ rem. The limits also apply to any type or combination of types of ionizing radiation dose, that is, α, β, γ, neutron, etc., received by each organ of the body.

When operating in conformance with the recommendations of the NCRP and the ICRP, the basic standard for exposure to ionizing radiation is the dose rate or dose that may be received by the organs of the body. Sometimes for

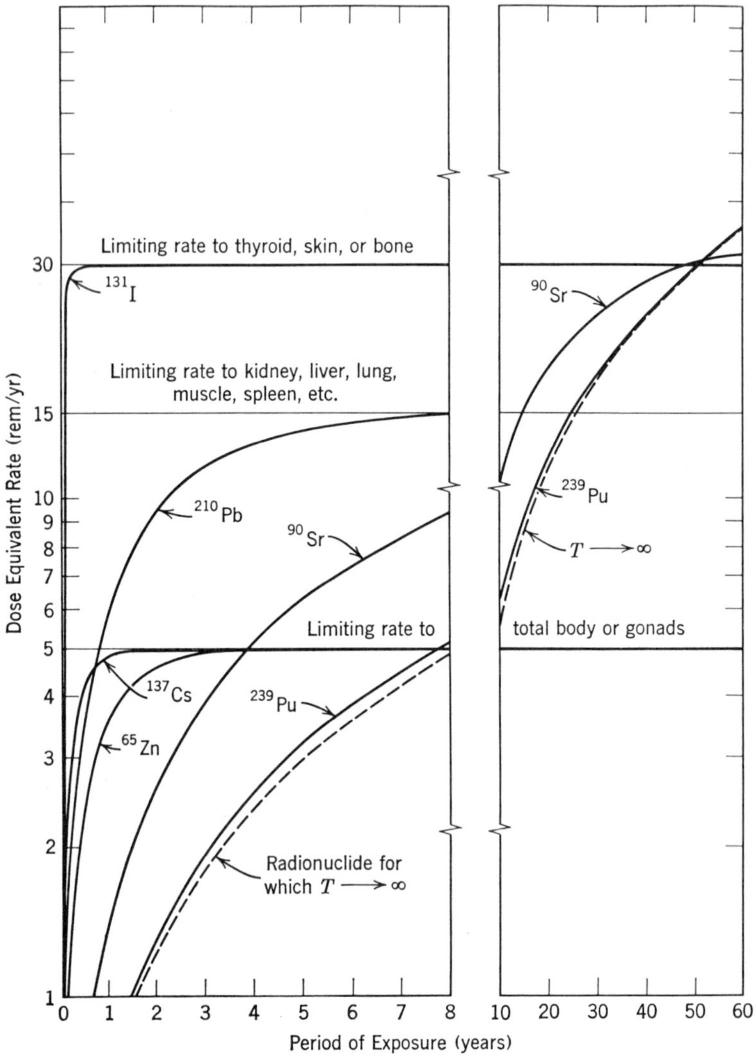

Fig. 14-2. Plot of dose equivalent versus years of exposure at the constant level of MPC.

practical reasons this limiting dose rate or dose is controlled by secondary standards such as the maximum permissible concentration (MPC) of the radionuclide in air, water, and food; by the maximum permissible levels of surface contamination; or by limits on the working time.

For internal dose, the pattern of maximum permissible continuous exposure is indicated by a few illustrative samples shown in Fig. 14-2. As explained in Chapter 10, the MPC values are calculated on the assumption that the annual permissible dose equivalent rates R_{50} shown in column 3 of Table 14-3 will be reached by the standard man only after an occupational exposure at the constant MPC value lasting for 50 years. Although an equilibrium condition (that is, constant dose rate) theoretically is never reached with continuous deposition of a radionuclide in a body organ, for practical purposes about 99%, or effective, equilibrium is reached after a period of body intake t corresponding to six effective half-lives (that is, if $t/T = 6$, $1 - e^{-0.693t/T} = 0.984$). Thus a radionuclide such as ^{131}I with a radioactive half-life $T_r = 8$ days and a biological half-life $T_b = 138$ days in the thyroid (effective half-life $T = 7.6$ days), will reach effective equilibrium in the thyroid at the constant dose equivalent rate $R_{50} = 30$ rem/yr after an occupational exposure at the MPC lasting for 6 or 7 weeks. For a radionuclide such as ^{137}Cs, which is rather uniformly distributed throughout the body, the value $T_r = 26.6$ years is long, but the value $T_b = 70$ days in the total body is relatively short, so that the resulting effective half-life T is 70 days. Thus effective equilibrium at the constant dose equivalent rate $R_{50} =$ 5 rem/yr is reached in the total body after a little more than a year. For ^{65}Zn, $T_r = 245$ days, $T_b = 933$ days for the total body, and $T = 194$ days, so that effective equilibrium at the constant dose equivalent rate $R_{50} =$ 5 rem/yr is reached in the total body in about 3 years. For most organs of the body, the limiting rate R_{50} is 15 rem/yr. The isotope ^{210}Pb, with the kidney as the critical body organ, is typical of this class. In this case, $T_r = 19.4$ years, $T_b = 531$ days in kidney, and $T = 494$ days are all of intermediate length, so that an exposure period of about 8 years is required for effective equilibrium at the constant dose equivalent rate $R_{50} = 15$ rem/yr to be reached. For bone seekers like ^{90}Sr and ^{239}Pu, the limiting dose rate is set by comparison with ^{226}Ra in such a way that the dose equivalent rate following occupational exposure at the MPC for 50 years $R_{50} = 30$ rem/yr.

As indicated in Fig. 14-2 and in Table 14-6, the bone burden or the corresponding dose equivalent rate from ^{90}Sr with $T_r = 27.7$ years, $T_b = 50$ years in bone, and $T = 17.5$ years just reaches the limiting value of $R_{50} = 30$ rem/yr and only $100\,(1 - e^{-0.693t/T}) = 86\%$ of equilibrium after 50 years of occupational exposure at the MPC. Effective equilibrium would not be reached except following an unlikely occupational exposure lasting 6×17.5 years or more than 100 years, at which time the constant dose equivalent rate would be $30/0.86 \approx 35$ rem/yr. Similarly, the bone burden or the corresponding dose equivalent rate for ^{239}Pu with $T_r = 24{,}360$ years, $T_b = 200$ years in bone, and $T = 197$ years reaches the limiting value $R_{50} = 30$ rem/yr, but only 16% of equilibrium after 50 years of occupational exposure at the MPC. Here effective equilibrium would be reached only after an occupational exposure lasting 6×197 years or about 1200 years, at which time the constant dose equivalent rate would be $30/0.16 \approx 190$ rem/yr.

Table 14-6 Values of Dose Equivalent Rate and Integrated Dose Equivalent Following Continuous Exposure at Occupational MPC Level

Exposure Period	6 Months				5 Years			
Radionuclide and Critical Body Organ	Rate (rem/yr)	% R_{50}	Dose (rem)	% MPAD	Rate (rem/yr)	% R_{50}	Dose (rem)	% MPAD
^{131}I: Thyroid	30	100	14	94	30	100	149	99
^{137}Cs: Total body	4.2	84	1.3	54	5	100	24	94
^{210}Pb: Kidney	3.4	23	0.88	12	14	92	48	64
^{65}Zn: Total body	2.4	48	0.66	27	5	100	21	85
^{90}Sr: Bone	0.68	2.3	0.17	1.1	6.2	21	16	11
^{239}Pu: Bone	0.33	1.1	0.082	0.54	3.2	11	8	5.3

Exposure Period	50 Years				70 Years			
Radionuclide and Critical Body Organ	Rate (rem/yr)	% R_{50}	Dose (rem)	% MPAD	Rate (rem/yr)	% R_{50}	Dose (rem)	% MPAD
^{131}I: Thyroid	30	100	1.499×10^3	99.9	30	100	2.099×10^3	99.96
^{137}Cs: Total body	5	100	249	99	5	100	349	99.6
^{210}Pb: Kidney	15	100	721	96	15	100	1.0×10^3	97
^{65}Zn: Total body	5	100	246	98	5	100	346	99
^{90}Sr: Bone	30	100	983	66	33	109	1.6×10^3	77
^{239}Pu: Bone	30	100	772	51	41	135	1.5×10^3	71

Note: The values of dose equivalent rate are given in rem/year at the end of the indicated exposure periods. The % R_{50} is the percent of the values of maximum permissible rate as given in Column 3 of Table 14-3. The values of accumulated dose equivalent are given in rem at the end of the indicated exposure periods. The % MPAD is the percent the accumulated dose equivalent is of the maximum permissible accumulated dose equivalent as given by the equation $5(N - 18)$ for the total body and by the product of the exposure time by the permissible average annual dose equivalent rate ($t\, R_{50}$) for the other organs assuming that the occupational exposures began after age 18.

Thus, as shown in Table 14-6, the patterns of dose rate and integrated dose as a function of period of exposure are quite different for the various radionuclides, although the dose equivalent rate is exactly 100% of the annual permissible value R_{50} in every instance at the end of the 50th year of exposures at the MPC.

It is to be emphasized that for radionuclides, such as ^{131}I, with short effective half-lives, 100% of R_{50} and 100% of maximum permissible accumulated dose, MPAD $= t\, R_{50}$, are reached effectively in a few months, whereas for the radionuclides such as ^{90}Sr and ^{239}Pu with long effective half-lives, 100% of R_{50} is reached only after an occupational exposure at the MPC for 50 years, and after 70 years such exposure has reached only 77% of MPAD for ^{90}Sr and 71% of MPAD for ^{239}Pu. The fact that the dose equivalent rate exceeds R_{50} when persons are occupationally exposed for more than 50 years at the MPC of a radionuclide with a long effective half-life T is no cause for alarm, however, since few persons, if any, will be so exposed. Even if they were, the accumulated dose equivalent does not reach 100% of MPAD until after a total occupational exposure lasting 183 years for ^{90}Sr and 103 years for ^{239}Pu and 100 years for a radionuclide for which $T \rightarrow \infty$. It can be shown that for a radionuclide of any value of T the dose equivalent rate $R(t)$ as a result of exposure period t is

$$R(t) = \frac{R_{50}(1 - e^{-0.693t/T})}{1 - e^{-0.693 \times 50/T}} \qquad (14\text{-}3)$$

$$R(t) \geq R_{50}\frac{t}{50} \quad \text{if} \quad t < 50 \qquad R(t) \leq R_{50}\frac{t}{50} \quad \text{if} \quad t > 50 \qquad (14\text{-}4)$$

$$R(t) = R_{50} \quad \text{if} \quad t = 50 \quad \text{and} \quad R(t) = R_{50}\frac{t}{50} \quad \text{if} \quad T \rightarrow \infty$$

The accumulated dose equivalent $D(t)$ is

$$D(t) = \int_0^t R(t)\, dt = \frac{R_{50}[t + (T/0.693)e^{-0.693t/T} - T/0.693]}{1 - e^{-0.693 \times 50/T}} \qquad (14\text{-}5)$$

or

$$D(t) \quad \geq \frac{R_{50}\, t^2}{100} \quad \text{but} \quad \leq R_{50}\, t \quad \text{if} \quad t < 50$$

$$D(t) \quad \text{may be} \geq \text{or} < \frac{R_{50}\, t^2}{100} \quad \text{if} \quad t > 50 \qquad (14\text{-}6)$$

$$D(t) \geq 25\, R_{50} \quad \text{but} \quad \leq 50\, R_{50} \quad \text{if} \quad t = 50$$

and

$$D(t) = \frac{R_{50}\, t^2}{100} \quad \text{if} \quad T \rightarrow \infty$$

Thus, when $t = 50$ years of occupational exposure at the MPC, the limit of accumulated dose is

$$D(t) = 25\ \mathrm{R}_{50} \quad \text{if} \quad T \to \infty \quad \text{and} \quad D(t) = 50\ \mathrm{R}_{50} \quad \text{if} \quad T \to 0 \tag{14-7}$$

Thus, during a 50-year exposure period, there is a built-in safety factor which approaches 2 for continuous exposure to the very long-lived radionuclides and which drops down to 1 for the short-lived radionuclides. In practice, however, this factor of safety of 2 is applicable in very few cases because most persons receiving internal dose from occupational exposure to the radionuclides of long effective half-life are exposed for only a few years while they are still relatively young employees. Figure 14-3 illustrates the case where the exposure period τ lasted for five years. The dashed curves indicate the dose equivalent rate as a result of exposure for the period τ = five years and the solid curves indicate the dose equivalent rate had there been continuous exposure for 50 or more years. The 50-year accumulated dose equivalents to the bone from the five-year intakes of ^{144}Ce (T = 243 days), ^{90}Sr, ^{239}Pu, and a hypothetical radionuclide for which $T \to \infty$ are 150 rem, 147.6 rem, 143.1 rem, and 142.5 rem, respectively, so that the safety factor has disappeared, even for a radionuclide for which $T \to \infty$.

As explained previously, the ICRP[8] has specified that in a quarter of a year up to one-half of the maximum permissible dose commitment may be accumulated. In the case of external exposure and/or internal exposure to the radionuclides having short effective half-lives T this corresponds to a quarterly exposure at twice the usual maximum permissible occupational rate R_{50} or to receiving a dose equivalent of $\mathrm{R}_{50}/2$ rem in 13 weeks. For internal exposure to the radionuclides having a long effective half-life T (or for internal exposure to radionuclides of any effective half-life T), this corresponds to the intake of the radionuclide at the MPC for half a year or an intake at twice the MPC for one-fourth year. Such exposures correspond to accumulated dose equivalents over a 50-year period that are numerically equal to $\mathrm{R}_{50}/2$. These exposures may be taken in any pattern during a given quarter interval, that is, the dose may be received in a microsecond, in a minute, in an hour, from continuous exposure, etc. Although it is recognized that there is a rate effect which influences the radiation damage or which should be considered in assigning the quality factors at high exposure rates, the ICRP considers that for lack of detailed information and for practical reasons it is not necessary to complicate the maximum permissible exposure pattern at these relatively low occupational levels by placing restrictions on the dose equivalent rate. Also, for practical purposes it was specified that the calendar quarter could be used as the quarter interval, provided an effort was made to avoid a high single dose equivalent at the beginning of one quarter immediately following a high single

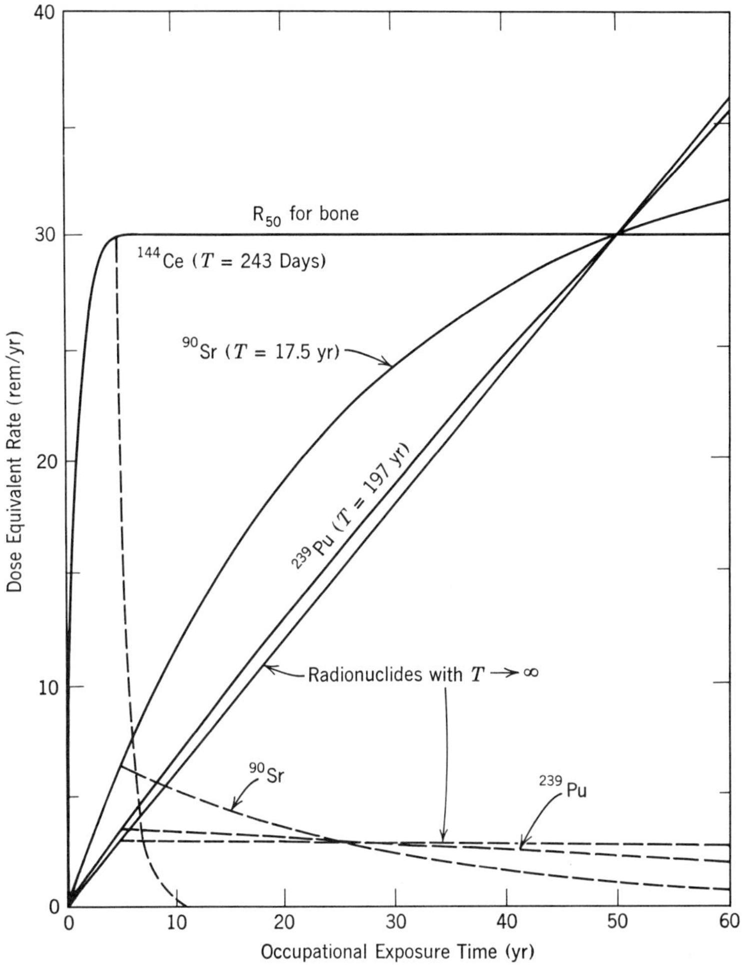

Fig. 14-3. Dose equivalent rate versus occupational exposure time.

dose equivalent at the end of the previous quarter, that is, two quarter doses over a period of a few days would not be permitted.

Figure 14-4 shows the maximum permissible single or quarterly exposure pattern for ^{137}Cs. In this case, a single intake has resulted in a body burden of ^{137}Cs which delivers a dose equivalent during the following t years. This dose from $t = 0$ to 50 years may be represented by the area under the dose-rate curve A. The ICRP recommendations specify that this maximum permissible single or quarterly intake of a radionuclide is to correspond to an exposure for half a year at the occupational MPC or for a quarter of a year at twice the occupational MPC. Such an exposure would deliver a dose equivalent that is represented by the area under curve B from $t = 0$ to 50 years. (Actually, for an exposure period $\tau = 0.5$ year, the area under curve B should be integrated from $t = 0$ to $t = 50 + \tau/2 = 50 + 0.5/2 = 50.25$ years.) It can be shown

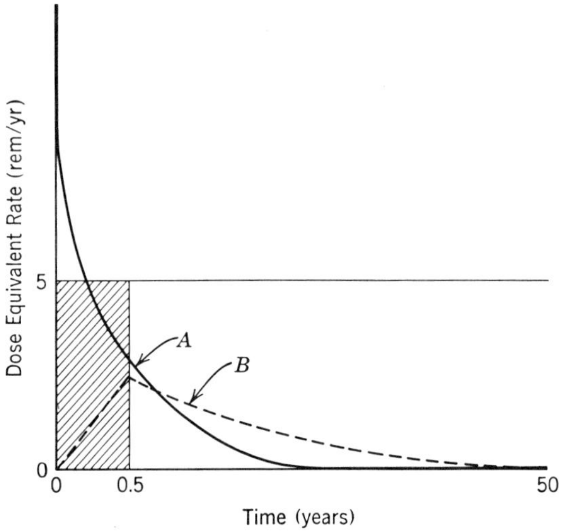

Fig. 14-4. Curve *A*: Dose equivalent for total body and gonads from single intake of ^{137}Cs at zero time. Curve *B*: dose equivalent from intake of ^{137}Cs for $\frac{1}{2}$ year at the MPC. If the area under curve *A* is equal to area under curve *B*, curve *A* corresponds to a maximum permissible quarterly or single intake.

that for such a maximum permissible single or quarterly intake of any radionuclide the area under curve A equals area under curve B equals area under the shaded rectangle drawn in Figure 14-4, and this area, in turn, is equal numerically to τR_{50}. Thus, as indicated in Table 14-7, the maximum permissible dose equivalent from a single or quarterly exposure is $R_{50}/2$, that is, $\frac{5}{2} \approx 3$ rem to total body, gonads and red marrow, $\frac{30}{2} = 15$ rem to thyroid, skin or bone, and $\frac{15}{2} \approx 8$ rem to the other internal organs.

14-8 PLANNED SPECIAL EXPOSURES AND EMERGENCY EXPOSURES

There may be infrequent occasions when planned emergency operations are necessary for protection of personnel or the saving of valuable property. Such occasions make it essential that exposures in excess of the maximum permissible occupational exposures be permitted. The ICRP[8] has recognized the need for such levels of infrequent occupational exposure. This planned special exposure (as recommended by the ICRP[9]) follows the same pattern

Table 14-7 Occupational Exposure Levels[a] Recommended by ICRP

Categories of Exposure	Maximum Dose Equivalent for External Exposure and/or Internal Exposure to Radionuclides With a Short Effective Half-Life (rem in 13 wk)	(rem in one yr)	Dose Commitments Dose Equivalent Integrated Over a 50-Year[d] Period Resulting From a Permissible Quarterly Dose Equivalent or From a Planned Special Exposure (rem in 50 yr[d])	Maximum Permissible Intake of any Radionuclide for Single, Quarterly, or Planned Exposure Corresponds to an Intake: for 13 wk at	at the MPC for
A—Maximum permissible dose equivalent for routine applications	$R_{50}/4$	R_{50}	$R_{50}/4$	1 × MPC	$\frac{1}{4}$ yr
B—Maximum permissible dose equivalent for single exposures or for exposures on a quarterly basis	$R_{50}/2$[b]	2 R_{50}[b] to total body, gonads, or red bone marrow if $5(N-18)$ is not exceeded. R_{50}[b] to all other organs	$R_{50}/2$	2 × MPC	$\frac{1}{2}$ yr
C—Planned special exposures[c]	2 R_{50} committed in any single event to any body organ		2 R_{50}	8 × MPC	2 yr
D—Summation of all planned special exposures in a lifetime[c]	5 R_{50} committed in a lifetime to any body organ from planned special exposures		5 R_{50}	20 × MPC	5 yr

[a] Exposure levels are given in terms of the annual permissible dose equivalent R_{50}. Values of R_{50} for individual body organs in various radiosensitivity organ groups are as follows:

Group (1) for gonads, total body, and red bone marrow $R_{50} = 5$, $R_{50}/2 \rightarrow 3$, $2\,R_{50} \rightarrow 12$ and $5\,R_{50} = 25$ rem.

Group (2) for thyroid, skin, and bone $R_{50} = 30$, $R_{50}/2 = 15$, $2\,R_{50} = 60$ and $5\,R_{50} = 150$ rem.

Group (3) for hands, forearms, feet, and ankles $R_{50} = 75$, $R_{50}/2 \rightarrow 38$, $2\,R_{50} = 150$ and $5\,R_{50} = 375$ rem.

Group (4) for all other body organs (including the lenses of the eyes) $R_{50} = 15$, $R_{50}/2 \rightarrow 8$, $2\,R_{50} = 30$ and $5\,R_{50} = 75$ rem.

[b] The ICRP states that it would be undesirable to repeat this quarterly dose equivalent of $R_{50}/2$ at close intervals but to provide flexibility, it is permitted on infrequent occasions to receive 2 R_{50} in a year to the gonads, total body, and red bone marrow, provided the accumulated dose equivalent of $5(N - 18)$ rem is not exceeded. Ordinarily, the annual dose equivalent R_{50} should not be exceeded for exposure to any of the body organs.

[c] Planned special exposures are not permitted if a single exposure in excess of $R_{50}/2$ has been received in the previous 12 months or if at any time the worker has received an abnormal exposure in excess of 5 R_{50}. Planned special exposures are not permitted to women of reproductive capacity. They are not permitted to gonads, total body, or red bone marrow if as a consequence $5(N - 18)$ rem is exceeded.

[d] Actually, over a period of $50 + \tau/2$ years for external and/or internal exposures to radionuclides of any effective half-life, τ is period of intake of a radionuclide.

as that for single or quarterly exposure described above except in this case emergency work involving exposure above permissible limits to radioactive materials shall be planned on the basis that the total dose equivalent does not exceed twice the maximum permissible dose commitment. For external exposure and/or internal exposure to the radionuclides having short effective half-lives T this corresponds to a quarterly exposure at eight times the usual maximum permissible occupational rate R_{50} or to receiving a dose equivalent of 2 R_{50} rem in 13 weeks. For internal exposure to radionuclides having a long effective half-life T (or for internal exposure to radionuclides of any effective half-life T), this corresponds to the intake of the radionuclide at the MPC for two years or intake at eight times the MPC for one-fourth year. Such exposure corresponds to an accumulated dose equivalent over a 50-year period (actually $50 + \tau/2 = 51$ years) that numerically is equal to 2 R_{50}. In this case (compare Fig. 14-4), the area under the single exposure curve A may equal but must not exceed the area under the curve B for intake at the MPC over a period of $\tau =$ two years and integrated over a 51-year period. These areas, in turn, must not exceed the area under the shaded rectangle of height R_{50} and width $\tau =$ two years (instead of width $\tau = 0.5$ year as was the case for single or quarterly exposure illustrated specifically in Fig. 14-4). Thus, as indicated in Table 14-7, the maximum permissible planned special exposure corresponds to a dose equivalent over a 51-year period of $\tau\ R_{50} = 2\ R_{50}$, that is, 12 rem to total body, gonads, red marrow, 60 rem to thyroid, skin, or bone, and 30 rem to other internal organs.

The ICRP[8] permits planned special exposures as long as the total of such doses received during the lifetime of an occupational worker is no more than five times the maximum permissible dose commitment. For external exposure and/or internal exposure to radionuclides having a short effective half-life T this corresponds to a quarterly exposure at 20 times the usual maximum permissible occupational rate R_{50} or to receiving a dose equivalent of 5 R_{50} rem in 13 weeks. For internal exposure to the radionuclides having a long effective half-life T (or for internal exposure to radionuclides of any effective half-life T), this corresponds to intake of the radionuclide at the MPC for five years or intake at 20 times the MPC for one-fourth year. Such exposures correspond to accumulated dose equivalents over a 50-year period (actually 52.5 years) that are numerically equal to 5 R_{50}. Thus the maximum dose accumulated from planned special exposures in one's lifetime corresponds to a dose equivalent over a 52.5-year period of 25 rem to total body, gonads, red bone marrow, etc.; 150 rem to thyroid, skin, or bone; 75 rem to other internal organs, and 375 rem to hands, forearms, feet, and ankles. This 25 rem limit to total body is recommended also by the NCRP, but as a once-in-a-lifetime exposure. The NCRP further limits application of this dose equivalent of 25 rem once-in-a-lifetime to external exposure to X- and γ-radiation of energy less than 3 MeV. The NCRP states that although two exposures to the total body of 12.5 rem separated by a short or long interval might be treated as a single "accidental dose of 25 rem once in a lifetime," for practical purposes of bookkeeping and to protect the worker that may move from one place of employment to another, the "once in a lifetime" refers only to "one episode in a lifetime." The episode may be any total-body dose equivalent greater than 12 and up to 25 rem and may have a duration of as much as one month. The NCRP, in addition, specifies a local exposure once in a lifetime resulting in a total dose equivalent of 125 rem to the hands, forearms, feet, and ankles. This may be compared with the ICRP planned special single exposure of 2 $R_{50} = 150$ rem or a total of 5 $R_{50} = 375$ rem in a lifetime. These recommendations provide some leeway to the most difficult problem of accurately assessing the surface dose to the hands that frequently are contaminated with radioactive materials and thus are in direct contact with the source.

The "25 rem episode once in a lifetime" ostensibly offers considerable flexibility to the employee in occupational work with sources of ionizing radiation. In practice, however, since this is permitted only once in a lifetime, the employee who makes use of the provision no longer has the flexibility of other employees and is never again permitted such an episode. Hereafter, his work must be so adjusted or, if necessary, so modified that he has an unusually low probability of a recurring high exposure. The NCRP states "accordingly, it is generally desirable to institute compensatory measures in such cases."

The ICRP specifies[8] that planned special exposures should not be permitted if the addition of the intended dose equivalent to the worker's accumulated dose exceeds the amount determined by the formula for total-body, gonad, and red marrow exposure and, presumably, also for the other organs if the accumulated dose exceeds $t\ R_{50}$ in which t = years of radiation exposure work; or if the worker has received a single exposure in excess of the quarterly limit in the previous 12 months; or if the worker has previously received abnormal exposures in excess of five times the average permissible annual dose commitment. The ICRP specifies that special planned exposures should not be permitted to women of reproductive capacity and that all such doses should be recorded and examined in respect to the above listed restrictions. However, any excess over the recommended normal limits should not necessarily serve as a reason to exclude a worker from his usual employment. Finally, the ICRP "emphasizes" that doses or intakes corresponding to dose commitments in excess of $R_{50}/4$ from occupational exposures during a 13-week period are only justified when alternative techniques, which do not involve such exposure of workers, are either unavailable or impractical.

It should be noted that Fig. 14-4 can be used to illustrate the maximum permissible single or quarterly exposure, the maximum permissible planned special exposure, and the total of lifetime planned exposures. For these three cases, the base of the shaded rectangle τ in Fig. 14-4 should be 0.5 year, two years, and five years, respectively, over which the occupational worker is exposed at the MPC. Also, in these situations the resultant dose commitments in $50 + \tau/2$ years are $R_{50}/2$, $2\ R_{50}$, and $5\ R_{50}$, respectively. In every case, the permissible exposure corresponds to the area under curve *A* equals area under curve *B* equals area $\tau\ R_{50}$ of the shaded rectangle.

It might be unfortunate to discontinue the discussion of planned special exposures and emergency exposures without emphasizing that, during certain emergencies, high exposures must be taken on an individual basis and these must be taken while quickly weighing the benefits against the hazards. Although the ICRP[8] makes a distinction between "planned special exposures" and "accidental" exposures, those who have had experience with radiation emergencies or accidents know that there is seldom, if ever, a sharp distinction between high exposure that is planned and exposure that is accidental. This is true because most planned high exposures are received during emergency operations associated with an accident—the cleaning up of a spill, the repair of a radiation facility, the rescue of an injured person following a reactor incident, etc. The differences are a matter of degree.

For example, one plans to receive no more than 3 rem during a cleanup or rescue operation but, in fact, may receive several times this planned amount. If one is faced with the probability of receiving a high exposure in a rescue operation or in performing a function essential to national defense, he must take certain calculated risks and the greater the urgency and speed with which decisions must be made, the greater the uncertainties in the dose estimates. Although no official permissible levels are specified for this, it is suggested that a person should not hesitate to receive 100 rem of total-body dose equivalent in the saving of a life. Beyond this level the persons on the rescue operation should judge the risks and possible benefits of the mission very carefully by keeping in mind the probabilities of large errors in estimating the potential exposure. For example, a person may decide to take 300 rem in order to drag from a highly contaminated area a casualty that has received 500 rem, but the decision could be unfortunate if both of these dose estimates were low by a factor of 2. On the other hand, a dose equivalent of even 100 ± 50 rem probably would not be warranted to "rescue" a person who had received a lethal dose equivalent of 2000 ± 1000 rem from total-body exposure.

In internal exposures during emergency operations, it should be remembered that an exposure for *H* hours to an MPC (based on occupational exposure 40 hr/wk) corresponds to an integrated dose equivalent to the critical body organ during the following 50 years of $HR_{50}/(40 \times 52)$ rem. Thus a three-hour exposure to 347 times the MPC (for 40 hr/wk) would correspond to a maximum permissible single dose equivalent of $R_{50}/2$ rem which the person would receive to the critical body organ during the following 50 years. Also, the particular radioisotope, its chemical form, and its initial deposition in the pulmonary system and GI tract must be taken into account in estimating emergency exposures. For example, in an extreme emergency an exposure 100 times the one mentioned above (that is, three hours at 3.5×10^4 times the MPC) might be acceptable for ^{90}Sr but not for ^{24}Na. Although in these cases the 50-year dose equivalents would be 1500 rem or 300 rad to bone and 250 rem to total body, respectively, the dose equivalents for the first three days would be about $\frac{1}{2}$ rem to bone and 250 rem to total body, respectively. This 250 rem to the total body in three days (most of which would be received the first day), plus a probable much larger dose equivalent to the GI tract and/ or lungs, could be lethal, whereas an exposure of $\frac{1}{2}$ rem in three days, 50 rem the first year, and a total of 1500 rem over 50 years might cause leukemia or bone cancer over the next 5 to 30 years, but would not cause early death. Added caution should be exercised when contemplating high internal emergency exposures, not only because of the large probable errors in estimating such resulting dose equivalents to the critical body organs for chronic exposure but also because of the risk of extremely high early dose equivalents to the upper respiratory tract and to the GI tract.

The ICRP states that all doses received under abnormal circumstances should be recorded together and clearly distinguished from normal exposures. In such cases, as much nonroutine dose data and related information (for example, spectral distribution, angular distribution, quality factor, peak doses in body, factors responsible for the exposure, etc.) as are available should be recorded and carefully evaluated. The ICRP states further that if the dose or the intake of radioactive material exceeds twice the annual limit (that is, exceeds the planned special exposure), the situation should be investigated by a competent medical authority and, of course, by a qualified health physicist. The worker may still be allowed to continue his routine work if there is no objection from the medical standpoint, due account being given to his previous health, age, and special skills, as well as his social and economic responsibilities.

The ICRP recommends that all persons doing work in certain areas or under such conditions that they "might exceed" an annual dose equivalent of 0.3 R_{50} should be classed as radiation workers in the sense that they are subjected to personnel monitoring and to special health physics surveillance. It is emphasized that this does not mean a person is classed as a radiation worker after he has or is almost certain to receive an annual exposure $\geq 0.3\ R_{50}$, but rather where there is a reasonable chance that he will receive such an exposure. This means, of course, that in a well-conducted operation most radiation workers who wear personnel monitoring meters will receive less than 0.3 R_{50}/yr or even less than 0.1 R_{50}/yr but a few will receive x R_{50}/yr in which $x \geq 0.3$ but ≤ 1. Employees not classed as radiation workers should not receive in excess of 0.3 R_{50} per year to any body organ as a result of their work, barring the exceptional serious radiation accident. If an accident occurs, it may happen, too, that some members of the population-at-large will exceed exposures of 0.1 R_{50} per year, but, again, such an accident must be a most rare event. The goal is to avoid all unnecessary radiation exposure, to receive exposure even within the limits discussed above when and only when the expected benefits exceed the likely harmful consequences of radiation exposure. In every instance one must learn to respect rather than fear ionizing radiation, to use it as a dangerous and most powerful tool for the good of mankind.

REFERENCES

1. *Radiation Protection Guidance for Federal Agencies*, Federal Radiation Council Rept. No. 1, Washington, 1960.

2. *Selected Radiation Protection Guides*, Federal Radiation Council Rept. No. 2, Washington, 1961.

3. Federal Radiation Council, *Federal Register*, p. 6953 (May 22, 1965).

4. "Report on 1956 Amendments to the Recommendations of the International Commission on Radiological Protection (ICRP)," *Radiology* **70,** 261 (1958).

5. "Recommendations of the International Commission on Radiological Protection," *ICRP Publ. 1*, Pergamon Press, London, 1959.

6. "Report on Decisions at the 1959 Meeting of the International Commission on Radiological Protection (ICRP)," Addendum to *ICRP Publ. 15*, *Radiology* **74,** 116 (1960).

7. "Recommendations of the International Commission on Radiological Protection," *ICRP Publ. 6*, Pergamon Press, London, 1964.

8. "Recommendations of the International Commission on Radiological Protection," *ICRP Publ. 9*, Pergamon Press, London, 1966.

9. "Recommendations of the International Commission on Radiological Protection, Report of Committee II on Permissible Dose for Internal Radiation," *ICRP Publ. 2*, Pergamon Press, London, 1959, and *Health Phys.* **3,** 1 (1960).

10. "Report of Committee III on Protection Against X-Rays Up to Energies of 3 MeV and Beta- and Gamma-Rays from Sealed Sources," *ICRP Publ. 3*, Pergamon Press, London, 1960.

11. "Principles of Environmental Monitoring Related to the Handling of Radioactive Materials," *ICRP Publ. 7*, Pergamon Press, London, 1965.

12. Morgan, K. Z., *Science* **139,** 565 (1963).

13. "Safe Handling of Radioactive Luminous Compounds," *Natl. Bur. Std.* (*U.S.*) *Handbook* **H-27** (1941).

14. "Radium Protection for Amounts up to 300 Milligrams, *Natl. Bur. Std.* (*U.S.*) *Handbook* **18,** [published for Advisory Committee on X-Ray and Radium (now NCRP)] (1934).

15. "Safe Handling of Radioactive Isotopes," *Natl. Bur. Std. (U.S.) Handbook* **42** (1949).

16. "Recommendations of the International Commission on Radiological Protection and of the International Commission on Radiological Units—1950," *Natl. Bur. Std. (U.S.) Handbook* **47** (1951).

17. "Maximum Permissible Amounts of Radioisotopes in the Human Body and Maximum Permissible Concentrations in Air and Water," *Natl. Bur. Std. (U.S.) Handbook* **52** (1953).

18. "Permissible Dose from External Sources of Ionizing Radiation," *Natl. Bur. Std. (U.S.) Handbook* **59** (1954).

19. "Maximum Permissible Radiation Exposures to Man," *Suppl. Natl. Bur. Std. (U.S.) Handbook* **59** (1957).

20. "Maximum Permissible Radiation Exposures to Man," *Suppl. Natl. Bur. Std. (U.S.) Handbook* **59** (1958).

21. "Maximum Permissible Body Burden and Maximum Permissible Concentrations of Radionuclides in Air and in Water for Occuptional Exposure," *Natl. Bur. Std. (U.S.) Handbook* **69** (1959).

22. "Somatic Radiation Dose for the General Population: Ad Hoc Committee Report of the National Committee on Radiation Protection and Measurements," May 6. 1959, *Science* **131,** 482 (1960).

23. "National Committee on Radiation Protection and Measurements," *Radiology* **75,** 122 (1960).

24. Stone, R. S., *Radiology* **58,** 639 (1952).

25. Taylor, L. S., *Health Phys.* **1,** 3 (1958).

26. Taylor, L. S., *Health Phys.* **1,** 306 (1958).

27. Taylor, L. S., *Health Phys.* **1,** 97 (1958).

28. *Principles of Environmental Monitoring Related to the Handling of Radioactive Materials*, Committee 4 Report of the International Commission on Radiological Protection, Publication 7, Pergamon Press, London, 1965.

PROBLEMS

1. Give a brief history of the organization of the ICRP and the NCRP.

2. Give the principal reasons for the reductions in the ICRP and NCRP permissible levels of exposure to ionizing radiation.

3. Define "critical body organ" as used by the ICRP for internal dose.

4. Name three types of damage from ionizing radiation that are commonly considered to vary monotonically with the dose equivalent received by the total body.

5. Explain why, in general, the quality factor (QF) of neutron radiation increases as the dose rate decreases.

6. Derive the equation for the dose equivalent D (rem) delivered in time t (days) to the critical body organ of mass m (grams) resulting from I_0 (μc) of an isotope that delivers to the organ $\sum E(\mathrm{QF})n$ MeV per disintegration when taken into the body at time $t = 0$. Assume that fraction f of the isotope arrives in the organ of mass m (grams) and the effective half-life in the critical organ is T (days).

7. Assuming biological elimination of a radionuclide from an organ to be exponential, find the biological half-life T_b if the radioactive half-life T_r is 50 years and the effective half-life T is 40 years. Show why T can never be greater than T_r or T_b.

8. Discuss the maximum permissible dose equivalent for occupational exposure in the so-called radiation industries with reference to (*a*) probability of severe somatic and genetic injury, (*b*) probability of some biological effect on the exposed person, and (*c*) damage in other common industries.

9. Name two types of radiation damage that can be expressed as having a threshold dose below which the radiation damage is not evident.

10. Why should the maximum permissible dose levels for the population-at-large be set at values which are less than those for the occupational worker? Give at least six reasons.

11. Why is it important that maximum permissible dose levels be set neither too low nor too high?

12. Discuss why there is a need for maximum permissible exposure levels as well as perhaps a need for exposure guides.

13. Discuss some of the problems involved in balancing the risks of exposure to ionizing radiation against the benefits in its use.

14. Values of maximum permissible flux as given by the ICRP are usually calculated with reference to what location in the body?

15. Which basic radiation standard or exposure level that was set in 1941 is still in use today (1967)?

16. Why should, in general, larger exposures be permitted to individual organs of the body than to the total body? What exception does the ICRP make to this rule?

17. At the present time (1967), what is the annual permissible dose equivalent of the ICRP for the occupational worker for liver, kidney, and the lenses of the eyes?

18. Answer problem 17 with reference to the ICRP annual permissible absorbed dose for the occupational worker for fast neutrons for which QF = 10.

• **19.** An atomic energy worker (age 35) has received 83 rem of occupational dose equivalent to the total body. How much additional exposure may he receive in a single day and in a 13-week period until he is age 36? Use the ICRP recommendations.

20. To what does "blood-forming organs" as used by the ICRP refer?

21. Do the annual permissible occupational dose equivalent values of the ICRP include medical exposure or background dose?

22. Would the present (1967) ICRP recommendations for routine occupational exposure permit an employee to receive, during 13 weeks, dose equivalents of 8 rem to the lenses of the eyes, 15 rem to the skin, and 8 rem to the kidney? Indicate whether these dose equivalents would be permitted individually to the indicated organs and/or simultaneously. If not, indicate what the limiting exposures would be in the various cases.

23. What are the four principal ICRP regulations relating to permissible dose that limit exposure of the occupational worker to ionizing radiation?

24. Is 12 rem/yr to the gonads of an occupational worker permitted? Explain.

25. In what respects may the present dose limits set by the ICRP for the occupational worker seem to be too high?

26. Define an annual "dose commitment" as the ICRP relates it to exposure to the radionuclides with long effective half-life.

27. Is it permissible to apply the dose commitment concept as used by the ICRP to radionuclides with a short effective half-life?

28. Why would it not be acceptable to have an employee ingest such an amount of a radionuclide having a long effective half-life in bone that the dose equivalent to the bone during the first year of exposure is 30 rem?

29. Show that the integrated dose equivalent delivered to the critical body organ in 50 years as a result of exposure at the MPC for one year is approximately equal numerically to the appropriate annual permissible dose equivalent R_{50} as listed in column 3 of Table 14-3.

30. Repeat derivation from problem 29 but integrate the dose equivalent over a period $50 + \tau/2$ years in which τ is the period of intake. Show which integration more nearly corresponds numerically to the annual permissible dose equivalent.

31. Give values recommended by the ICRP for the relative hazard factor n for the following situations: (*a*) α-radiation from ^{241}Am in bone, (*b*) γ-radiation in bone, (*c*) α-radiation from ^{226}Ra in bone, (*d*) fission recoil radiation in liver, (*e*) β-radiation in muscle.

32. What is the magnitude of the lens modifying factor for fast neutron irradiation? What is the operational value of the quality factor QF in this case? What is the over-all modifying factor?

33. When choosing MPC values acceptable to the ICRP for the disposal of radioactive material into the environment, is the permissible value simply $\frac{1}{10}$ of the MPC level for occupational workers? Explain to what permissible occupational value the reduction factor(s) should be applied and to what group it (they) should be applied.

34. Give the three principal ICRP rules that are to be applied in reference to reducing the permissible external dose equivalent rate for employees who have accumulated radioactive material in their bodies.

35. Using the earlier ICRP 1959 recommendation, what is the acceptable size of the population comprising "special groups" (that is, group B) in a country with a total population of 10 million?

36. What are the primary standards of the ICRP and the NCRP for controlling radiation exposure? Are the MPC values primary standards?

37. How long would a person have to be exposed at the MPC to reach 99% of equilibrium in the skeleton for a radionuclide having a biological half-life T_b in the skeleton of 10^4 years and a radioactive half-life T_r of 2×10^4 years? Assume the person is a standard man.

38. A person is occupationally exposed at the MPC of a radionuclide (that is, exposed continually 40 hr/wk at the MPC applying to 40 hr/wk exposure). When will the dose rate in the critical body organ reach the limiting annual dose rate listed in column 3 of Table 14-3? When will the dose rate in the critical body organ reach $\frac{1}{2}$ R_{50} if the effective half-life is $T = 100$ years? Assume the person is a standard man.

39. How long would a person have to be exposed to ^{131}I at the MPC to reach 50% equilibrium in the thyroid? What would his dose equivalent rate be in the thyroid at 50% of equilibrium? Assume the person is a standard man and that the effective half-life $T = 7.6$ days in this case. Repeat this problem for ^{239}Pu for which $T = 197$ years.

40. What is the dose equivalent rate to bone for occupational exposure at the MPC for 25 years? Assume here that bone is the critical body organ and that the effective half-life T approaches an infinite value. Assume the person is a standard man. Repeat this problem for $T = 10$ years.

41. What are the accumulated dose equivalents in the 25-year period for the two parts of problem 40?

42. A person is occupationally exposed for a period $\tau = 10$ years at the MPC of a radionuclide having an effective half-life in the kidney of 100 years, and then has no further occupational exposure (intake) of the radionuclide. What is the integrated dose to the kidney 60 years after the intake of the radionuclide began? Assume the standard man.

43. Give the approximate dose equivalent integrated from $t = 0$ to 50.25 years for the permissible single or quarterly exposure of the occupational worker to a radionuclide for which bone is the critical body organ. Give the value of this integrated dose equivalent when one of the following is the critical body organ: kidneys, muscle, lungs, gonads, thyroids, or lenses of the eyes.

44. A person is continually exposed at the occupational MPC (that is, 40 hr/wk at the MPC applying to 40 hr/wk exposure) of a radionuclide with an effective half-life $T = 20$ years. When will the accumulated dose equal the MPAD?

45. In problem 44 find when the accumulated dose from this radionuclide is equal to the dose that would have been accumulated from a radionuclide with an effective half life $T \to \infty$.

46. Prove that the integrated dose $D(t) = R_{50}\, t^2/100$ for a radionuclide for which $T \to \infty$.

47. What is the effective half-life T of a radionuclide which reaches effective equilibrium (for example, $t/T = 6$) when the dose rate $R(t) = 2\,R_{50}$? When is this effective equilibrium reached?

48. Under what conditions and to whom are planned special exposures permitted? Explain.

49. Derive an equation for the energy E_r of a recoil atom in terms of the energy E_α of an α-particle emitted from an atom of atomic mass number A.

50. What approximate dose equivalent would be received in the 13-week period following a maximum planned special exposure to a radionuclide with an effective half-life of one day if the thyroid were the critical organ? What would be the approximate dose equivalent over the following 50 years as a result of this single planned special exposure?

51. What approximate dose equivalent would be received in the 13-week period following a maximum planned special exposure to a radionuclide with an effective half-life of 40 years if bone were the critical organ? What would be the approximate dose equivalent over the following 50 years as a result of this exposure?

52. How does one determine what dose should be permitted to rescue personnel entering the area of an accident where the dose rate is very high? Explain.

53. If the MPC for a radionuclide is 10^{-7} $\mu c/cm^3$ of air for occupational exposure (40 hr/wk), what air concentration of the radionuclide would be the maximum permitted in one hour of ordinary radiation exposure if there will be no further exposures for 13 weeks? What would be the permissible concentration of this radionuclide for one hour for a planned special exposure?

54. Repeat problem 53 if liver is the critical body organ for the radionuclide and if the dose equivalent rate from an external high-energy γ-emitting source is 3 rem/hr.

55. In what ways must the particular chemical compound of a radionuclide be taken into consideration in determining the permissible air concentration over a given exposure period? Consider both the radiation and toxicity problem.

56. At what level of exposure to ionizing radiation does the ICRP recommend that occupational workers be required to wear personnel monitoring meters? Discuss.

CHAPTER 15

Health Physics Instruments

E. D. Gupton and D. M. Davis

15-1 CHARACTERISTICS OF INSTRUMENTS IN USE

Human senses do not respond to ionizing radiation, so that some form of instrumentation must be used for detection and measurement. Qualitative as well as quantitative measurements are needed, because the hazard of radiation to man depends on the type of radiation and its energy spectrum, as well as the quantity to which he is exposed. This chapter discusses health physics instruments from the standpoint of personnel dosimetry. The operation of many of the instruments described depends upon the physical principles discussed in Chapters 4 to 7.

Standard types of instrumentation are now available for the assessment of radiation hazards associated with naturally occurring and man-made radionuclides, and with radiation from X-ray and accelerator equipment, except for very high-energy accelerators ($\geqslant$1 GeV). Dosimetric techniques for monitoring high-energy accelerators are still in a developmental stage. Typical types of radiation and the energy ranges in which they occur are shown in Table 15-1.

The development of health physics instruments is complicated by two factors. First, the interaction of radiation with various materials which may be used as detecting devices is usually not the same as the interaction of the radiation with human tissue. This means that a detector may respond in a different manner than tissue to the same radiation. Second, the damage resulting from a given absorbed dose in human tissue may be quite different for different tissues. It is necessary, then, to measure radiation with an instrument and predict its relationship with a human being.

Health physics instruments are used to monitor radiation dose and dose rates, air-borne and surface contamination, environmental radioactivity, waste

Table 15-1 Ionizing Radiation and Common Energy Ranges

Radiation	Energy Range
Alpha particles (α)	Up to ~7 MeV
Beta particles (β)	Up to ~5 MeV
Gamma rays (γ), nuclear decay	Up to ~3 MeV
Gamma rays (γ), fission	Up to ~10 MeV
Neutrons:	
Thermal (N_{th})	~0.025 eV
Intermediate	>Thermal to ~0.1 MeV
Fast (N_f)	>0.1 MeV
Protons	Function of accelerator type and energy range
Electrons	Function of accelerator type and energy range

effluent, and in other applications. An *ideal* instrument for practical use would have a number of the following characteristics.

1. Measure dose, or dose rate, in units directly applicable to the tissue of concern.
2. Respond to only one kind of radiation at a time.
3. Have the desired sensitivity and accuracy, independent of the energy of the radiation.
4. Be free of calibration and zero drift within reasonable times.
5. Be free of extraneous effects from gravitational forces, temperature, humidity, chemical vapors, electrical and magnetic fields, shock, and so on.
6. Have a short response or readout time.
7. Indicate the occurrence of full- or off-scale readings if its range is exceeded.
8. Have a means of field-checking its operability.
9. Have a means of indicating when it is inoperable.
10. Be small, light, and rugged, particularly if it is a portable instrument.
11. Be relatively trouble-free and require only infrequent, simple repair and maintenance.

Many of these features are available, although some only at considerable cost. A few (such as items 1, 2, and 3) are not available, at this time, at any cost.

Since no single instrument at present has all the properties named, different types of instruments are used in various applications to obtain the desired ranges, sensitivities, and monitoring characteristics for different forms of radiation hazards. The types of instruments most commonly used include the following.

1. Dose and dose-rate monitors for photons or neutrons in fixed locations in work area. These are usually equipped with alarm signals and/or relays,

Fig. 15-1. Monitron, γ-radiation monitor.

and are operated by power directly from the a-c line. An instrument of this type is shown in Fig. 15-1.

2. Air monitors for detecting air-borne (α and β) radionuclides. These monitors are usually equipped with recorders and alarm signals, with power also from the a-c line. They may be used to monitor work and environmental areas and regions through which gaseous effluents pass (Fig. 15-2).

Fig. 15-2. Continuous air monitor for beta particulates.

3. Water monitors for estimating the specific activity of radionuclides in liquid wastes and environmental water sources. These are usually equipped with recorders and alarm signals and are a-c line powered (Fig. 15-3).

4. Personnel contamination monitors for detecting radioactive materials on the clothing or body surfaces of persons. These are usually located at the exits of areas in which such contamination is likely; they are usually equipped with alarm signals and are a-c line powered (Fig. 15-4).

Fig. 15-3. Geiger tube probe for water monitor.

5. Portable, battery-powered instruments for dose rate measurements, surface contamination analysis, and special analyses of radiation hazards. Most of these provide information by means of an ammeter with the readout units designated thereon (Fig. 15-5).

6. Personnel monitoring instruments are carried by the person being monitored. These may be either dose-rate warning devices or devices that are read for integrated dose. The latter type often requires auxiliary equipment and specialized techniques for dose interpretation. Devices of both types are usually small and light and can be attached to the user's clothing or carried in his pocket (Fig. 15-6).

7. Specialized systems, such as those for γ-spectrometry, neutron activation and analysis, and radioactive sample counting and analysis. Figure 15-7 shows a neutron threshold detector unit.

Note that some instruments provide a measurement of the *quantity* of radiation, or radiation dose, whereas others only detect the presence of radiation or at most indicate the relative quantity. An example of the first type is an instrument with an ionization chamber for the measurement of air-equivalent γ-radiation doses, such as the one shown in Fig. 15-8. An example of the second type is an instrument with a Geiger-Müller counter for β-γ surface contamination monitoring, such as shown in Fig. 15-9.

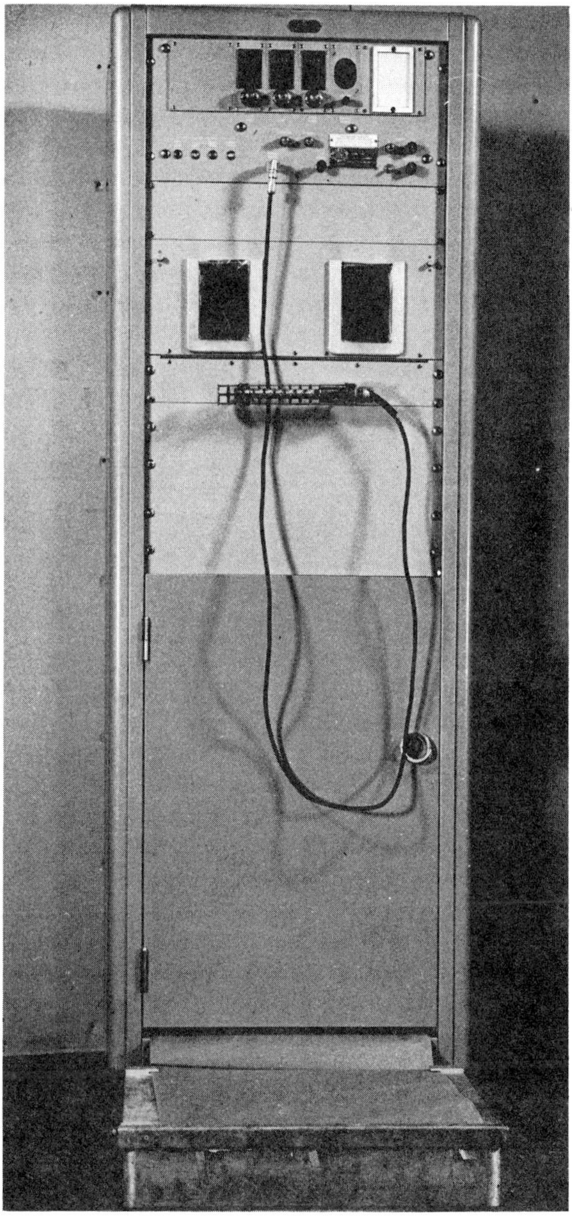

Fig. 15-4. Hand, foot, and clothing monitor for beta-gamma radiation.

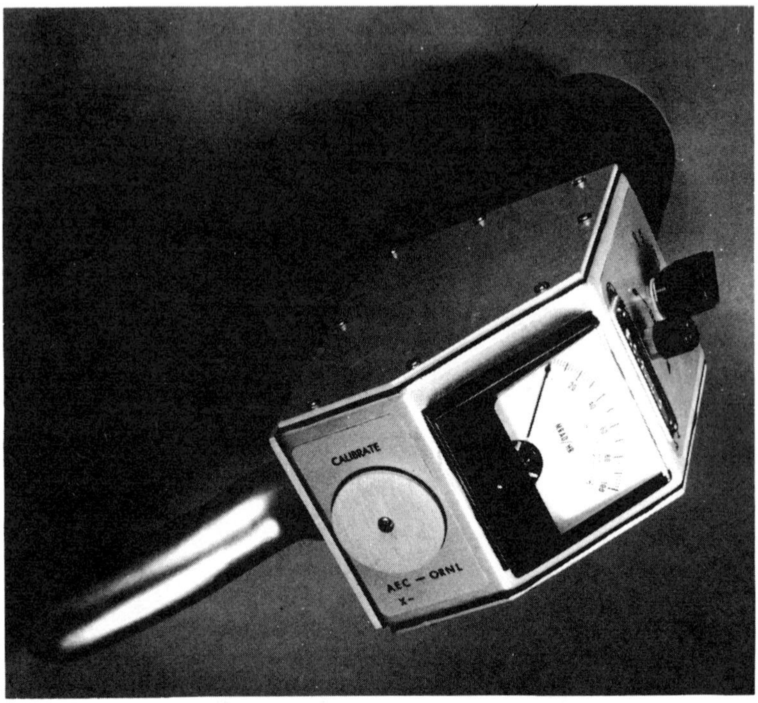

Fig. 15-5. Portable survey meter for beta-gamma radiation (Cutie Pie).

15-2 DETECTORS

An instrument or other system for radiation monitoring has at least two main components: a detector, in which some effect is produced by the radiation, and a means of readout or interpretation of the magnitude or degree of this effect.

The most important radiation detectors for monitoring purposes are: (1) ionization chambers, (2) Geiger-Müller (G-M) tubes, (3) proportional counters, (4) scintillation detectors (phosphor, crystal, liquid, gas), (5) solid-state detectors (semiconductor, luminescent, fluorescent), (6) induced radiation or fission materials, (7) chemical compounds, and (8) photographic emulsions. The selection of a detector for a given application is influenced by the kind and quantity of radiation to be dealt with, its energy spectrum, the degree of accuracy wanted, the inherent absorber that may be acceptable, the practical size and weight of the detector, and its resistance to deleterious effects caused by the environment in which it is to be used.

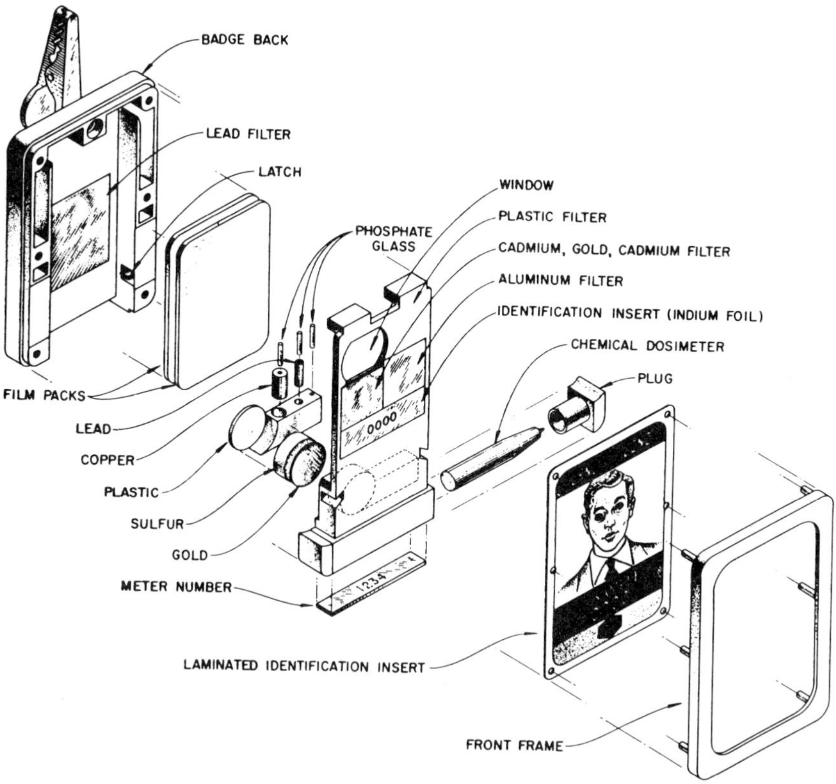

Fig. 15-6. Personal badge-dosimeter.

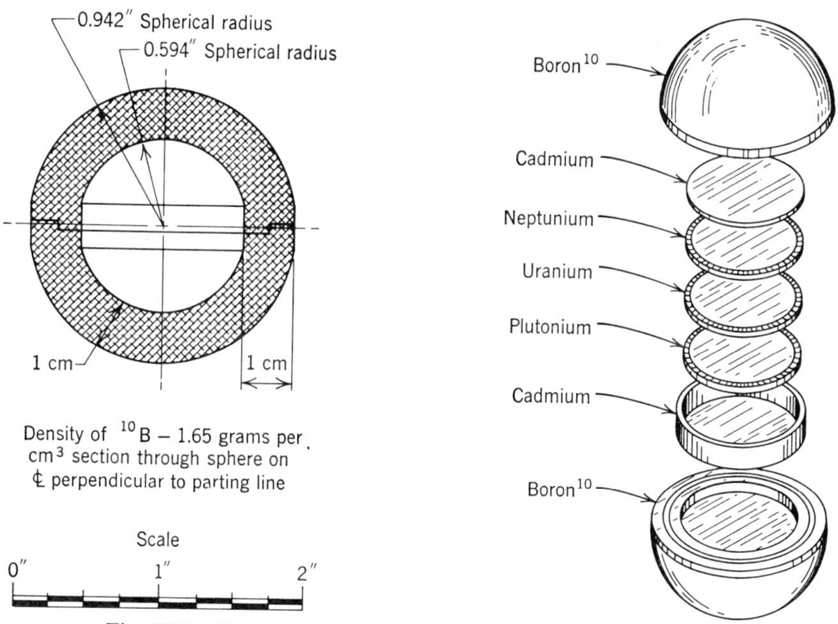

Fig. 15-7. Neutron threshold detector assembly and components.

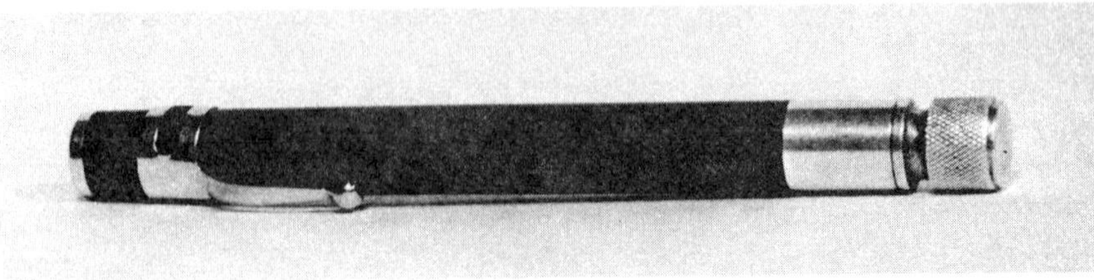

Fig. 15-8. Pocket ion-chamber dosimeter.

We now consider how these detectors have been put into use. More detailed information on the ionization and chemical reactions produced by radiation is given in Chapters 5 and 7.

Ionization Chambers (Sections 5-3 and 5-4)

These are used mainly to measure dose and dose rate (in the media of which the chambers are composed) from β-, γ-, and X-radiations, and in some

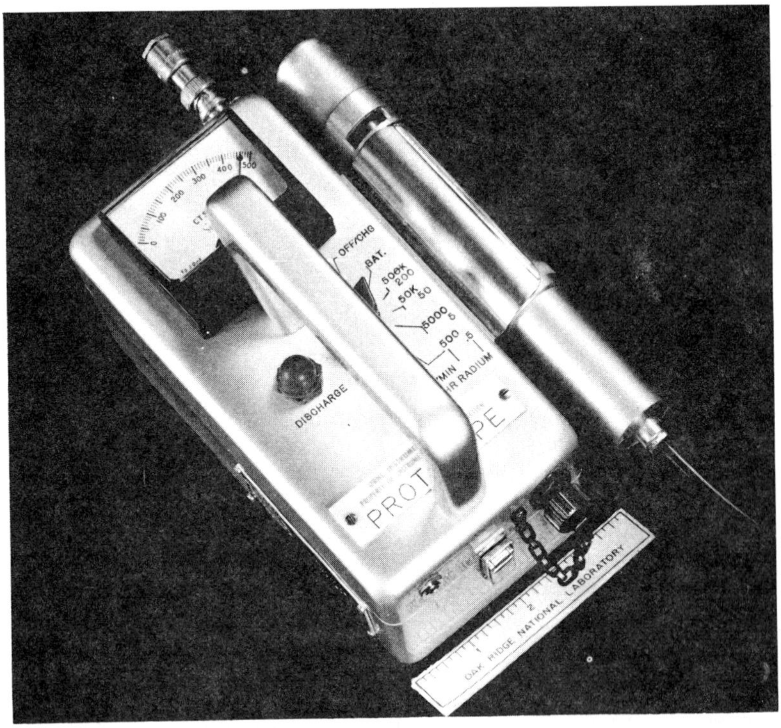

Fig. 15-9. Geiger-Müller survey meter.

applications for measurements of other radiation. The geometric configuration of the chambers may be that of concentric cylinders, parallel plates, concentric spheres, or rectangular boxes. The electrodes must be well insulated from each other to minimize leakage current. The potential difference across the electrodes must be sufficient to collect all the ion pairs formed within the chamber.

Practical electronic circuitry and insulation for field-type, rate-measuring instruments are limited to input resistance not greater than 10^{12} ohms. Electrometer systems of the vibrating reed or capacitor type can measure currents of less than 10^{-15} amp, but the simple d-c electrometers for field use require a signal of about 0.1 volt for a minimum accurate measurement.

The volume of an ionization chamber depends on the doses or dose rates for which it is to be used and on the sensitivity of the associated readout system. Very sensitive systems may require the use of large or highly pressurized chambers. Ideally, the chamber should be small enough so that the intensity of the incident radiation is not significantly diminished in traversing it; otherwise the energy flux will be nonuniform within the chamber, with subsequent uncertainties in measurement. Also, the walls of the chamber should not significantly attenuate the radiation, unless such attenuation is intended.

Most ionization chambers for monitoring instruments are unsealed and air-filled. However, many types are sealed and may be filled with air, nitrogen, carbon dioxide, helium, or other gases at pressures above or below atmospheric. For health physics use, the chamber walls, internal components, and gaseous filling are usually either air-equivalent or tissue-equivalent.

Ionization chambers are usually operated in one of two modes for radiation monitoring. A voltage may be applied continuously to the collecting electrodes so that the quantity of charge or ionization current may be measured as it is produced; or the chamber may be used as a condenser and removed from the voltage source after charging; having been exposed to radiation, the charge difference is then measured.

Geiger-Müller Counters (Section 5-6)

Geiger-Müller counters are normally used for detecting ionizing events which take place within the sensitive volume of the counter. These ionizing events are transformed into output voltage pulses or counts if the counter is operated within the Geiger voltage region with adequate quenching (prevention of multiple discharges). Less commonly, G-M counters may be used in the current-ouptut mode of operation.

The G-M counters for use in practical instruments are self-quenching. Halogen-quenched counters are the type most commonly used, because the quenching action does not consume the quenching agent, the output voltage pulse is relatively larger, and the operating voltage is relatively lower than that for organic-quenched counters.

Geiger-Müller counters are available in a wide variety of shapes, sizes, wall materials, wall thicknesses, and operating voltages. The majority of those used for health physics instrumentation are cylindrical, with the wall of the cylinder serving as the cathode and a fine, axially mounted wire as the anode. They have glass, aluminum, or stainless steel shells or walls, and operate at a voltage level between 700 and 900 volts. The diameter is in the range 0.5 to 1 inch and the length within the range 4 to 6 inches. They may be halogen or alcohol quenched, and the wall thickness is usually about 30 mg/cm^2. Much use is made also of mica window counters, which are available with window diameters of about 0.25 to 1.0 inch and thicknesses of about 1.5 to 5 mg/cm^2. These "end-window" counters are usually employed for β-particle detection.

Except for counting losses due to "dead time" and coincident events, pulse-operated G-M counters produce a voltage pulse for each ionizing event which occurs in the sensitive volume of the counter independently of the quantity of energy associated with the event. Such counters therefore are not capable of measuring dose rates *per se*, but may only be used in measurements of the relative intensity of radiation.

Proportional Counters (Section 5-5)

These are used to detect one type of radiation in the presence of other types or to obtain output signals greater than those obtainable with ion chambers of equivalent size. Because the electrical pulse output from such a counter is proportional to the quantity of ionization which is produced by an event in the sensitive volume of the counter, pulse-height discrimination may be used to count selectively the pulses due to particles with a given degree of specific ionization. Higher and more critically selected voltages are required, in general, for the operation of proportional counters than for the operation of other commonly used detectors.

Proportional counters may be used either to detect events or to measure absorbed energy (dose), because the output pulse is directly proportional to the energy released in the sensitive volume of the counter (even though gas amplification produces an output pulse larger than that produced by the ionizing event).

A wide variety of gases and gas mixtures at various pressures is used in proportional counters. They may be operated at, above, or below atmospheric pressure. The counters can be sealed to prevent gas leakage or connected with a supply of gas that continuously flows through the counter.

Shapes are usually either cylinders, spheres, rectangular boxes, or hemispheres.

Proportional counters are most widely used for the detection of α-particles, neutrons, and protons. Some have a means of introducing the radiation source into the sensitive volume of the detector, but most detect radiation from sources external to the counter. Very thin counter walls or windows must be used for the transmission of α-particles and protons.

Scintillation Counters (*Section 7-5*)

This type is a combination of a photomultiplier tube with a scintillating material, which may be a crystal or other phosphor (solid, liquid, or gas).

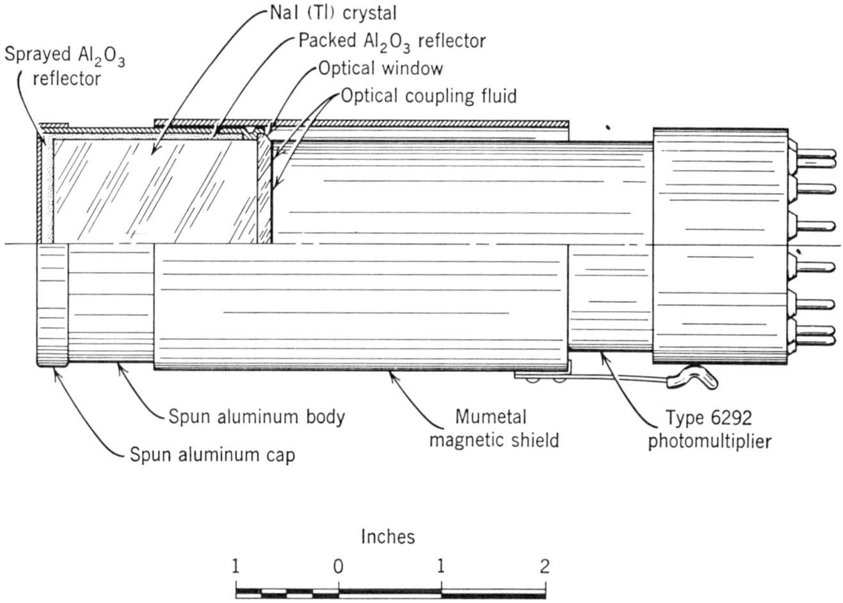

Fig. 15-10. Gamma scintillation counter assembly.

A typical arrangement is shown in Fig. 15-10. Light pulses produced in the scintillator by radiation release photoelectrons from the cathode of the photomultiplier tube, which then initiate pulses of current that can be counted. The light-production efficiency of scintillators depends on many characteristics of the material and the incident radiation, and the choice of a particular scintillator is based on these factors. Scintillation counters are available for detecting all of the radiations shown in Table 15-1. The most common ones for health physics monitoring are α-counters with ZnS phosphors and γ-counters with NaI crystals.

In many respects the scintillation counter is similar to the proportional counter. Relatively high voltages are required and the magnitude of the output signal is directly proportional to the energy absorbed in the detecting medium and to the operating voltage of the detecting system. Both types of counters are affected by shock and temperature extremes more than some other detectors and they are relatively costly.

Semiconductors

Semiconductors of either the *p-n* junction or surface-barrier types can be used as radiation detectors. They are usually specially prepared from thin wafers machined from germanium or silicon crystals. They are most useful for measurements of the energies of particles of high specific ionization, and in particular for α-particle energy analysis. They can be used for detecting heavy particles such as α-particles, protons, and fission fragments in the presence of other radiations.

The use of semiconductor detectors in field work has been limited because they have been available only with usable areas not much larger than a few square centimeters. However, when small size is not a disadvantage—and in some cases it may be a distinct advantage—a semiconductor may be incorporated into a small, compact detector assembly.

Solid State Detectors

Solid state is a term used in the literature in a number of ways to classify a variety of detectors. We shall consider solid-state detectors as homogeneous solids in which changes produced in the solid by radiation are stored for later readout in comparison with a calibrated standard. In practice, only small sizes (a few millimeters in both diameter and length) are required for most health physics dosimetric applications.

The most widely used solid state detectors are silver metaphosphate glass, calcium fluoride, and lithium fluoride. The amount of energy stored in the glass from absorbed radiation is inferred from the quantity of light emitted from the glass when it later undergoes ultraviolet irradiation at readout. The fluorides, which are used in a finely powdered form, release stored energy as light when heated. The readout operation does not modify the energy states in the glass, but those in the fluorides are returned to or near the ground state. Sensitivity of solid-state detectors is comparable to that of photographic emulsions as described below.

Chemical Dosimeters (Section 7-3)

These are systems in which measurable chemical changes are produced by ionizing radiation. Most chemical systems of practical size are useful only for

γ-doses of hundreds to millions of rad. However, small volumes (several milliliters) containing chloroform, water, and a dye indicator can be used for γ-doses in the range from a few rad to several thousand rad. Radiation produces acids in this system, the amount of which can be determined from visible color changes or, more accurately, by titration or pH readings.

Photographic Emulsions (Section 7-2)

In photographic emulsions ionizing radiation produces a latent image of developable grains. The effect of radiation appears as a blackening in the film when it is developed. Two types of emulsions are generally used for monitoring purposes. One is an X-ray type film in which radiation produces a generalized blackening, the extent of which is proportional to the dose absorbed in the film. The other is a nuclear track emulsion in which charged particles produce a track (or a linear series of developable grains).

The films most widely used for personnel monitoring are in dental-sized packets (about $1\frac{1}{4} \times 1\frac{5}{8}$ inches) and consist of an acetate or plastic base that is coated on one or both sides with gelatin containing silver bromide crystals. The films are contained in a light-tight wrapper. A film packet consists of one or more films and the wrapper.

Dosimetry with photographic emulsions is complicated by several circumstances. Although the blackening in a film is proportional to the dose of radiation absorbed in the film, the dose absorbed in the film is not in the same ratio to the dose absorbed in air or soft tissue for all energies of radiation. Therefore the response of the film is energy dependent as a measure of dose absorbed in air or tissue, particularly for low-energy photon radiation. Various systems of absorbers or filters are used for overcoming this energy dependency. Films can be affected adversely for monitoring purposes by a number of agents, such as heat, light, moisture, chemical vapors, age, and processing errors.

Dose equivalent ranges for which films may be used go from a few mrad to several hundred rad of photon and β-radiation. Special techniques may be used to extend this range or to interpret the effect of neutron radiation in the X-ray type emulsion. Nuclear track films are commonly used to determine fast neutron doses from a few mrad to about ten rad. Here, protons in the emulsion arising from collisions with the incident neutrons produce tracks which are measured. Neutrons are uncharged and hence leave no track in photographic emulsions. The principal reactions with neutrons are elastic scattering with hydrogen and, to a lesser degree, the $^{14}N(n, p)^{14}C$ reaction with thermal neutrons. The energy spectrum of the incident neutrons is inferred from the track-length spectrum of the recoil protons, and the flux of incident neutrons is determined from the number of tracks.

The emulsion response itself is not significantly energy dependent for most β-radiation. However, an object such as a wrapper, located between the emulsion and the incident β-radiation, will absorb β-particles, and so the β-dose in an emulsion may be much less than the dose at the surface of the film container. In the nuclear track emulsion, recoil protons from neutrons of energy less than about 0.5 MeV do not produce easily definable tracks in the emulsion, and so the film is not useful for monitoring neutrons, other than thermal, below that energy. Any material used to contain or support the emulsion may alter significantly the energy spectrum of the neutrons which are to be monitored.

Other Detectors

Other detectors, not included in the preceding descriptions, are also sometimes used in health physics work. Some detectors make use of changes produced in atomic nuclei, for example, the absorption of thermal neutrons resulting in induced radioactivity of indium. A small piece of indium included in a γ-film packet can be monitored with a G-M counter to detect whether it has been exposed to thermal neutrons.

Other examples include detectors that respond to neutrons with energies above various thresholds for nuclear reactions, such as fission, neutron activation, and nuclear transmutation. By combining a number of materials, like the system shown in Fig. 15-7, these threshold detectors give information about the incident neutron energy spectrum as well as flux (see Section 7-8).

15-3 COMPONENT INSTRUMENTATION

A complete health physics instrument consists of one or more detectors and some combination of electronic components. Except for the specialized equipment used for reading films, solid-state devices, and chemical systems, most instruments have a combination of rate or integrating circuitry and readout. Rate circuitry may be designed for either pulses or ion current produced in the detector, and the readout units may be counts per unit time, radiation exposure per unit time, or dose per unit time. Integrating circuitry may be designed for total counts, exposure, or dose. The readout for rate systems may be panel meters, recorders, and audible or visual signals which vary in intensity or frequency. The readout for integrating devices may be panel meters, scaling lights, registers, or electrostatic electrometers. A typical circuit for one of the more common types of health physics instruments (Cutie Pie) is shown in Fig. 15-11.

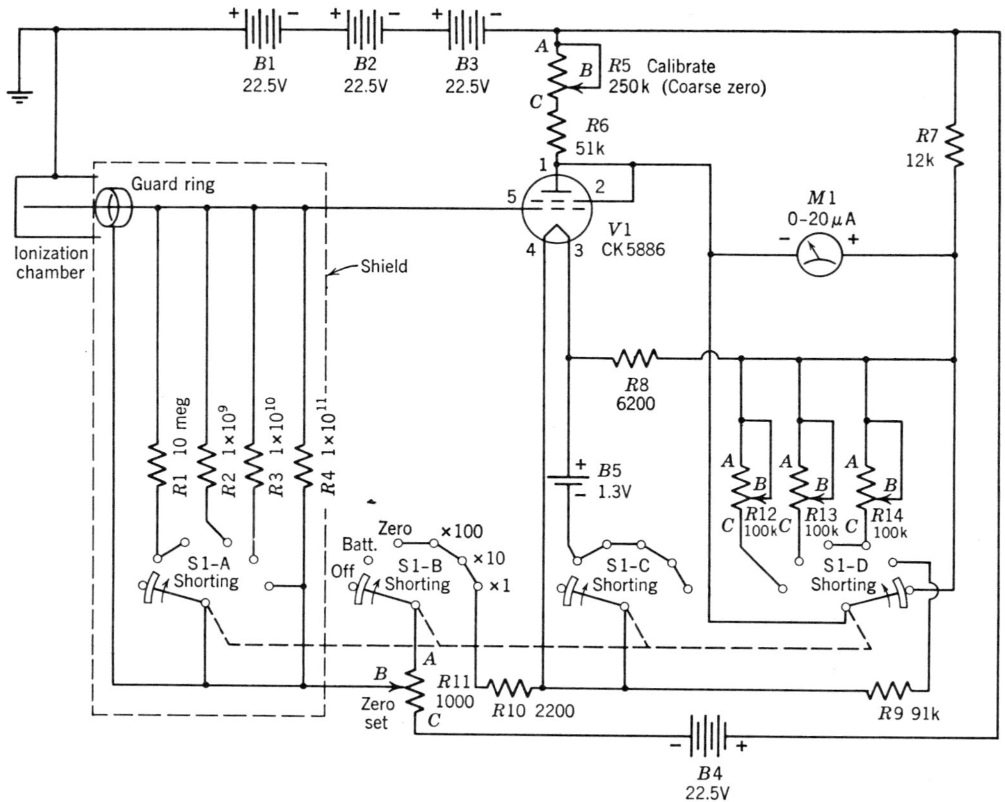

Fig. 15-11. Ionization chamber survey meter—Cutie Pie Mark V wiring diagram.

Electrical power is usually provided from batteries or common 110 volt a-c sources. In both cases, many of the voltages in an instrument usually need to be well regulated, particularly in proportional counters, photomultipliers, some transistors, some G-M counters, d-c amplifiers, and discriminator circuits. Thus voltage regulation may add materially to the complexity of an instrument.

The output information from a monitoring instrument can be relayed or telemetered to remote meters, recorders, and annunciator signals, and may be used to initiate modification or stoppage of operations and processes.

In addition to the detector and electronic circuitry, pumps, blowers, radiation shielding, and other mechanical equipment may be incorporated in a complete instrument or instrumentation system. For example, some β-air monitors have a shielded detector, a filter-paper drive-mechanism, timers, power supply, rate meter, alarms and relays, vacuum pump, and recorder mounted in a single cabinet unit, having a total weight of several hundred pounds.

The weights of other instruments vary. A pocket-type radiation monitor may weigh only a few ounces, complete with detector, battery, case, circuitry, and readout or signaling device. A film meter may weigh a few grams, and an *in vivo* γ-spectrometer with shielding may weigh more than a hundred tons. Most of the portable, battery-operated instruments for monitoring weigh less than five pounds. The instruments listed in Tables 15-2, 15-3, and 15-4 are typical of those in common use at Oak Ridge National Laboratory about 1964.

15-4 CALIBRATION AND MAINTENANCE

Inherent sources of error exist in any measurement made with radiation instruments due, for example, to the limited degree of certainty with which a meter dial can be read. However, even the most carefully made measurement may be seriously in error if the instrument is not properly calibrated, adjusted, and serviced. Radiation calibration should be performed often enough to prevent errors in measurements with an otherwise operable instrument. The required frequency of calibration depends upon the calibration stability of the instrument, and this is usually determined from experience and tests with the instrument. In the routine use of health physics instruments, a regular maintenance schedule should be established for each type of instrument, and recalibration should follow any maintenance or repair operation which may have affected the calibration.

Radiation monitoring instruments should be checked routinely for over-all operability both in fields of radiation to which they are to respond and in

Table 15-2—Portable Personnel Monitoring Instruments

Instrument	Detector	Radiation[a] Detected	Range (Nominal)	Application	Remarks
Film meter	Photographic emulsions	γ, β N_f, N_{th}	0.1–1,000 rad of mixed radiations	Measurement of integrated dose to the meter user	Orientation of film during exposure is important
Pocket chamber (indirect reading)	Ionization chamber (air)	γ	5–200 mR	Measurement of daily gamma exposure	Relatively energy independent for γ; charger-reader required
Pocket chamber (direct reading)	Ionization chamber (air)	γ	5–200 mR; available with higher ranges	Visual check on gamma exposure	Position of electrometer fiber read through magnifying lens
Personnel radiation monitor	G-M tube	γ, X high-level β	Maximum audible warning at ~0.5 R/hr	Audible warning of radiation field	Signal frequency proportional to radiation intensity; available in higher rate ranges
Chemical dosimeter	Tetrachloroethylene + pH indicator	γ	5 rad to 2×10^6 rad	Measure gamma component of a mixed radiation field	Read by titration, measurement of conductivity change, measurement of pH colorimetrically or electrometrically
Glass dosimeter	Metaphosphate glass containing silver	γ	0.005 to several thousand rad or more	Dose measurement of γ exposure over a wide range	Calibrated reader required

[a] Symbols defined in Table 15-1.

Table 15-3—Portable Survey Instruments (Battery Powered)

Instrument	Detector	Radiation[a] Detected	Range (Nominal)	Application	Remarks
Cutie pie survey meter	Ionization chamber (air)	γ, X	5–10,000 mrad/hr	Dose-rate meter for γ and X (0.008–2 MeV) within 10%	Most widely used instrument for these measurements
		High-energy β		Measures with at least 50% efficiency the externally hazardous betas	
Juno survey meter	Ionization chamber (air)	α	Three scales: 10,000, 100,000, and 1,000,000 d/min	Dose-rate meter for γ and β; relative-intensity meter for α	Maximum error 10% full scale; manually positioned shields used
		γ	Three scales: 50, 500, and 5000 mR/hr		
		β			
Geiger-Müller survey meter	G-M tube	$\beta > 0.2$ MeV, γ	Three scales: $\times 1$, $\times 10$, $\times 100$; $\times 1$ may be 600–800 counts/min full scale	Detection instrument for $\beta > 0.2$ MeV and γ. Rate meter and audible pulse	Energy dependent, should be used with earphones for faster response; sliding shield for β-γ discrimination. Some models saturate above 50–100 mR/hr and will not indicate higher dose rates. Insensitive to low β-energies, unless equipped with thin window counter

Alpha proportional counter, gas flow	Proportional counter (propane)	α	May detect as little as 50 d/min α in presence of 1 rad/hr γ; range to 500,000 d/min	Discriminates between α and β-γ	More stable than air proportional counter
Alpha scintillation counter	Scintillation (phosphor and photomultiplier)	α	To $\sim$1000 cpm with $\sim$15% geometry	Assay of α emitters; registers accumulated counts; register and audible signal	Simpler to operate and require less maintenance than proportional counters
Disk air sampler	None	α, β, γ, filter later counted		Air drawn through filter by a-c operated blower	Collection time and airflow rate should be noted
Thermal neutron proportional counter	(BF_3 enriched in ^{10}B) proportional counter, gas	N_{th}	20–20,000 $N_{th}/cm^2 \cdot sec$	Can discriminate against intense γ-radiation (measure 200 $N_{th}/cm^2 \cdot sec$ in field of 100 R/hr γ)	Employs $^{10}B + n \rightarrow {}^7Li + \alpha$ reaction
Fast neutron proportional counter	Proportional counter, gas	N_f	2–2500 mrem/hr	Measures neutron (0.15–10 MeV) dose rate within $\pm$25%. Rate reading unaffected by γ dose rate of 50 R/hr	Tissue-equivalent walls and gas, proton recoil
Gamma scintillation counter	NaI crystal	γ	Low-level	Used for very low-level γ-monitoring, 0.001–1 mR/hr	Sensitivity dependent on size of crystal; energy dependent
Beta scintillation counter	Phosphor	β	—	Very few applications in portable instruments	G-M tube generally preferable
Neutron scintillation counter	ZnS(Ag), molded in Lucite	N_f	—	Fast neutron detection where dose rate is not required	

[a] Symbols defined in Table 15-1

Table 15-4—Nonportable Monitoring Instruments (a-c Powered)

Instrument	Detector	Radiation Detected	Range (Nominal)	Application	Remarks
Continuous β-γ particulate air monitor	G-M tube (shielded)	β, γ	0–5000 cm	Continuous recording of β-γ particulate radiation; light and audible alarms for preset levels	Count-rate and strip-chart recorder incorporated; does not distinguish between β and γ
Continuous α-particulate air monitor	ZnS(Ag) for α	α, β	0–1000 cpm	Continuous recording of α-particulate radiation; light and audible alarms for preset levels	Radon tends to "mask" more hazardous long-lived emitters
Continuous gaseous air monitor	Ionization chamber (gas-flow, shielded or compensated)	All	1/100–10 × MPC for tritium	Tritium monitor	Air is filtered and electrostatically precipitated prior to entering chamber
Fast neutron dosimeter (Radsan)	Proportional counter lined with polyethylene and filled with ethylene	N_f	From 1/10 maximum permissible exposure; measures first-collision dose	Fast neutron dosimetry; insensitive to γ less than 5 R/hr	Can be checked with internal α-source; tissue-equivalent chamber
Monitron	Ionization chamber	γ if chamber is coated with carbon only, γ and N_{th} if coated with ^{10}B-enriched boron	To 125 mR/hr	Dose-rate meter for γ background monitoring; measures only the relative intensities of N_{th}	Has adjustable alarm setpoint

Neutron monitor	Moderated BF_3, proportional counter	Neutrons	To about 20 mrem/hr	Approximate dose and dose rate from neutrons; detects short-duration bursts of neutrons	Alarm provided for bursts, dose rate and integrated dose; energy dependent
Hand and foot monitor	Halogen-quenched G-M tube	β, γ	Low-level	Simultaneous detection of β- and γ-contamination of hands and shoes	Has auxiliary probe for monitoring clothing
Water effluent monitor	Cluster of G-M tubes or NaI scintillator	γ, β	—	Monitoring water wastes or coolants; may be connected to rate meter, recorder, and alarm systems, or to check and diversion valves for control of water flow	Thin films of water (or other material) will absorb α and low-energy β; monitoring for any radiation may be complicated by silt, algae, radioactive contamination of the detector, variability in water flow, and surface levels
Portal monitors	Five or more G-M tubes	β, γ		Monitoring exits from areas of suspected contamination	Slow passage through narrow portal required for maximum instrument response
Lab monitor, α	Scintillation	α	5 ranges: 250–25,000 cpm full scale	Monitoring personnel, clothing, and equipment for α-contamination	Has audible count signal and adjustable alarm point
Lab monitor, β	G-M tube	β, γ	5 ranges, 250–25,000 cpm full scale	Monitoring personnel, clothing, and equipment for β-contamination and for background monitoring	Has audible count signal and adjustable alarm point

fields against which they are to discriminate. A daily operational check of instruments that provide continuous monitoring and alarm signals is desirable. Portable instruments used for dose rate and contamination measurements should be checked before each use.

The most widely used sources for calibration are sealed γ-ray emitters, sealed (α, n) systems, and α-emitters on metal disks. The ranges of energies and source strengths of readily available calibration sources are limited by practical considerations of short half-lives, bulk and containment of sources. and their standardization. This applies particularly to sealed neutron sources. Available sources of one type or another are adequate for most routine calibrations except β-dose rates. Most sources are not, however, adequate for energy-dependence evaluation.

Specialized sources, materials, techniques, and equipment are required for energy dependence studies and β-dose rate calibration. This may involve calibrated X-ray equipment, accelerators, high-voltage machines, and radioisotope sources.

BIBLIOGRAPHY

Ehrlich, M., "The Use of Film Badges for Personnel Monitoring," International Atomic Energy Agency, Vienna, 1962.

Handloser, J. S., *Health Physics Instrumentation*, Pergamon Press, New York, 1959.

Hine, G. J., and G. L. Brownell, eds., *Radiation Dosimetry*, Academic Press, New York, 1956.

O'Kelley, G. D., *Detection and Measurement of Nuclear Radiation*, NAS-NS 3105 Office of Technical Services, Dept. of Commerce, Washington, 1962.

Price, W. J., *Nuclear Radiation Detection*, McGraw-Hill, New York, 1958.

Renne, H. S., *How to Detect and Measure Radiation*, Howard W. Sams and Co., Inc., Indianapolis, 1963.

Snell, A. H., ed., *Nuclear Instruments and Their Uses*, John Wiley and Sons, Inc., New York, 1962, Vol. I.

"Measurement of Neutron Flux and Spectra for Physical and Biological Applications," *Natl. Bur. Std. (U.S.) Handbook* **72** (1960).

"Measurement of Absorbed Dose of Neutrons, and of Mixtures of Neutrons and Gamma Rays," *Natl. Bur. Std. (U.S.) Handbook* **75** (1961).

PROBLEMS

1. Why may not a stainless-steel walled, helium filled, ionization chamber be used for health physics dosimetry of photon radiation?

2. It is desired to have an air ionization chamber-electrometer instrument indicate, at standard temperature and pressure, a full-scale reading for an exposure rate of 1 R/hr. If the electrometer input sensitivity is 0.1 volt for a full-scale reading of the meter, and if the volume of the ionization chamber is one liter, what should be the approximate value of the electrometer input resistor?

3. The ionization chamber of the instrument in problem 2 is spherical. For accurate measurements, the distance between a small-diameter source and the center of the chamber should be no less than five times the diameter of the chamber, and at least one-tenth of a full-scale reading is required. What is the least amount (millicuries) of ^{60}Co that the source must contain for an accurate measurement to be obtained?

4. A miniaturized, battery-operated electrometer that would have in input sensitivity for accurate reading of 10^{-13} amp can be made about the size of a cigarette package. What is the practical limitation which would prevent its being used for a pocket-size, exposure-rate instrument for measurements as low as 1 mR/hr?

5. Why may not a G-M counter with a 30 mg/cm^2 stainless steel cathode (outer shell) be used for (*a*) measurements of exposure rates from photons of unknown energies; (*b*) detection of β-particles of unknown energies; and (*c*) detection of α-particles?

6. A G-M counter instrument is to be used to locate a 10 mc source of ^{60}Co which has been lost in high grass. Neglecting scattering and absorption, within what minimum distance can the source be located if the counter has a cross-sectional area of 10 cm^2 and a γ-detection efficiency of 0.01, and if the instrument has a maximum counting rate of 50,000 cpm.?

7. Why are intermediate energy neutrons usually monitored with instruments which detect only thermal neutrons?

8. A filter-type air monitor is used for α-air particulate monitoring. If the monitor indicates 1 cpm for each 3 dpm (disintegrations per minute) from the material in the air, how many times MPC_a (averaged over the sampling interval) is equivalent to 100 cpm under the conditions that $MPC_a = 10^{-12}$ μc/cm^3, sampling interval is one hour, and sampling rate is 10^5 cm^3/min?

9. A radiation monitoring operation is to be performed to determine surface contamination levels in a laboratory where radium bromide has leaked from a source. It is necessary to locate and evaluate, in terms of disintegrations per minute per 100 cm^2, α-particle emission rates in excess of 30 dpm per 100 cm^2. What type or types of instruments should be used and why?

10. What types of instruments and personnel monitoring devices might be required for equipping a facility which separates kilogram quantities of ^{239}Pu from spent reactor fuels?

CHAPTER 16

Prevention of Criticality Accidents

Dixon Callihan

16-1 INTRODUCTION

The practicing health physicist is confronted with problems of many types and of many origins. Some of these are associated in routine procedures with sources of radioactivity having known characteristics and predictable behavior, exemplified by tracer and medical uses of radioisotopes, by industrial-scale processes of natural radioactive substances, and even by the operation of nuclear reactors in their designed manner. The health physicist must also be prepared for emergencies in any of these operations which can amplify significantly the hazard known to be normally present. Accidental dispersal of radioisotopes and an uncontrolled surge in the power of a nuclear reactor are examples of these occurrences. It is the purpose of this chapter to review briefly still another problem which is, in fact, almost insidious. This problem arises when fissionable materials are accidentally brought together in sufficient quantity to achieve criticality. The ensuing nuclear chain reaction can release lethal quantities of radiation. It is possible that the reaction may generate sufficient energy to damage equipment and to disperse radioactive contamination in amounts that prohibit access to an area.

The problem of criticality accidents is potentially present in all operations with fissionable materials which, in this discussion, are exemplified by chemical and metallurgical processes and by shipment and storage of these materials. We will treat the bases for accident control rather than any detailed safety analysis of complex industrial-scale processes. Accident controls must be carefully and thoroughly developed and enforced in industrial practices to prevent establishing nuclear chain reactions.

In order to gain insight into the manner in which nuclear accidents can be prevented, it is necessary to describe the conditions whereby nuclear chain reactions develop. Sections 16-2 through 16-5 review the necessary physical principles, and Section 16-6 shows how the principles are applied to the prevention of criticality accidents. Administrative problems and operating experience are discussed in the concluding sections.

16-2 NUCLEAR STRUCTURE AND ISOTOPES

The nuclei of atoms are composed of neutrons and protons, the number of protons distinguishing one chemical element from another and the number of neutrons distinguishing one species, or isotope, of a particular element from another. The number of protons in a nucleus is the atomic number Z of the element and the total number of protons and neutrons is the mass number A of the nucleus. Since the neutron is electrically neutral, the positive charge carried by the nucleus is numerically equal to Z units of the electronic charge. One atomic mass unit (amu) is defined as one-twelfth the mass of a single atom of the carbon isotope ^{12}C ($Z = 6$, $A = 12$): 1 amu = 1.657×10^{-24} gm = 931.8 MeV. On the atomic mass scale the mass of the proton is 1.007277 amu and that of the neutron 1.008665 amu.

Some 300 stable isotopes exist among the approximately 90 chemical elements found in nature. In the stable nuclei of some elements, such as gold and fluorine, the ratio of neutrons to protons is unique and only a single isotope exists naturally. In the stable nuclei of other elements the number of neutrons varies. For example, tin has ten stable isotopes.

About 50 isotopes found in nature are radioactive and disintegrate spontaneously. A large portion of the naturally occurring radioactive isotopes are found in the decay chains described in Chapter 1. For example, the two most abundant uranium isotopes decay by emission of an α-particle according to the scheme

$$\begin{aligned} {}^{238}_{92}\mathrm{U} &\rightarrow {}^{4}_{2}\mathrm{He} + {}^{234}_{90}\mathrm{Th} \\ {}^{235}_{92}\mathrm{U} &\rightarrow {}^{4}_{2}\mathrm{He} + {}^{231}_{90}\mathrm{Th} \end{aligned} \tag{16-1}$$

A naturally occurring radioactive isotope of relatively low mass is ^{40}K, which emits a negative β-particle

$${}^{40}_{19}\mathrm{K} \rightarrow {}^{0}_{-1}\beta + {}^{40}_{20}\mathrm{Ca} \tag{16-2}$$

The emission of penetrating electromagnetic radiation from the nucleus in the form of γ-rays usually follows radioactive decay.

16-3 NUCLEAR REACTIONS

Some 1000 radioactive isotopes have been produced artificially by nuclear reactions. For example, the compound nucleus of boron with mass number

11, formed by neutron bombardment of ^{10}B, is unstable and decays almost immediately into ^{7}Li and an α-particle in the following way:

$$^{10}_{5}\text{B} + ^{1}_{0}\text{n} \rightarrow [^{11}_{5}\text{B}] \rightarrow ^{4}_{2}\text{He} + ^{7}_{3}\text{Li} \qquad (16\text{-}3)$$

The resulting energetic α-particle readily ionizes a gas and the reaction is, indeed, commonly used in the detection of neutrons (see Chapter 15).

In another example the reaction between two deuterons gives

$$^{2}_{1}\text{H} + ^{2}_{1}\text{H} \rightarrow [^{4}_{2}\text{He}] \rightarrow ^{3}_{1}\text{H} + ^{1}_{1}\text{H} \qquad (16\text{-}4)$$

The product nucleus $^{3}_{1}$H, the hydrogen isotope called tritium, is itself radioactive and decays by emission of a negative β-particle,

$$^{3}_{1}\text{H} \rightarrow ^{\;\,0}_{-1}\beta + ^{3}_{2}\text{He} \qquad (16\text{-}5)$$

with a half-life of about 12 years.

16-4 FISSION

In addition to the natural radioactivity illustrated by the α-particle emission by uranium described in Eq. (16-1), many of the heavier elements have the property, after appropriate excitation, of disintegrating into two fragments of nearly equal mass accompanied by lighter particles and by relatively large quantities of energy. This type of disintegration is called fission. In some nuclei the stimulus for fission is naturally present, causing spontaneous fission although at a very low rate. Usually, however, fission occurs following the capture of a photon or a particle by the nucleus. The particle inaugurating fission is often a neutron. Nuclei capable of fission are said to be fissionable; a special species capable of fission even though the kinetic energy of the incident particle (a neutron) is essentially zero is said to be fissile.

As an example of fission consider the capture of a thermal neutron by a ^{235}U nucleus. The disintegration of the compound nucleus, ^{236}U, may occur in several ways illustrated by the following reactions:

$$\begin{aligned} ^{235}_{92}\text{U} + ^{1}_{0}\text{n} \rightarrow [^{236}_{92}\text{U}] &\rightarrow ^{235}_{92}\text{U} + ^{1}_{0}\text{n} \\ &\rightarrow ^{236}_{92}\text{U} + \gamma \qquad (16\text{-}6) \\ &\rightarrow ^{141}_{56}\text{Ba} + ^{92}_{36}\text{Kr} + 3^{1}_{0}\text{n} + \gamma \end{aligned}$$

In the first reaction, called scattering, the neutron is reemitted; in the second, called radiative capture, the excited ^{236}U nucleus loses energy through the emission of a γ-ray and thereafter behaves as a naturally radioactive uranium nucleus; in the third, called fission, a barium and a krypton nucleus together with three neutrons and some γ-rays are emitted. The latter process is described by saying that a ^{235}U nucleus fissions as a result of neutron absorption and the example is only one of about 30 observed modes. Although as many as seven neutrons are emitted in one of the modes, the average is about 2.5. The average fission neutron production by the fissile nuclei ^{233}U and ^{239}Pu is about 2.5 and nearly three, respectively.

The uranium isotopes of mass 233 and 235 and the plutonium isotope of mass 239 are the only ones considered important at this time in protecting against criticality accidents. Although other heavy isotopes fission, either spontaneously or following absorption of a neutron or other elementary particle, criticality accidents with them are highly unlikely because of the type and energy of the particles producing fission or because the quantities of the isotopes presently in existence are too small to present risks of this type. The reader is alerted to the potential risks associated with ^{238}Pu, ^{244}Cm, ^{245}Cm, and other fissionable isotopes when they are produced in quantity.

Fission Energy

The emission of a relatively large quantity of energy accompanies fission. Nearly 200 MeV per fissioning nucleus are released in the case of ^{235}U, compared to a few electron volts per atom in a chemical reaction. Fission energy is divided among the products in the manner shown in Table 16-1, a preponderant amount appearing as the kinetic energy of the recoiling heavy fission fragments.

The origin of the energy released in fission can be considered qualitatively in the following manner. Many experimental measurements have shown that the masses of nuclei are less than the sums of the masses of their constituent neutrons and protons. This difference m, called the mass defect, corresponds

Table 16-1 Typical Distribution of Fission Energy among Products[26]

Source of Energy	Quantity, MeV
Kinetic energy of fission fragments	167
Prompt γ-rays	6
Kinetic energy of neutrons	5
Fission product decay	
γ-rays	6
β-particles	8
Neutrinos	12
Total	204

to a quantity of energy given by the well-known relation $E = mc^2$, where E and m are expressed in ergs and grams, respectively, and c is the velocity of light, 3×10^{10} cm/sec. This is the energy which would be released in assembling neutrons and protons to form a nucleus and, conversely, the energy required to disintegrate a nucleus into its primary constituents. In the second concept it is designated as a binding energy and is a measure of the tightness with which the primary particles are held together. An inspection of these properties of the elements shows that the binding energy per primary nuclear particle (proton or neutron) is greatest for nuclei of intermediate mass, that is, in the mass number range of 50 to 120, whereas the binding energy per nucleon for the nuclei of higher mass number is somewhat less. More energy would be released, therefore, in assembling fission products from free protons and neutrons than in assembling a ^{235}U nucleus. A net release of energy occurs, then, as the more massive nucleus is broken into two nearly equal parts. A comparison of the masses of the particles in the last reaction, the third of Eq. (16-6), shows that the sum of those on the right is about 0.2 amu less than the ^{235}U mass plus the neutron mass.

Delayed Neutrons

More than 99% of the neutrons produced in fission arise from the disintegration of the compound nucleus and make their appearance within the order of 10^{-12} sec after the fission is inaugurated. The remaining neutrons (0.65% of those from ^{235}U fission, 0.27% from ^{233}U, and 0.21% from ^{239}Pu) result from the radioactive disintegrations of the unstable fission products having intermediate mass.* Since these disintegrations are statistical and occur with definite half-lives, the resulting neutrons appear, on the average, at discrete times following the fissions responsible for them. There are at least six groups of these neutrons corresponding to six precursors with half-lives ranging from 0.2 to 56 sec. These neutrons are called delayed neutrons and are of utmost importance in the control of nuclear chain reactions. Their effect is to increase the apparent neutron lifetime from something like 10^{-4} sec to the order of 0.1 sec, and they have a significant influence on the time behavior of the neutron population in a nuclear chain reacting system.

Nuclear Chain Reactions

In the absence of nonfission-producing absorptions and of other losses of neutrons, a yield of one neutron from fission is adequate for establishing a

* Energy differences between prompt and delayed neutrons and the properties of the environment in which they are produced determine an "effectiveness" of the delayed neutrons so that they, in fact, behave as though they were in greater proportion. Effective delayed neutron percentages may be as large as 0.9% or more.

nuclear chain reaction since that neutron will produce another fission. Actually, of course, for chain reactions to develop, the neutron yield must be greater than unity in order to counter losses of neutrons through absorptions (even in the fissionable material itself) which do not cause fission and through neutron leakage from the fissionable material. It is also necessary that the fission-produced neutrons have an energy at least equal to that with which they, in turn, may cause fission.

The uranium isotopes having a mass of 233 and 235 and the plutonium isotope of mass 239 have combinations of nuclear properties conducive to the establishment of chain reactions. On the average, these nuclei, upon fission, produce more than two neutrons. Although these fission neutrons are produced with high kinetic energy ($\approx$2 MeV), they can be readily degraded in energy, or moderated, through collision with other nuclei until they reach thermal energies ($\approx$0.025 eV), where fission cross sections of these isotopes are largest. In contrast, ^{238}U is also fissionable and yields, on the average, more than two neutrons per fission. However, neutrons of about 1 MeV energy are required to cause fission of ^{238}U. Since this energy is approximately the same as that of the fission-produced neutrons at birth, the probability of their capture by ^{238}U before they lose too much energy to initiate fission is small. For this reason ^{238}U will not support a chain reaction.

The concept of neutron economy or neutron conservation from generation to generation is facilitated by defining a quantity called the neutron multiplication factor, usually denoted by k. This factor is the ratio of the neutron population of one generation to that of the generation immediately preceding. It will be shown later that k is characteristic of the combined properties of the fissionable and associated materials, as well as of their shape, size, and arrangement. To maintain a chain reaction in equilibrium, that is, to establish a critical system, the value of k must be unity. If k is less than unity, the system is subcritical and the neutron population will decrease; if k is greater than unity, the system is supercritical and the neutron population will increase.

The time dependence of the neutron population on k can be shown qualitatively in the following simplified manner. If n is the neutron density at time t, the density in the next generation, at time $t + L$, is kn, L being an average neutron generation time, if we consider the fact that the production of some neutrons is delayed. The time rate of change in neutron density or population is

$$\frac{dn}{dt} = \frac{kn - n}{L} = \frac{n(k-1)}{L} \tag{16-7}$$

The solution of this equation is

$$\frac{n}{n_0} = \exp \frac{(k-1)t}{L} \tag{16-8}$$

where n_0 is the neutron density at time $t = 0$. By defining $L/(k - 1)$ as the period T of the chain reaction, Eq. (16-8) may be written

$$\frac{n}{n_0} = e^{t/T} \tag{16-9}$$

The period is thus the time required for the neutron population to change by a factor e, the logarithmic base. Equation (16-9) shows that the behavior of n with time is exponential. When k becomes greater than unity, the neutron population, and hence the fission rate and power developed, increase rapidly. Assume, for example, that k is 1.0005 and that the mean neutron generation time is 0.1 sec; then after 1 sec the neutron population will have increased about 0.5%. If there were no delayed neutrons, L would be about 10^{-4} sec and the neutron population would increase 150-fold in 1 sec.

If the population of all neutrons in a chain reacting system, including the delayed neutrons, remains constant in time, the system is delayed critical and is controllable. If the delayed neutrons could be removed from this critical system in some manner and not allowed to propagate, the system would become subcritical. A change in some property of the system, such as the mass, for example, must be made to reestablish criticality with prompt neutrons alone. Such a prompt-critical system cannot be controlled because of the short mean life of prompt neutrons. In the actual case, the neutron population rises rapidly in a prompt critical system because of the delayed-neutrons. Since, in its usual usage, a multiplication factor k equal to unity signifies delayed critical, prompt criticality ensues when k is increased by an amount equal to the fraction of delayed neutrons. It is apparent that the reactivity margin between delayed- and uncontrollable prompt-critical conditions is exceedingly small.

Cross Sections

Consider a thin slab of material placed normal to a stream of neutrons. If N is the number of nuclei per unit volume of the material and dx is the thickness of the slab, $N\,dx$ is the number of nuclei per unit area of the slab. If there are n_0 neutrons per unit volume of this idealized, monoenergetic neutron stream and v is their velocity, n_0v is the number of neutrons incident upon unit area of the slab per unit time. (The quantity nv is often referred to as the neutron flux, ϕ, and, in a more general sense, as the neutron track length. In the latter connotation it is the total distance traveled per unit time by all the neutrons in the unit volume.) The number of events per unit time per unit beam area is proportional to the product of the number of potentially interacting particles and the number of potentially interacting nuclei. That is,

$$\text{Number of reactions per unit area per unit time} = \sigma n_0 v N\, dx \quad (16\text{-}10)$$

where σ is a proportionality constant, called the microscopic cross section, expressed in units of area. Stated differently, the cross section σ is the ratio of the fraction of the target nuclei that interact to the total number of incident neutrons available for interaction; it refers to a single event involving a single nucleus. A typical value for σ describing reactions of neutrons with matter is the order of 10^{-24} cm², or one barn. Solution of Eq. (16-10) for neutron absorption yields the well-known exponential decrease in intensity as a function of x. Cross sections describe each of the possible reactions of neutrons with nuclei: absorption, scattering, capture, and fission.

Table 16-2 Thermal-Neutron Properties of Fissile Isotopes[a]

	^{235}U	^{233}U	^{239}Pu
Absorption cross section, σ_a (barns)	678	573	1014
Fission cross section, σ_f (barns)	577	525	741
Capture cross section, σ_c (barns)	101	49	286
Scattering cross section, σ_s (barns)	15	13	10
Neutron yield per fission, ν	2.43	2.50	2.89
Neutron yield per neutron absorption, η	2.08	2.29	2.12

[a] Neutron velocity = 2200 m/sec; energy = 0.025 eV.

A related quantity is the macroscopic cross section, $\Sigma = \sigma N$, defined as the product of σ and N the number of nuclei per unit volume. The macroscopic cross section has the units of reciprocal length and is a property of a material rather than of a single nucleus.

Experiment shows that microscopic cross sections are strong and complex functions of the energy of the incident particles. In general, cross sections decrease as the neutron energy increases, following, in the simplest case, the relation $\sigma \propto 1/v$, where v is the neutron volocity. There are few clear-cut examples of this relation, the most common one being the neutron absorption cross section of boron. As a rule, there are many resonances superposed on the simple $1/v$ relation. These resonances represent high probabilities of events at particular energies and may have values of the order of 10^4 barns.

Table 16-2 shows some thermal-neutron properties of the three abundant fissile nuclei. It should be remembered that the values, like many physical properties, are subject to change with improved experiments.*

* The standard reference for neutron cross sections is Reference 13. The values in Table 16-2 were taken from Supplement No. 2 of the Second Edition, 1965.

16-5 NUCLEAR CHAIN REACTING SYSTEMS*

Infinite Homogeneous Medium

Consider a homogeneous medium of unlimited size characterized by a uniform mixture of fissionable atoms and other neutron absorbing and scattering elements. In scattering, a neutron generally loses some of its kinetic energy to the scattering nucleus. Thus, after a number of collisions, a fast neutron will be slowed down or moderated to a thermal neutron, unless, of course, it is absorbed before becoming thermalized. The average energy decrease per collision is greater with nuclei more nearly equal in mass to the neutron. Hydrogen is one of the most effective moderators since the proton and neutron have approximately equal masses. Other low-mass-numbered elements of low absorption cross section, such as deuterium, beryllium, and carbon, are also good moderators.

The cycle of events to be considered in this. Fission neutrons are born with an energy of about 2 MeV. In subsequent collisions with nuclei, these fast neutrons are either scattered with loss of kinetic energy or are absorbed. Some of these fast-neutron absorptions may result in "fast" fission. Other neutrons survive thermalization and continue to collide with nuclei until they are absorbed either by fissionable or by other nuclei. Some of the absorptions by fissionable nuclei result in fission and the chain is continued.†

If N_1 is the number of fission neutrons born in the first generation and if p_1 is the probability that some of these will produce fission before they are thermalized, the number of second generation fission neutrons arising from this source is

$$N_1 p_1 \nu \tag{16-11}$$

where ν is the number of neutrons born per fission. On the other hand, if p is the probability that neutrons will survive until they are thermalized, if p_2 is the probability that they will then be absorbed by a fissionable nucleus, and, furthermore, if p_3 is the probability that an absorbed thermal neutron produces fission, then the over-all probability that a fast neutron will slow down and produce fission as a thermal neutron is given by pp_2p_3. The number of neutrons accomplishing this is $N_1pp_2p_3$ and hence the number of second generation neutrons so produced is

$$N_1 p p_2 p_3 \nu \tag{16-12}$$

* For more complete considerations of this topic the reader is referred to treatises on reactor theory such as References 8 and 26. The discussion in this section is based in part upon these sources, and their aid is acknowledged together with that of R. K. Osborn.[17]

† In the medium of infinite extent being considered here, no neutrons are lost by leakage from the medium.

The total second generation population is the sum of this quantity and the number of neutrons produced by fast fission given by Eq. (16-11):

$$N_2 = N_1 p_1 \nu + N_1 p p_2 p_3 \nu \qquad (16\text{-}13)$$

which can be rewritten as

$$\frac{N_2}{N_1} = \nu p p_2 p_3 \left(\frac{p_1 + p p_2 p_3}{p p_2 p_3} \right) \qquad (16\text{-}14)$$

The term in parentheses is the ratio of the number of fissions produced by fast and thermal neutrons to the number produced by thermal neutrons alone and is the fast fission factor ϵ.

The quantity p is the probability that neutrons will be thermalized, that is, that they will escape capture at high and intermediate energies. It is referred to as the resonance escape probability since many of the absorptions in this energy region are resonance phenomena.

The probability p_2 that a thermal neutron will be absorbed by a fissionable nucleus is called the thermal utilization f. It is expressed in terms of macroscopic absorption cross sections as

$$p_2 = f = \frac{\Sigma_a^F}{\Sigma_a^F + \Sigma_a^O} \qquad (16\text{-}15)$$

Here the subscript a denotes absorption and the superscripts F and O denote the fissionable and all other materials, respectively.

The probability p_3 that a thermal neutron absorbed by a fissionable nucleus will produce fission is

$$p_3 = \frac{\Sigma_f^F}{\Sigma_a^F} \qquad (16\text{-}16)$$

where the subscript f denotes fission. Multiplication by ν gives the yield of fission neutrons per neutron absorbed in the fissionable material,

$$\eta = \frac{\nu \Sigma_f^F}{\Sigma_a^F} \qquad (16\text{-}17)$$

which is an index of its efficiency.

The left side of Eq. (16-14) is the neutron multiplication factor, k, so that the equation can now be rewritten as

$$k_\infty = \eta f p \epsilon \qquad (16\text{-}18)$$

where the subscript ∞ refers to a medium of infinite extent, from which there is no neutron leakage. It should be noted that Eq. (16-18) is independent of absolute values of neutron density or flux and refers only to the ratio of these quantities in successive generations of fission-produced neutrons.

The values of p and ϵ depend upon the fission, absorption, and scattering properties of the constituents of the medium for neutrons of high and intermediate energies, and upon the number of collisions the neutrons make as their energy is reduced. Some insight into their meaning can be obtained from the following qualitative argument. The average number of collisions experienced by a neutron in traversing an energy interval dE' at some energy E' is

$$\frac{dE'}{\xi E'} \tag{16-19}$$

where ξ is the logarithmic energy decrement, that is, the change in the logarithm of the energy per collision during slowing down. This quantity depends on the masses of the colliding nuclei. An estimate of the probability that a neutron will be absorbed in dE' is

$$\frac{\Sigma_a}{\Sigma_a + \Sigma_s}\left(\frac{dE'}{\xi E'}\right) \tag{16-20}$$

where cross sections are evaluated at E'. An estimate of the probability that an absorption will occur within the energy range between that of fission neutrons at birth E_0 and thermal energy E_{th} is

$$\int_{E_{\text{th}}}^{E_0} \frac{\Sigma_a}{\Sigma_a + \Sigma_s}\left(\frac{dE'}{\xi E'}\right) \tag{16-21}$$

and the probability that the neutron will survive this interval is

$$1 - \int_{E_{\text{th}}}^{E_0} \frac{\Sigma_a}{\Sigma_a + \Sigma_s}\left(\frac{dE'}{\xi E'}\right) \tag{16-22}$$

For absorbers with particular properties a more exact treatment gives for the latter probability

$$p = \exp\left[-\int_{E_{\text{th}}}^{E_0} \frac{\Sigma_a}{\Sigma_a + \Sigma_s}\left(\frac{dE'}{\xi E'}\right)\right] \tag{16-23}$$

The fast fission factor is the ratio of the number of neutrons produced in all fissions to the number produced in fissions by thermal neutrons only,

$$\epsilon = \frac{N_1 \eta f p + \text{neutrons produced by fast neutrons}}{N_1 \eta f p} \tag{16-24}$$

The number of fissions produced by neutrons of energy in some intermediate interval, dE', is the product of (1) the number of fission-produced neutrons, (2) the probability of their survival to energy E', (3) the number of collisions or events experienced in interval dE', and (4) the probability that one of these events will be a fission. The number of fissions in the whole energy region above thermal is the integral of this multiple product. This product, multiplied by ν gives the expression:

Number of neutrons produced in fast fissions

$$= N_1\nu \int_{E_{\mathrm{th}}}^{E_0} p(E_0 \to E')\left(\frac{dE'}{\xi E'}\right)\frac{\Sigma_f}{\Sigma_a + \Sigma_s} \tag{16-25}$$

where $p(E_0 \to E')$ is the resonance escape probability to energy E'; $dE'/\xi E'$ is the number of collisions experienced by the neutron within energy interval dE'; and the macroscopic fission, absorption, and scattering cross sections, Σ_f, Σ_a, and Σ_s are evaluated at E'. (The number of neutrons produced per fission, ν, is slightly energy dependent.)

Although the value of the product $p\epsilon$ for homogeneous hydrogen-moderated ^{235}U is not greatly different from unity, it will be pointed out later that these terms do make a significant contribution to the reactivity of a strongly heterogeneous array of fissionable and moderating materials.

Finite Homogeneous Medium

Consider a finite volume of the material just discussed in which the events in a neutron's lifetime are the same as before, with an added probability now that they can escape from the volume. Let the volume be located in a vacuum so that there is no return of the neutrons that escape. Neutrons which are in the process of being slowed as well as thermal neutrons can be lost from the volume by leakage. The behavior of thermal neutrons will be first discussed from the viewpoint of elementary diffusion theory with reference, for analogy, to diffusion in gases. Later the effect of leakage of fast neutrons on only the multiplication constant will be noted.

A concentration gradient in a system of particles (neutrons or gas molecules, for example) gives rise to a transport of the particles in a direction opposite to that in which the gradient increases. This transport may be likened to a current which can be expressed by

$$\mathbf{J} = -D \operatorname{grad} \phi \tag{16-26}$$

where J is expressed as particles per unit area per unit time, D is the diffusion constant in units of length, ϕ is the neutron flux expressed as the product of the number of particles per unit volume and the average distance each travels per unit time, $\phi = nv$. The minus sign enters because the current and the increase in concentration are oppositely directed. The relationship, Eq. (16-26), is sometimes called Fick's Law of Diffusion.

The rate of change of neutron density within a volume V can be obtained in the following manner. Consider an increment of volume, $dx\, dy\, dz = dV$, located in a neutron field in which there is a gradient. The flow in the x

direction into dV is $J_x\, dy\, dz$ and the flow out of dV is $J_{(x+dx)}\, dy\, dz$ where the subscripts denote the coordinates at which J is measured. The net loss of neutrons in the x direction from dV per unit time is

$$\begin{aligned} [J_{(x+dx)} - J_x]\, dy\, dz &= -D\left[\left(\frac{\partial \phi}{\partial x}\right)_{(x+dx)} - \left(\frac{\partial \phi}{\partial x}\right)_x\right] dy\, dz \\ &= -D\frac{\partial^2 \phi}{\partial x^2}\, dx\, dy\, dz \\ &= -D\frac{\partial^2 \phi}{\partial x^2}\, dV \end{aligned} \qquad (16\text{-}27)$$

By similar reasoning, the leakage per unit time along the other coordinates is identical in form. The net thermal (monoenergetic) neutron leakage per unit volume is therefore

$$-D\left(\frac{\partial^2 \phi}{\partial x^2} + \frac{\partial^2 \phi}{\partial y^2} + \frac{\partial^2 \phi}{\partial z^2}\right) = -D\nabla^2\phi \qquad (16\text{-}28)$$

There are three terms in the neutron balance equation describing the critical condition in the finite homogeneous medium, if, for the moment, the mechanism whereby fission neutrons are thermalized is neglected:

$$\text{production} = \text{absorption} + \text{leakage}$$

$$\nu\Sigma_f\phi = \Sigma_a\phi - D\nabla^2\phi \qquad (16\text{-}29)$$

This expression can be rewritten

$$\nabla^2\phi + \frac{\nu\Sigma_f - \Sigma_a}{D}\phi = 0 \qquad (16\text{-}30)$$

or

$$\nabla^2\phi + B^2\phi = 0 \qquad (16\text{-}31)$$

where

$$B^2 = \frac{\nu\Sigma_f - \Sigma_a}{D} \qquad (16\text{-}32)$$

The quantity B^2 is referred to as the material buckling. We shall see later how the material buckling in a critical system is related to the thermal neutron distribution.

From Eqs. (16-29) and (16-32) the thermal neutron leakage is given by

$$\text{leakage} = (\nu\Sigma_f - \Sigma_a)\phi = \phi D B^2 \qquad (16\text{-}33)$$

The probability that leakage will occur is the ratio of the leakage to the disappearance of neutrons from the system by all methods, leakage plus absorption. The nonleakage probability is therefore

$$1 - \frac{\phi\, DB^2}{\phi\, DB^2 + \phi\Sigma_a} = \frac{\Sigma_a}{DB^2 + \Sigma_a} = \frac{1}{1 + L^2B^2} \tag{16-34}$$

where $L^2 = D/\Sigma_a$ is the square of the thermal diffusion length.

It can be shown that

$$L^2 = \tfrac{1}{6}\, \overline{r^2} \tag{16-35}$$

where $\overline{r^2}$ is the mean square (crow-flight) distance a neutron travels after thermalization before it is absorbed.

In considering the leakage of fast neutrons, it must be noted that the actual slowing down process is stepwise rather than continuous. A neutron loses some energy in a single collision, then travels with constant velocity until the next collision, and so on. The magnitude and frequency of these energy decrements depend upon the masses of the scattering nuclei and upon the neutron velocity and the distance of travel between events. It is apparent that the rate of leakage during the slowing down process depends upon these slowing down properties. If the distance traveled during slowing down is large, there is a large probability of leakage before thermalization. The spatial distribution of these neutrons in the process of being slowed down is therefore an important consideration.

Fermi assumed that the actual stepwise slowing down process of a neutron can be replaced by one that is continuous in time. Relations for the distribution of neutrons in both space and energy can then be derived in terms of a property of the medium called the Fermi age. It should be pointed out, however, that the Fermi age τ is not a unit of time, but has the dimensions of length squared. The relation between the age and the distance r between the point of neutron birth and the point at which the neutron is thermalized is

$$\tau = \tfrac{1}{6}\, \overline{r^2} \tag{16-36}$$

It can be shown that the probability of neutrons at intermediate energies not leaking from a medium of finite dimensions is $e^{-\tau B^2}$; when B^2 is small, this expression can be written as

$$1 + \tau B^2 \tag{16-37}$$

The continuous slowing down model of neutron energy degradation gives a poorer representation for collisions with light nuclei, such as hydrogen, than with heavy ones.

Equation (16-18) expressed the neutron multiplication factor k in terms of the neutron yield in fission and the probabilities of the various events in the neutron lifetime, provided the medium was boundless. Incorporation of the nonleakage probabilities Eqs. (16-34) and (16-37) in Eq. (16-18) yields

$$k_{\text{eff}} = \frac{nfp\epsilon}{(1 + L^2B^2)(1 + \tau B^2)} \tag{16-38}$$

as the expression for the effective multiplication factor of a finite system.

To obtain a concept of the spatial distribution of thermal (monoenergetic) neutrons, we note that Eq. (16-31) describes not only the neutron balance in a finite, unreflected critical system but that it also expresses the steady-state

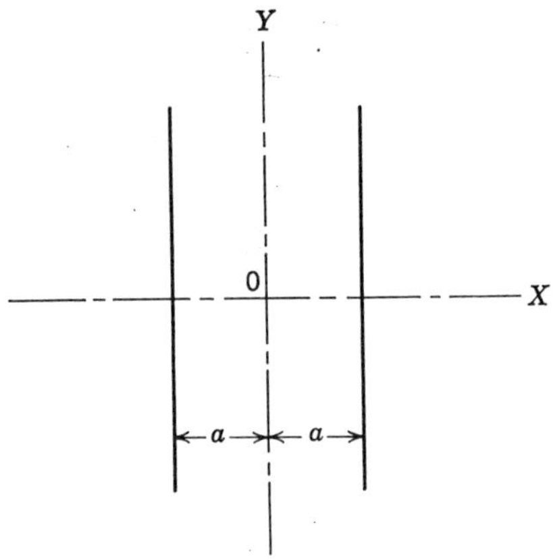

Fig. 16-1

spatial distribution of neutrons provided the terms are evaluated. For a slab-shaped medium infinite in extent in two dimensions and of half-thickness a (Fig. 16-1), Eq. (16-31) can be written

$$\frac{d^2\phi}{dx^2} + B_x^2\phi = 0 \tag{16-39}$$

where the subscript x refers to the distribution of neutrons associated with the x direction. The general solution to the homogeneous equation (16-39) is

$$\phi = C_1 \sin B_x x + C_2 \cos B_x x \tag{16-40}$$

where C_1 and C_2 are arbitrary constants.

Two boundary conditions are needed to determine these constants. First, since the yz plane is a plane of symmetry of the flux distribution, C_1 must be zero. Second, the flux must effectively vanish at some distance, δ, called the extrapolation distance, outside the boundary of the medium. The extrapolation distance is a mathematical convenience rather than a true physical quantity; it has a value of no more than a few centimeters, although neutrons can be detected at distances of many meters. This boundary condition gives

$$\phi = 0 = C_2 \cos B_x(a + \delta) \tag{16-41}$$

so that $B_x(a + \delta) = n(\pi/2)$ where n is any odd integer.

The solution of Eq. (16-40) now becomes

$$\phi = C_2 \cos B_x x = C_2 \cos \frac{n\pi}{2(a + \delta)} x \tag{16-42}$$

The simplest, or fundamental, distribution occurs for $n = 1$, which gives, from Eq. (16-42)

$$B_x^{\ 2} = \left[\frac{\pi}{2(a + \delta)}\right]^2 \tag{16-43}$$

This function of the finite dimension of the critical slab is called the geometric buckling.

If the other dimensions of the slab were finite, the geometric buckling would be

$$B^2 = \left[\frac{\pi}{2(a + \delta)}\right]^2 + \left[\frac{\pi}{2(b + \delta)}\right]^2 + \left[\frac{\pi}{2(c + \delta)}\right]^2 \tag{16-44}$$

which describes a parallelepiped of dimensions $2a$, $2b$, and $2c$.

By similar arguments, it can be shown that for a sphere of radius r, the buckling is

$$B^2 = \left[\frac{\pi}{r + \delta}\right]^2 \tag{16-45}$$

and for a right circular cylinder of radius r and half-height h the buckling is

$$B^2 = \left[\frac{\alpha}{r + \delta}\right]^2 + \left[\frac{\pi}{2(h + \delta)}\right]^2 \tag{16-46}$$

where $\alpha = 2.405$ is the first zero of the Bessel function J_0.

It is seen in the preceding discussion that the buckling B^2 has two rather different physical meanings. In Eq. (16-32) it is a property of the material comprising the chain reacting system and in this connotation it is usually called the material buckling. In Eq. (16-42), B appears in the argument of the cosine as a measure of the shape or "buckling" of the spatial neutron flux distribution. The square of B is called the geometric buckling. In both usages of B^2, reference was made to a critical system, one sustaining a nuclear chain reaction, which establishes the criterion that for the multiplication factor to be unity, the material and geometric buckling must be equal.

This criterion is of practical importance in the following way. The material buckling depends only upon the properties of the material and is independent of geometry. If therefore the critical dimensions of a particular fissionable material in some geometrical configuration are known—for example, if the height and diameter of a cylinder of an aqueous solution of a ^{235}U salt are known—then the critical dimensions of the material in another geometry can, in principle, be calculated by equating the geometric bucklings of the two volumes. The critical radius of a sphere in this case can be obtained by equating expressions (16-45) and (16-46).

A difficulty in this procedure arises from the inexactness with which the extrapolation distance is usually known. Precise evaluation from measurements of neutron flux distributions is difficult and the results of calculations from the properties of the material are uncertain. The method does, however, yield useful approximations.

The foregoing, based on diffusion theory, does not describe the behavior of neutrons near boundaries or near other perturbations in a medium, and its application to volumes of small dimensions is unreliable. The results also rest on an assumed isotropic scattering of the neutrons by the moderating nuclei as well as other simplifications introduced by the model of continuous slowing down. Comparison of the results of diffusion theory methods of analysis with those from experiments have shown fair agreement, particularly when the fissionable material is well moderated, as in a dilute aqueous solution.

Analyses have been made by more elaborate diffusion theory methods than those described here, although, in general, the above results are not strikingly improved. For example, the energy interval between 2 MeV and thermal energy can be divided into many steps in order to better evaluate appropriate cross sections. For each of these subintervals a diffusion equation can be written with terms describing absorption and leakage and the loss and gain of neutrons from and to that energy interval by scattering. These equations can be solved simultaneously by electronic computers.

Diffusion theory treats only neutron densities, fluxes, and currents—not their kinetic energy or the direction of their motion. These latter quantities, together with the properties described in diffusion theory, are included in analyses of neutron behavior by transport theory. In an even more sophisticated method, called the Monte Carlo method, the distribution of successive events in the lifetime of a neutron are followed and, provided sufficient histories are included to give the results statistical significance, the neutron behavior from generation to generation can be established. In addition to the multiplication factor, calculations of this kind, using high-speed electronic computers, give information on the spectral and spatial distribution of neutrons and their mean life-span. The agreement between the experimental values of critical dimensions and those calculated by these methods is very much better than that obtained with simple diffusion theory. The reader is referred to reactor physics texts for treatment of these topics.

In addition to inadequacies in any calculational model, there are also large uncertainties in the knowledge of basic physical constants, primarily cross sections, which must be inserted in the analyses to yield numerical results.

Heterogeneous Systems

A core of fissionable material surrounded by a substance which scatters and moderates neutrons without excessive absorption is a simple example of a heterogeneous system. There are several effects of this external medium, called a neutron reflector, on the neutron balance within the core. Whereas in the absence of a reflector, any neutron reaching the boundary of the core was irretrievably lost, any leakage neutron now has some probability of being returned. The neutron which is returned has, on the average, a lower energy than that with which it left, particularly if the reflector has a low nuclear mass. Reflected neutrons have therefore a greater probability of producing fission than they had at the time of leakage.

These effects reduce the critical parameters by amounts which vary widely with the reflector material and its geometry. As one example, consider a long critical cylinder of ^{235}U solution which is on the order of 10 cm in radius. Surrounding the cylinder with water makes the cylinder supercritical because of the decreased leakage. To return the system to criticality requires a decrease in the cylinder radius by a few centimeters. This change in radius is sometimes called the reflector savings.

In another type of heterogeneous system the fissionable material and the moderator occupy discrete volumes, which may be exemplified by a lattice of uranium metal rods in water. The effect of this distribution upon the reactivity depends in a complex way on the properties of materials. In general, the critical mass of uranium highly enriched in ^{235}U is minimal when the uranium and the moderator are homogeneously mixed. Any lumping of the uranium increases the critical mass provided, of course, other properties of the mixture are unchanged.

The situation is reversed, however, with natural uranium or uranium enriched to only a few percent in ^{235}U. The more abundant ^{238}U does not fission upon absorption of neutrons having energy less than $\sim$1 MeV, and has a strong neutron absorption resonance in the electron-volt region. This means that if neutrons collide with ^{238}U nuclei as they pass this lower energy in being slowed down, there is a high probability of their capture and disappearance from the chain. If, however, this uranium of low ^{235}U content is heterogeneously distributed in water in such a manner that the energy of the neutrons is reduced below that of this ^{238}U resonance while they are in the water, the ^{238}U nonfission absorption is significantly lessened. In other words, the value of the resonance escape probability p is increased by latticing. The critical mass of a lattice of low-enriched uranium in water is much lower than that of a homogeneous mixture of the same materials in the same proportions.

A number of large reactors have operated with natural uranium arranged in the form of a lattice of uranium "slugs" embedded in a suitable moderator such as graphite.

16-6 PREVENTION OF ACCIDENTAL CRITICALITY

Introduction

Consideration is now given to the question of how a neutron chain reaction can be prevented in fissionable materials that are being processed, stored, or transported on an industrial scale. The question may be divided into several parts:

First, there are purely scientific problems connected with the conditions needed for a chain reaction; these were considered in Section 16-5. If a definitive analytical model, accurate cross sections, and other nuclear data were available, it would be possible to calculate critical quantities. At the present time, however, the data are not sufficiently complete and the theoretical methods are not sufficiently precise to permit adequate calculation of critical masses except with simple materials in simple geometries without neutron absorbers. It is necessary to rely primarily on experimental measurements of critical conditions and perhaps make short extensions of these by theory.

Second, there are problems of an engineering type. In some processes it is not only necessary to determine in detail the normal physical conditions but also very important to know those unusual conditions which are physically possible within the process and which may be favorable for chain reactions.

Third, work on an industrial scale involves men and equipment, which entails certain administrative problems. In considering possible events that may lead to dangerous configurations of fissionable material, it is necessary to know the rules under which men operate the process equipment, what violations of procedures are possible, and what physical controls exist to minimize violations. It is only with such knowledge that a careful administrative system of routine checks can be prescribed and carried out effectively.

The approach to nuclear criticality problems in an industrial plant can be described succinctly as follows. With guidance from experimentally determined critical parameters,* a detailed study can be made of the equipment and

* Many data describing critical systems have been compiled.[22] Reference 16 gives values for the dimensions of fissile materials which may be singly applied to guide safety in operations. Many of the more restrictive limits may be relaxed by a detailed study of particular operations by knowledgeable and experienced individuals.

conditions in which the fissionable material is processed, and a safe practical distribution of mass throughout the plant can be recommended. Nuclear safety operating rules should be formulated in detail and an administrative system set up to enforce them rigorously. The aim of such a program is to attain a high degree of assurance that chain reactions will not occur yet to operate the plant efficiently and economically.

The Basic Parameters

An examination of the expression Eq. (16-38) for the neutron multiplication factor shows its dependence on a number of individual properties and dimensions of a potentially critical volume or core. Since each of the quantities in the numerator of Eq. (16-38) is a function of one or more macroscopic cross sections which, in turn, depend upon nuclear densities and microscopic cross sections, it follows that k_{eff} is determined partly by the nuclear properties of the constituents and their concentration. The dimensions and shape of the core, which appear in the expression for the buckling, determine the loss of neutrons by leakage. In the case of a reflected core the dimensions in the expression for the buckling include a quantity called the reflector savings which depends on the nuclear properties of both the reflector and the core. The reflector savings is the amount by which a critical physical dimension of an unreflected core—a sphere radius or a slab half-thickness, for example—is reduced by the addition of the reflector. The dimensions, minus the appropriate extrapolation distance or reflector savings, and the density of the fissile material fix the critical mass. The presence of neutron absorbing materials appears in the thermal utilization f. The properties of ^{238}U, important when the uranium is only partly enriched in the ^{235}U isotope, are incorporated in η, latticing influence, and resonance escape probability p. It is thus apparent that control of a number of parameters, either singly or in combination, can restrict the establishment of criticality. These will be considered now in somewhat more detail.

It is convenient to select a single property in a system of fissionable material as being more basic than others, and to consider the manner in which its value is varied by other factors. If we select the critical mass, then one factor of importance in determining its value is the leakage of neutrons that would otherwise produce fissions. It has been pointed out that the leakage depends on the shape, size, and composition of the system and on the neutron-reflecting properties of surrounding materials. It is possible, for example, to specify dimensions of a container of solution that gives sufficient leakage, by virtue of a large surface-to-volume ratio, to prevent a chain reaction, regardless of the quantity or quality of material contained. If the container is encased in a cooling jacket, or is near other process equipment or structural materials which can reflect neutrons, its dimensions must be reduced because of these conditions. In the problems typically encountered in industry, it is assumed that natural water, concrete, graphite, and steel act as reflector materials. Although more effective reflectors are known (for example, heavy water and beryllium), they are uncommon in processing plants.

The value of the critical mass is also extremely sensitive to the presence of neutron moderating elements, particularly hydrogen, mixed with the fissionable material. The ratio of the number of hydrogen atoms to the number of fissionable atoms may range from zero for uranium metal or a dry unhydrated salt to several thousand for dilute aqueous solutions. The corresponding critical mass may vary from a few tens of kilograms of metal, through a minimum of a few hundred grams in an aqueous solution of optimum concentration, to indefinitely large amounts in dilute solutions in which the low fuel density and the neutron absorption by hydrogen make chain reactions impossible. In this latter limit, nuclear safety is assured by the chemical concentration alone.

As a rule, the critical mass of a fissionable material associated with a moderator is minimal when the two are intimately mixed, as in an aqueous solution. As pointed out, however, uranium containing a few percent ^{235}U is an exception to this generalization, a lower critical mass resulting when it is latticed in a moderator. The maximum enrichment of the uranium at which latticing does reduce the critical mass is not known exactly, although it is probably between 3% and 5% ^{235}U by weight.

It has not been unequivocally established that natural uranium latticed in water will always be subcritical. If, however, these materials can constitute a critical array, the mass of uranium required is greater than amounts usually encountered in any industrial processes and, in addition, it would be necessary for the dimensions of the uranium units to be carefully selected and for them to be precisely latticed in the water. Homogeneous mixtures of natural uranium and water in any proportion, however, cannot be made critical because of the resonance absorption in ^{238}U. In fact, it has been shown that for such a mixture to be critical, the ^{235}U content of the uranium must be about 1%.* This property of homogeneous, hydrogen-moderated, slightly enriched uranium, whereby it cannot be made critical, provides a convenient means of assuring safety in some industrial processes.

The critical mass of a fissionable isotope also depends upon its density, or concentration, in homogeneous mixtures. The actual dependence can be specified quantitatively only in special cases, but the critical mass always increases as the density decreases. The critical mass of a sphere of ^{239}Pu metal, for example, is less than that of a spherical volume of dry ^{239}Pu sawdust. The critical mass of ^{235}U in an aqueous solution is greater than that

* The ^{235}U content of natural uranium is 0.72%.

in a homogeneous aqueous slurry of high density UO_2 of the same hydrogen-to-^{235}U ratio, because the mass of U^{235} per unit volume is greater in the slurry.

Neutron absorbing materials, such as cadmium and boron, which have relatively high thermal neutron absorption cross sections, when distributed within the fissionable material can render equipment and processes safe. Adequate experimental data must assure mechanical and chemical stability of the absorber in the process. The inclusion of solid absorbers in the construction and assembly of equipment is often recommended; the use of solutions of neutron absorbers as components of process streams is inherently less acceptable because of the administrative control required to assure their presence. Caution must be exercised, however, in placing neutron absorbing materials on the outside of vessels containing fissionable materials. For example, if vessels surrounded by a thin layer of cadmium are, in turn, surrounded by water, the cadmium is very effective in increasing the mass in the vessel required for criticality, since the cadmium prevents neutrons thermalized in water from reentering the vessel. In the absence of the external water, however, the use of cadmium alone would decrease the critical mass in the vessel because the cadmium would serve merely as a neutron reflector.

The presence of nitrogen in nitrate solutions often used in chemical processing and the presence of ^{240}Pu, a predominantly thermal-neutron absorbing impurity in plutonium solutions, increase the margin of safety of many operations. In plutonium processes containing little or no hydrogen or other moderating nuclei, however, the neutron population is essentially fast and ^{240}Pu is not effective as a parasitic neutron absorber. Small amounts of ^{241}Pu, an isotope readily fissionable by thermal neutrons, can be treated like ^{239}Pu.

Most homogeneous systems of fissionable material have negative temperature coefficients of reactivity due to a density change (including the formation of vapors in liquid systems) and to changes in neutron energy distribution. Although this property is important in reactor design, where it facilitates shutdown in case of a power excursion, it does not contribute to the prevention of such excursions. Much damage can occur before high temperature begins to control a reaction initiated at a low temperature. The magnitude of the temperature coefficient depends on the material, the geometry of the system, and on the range of the temperature change.

In summary, the basic parameters that govern nuclear chain reactions in fissionable materials are the mass, volume, neutron leakage (established primarily by the shape of the volume and the reflector properties of the environment), neutron moderation, the presence of neutron absorbers, and the spatial distribution of the fissionable elements, including their distribution within the moderator.

Specifications of Single Parameters

Table 16-3 gives values of some of the basic parameters describing the readily fissionable isotopes both in an aqueous solution and as a metal containing no internal neutron moderating material. For every value, however, a thick hydrogenous neutron reflector (or its equivalent) is assumed to be present. The quantities in the columns disignated "Recommended" are those suggested for application in the control of nuclear safety; these are so selected that any one, applied singly, will assure safety regardless of other properties or quantities of the material in question. Aggregates of solids, such as bundles of rods and accumulations of pellets, which may become submerged, are specifically excluded. The best estimates of the minimum critical value of each of these controlling parameters, with all others optimized and with an effectively infinite hydrogenous reflector, are also tabulated. These allow an approximate evaluation of the safety factors contained in the recommended values. The safety factors reflect somewhat the uncertainties in the experimental data. The values for solutions are based on optimum chemical concentration. The minimum critical mass and the minimum critical volume do not occur at the same chemical concentration. The two sets of values given for plutonium metal describe the α-phase, having a density of 19.60 gm/cm^3, and the δ-phase, having a density of 15.65 gm/cm^3, respectively. Safety factors appropriate to uncertainties in sampling and analyses should be applied to the recommended values of the chemical concentrations and of the ^{235}U enrichment of homogeneous moderated uranium. If two or more of the basic parameters can be controlled simultaneously, none needs to be as stringent as that recommended in the table.

Arrays of Units

The preceding comments refer to a single unit of fissile material isolated from other units. If, however, the units are not isolated, the exchange of neutrons among those in the group must be considered. Specification of adequate separation of the units and the method of assuring that the specification will not be violated are factors which must receive careful attention, both in the design of plant facilities and, particularly, in storage and transportation. Compactness of storage and shipping arrays, usually desirable in normal industrial methods, is often impossible with fissionable materials.

The probability of neutron interaction, and hence its effect on criticality of an array, is dependent upon such goemetrical factors as the size, shape, and separation of the units, as well as on the over-all size and shape of the array itself. The effects of materials which may be intermingled among the units of an array or which may surround the array (such as the concrete walls of a

Table 16-3 Values of Some Basic Nuclear Safety Parameters

Parameter	^{235}U		^{233}U		^{239}Pu	
	Recommended	Minimum Critical	Recommended	Minimum Critical	Recommended	Minimum Critical
Mass, kilograms						
Aqueous solution	0.35	0.82	0.25	0.59	0.22	0.51
Metal	10.0	22.8	3.2	7.5	2.6	5.6 (α-phase)
					3.5	7.6 (δ-phase)
Diameter of infinite cylinder, inches						
Aqueous solution	5.0	5.4	3.7	4.4	4.7	5.8
Metal	2.7	3.1	1.7	1.9	1.4	1.7 (α-phase)
					1.8	2.1 (δ-phase)
Thickness of infinite slab, inches						
Aqueous solution	1.5	1.7	0.8	1.2	1.2	1.8
Metal	0.5	0.60	0.2	0.3	0.2	0.3 (α-phase)
					0.3	0.4 (δ-phase)
Volume of spherical aqueous solution, liters	4.8	6.3	2.3	3.3	4.4	6.9
Chemical concentration of aqueous solution, gram (of isotope)/liter	10.8	12.1	10.0	11.2	6.5	7.3
^{235}U Enrichment of homogeneous hydrogen-moderated uranium, weight %	0.95	1.0	—	—	—	—

storage vault) are also important. An interacting array of closely spaced units that is critical when submerged may become subcritical if the water is removed. Conversely, an interacting array of more widely spaced units that is subcritical when submerged may actually become critical when the water is removed. In the latter case, the water, as a neutron absorber, isolates the units. The approach to these problems has been essentially emprircal.

Design Criteria

The inclusion of safety features in equipment design rather than in operation is preferred. This practice reduces the chance that a violation of an operating procedure will lead to a criticality accident. Control of safety through limitations imposed on the mass of material or the chemical concentration, for example, is less certain than control by features embodied in the design of the equipment which may include, in addition to shape and size, the presence of neutron absorbers which are mechanically stable and chemically compatible with process materials.

In general, operations with uranium of limited ^{235}U enrichment can be carried out in equipment larger than that recommended for pure ^{235}U. Assurance of this enrichment control combined with appropriately sized vessels is another example of a practical combination of features to produce the over-all safety and economy of an operation.

The safety controls designed into a process should be adequate to allow the occurrence of at least two unlikely, independent, and concurrent changes in one or more of the conditions specified as essential before a nuclear accident is imminent.

In practice, it has been customary to assume operating conditions to be more severe than they probably will be. Piping, for example, is often designed on the assumption that it may become surrounded by a thick layer of water as perhaps it would as a result of the rupture of a water main and the stoppage of drains. A more important reason for conservative designs is the unknown neutron-reflecting properties of nearby concrete walls and floors, of neighboring water lines and process vessels, and of personnel.

Instrumentation

Although radiation-detecting instrumentation is useful in warning of impending hazard, there are practical limitations to its use. An approach to a chain reaction is manifested by the multiplication of the ambient neutron field. There are three components of such a multiplication-measuring system: the source of neutrons, the detector, and the multiplying medium. It is desirable to have the source neutrons distributed within the process material in order that any multiplication may be more readily observed. With some process materials this condition is achieved by spontaneous fissions or by (α, n) reactions in the material itself. The latter reaction occurs, for example, with oxygen in an aqueous solution of plutonium. In other instances, the source may be an encapsulated mixture of beryllium with plutonium or with polonium, placed adjacent to or within process vessels. Multiplication by the fissionable material of neutrons from any of these sources may establish in a neutron-sensitive device a signal which depends upon the multiplication by fissionable material. However, unless the three components are carefully located with respect to each other, the signal may not reach a significant strength until the system becomes supercritical, and then the time rate of change of the radiation level will increase rapidly.

Properties of fissionable isotopes, or of other materials closely associated with fissionable isotopes in chemical processes, can be utilized in indirect methods for criticality control. For example, accumulations of ^{235}U can be detected through measurement of its characteristic γ-radiation before the accumulations become dangerously large. As another example, the absorption by the heavy elements of γ-rays directed through a process stream depends on the chemical concentration of the solution and can be used for concentration control. In a third case, the isotope ^{240}Pu, which has a high spontaneous fission rate, usually accompanies ^{239}Pu in some proportion that is characteristic of the material history. The neutron background in a plutonium process is, therefore, a measure of the ^{239}Pu concentration, and an increase from an established background can signal an abnormal condition in the process stream. All of these indirect methods of safety control are empirical and are based on the calibration of appropriate instruments.

Instrumentation to indicate the radiation hazard existing after the occurrence of an accident should be installed in all operations with potentially critical quantities of fissionable materials.

16-7 ADMINISTRATION OF NUCLEAR SAFETY

Detailed administrative controls of nuclear safety must be established for each organization operating with fissionable material in quantity. Those installations having continuous problems as a consequence of their inventory, or because of frequent alterations in their process, generally assign to a staff group the responsibility for advising design and operating personnel in these matters. The infrequent problems of facilities processing only small amounts of material have often been referred to qualified consultants from other organizations.

The responsibility for nuclear safety must be clearly defined. In some organizations individuals directing activities which may involve nuclear risks are responsible for nuclear safety controls to the same extent that they are responsible for research, design, maintenance, and operation. Guidance in this responsibility is usually obtained from personnel familiar with potential hazards and methods of their control; formal approval of processes and designs by an authorized group should be required in order to establish these responsibilities.

16-8 NUCLEAR ACCIDENT EXPERIENCE

Introduction

The consequences of a nuclear accident cannot be predicted with exactness since the intensity will depend not only upon the characteristics of the material and the manner in which it is made critical, but also upon the immediate environment in which the accident occurs. It is fortunate that to date the accident experience is too small to allow formulation of other than generalized expectations. That rates of energy release from critical accumulations of fissile isotopes mixed with hydrogenous or other moderating materials will exceed those from typical steam explosions is believed to be highly improbable. On the other hand, the rapid consolidation of a number of pieces of ^{235}U metal due, say, to the collapse of shelving, could yield a power release equivalent to that from the detonation of a quantity of high explosive.

Six nuclear accidents are recorded as having occurred in chemical-plant operations and fifteen or more have occurred in the performance of critical experiments. Several of these accidents have been analyzed by Stratton.[24] Two fatalities and a few other significant personnel exposures resulted from the chemical-plant accidents. Although most of the accidents in critical assemblies have occurred in laboratories designed with adequate shielding to protect the experimenters, a number of exposures of varying severity, including three fatalities, have resulted; the fatalities occurred in laboratories where no shielding was provided. In addition, a steam explosion, brought about by an inadvertent addition of reactivity to an experimental nuclear reactor, caused the death of three individuals who probably also received a lethal radiation dose. Only a brief summary of some of these accidents will be included here since all have been reported in the literature.

Accidents in Chemical Processes

Perhaps of greatest interest to the reader are the accidents in process operations since the context of this chapter is directed to industrial operations with fissile materials and not to nuclear reactors or to experiments in which nuclear chain reactions are purposely established. The first recorded fatality[18,20] stemmed from a mishap in a plutonium recovery operation preparatory to making an inventory of fissile materal at the Los Alamos Scientific Laboratory in December 1958. About 3 kg of plutonium was inadvertently accumulated in a 225-gallon, 58-inches-diameter solvent-treating cylindrical tank together with aqueous and organic reagents. The quantity of organic solvent, tributyl phosphate, and its affinity for plutonium resulted in the formation of a slab-shaped layer of liquid, relatively rich in plutonium, of sufficient dimensions and concentration to be only slightly subcritical. Action of a stirrer, started by an operator at the tank, caused an immediate relative displacement of the immiscible liquids which thickened the organic layer sufficiently to initiate a chain reaction. Continued operation of the stirrer, with some additional mixing of the solution by the energy released from the nuclear reaction, distributed the plutonium throughout a larger subcritical volume, limiting the energy release apparently to a single burst of 10^{17} fissions, equivalent to about 1 kw-hr. The operator received a dose equivalent of the order of 10^4 rem and survived only about 36 hours. The pressure developed in the reaction was insufficient to rupture the closed tank although the shock displaced it horizontally about $\frac{3}{8}$ in. at its supports. There was no dispersal of plutonium outside of the system.

The other fatality occurred in a uranium-scrap recovery plant operated by the United Nuclear Corporation[28] in July 1964. In the pertinent phase of the operation, dilute uranium solution was treated in a tank 18 in. in diameter and 26 in. high. The solution was transferred to the tank from storage cylinders having a diameter such that uranium of any concentration was subcritical. Two factors contributed to the accident—one was a recent procedural change, whereby the large vessel was used for removal of uranium from trichlorethane taken from stripping columns; this change was not generally known throughout the organization. The other factor arose from some operational difficulties leading to storage of concentrated uranium solution in at least one limited-diameter cylinder usually containing the dilute solution and stored at the same location. An operator, cognizant of the procedural change but without realizing that a solution of greater than usual concentration was present, poured the contents of the cylinder into the larger vessel where the solution became critical. The operator received a dose estimated at 12,000 rad and survived 49 hours. The occurrence was compounded by some inadequately considered countermeasures by supervisory personnel whereby they received significant but nonfatal radiation exposures from a second excursion initiated by stopping the stirrer in the tank. Actions by the supervisors ultimately transferred the solution to vessels of limited geometry.

Another industrial nuclear accident occurred in mid-1958 at the Y-12 Plant in Oak Ridge. The accident is described in References 21, 7, 15, 12, and 1.

This accident also took place in a salvage process involving ^{235}U at a time when normal production procedures were interrupted in order to make a material inventory. One section of a chemical-operations complex had been restarted while another, downstream, was being cleaned and reassembled. A solution having a uranium concentration of about 50 gm/liter, which had fortuitously accumulated in some restricted-geometry equipment, was subsequently drained into a standard 55-gallon drum intended only for water used in leak-testing the reassembled equipment. The solution was followed into the drum by the water. About 2.5 kg of ^{235}U became critical in an initial volume of about 50 liters, and remained critical for some 20 minutes until dilution of the solution by the continuing inflow of water terminated the reaction. During this interval about 1.3×10^{18} fissions occurred. Personnel, all of whom were at least 3 ft from the source, evacuated expeditiously as a result of automatically sounding alarms. All personnel received dose equivalents of less than 500 rem which resulted from exposure to only the initial portion of the energy release.

In October 1959 an accidental excursion in the Idaho Chemical Processing Plant at the National Reactor Testing Station resulted from the accidental transfer of enriched uranium solution from restricted-geometry equipment into a large waste-storage tank.[9,25] The transfer was caused by an abnormal pneumatic pressure, arising from the maloperation of air spargers in the storage system, which initiated a siphon action. Approximately 200 liters of solution, at a uranium concentration of 170 gm/liter, drained into a 5000-gallon tank where it mixed with about 600 liters of very dilute solution. The uranium was enriched to about 90% in ^{235}U. At some stage of mixing, the solution became critical and remained so for an undetermined time, generating about 4×10^{19} fissions. Gaseous and airborne contamination was apparently forced from vent lines and drain connections by some mechanism actuated by the released energy. The occurrence was signaled by the response of radiation monitors to this contamination. The mechanism by which the reaction was terminated is not entirely known. At its conclusion, however, the solution was found distributed between two tanks, the distribution apparently having been caused by pressure developed in the excursion. Since the waste tank is 50 ft below grade and is covered by a 4-ft layer of concrete, personnel exposures were primarily from airborne radioactivity. External dose equivalents of personnel, due to both β- and γ-rays, did not exceed 50 rem. There were no significant neutron exposures or internal doses from inhalation, and only slight damage occurred to the equipment.

A second nuclear incident occurred in the Idaho Chemical Processing Plant[19] in January 1961 whereby 20 MWsec of energy (6×10^{17} fissions) were released. Enriched uranium solution, at a concentration of approximately 200 gm/liter, was raised, apparently by air pressure, from the limited-geometry section of an evaporator into the 24-inches-diameter vapor-disengagement section. This equipment, like that in the preceding incident, was enclosed in heavy shielding and no significant exposure of personnel to penetrating radiation resulted from the excursion. Control of gaseous fission products prevented internal exposures. There was no equipment damage or material loss.

Criticality was accidentally achieved in unshielded plutonium-recovery equipment at the Hanford Atomic Products Operation in April 1962.[6] About 45 liters of solution, containing 1400 to 1500 gm of plutonium, which had been separated from the uranium and fission products with which it was associated earlier, was drawn by vacuum into an 18-inches-diameter cylindrical vessel where it became supercritical. Oscillations between supercritical and subcritical apparently followed during the succeeding 37 hours and generated 8×10^{17} fissions. Dosimetric observations showed that three employees near the site of the accident received dose equivalents from neutron and γ-radiation totaling 110, 43, and 19 rem, respectively. No clinical symptoms attributable to these exposures have been observed. None of the 19 other persons in the building received more than 2 rem. These exposures are not inconsistent with the plausible power-versus-time pattern of the excursion which was developed from the nuclear properties of the materials concerned and an estimate of their quantity and their location. The power excursion caused no rupture of process lines or damage to equipment, and the plutonium contamination problem, which is normally present in these operations and requires that the equipment be provided with containment barriers, was in no way aggravated by the accident. Subcriticality was reestablished probably owing to the reduction of the volume of solution by boiling and evaporation under the reduced pressure above the surface. There was evidence that some of the plutonium, in an organic solvent of relatively high density, flowed into a tube attached to the bottom of the cylinder thereby contributing to the shutdown. It will be noted that this occurrence, like the ones previously mentioned, was during salvage operations and points to the need for unusually careful control of fissile materials in areas devoted to salvage. The extended time the solution remained critical, until a set of favorable conditions ensued, is an interesting and instructive observation.

Accidents in Critical Experiments

Two experimenters were killed by the radiation arising from accidents involving supercritical assemblies at Los Alamos in 1945 and 1946. These accidents resulted from errors in judgment during the manipulation by hand of components of the assemblies.[10,11]

Several individuals received radiation doses as a consequence of an accident which occurred during the performance of critical experiments in October

1958 at the Boris Kidric Institute, Yugoslavia, near Belgrade.[3,4,23] One fatality resulted from these exposures. The critical assembly consisted of a lattice of natural uranium which was made critical by the addition of D_2O. In the operation in which the exposures occurred, the heavy water was apparently added accidentally by a means not clearly reported and without the knowledge of eight persons located 10 to 25 ft from the assembly. The system remained critical for about 10 minutes and generated 2.4×10^{18} fissions.

All of the other excursions in critical assembly work in the United States* occurred in laboratories equipped for remote control operation and provided with shielding for the protection of experimenters. No fatalities have resulted and in only one accident,[2] where a control element was inadvertently removed by hand from a water-moderated and -reflected lattice of fissile material, was there significant personnel exposure. In all cases some shutdown device functioned as designed and the liberated energy was limited to that arising from the order of 10^{17} fissions. This limitation was probably first imposed by density and temperature changes in the assembly brought about by the excursion itself. The shutdown by some mechanical means prevented a recurrence of a supercritical condition.

Termination of Accidental Excursions

It is not possible to establish a pattern by which a critical accident subsides because the mechanism usually depends upon the type and cause of the accident. Perhaps the method most expected is the self-dispersal of the fissionable material, although it has not always been the effective one in past experience. The following discussion of hypothetical accidents will illustrate other stopping mechanisms.

The transfer of an aqueous solution from restrictive geometry into one allowing prompt criticality results, first, in the generation of thermal and radiant energy. The thermal expansion of the liquid and the formation of bubbles of vapor by radiolytic decomposition of the water or by boiling decreases the density and increases the neutron leakage sufficiently to make the volume subcritical, although the pressure may not be great enough to rupture the vessel and disperse the liquid. Some cooling occurs upon expansion and evaporation. The bubbles collapse and the reactivity increases, perhaps even to prompt criticality again, and the cycle repeats itself. In the process, however, some liquid is lost from the system, and eventually the chemical concentration is increased to a value which will preclude criticality in the geometry available. Many minutes or even hours could be required for

* Several of these accidents have been described by Stratton.[24] More recent ones are described in References 5 and 14.

termination by this means, which would involve a correspondingly large release of energy. This was the situation in the chemical operations accident at Hanford.

An alternate shutdown mechanism, in contrast to the above, is dilution of the solution to the point where the increased neutron absorption by hydrogen and decrease in the fissionable material density preclude a chain reaction. This, too, may require an extended time. This was the way in which the Y-12 excursion was terminated.

The settling of a precipitate, formed in a subcritical solution, into an appropriate density-geometry condition may induce criticality. In this case the first energy pulse will probably redisperse the solids to a lower over-all density not capable of supporting a chain reaction. This system, too, may cycle again and again until some more permanent condition ensues. The ultimate action of the stirrer in the Los Alamos plutonium-solution accident was such a dispersal.

Accidents involving fissionable isotopes in metallic form or even as salts may be very severe because no permanent shutdown mechanism other than dispersal induced by self-generated energy is readily available.

It is difficult to predict the effect of the worst, yet realistic, accident which might occur in a process operation and it is emphasized that the limited experience may not be typical of expectations. The consequence of each of the accidents, except for fortunate conditions, could have been much more severe. Apparently thermal expansion or bubble formation in solutions sets a limit of something like 10^{12} fissions/cm^3 in the first power surge. If an accident involved solids, however, and consisted of dropping several only slightly subcritical pieces of metal into an appropriate configuration, the energy release could be more severe.

REFERENCES

1. Andrews, G. A., B. W. Sitterson, A. L. Kretchmar, and M. Brucer, *Health Phys.* **2,** 134 1959).

2. Brittan, R. O., R. J. Hasterlik, L. D. Marinelli, and F. W. Thalgott, *Technical Review of ZPR-I Accidental Transient—The Power Excursion, Exposures, and Clinical Data*, Argonne National Laboratory Report ANL-4971 (1953).

3. Yugoslavian Criticality Accident, October 15, 1958, *Nucleonics* **17,** No. 4, 106 (1959).

4. Callihan, Dixon, *Nucl. Safety* **1,** No. 1, 38 (1959).

5. Callihan, Dixon, *Criticality Excursion of November 10, 1961*, Oak Ridge National Laboratory Report TM-139 (1962).

6. Callihan, Dixon, *Nucl. Safety* **4,** No. 4, 136 (1963).

7. Callihan, Dixon, and J. T. Thomas, *Health Phys.* **1,** 363 (1959).

8. Glasstone, S., and M. C. Edlund, *The Elements of Nuclear Reactor Theory*, Van Nostrand, New York, 1952.

9. Ginkel, William L., C. W. Bills, A. O. Dodd, K. K. Kennedy, and F. H. Tingey, *Nuclear Incident at the Idaho Chemical Processing Plant on October 16, 1959*, Idaho Operations Office Report IDO-10035 (1960).

10. Hempelmann, L. H., and H. Lisco, *The Acute Radiation Syndrome*, Los Alamos Scientific Laboratory Report LA-1095 (1950), Classified.

11. Hoffman, J. G., *Radiation Dose in the Pajarito Accident of May 21, 1946*, Los Alamos Scientific Laboratory Report LA-687 (1948).

12. Hurst, G. S., R. H. Ritchie, and L. C. Emerson, *Health Phys.* **2,** 121 (1959).

13. Hughes, D. J., and R. B. Schwartz, *Neutron Cross Sections*, Brookhaven National Laboratory Report BNL-325. Revisions are published frequently.

14. Kathren, R. L., W. C. Day, D. H. Denham, and J. L. Brown, *Health Phys.* **10,** 183 (1964).

15. McLendon, J. D., *Health Phys.* **2,** 21 (1959).

16. *Nuclear Safety Guide*, USAEC Report TID-7016, Rev. 1, Office of Technical Services, Department of Commerce, Washington, D.C. (1961).

17. Osborn, R. K., in *Studies in Nuclear Safety*, Oak Ridge Gaseous Diffusion Plant Report K-1380 (1957).

18. Paxton, H. C., R. D. Baker, W. J. Maraman, and Roy Reider, *Nuclear-Critical Accident at the Los Alamos Scientific Laboratory on December 30, 1958*, Los Alamos Scientific Laboratory Report LAMS-2293 (1959).

19. Paulus, R. C., A. O. Dodd, K. K. Kennedy, F. H. Tingey, and F. M. Warzel, *Nuclear Incident at the Idaho Chemical Processing Plant on January 25, 1961*, Idaho Operations Office Report IDO-10036 (1961).

20. Paxton, H. C., R. D. Baker, W. J. Maraman, and Roy Reider, *Nucleonics* **17,** No. 4, 107 (1959).

21. Patton, F. S., J. C. Bailey, Dixon Callihan, J. M. Googin, G. R. Jasney, K. Z. Morgan, C. R. Sullivan, Jr., J. W. Wachter, and H. J. McAlduff, *Accident Radiation Excursion at the Y-12 Plant, June 16, 1958*, Y-12 Plant Report Y-1234 (1958).

22. Paxton, H. C., J. T. Thomas, Dixon Callihan, and E. B. Johnson, *Critical Dimensions of Systems Containing* U^{235}, Pu^{239}, and U^{233}, USAEC Report TID-7028, Office of Technical Services, Department of Commerce, Washington, D.C. (1964).

23. Savic, P. P., *Bull. Inst. Nucl. Sci. Boris Kidrich* **9,** 1 (1959).

24. Stratton, W. R., in *Criticality Control in Chemical and Metallurgical Plant*, p. 491, European Nuclear Energy Agency (1961).

25. Ullmann, J. W., and J. P. Nichols, *Nucl. Safety* **1,** No. 3, 75 (1960).

26. Weinberg, A. M., and E. P. Wigner, *The Physical Theory of Neutron Chain Reactors,* University of Chicago Press, Chicago, 1958.

27. Beck, C. K., D. Callihan, J. W. Morfitt, and R. L. Murray, *Critical Mass Studies, Part III*, Carbide and Carbon Chemicals Corporation (Oak Ridge Gaseous Diffusion Plant) Report K-343 (1949).

28. Nakache, F. R. and Mathew Shariro, *The Nuclear Aspects of the Accidental Criticality at Wood River Junction, Rhode Island*, USAEC Report TID-21995 (1964).

PROBLEMS

1. The following nuclear reaction takes place:

$$ {}_0^1n + {}_{92}^{235}U \rightarrow [{}_{92}^{236}U] \rightarrow {}_{53}^{139}I + {}_{39}^{95}Y + 2\,{}_0^1n $$

(*a*) Compute the energy release from this reaction.

(*b*) Compute the number of fissions per second necessary to produce one watt of power. For the purpose of this problem assume masses to be: $^{139}I = 138.955$; $^{95}Y = 94.945$; $^{235}U = 235.119$. Neglect kinetic energy of the incident neutrons.

2. Given an infinite and uniform dispersion of particulates of pure ^{233}U in water. Express the minimum critical concentration in grams of ^{233}U per liter. Assume all neutrons have thermal energies.

3. A critical sphere has a radius of 12 cm including the extrapolation distance. What is the critical thickness (including the extrapolation distance) of a slab of the same material which is infinite in two dimensions?

4. Write in words the neutron balance equation describing a finite, homogeneous, one-velocity critical system.

5. In a series of experiments with pure ^{235}U aqueous solutions having a concentration of 0.0764 gm $^{235}U/cm^3$ in unreflected cylindrical geometry the following critical dimensions were measured:

Diameter, cm	Height, cm
25.48	169.9
30.56	32.84
38.18	22.76

(*a*) Obtain a value of an extrapolation distance (to three significant figures) consistent with the three sets of data. (The "best" value is between 2 and 3 cm.)

(*b*) Calculate the physical dimensions of the following unreflected critical systems of this solution: (1) sphere; (2) cylinder infinitely long; (3) slab infinite in two dimensions.

6. Neglecting the effects of extraneous nuclei (O, S, N, etc), the thermal neutron diffusion length in hydrogenous solutions of ^{235}U salts, as a function of chemical concentration, is given by

$$L^2 = \frac{L_0^2}{1 + \dfrac{\Sigma_a(\mathrm{U})}{\Sigma_a(\mathrm{H})}}$$

where L_0 is the diffusion length in water and $\Sigma_a(\mathrm{U})$ and $\Sigma_a(\mathrm{H})$ are the thermal-neutron absorption cross sections of ^{235}U and hydrogen, respectively. A relation between UO_2F_2 solution density and concentration may be derived from the empirical fact that 1 gm UO_2F_2 displaces 0.133 gm H_2O upon dissolution.

(*a*) Use the migration area relation to calculate the critical radius of an unreflected spherical reactor of aqueous solutions of UO_2F_2 at each of the following concentrations (expressed as atomic ratios): $H{:}^{235}U = 500, 600, 700$. Take the extrapolation distance from problem 5 and the neutron age τ to be 33 cm², both constant for all concentrations.

(*b*) Estimate the minimum critical mass from a plot of the results of problem 6 (*a*) and the concentration.

(*c*) Calculate the critical mass at an $H{:}^{235}U = 500$ using an extrapolation distance of 4 cm.

7. From the experimental data in Table 16-4 evaluate:

(*a*) The minimum critical mass and minimum critical volume of $^{235}UO_2F_2$ aqueous solutions in water-reflected, cylindrical, steel vessels.

(*b*) The height-to-diameter ratio of the vessel in which the critical volume is a minimum.

(*c*) The chemical concentration of these solutions of minimum critical volume and mass.

(*d*) The ^{235}U mass contained in the minimum volume and the volume occupied by the solution with minimum mass.

8. Select the most appropriate of the three options given to complete the statement.

(*a*) An unreflected vessel filled with a ^{235}U solution is critical. Surrounding it with a sheet of cadmium, a good thermal neutron absorber, will make the solution (1) subcritical, (2) supercritical, (3) remain critical.

(*b*) Removal of an increment of volume from a critical system makes it subcritical. The ratio of the geometric and material bucklings of this subcritical system is (1) >1, (2) <1, (3) $=1$.

(*c*) If, instead, the system of problem 8(*b*) were made subcritical by adding a nonfissionable neutron absorber (without change in volume) the ratio of geometric and material bucklings would be (1) >1, (2) $=1$, (3) <1.

(*d*) The thickness of water which effectively shields one vessel from neutrons emitted by another is approximately (1) 10 cm, (2) 12 inches, (3) 90 inches.

(*e*) The "four factor" criticality formula, $k = \eta f p \epsilon$, is applicable to: (1) an unreflected spherical reactor with only thermal neutrons, (2) a spherical reactor surrounded by a finite reflector in which there are neutrons of all energies less than 10 MeV, (3) an infinitely large reactor in which fissions are produced by neutrons having various energies.

(*f*) The minimum critical mass of a solid piece of ^{235}U metal with water reflector is about (1) 50 kg, (2) 800 gm, (3) 20 kg.

(*g*) In a critical system (1) neutrons must be supplied from an external source, (2) the neutron density must be on the order of $10^6/cm^3$, (3) the neutron population remains constant with time.

(*h*) A homogeneous natural uranium-graphite array cannot be critical because of the (1) high absorption by ^{238}U of neutrons having intermediate energies, (2) high resonance escape probability, (3) low thermal neutron cross section of graphite.

(*i*) Given a finite homogeneous critical array operating entirely on thermal neutrons. The probability that a neutron will leak from the system is

$$(1)\quad \frac{1}{1 + L^2B^2}, \qquad (2)\quad \frac{1}{1 + \dfrac{1}{L^2B^2}}, \qquad (3)\quad \frac{1}{1 + \tau B^2}$$

where L is the diffusion length, τ is the "age," and B^2 is the buckling.

(*j*) A thin-walled aluminum cylinder is the containing vessel for a water-reflected and water-moderated critical reactor. A stainless steel shell $\frac{1}{16}$ inch thick is placed

Table 16-4 Critical Conditions for $^{235}UO_2F_2$ Aqueous Solution[a] in Steel Cylinders with Water Reflector

(From Ref. 27)

Cylinder Diameter (inch)	$H:^{235}U$ Atomic Ratio	$\frac{\text{gm } ^{235}U}{\text{gm sol'n}}$	Sol'n Density (gm/cm³)	$\frac{\text{gm } ^{235}U}{\text{cm}^3 \text{ Sol'n}}$	Critical Height (cm)	Critical Volume (liters)	Critical Mass (kg ^{235}U)
$6\frac{1}{2}$	31.6	0.385	1.88	0.724	49.0	10.49	7.59
	43.9	0.326	1.65	0.538	47.1	10.08	5.42
	62.7	0.264	1.50	0.396	47.9	10.25	4.06
	86.4	0.213	1.35	0.288	53.8	11.51	3.31
7	24.4	0.430	2.02	0.869	35.4	8.79	7.63
	31.6	0.385	1.88	0.724	34.0	8.44	6.11
	43.9	0.326	1.65	0.538	32.7	8.12	4.37
	61.1	0.268	1.47	0.394	32.6	8.09	3.19
	86.4	0.213	1.35	0.288	36.0	8.94	2.57
	123.2	0.164	1.25	0.205	42.3	10.50	2.15
	174	0.124	1.19	0.148	57.3	14.22	2.10
8	24.4	0.430	2.02	0.869	23.4	7.59	6.59
	26.2	0.418	1.98	0.827	23.5	7.62	6.30
	31.6	0.385	1.88	0.724	22.6	7.33	5.31
	43.9	0.326	1.65	0.538	21.9	7.10	3.82
	56.7	0.281	1.51	0.424	22.2	7.20	3.05
	58.8	0.275	1.51	0.415	20.8	6.74	2.80
	62.7	0.264	1.50	0.396	22.9	7.42	2.94
	86.4	0.213	1.35	0.288	23.8	7.72	2.22
	99.5	0.192	1.32	0.254	24.2	7.85	1.99
	123.2	0.164	1.25	0.205	26.0	8.43	1.73
	174	0.124	1.19	0.148	30.1	9.76	1.44
	192	0.115	1.17	0.134	29.4	9.53	1.28
	226	0.0995	1.15	0.114	36.3	11.77	1.34
	320	0.0732	1.10	0.0805	60.1	19.48	1.57
9	24.4	0.430	2.02	0.869	18.4	7.55	6.56
	31.6	0.385	1.88	0.724	18.1	7.43	5.38
	43.9	0.326	1.65	0.538	17.8	7.30	3.93
	62.7	0.264	1.50	0.396	18.0	7.39	2.93
	86.4	0.213	1.35	0.288	18.7	7.67	2.21
	123.2	0.164	1.25	0.205	19.9	8.16	1.67
	174	0.124	1.19	0.148	22.2	9.11	1.35
	226	0.0995	1.15	0.114	25.3	10.38	1.18
	320	0.0732	1.10	0.0805	33.0	13.54	1.09

[a] The ^{235}U content of the uranium was 93%.

around the reactor vessel. The reactor is now (1) supercritical, (2) subcritical, (3) unchanged.

(*k*) A 55-gallon drum is partially filled with water containing an unknown amount of uranium. In order to insure a good margin of safety you can (1) dilute with water, (2) boil off water, (3) transfer to smaller vessels.

(*l*) The age of neutrons is (1) the time required for a fast neutron to come to rest, (2) the average time required for a delayed neutron to be born, (3) directly proportional to the average square distance a neutron travels as its energy is decreased.

(*m*) The extrapolation distance (1) is the maximum distance traveled by a leakage neutron, (2) is a mathematical convenience used to apply boundary conditions in the solution of the equations describing the neutron flux distribution, (3) locates a point where the neutron flux is zero.

(*n*) The quantity not conserved in a nuclear reaction is the (1) kinetic energy, (2) nuclear charges, (3) total number of neutrons and protons.

(*o*) In a latticed array of solid fuel elements in water, the thermal neutron flux inside an element is (1) greater than, (2) equal to, (3) less than the thermal neutron flux in water.

(*p*) Of the three chemical elements described below the best neutron moderator has the properties

$$(1)\ \Sigma_a/\Sigma_s = 1,\ A = 1, \quad (2)\ \Sigma_s/\Sigma_a \gg 1,\ A = 1, \quad (3)\ \Sigma_a/\Sigma_s \ll 1,\ A \gg 1$$

(Σ_a and Σ_s are macroscopic absorption and scattering cross sections and A is the atomic number.)

(*q*) The thermal utilization is (1) the probability that a thermal neutron will produce a fission, (2) the ratio of the macroscopic thermal absorption cross section of the fissionable element to the macroscopic thermal cross section of all elements in the fuel, (3) σ_f/σ_a.

(*r*) The minimum critical mass of ^{235}U with a water moderator and reflector is about (1) 800 gm, (2) 350 gm, (3) 23 kg.

Author Index

*Asterisk denotes references that do not give initials.

Subject Index